Handbook of Research on Emerging Technologies for Electrical Power Planning, Analysis, and Optimization

Smita Shandilya
Sagar Institute of Research Technology & Science, India

Shishir Shandilya
Bansal Institute of Research & Technology, India

Tripta Thakur
Maulana Azad National Institute of Technology, India

Atulya K. Nagar
Liverpool Hope University, UK

A volume in the Advances in Computer and Electrical Engineering (ACEE) Book Series

ENGINEERING SCIENCE REFERENCE
An Imprint of IGI Global

Published in the United States of America by
Engineering Science Reference (an imprint of IGI Global)
701 E. Chocolate Avenue
Hershey PA, USA 17033
Tel: 717-533-8845
Fax: 717-533-8661
E-mail: cust@igi-global.com
Web site: http://www.igi-global.com

Library of Congress Cataloging-in-Publication Data

Names: Shandilya, Smita, 1982-
Title: Handbook of research on emerging technologies for electrical power planning, analysis, and optimization / Smita Shandilya, Shishir Shandilya, Tripta Thakur, and Atulya Nagar, editors.
Description: Hershey, PA : Engineering Science Reference, 2016. | Includes bibliographical references and index.
Identifiers: LCCN 2015048665| ISBN 9781466699113 (hardcover) | ISBN 9781466699120 (ebook)
Subjects: LCSH: Electric power production--Technological innovations. | Electric power consumption. | Electric utilities--Planning.
Classification: LCC TK1001 .H353 2016 | DDC 621.31--dc23 LC record available at http://lccn.loc.gov/2015048665

This book is published in the IGI Global book series Advances in Computer and Electrical Engineering (ACEE) (ISSN: 2327-039X; eISSN: 2327-0403)

British Cataloguing in Publication Data
A Cataloguing in Publication record for this book is available from the British Library.

For electronic access to this publication, please contact: eresources@igi-global.com.

Advances in Computer and Electrical Engineering (ACEE) Book Series

Srikanta Patnaik
SOA University, India

ISSN: 2327-039X
EISSN: 2327-0403

Mission

The fields of computer engineering and electrical engineering encompass a broad range of interdisciplinary topics allowing for expansive research developments across multiple fields. Research in these areas continues to develop and become increasingly important as computer and electrical systems have become an integral part of everyday life.

The **Advances in Computer and Electrical Engineering (ACEE) Book Series** aims to publish research on diverse topics pertaining to computer engineering and electrical engineering. **ACEE** encourages scholarly discourse on the latest applications, tools, and methodologies being implemented in the field for the design and development of computer and electrical systems.

Coverage

- Chip Design
- Circuit Analysis
- Computer Architecture
- Microprocessor Design
- Digital Electronics
- Sensor Technologies
- Computer science
- Algorithms
- Computer Hardware
- Optical Electronics

IGI Global is currently accepting manuscripts for publication within this series. To submit a proposal for a volume in this series, please contact our Acquisition Editors at Acquisitions@igi-global.com or visit: http://www.igi-global.com/publish/.

The Advances in Computer and Electrical Engineering (ACEE) Book Series (ISSN 2327-039X) is published by IGI Global, 701 E. Chocolate Avenue, Hershey, PA 17033-1240, USA, www.igi-global.com. This series is composed of titles available for purchase individually; each title is edited to be contextually exclusive from any other title within the series. For pricing and ordering information please visit http://www.igi-global.com/book-series/advances-computer-electrical-engineering/73675. Postmaster: Send all address changes to above address.

Titles in this Series

For a list of additional titles in this series, please visit: www.igi-global.com

Sustaining Power Resources through Energy Optimization and Engineering
Pandian Vasant (Universiti Teknologi PETRONAS, Malaysia) and Nikolai Voropai (Energy Systems Institute SB RAS, Russia)
Engineering Science Reference • copyright 2016 • 494pp • H/C (ISBN: 9781466697553) • US $215.00 (our price)

Environmental Impacts on Underground Power Distribution
Osama El-Sayed Gouda (Cairo University, Egypt)
Engineering Science Reference • copyright 2016 • 405pp • H/C (ISBN: 9781466665095) • US $225.00 (our price)

Handbook of Research on Advanced Computational Techniques for Simulation-Based Engineering
Pijush Samui (Centre for Disaster Mitigation and Management, VIT University, India)
Engineering Science Reference • copyright 2016 • 616pp • H/C (ISBN: 9781466694798) • US $310.00 (our price)

Reliability in Power Electronics and Electrical Machines Industrial Applications and Performance Models
Shahriyar Kaboli (Sharif University of Technology, Iran) and Hashem Oraee (Sharif University of Technology, Iran)
Engineering Science Reference • copyright 2016 • 549pp • H/C (ISBN: 9781466694293) • US $255.00 (our price)

Operation, Construction, and Functionality of Direct Current Machines
Muhammad Amin (COMSATS Institute of Information Technology, Wah Cantt, Pakistan) and Mubashir Husain Rehmani (COMSATS Institute of Information Technology, Wah Cantt, Pakistan)
Engineering Science Reference • copyright 2015 • 404pp • H/C (ISBN: 9781466684416) • US $200.00 (our price)

Performance Optimization Techniques in Analog, Mixed-Signal, and Radio-Frequency Circuit Design
Mourad Fakhfakh (University of Sfax, Tunisia) Esteban Tlelo-Cuautle (INAOE, Mexico) and Maria Helena Fino (New University of Lisbon, Portugal)
Engineering Science Reference • copyright 2015 • 464pp • H/C (ISBN: 9781466666276) • US $235.00 (our price)

Agile and Lean Service-Oriented Development Foundations, Theory, and Practice
Xiaofeng Wang (Free University of Bozen/Bolzano, Italy) Nour Ali (Lero- The Irish Software Engineering Research Centre, University of Limerick, Ireland) Isidro Ramos (Valencia University of Technology, Spain) and Richard Vidgen (Hull University Business School, UK)
Information Science Reference • copyright 2013 • 312pp • H/C (ISBN: 9781466625037) • US $195.00 (our price)

Electromagnetic Transients in Transformer and Rotating Machine Windings
Charles Q. Su (Charling Technology, Australia)
Engineering Science Reference • copyright 2013 • 586pp • H/C (ISBN: 9781466619210) • US $195.00 (our price)

To our little Master 'Samarth Koshagra Shandilya'

— S. S., S. K. S.

To my true inspirations – Shaivi and Shirish

— T. T.

To my beloved daughters – Kopal and Priyel

— A. K. N.

Editorial Advisory Board

List of Contributors

Table of Contents

Detailed Table of Contents

The identification and classification of electrical faults have a great importance in power system analysis. They help in the dimensioning and the adequate choice of electrical equipment, especially, for protective and interrupting devices. This chapter describes the various faults undergone by the power system and removes some ambiguities causing confusions and difficulties for the correct classification of faults. Once the faults terminologies are well understood and properly assessed; they can be used efficiently to develop enhanced algorithms dedicated to fault detection, classification, isolation and diagnostics.

Increased electricity demands and economic operation of large power systems in a deregulated environment with a limited investment in transmission expansion planning causes interconnected power grids to be operated closer to their stability limits. Meanwhile, the loads uncertainty will affect the static and dynamic stabilities. Therefore, if there is no emergency corrective control in time, occurrence of wide area contingency may lead to the catastrophic cascading outages. Studies show that many wide area blackouts which led to massive economic losses may have been prevented by a fast feasible controlled islanding decision making. This chapter introduces a novel computationally efficient approach for separating of bulk power system into several stable sections following a severe disturbance. The splitting strategy reduces the large initial search space to an interface boundary network considering coherency of synchronous generators and network graph simplification. Then, a novel islanding scenario generator algorithm denoted as BEM (Backward Elimination Method) based on PMEAs (Primary Maximum Expansion Areas) has been applied to generate all proper islanding solutions in the simplified network graph. The PPF (Probabilistic Power Flow) based on Newton-Raphson method and Q-V modal analysis has been used to evaluate the steady-state stability of created islands in each generated scenario. BICA (Binary Imperialistic Competitive Algorithm) has then been applied to minimize total load-generation mismatch considering integrity, voltage permitted range and steady-state voltage stability constraints. The best solution has then been applied to split the entire power network. A novel stochastic contingency analysis of islands based on PSVI (Probability of Static Voltage Instability) using MCS (Monte Carlo Simulation) and k-PEM (k-Point Estimate Method) have been proposed to identify the critical PQ buses and severe contingencies. In these approaches, the ITM (Inverse Transform Method) has been used to model uncertain loads with normal probability distribution function in optimal islanded power system. The robustness, effectiveness and capability of the proposed approaches have been validated on the New England 39-bus standard power system.

Asmaa Osamaa, Zagazig University, Egypt
Shaimaa Ahmed El-Said, Zagazig University, Egypt
Aboul Ella Hassanien, Cairo University, Egypt

Wireless sensor networks (WSNs), which normally consist of hundreds or thousands of sensor nodes each capable of sensing, processing, and transmitting environmental information, are deployed to monitor certain physical phenomena or to detect and track certain objects in an area of interests. The sensor nodes in WSN transmit data depending on local information and parameters such as signal strength, power consumption, location of data collection and accretion. Only reachable nodes are able to communicate with each other directly to collect and transmit data. The motes have limited energy resources along with constraints on its computational and storage capabilities. Thus, innovative techniques that eliminate energy inefficiencies that would shorten the lifetime of the network are highly required. Such constraints combined with a typical deployment of large number of sensor nodes pose many challenges to the design and management of WSNs and necessitate energy-awareness at all layers of the networking protocol stack. In this chapter, we present a survey of the state-of-the-art routing techniques in WSNs that take into consideration the energy issue.

Juan Aurelio Montero-Sousa, University of A Coruña, Spain
Luis Alfonso Fernández-Serantes, University of A Coruña, Spain
José-Luis Casteleiro-Roca, University of A Coruña, Spain
Xosé Manuel Vilar-Martínez, University of A Coruña, Spain
Jose Luis Calvo-Rolle, University of A Coruna, Spain

The successive energy crises, usually linked to the rising prices of oil, bring about new topics of the energy systems management in general terms. Over all, the electrical system is one of these cases. In addition, a greater concern for environmental issues has introduced, to a greater or lesser extent, the generation from renewable sources in the electrical system. In this context, the possibility of developing and using electricity storage systems would manage mismatches between generation and demand at electricity networks, making them more efficiently. In this research, we propose a number of possible strategies based on technical peak shaving and valley filling. The tool is used as energy storage systems in general terms, regardless of the accumulation technique used. The classification of strategies essentially serves two criteria: optimization service and increased profitability.

This chapter presents an unbalanced multi-phase optimal power flow (UMOPF) based planning approach to determine the optimum capacities of multiple distributed generation units in a distribution network. An adaptive weight particle swarm optimization algorithm is used to find the global optimum solution. To increase the efficiency of the proposed scheme, a co-simulation platform is developed. Since the proposed method is mainly based on the cost optimization, variations in loads and uncertainties within DG units are also taken into account to perform the analysis. An IEEE 123 node distribution system is used as a test distribution network which is unbalanced and multi-phase in nature, for the validation of the proposed scheme. The superiority of the proposed method is investigated through the comparisons of the results obtained that of a Genetic Algorithm based OPF method. This analysis also shows that the DG capacity planning considering annual load and generation uncertainties outperform the traditional well practised peak-load planning.

It is generally aware that real-time parameter estimation of A.C. induction machine is very important for the efficient operation of vector-controlled drives. However, most existing methods have not taken into account of the four imperfections within the system, i.e. the existence of severe power harmonics due to the PWM inverter drive, the imbalancing of the 3-phase supply from the inverter, the variation of supply frequency under load changing and acceleration/deceleration, and changes in resistance. In this example, we have developed a new method of parameter estimation based on real time data sampling of voltages and currents, no matter they are sinusoidal or not. Here, the assumption of a synchronous rotating speed of the stator flux is not critical. With the aid of hybrid genetic algorithm techniques, the model has been found useful for on-line speed/torque control in most field orientation control schemes as it is much easier to achieve a global minimum during the optimization process.

In this chapter, a Unified Power Quality Conditioning (UPQC) system has been presented for a two bus system. This is capable of compensating voltage and current disturbances simultaneously in a two bus system. In this topology one shunt voltage source converter (VSC) and two series VSC are present. All the converters are connected back to back on the dc side with a common dc link capacitor. Power transfer from one feeder to other feeder is made through this dc link during sag/swell and interruption. The performance of the UPQC and the proposed control algorithm has been validated in MATLAB/SIMULINK environment on a two bus system.

The authors in this chapter have put in efforts to throw light on optimization of power through the emerging technology- Smart Grid. The chapter has been divided in part which cover each section associated with the smart grid. The chapter starts with the knowhow of the system and goes till the implementation challenges. The authors have focused on the challenges associated, like cyber security, interconnection, power quality and many more. The challenges have been discussed with its significance. The most emerging technology to be harnessed to maximum, authors have also dealt with the possible solutions to the problems prevalent in the implementation of the smart grid technology. Generally the problem with the frame work has been with the integration and data handling. The chapter will help the researchers to further investigate the associated area of smart grid and find solutions to problems in smart grid.

Self-Excited Induction Generator (SEIG) offers many advantages such as low cost, simplicity, robust construction, self-protection against faults and maintenance free in today's renewable energy industry. However, the SEIG demands an external supply of reactive power to maintain the constant terminal voltage under the varying loading conditions, which limits the application of SEIG as a standalone power generator. The regulation of speed and voltage does not result in a satisfactory improvement although several studies have been emphasized on this topic in the past. To improve the performance of the SEIG system in isolated areas and to regulate the terminal voltage STATic COMpensator (STATCOM) has been modelled and developed in this dissertation. The STATCOM consists of AC inductors, a DC bus capacitor and solid-state self-commutating devices. The ratings of these components are quite important for designing and controlling of STATCOM to maintain the constant terminal voltage. The proposed generating system is modelled and simulated in MATLAB along with Simulink and sim power system block set toolboxes. The simulated results are presented to demonstrate the capability of an isolated power generating system for feeding three-phase resistive loads.

Issues related to power system voltage levels have become increasingly important issue during last two and half decades. In power networks, low voltage situations may result in the loss of stability, voltage collapse and eventually to cascading power outages. Large number of incidents of voltage collapse has been reported in different countries across the globe. A simple indicator that has the potential in real time, i.e. L indicator has been used to find voltage profile at different switching condition and simulated using ANN in network reconfiguration for the improvement of voltage stability. A method for improving voltage stability in a power network comprising of multiple lines and switches has been suggested in this chapter based on system reconfiguration approach. ANN based fast and efficient methodology has been developed to obtain the optimum switching combination to achieve best voltage stability. The proposed scheme has been tested on an IEEE 14-bus system.

The study of energy efficiency in buildings is an active field of research. Modeling and predicting energy related magnitudes leads to analyze electric power consumption and can achieve economical benefits. In this study, classical time series analysis and machine learning techniques, introducing clustering in some models, are applied to predict active power in buildings. The real data acquired corresponds to time, environmental and electrical data of 30 buildings belonging to the University of León (Spain). Firstly, we segmented buildings in terms of their energy consumption using principal component analysis. Afterwards, we applied state of the art machine learning methods and compare between them. Finally, we predicted daily electric power consumption profiles and compare them with actual data for different buildings. Our analysis shows that multilayer perceptrons have the lowest error followed by support vector regression and clustered extreme learning machines. We also analyze daily load profiles on weekdays and weekends for different buildings.

Building Automation Systems (BAS) have started attracting focus of global research community and demonstrating rapid growth potential due to their transformation from the legacy stand-alone security systems into intelligent computerized wired or wireless communication technology based solutions. Amid rising concerns over energy and security globally, BAS demonstrates commendable outperformance with promising research and commercial opportunities. Rising energy costs, stringent energy requirements and vibrant, but abnormal-unpredictable swings in environment, etc. all have contributed immensely as driving forces for BAS developments. This chapter presents the learning outcomes from author's research work and experiences in terms of instrumentation engineer's perspective towards the focused area nicely supported by literature review as well as practical hands-on exposure. The chapter judiciously puts forth historical and modern perspectives in overall development of BAS for reader's digest.

fault sections, and are the most important factors in decision making. During decision making, the most possible fault section is determined by maximum selection method. In this method most possible fault section is the one which is having highest membership grade. MATLAB code for the proposed scheme is developed and the results obtained in four cases for a power- system network.

A modern AC motor drive is a very intelligent system which covers a wide range of different electro technical apparatus and a wide scope of electrical engineering skills. A modern AC motor drive consists of four closely acting main parts: the AC machine, the power electronics, the motor control algorithm and the control hardware, i.e. the signal electronics. The advances in semiconductors and microelectronics have made the rapid development of AC motor drives possible. Semiconductors used in the switching converters provide the electric energy processing capability and microcontrollers and digital signal processors provide the data processing power for complex control algorithms.

Since the beginning of the nineties decade, many nations have undergone a change in the method of operations of the power network. The power sector was switched to open access market from vertically integrated market. The reasons for this change are many and they differ from nation to nation. The developing countries faced the problem of high demand growth along with irrational tariff policies and lack of efficiency in managing the power system. On the contrary, the developed countries restructured the power sector to provide electricity at lower prices and give the customers a number of operators to choose from. In short, the goal of restructuring is to encourage competition in the market along with providing consumers with more choices in purchasing economic energy and economic benefits. This might seem to be less complex but involves several issues which are discussed further. This chapter describes this restructuring of power sector with special reference to Indian scenario.

A discharge estimator is designed and analyzed in this chapter to estimate the discharge with voltage, frequency and current without prior knowledge of head. A control of discharge in submersible pump is achieved by varying the motor speed. In VSI fed induction motor the speed is varied by varying the switching frequency. The ratio of frequency to the applied voltage is maintained constant, to keep the air gap flux at a desired value. The development of discharge estimator, PID and Fuzzy controllers are discussed in this chapter.

The network is created through a mixture of IT, communication technology and electrical power systems. The smart grid means it is "smarter" in the sense it has its own intelligence hence a smarter power grid. The generated resources are being decentralized by the smart power grids. The major challenges with the smart grid include huge number of smart meters. Due to the large traffic and limited bandwidth, scheduling is very important. Traditional algorithms will maximize the throughput or minimize the average delay may not be valid in smart grid. Power grid plays an important role in datacenters which forms the backbone for the cloud computing infrastructure. The cloud computing and smart grid interacts through the datacenters. The load balancing and robustness can be achieved through this datacenter and cloud system. Many cloud computing features help to achieve this load balancing.

Foreword

With the increasing demand of electricity everywhere, there is an utmost need of systematic and strategic planning for the electrical generation, transmission and distribution systems. While a lot of research has already been carried-out separately in all these sectors, the analysis and optimisation are still less-focused. An efficient electrical system should not only cater the current electricity demands, but it should also be designed to accommodate more electric power to be transmitted in future, as the demand of electricity is ever-growing.

As the demand for efficient energy sources continues to grow around the globe, electrical systems are becoming more essential in an effort to meet these increased needs. As these systems are being utilised more frequently, it becomes imperative to find ways of optimising their overall function.

The Handbook of Research on Emerging Technologies for Electrical Power Planning, Analysis, and Optimisation features emergent methods and research in the systemic and strategic planning of energy usage. Highlighting theoretical perspectives and empirical research, this handbook is a comprehensive reference source for researchers, practitioners, students, and professionals interested in the current advancements and efficient use in power systems.

The present Handbook is targeted to researchers and practitioners in Electrical Power Analysis. This book includes the important and futuristic topics on Electrical Faults, Energy Loss, Energy Management, Load Uncertainty, Transmission & Distribution Planning and Advanced Power Analysis.

Atulya K. Nagar
Liverpool Hope University, UK

Atulya K. Nagar *holds the Foundation Chair as Professor of Mathematical Sciences at Liverpool Hope University where he is the Dean of Faculty of Science. Prof. Nagar has been the Head of Department of Mathematics and Computer Science at Liverpool Hope since January 2008. A mathematician by training, he possesses inter-disciplinary expertise in Nonlinear Mathematics, Natural Computing, Bio-Mathematics and Computational Biology, Operations Research, and Control Systems Engineering. He has an extensive background and experience of working in Universities in the UK and India. He has edited several volumes on Intelligent Systems, and Applied Mathematics. Prof. Nagar is the Editor-in-Chief of the International Journal of Artificial Intelligence and Soft Computing (IJAISC) and serves on editorial boards for a number of prestigious journals such as the Journal of Universal Computer Science (J.UCS). He has published with over 200 publications in prestigious publishing outlets and international journals. Prof. Nagar was awarded a prestigious Commonwealth Fellowship for pursuing Doctorate (DPhil) in Applied Non-Linear Mathematics from the University of York in year 1996.*

Preface

With growing population world-wide, the demand for various resources is increasing at a fast pace. Electrical power is one of such resources and its need is never-ending. It is also one of the basic necessities that are helpful in the economic development of a nation. Society today is highly dependent on electrical energy. From household chores, to agricultural and industrial production, electrical energy finds its uses in almost all activities of daily life. Moreover, with the help of new technologies, it is also possible now to convert electrical energy to other forms of energies such as light energy, mechanical energy, etc., thus making electrical energy more important in today's time.

Since it is not necessary that the electric energy will always be used at the same place where it is produced, long-distance transmission lines and distribution systems are required. But, the existing transmission and distribution systems are inefficient in catering to such humongous requirements. As the demand for electric power is growing, more and more load is being faced by the Electrical Power Transmission and Distribution Systems. It is therefore imperative to optimize the overall functionality of such systems.

The Handbook of Research on Emerging Technologies for Electrical Power Planning, Analysis, and Optimization features emergent methods and research in the systemic and strategic planning of energy usage.

The main characteristic of this handbook is to provide a single platform that is a collection of various chapters which deal with all necessary aspects of lossless and optimized energy transmission and distribution. From pros and cons of establishing smart grids, to minimizing energy loss planning and introducing soft computing technologies in building the transmission and distribution systems, this handbook discusses all features necessary for efficient and effective optimization of these systems.

Highlighting theoretical perspectives and empirical research, it is assumed that this handbook will prove to be a comprehensive reference source for researchers, practitioners, students, and professionals interested in the current advancements and efficient use in power systems.

The handbook can be considered as a catalog of 19 chapters presented by various authors across the globe. All the issues related to Electrical Energy Planning, its Transmission and Distribution are covered in following chapters as:

Chapter 1 - Electrical Faults in Power Systems: The chapter addresses the issues concerning the identification and classification of electrical faults. This chapter describes the various faults undergone by the power system and removes some ambiguities causing confusions and difficulties for the correct classification of faults.

Chapter 2 - Stochastic Contingency Analysis Based on Voltage Stability Assessment in Islanded Power System Considering Load Uncertainty Using MCS and k-PEM: This chapter introduces an optimized solution for separating bulk power system into several stable sections following a severe disturbance.

Chapter 3 - Energy-Efficient Routing Techniques for Wireless Sensors Networks: In this chapter, the authors have presented a survey of the widely used routing techniques in Wireless Sensor Networks that take into consideration the energy issue.

Chapter 4 - Energy Management Strategies to Improve Electrical Networks Using Storage Systems: With increasing concern of environmental issues, it has become important to use renewable resources for electric energy generation. Following the same concern, various possibilities have been proposed in this chapter which are based on technical peak shaving and valley filling.

Chapter 5 - Distributed Generation Capacity Planning for Distribution Networks to Minimize Energy Loss (An Unbalanced Multi-Phase Optimal Power Flow Based Approach): The chapter introduces an Unbalanced Multi-Phase Optimal Power Flow (UMOPF) approach that is used to determine the optimum capacities of multiple generation units in a distributed network. The results obtained are also compared against those that are produces using Genetic Algorithm based OPF method.

Chapter 6 - Hybridized Genetic Algorithm based machine parameters estimation for Direct Torque control of 3 Phase Motor for Wind Energy Systems: In this chapter, the authors have developed a new method of parameter estimation based on real time data sampling of voltages and currents. With the aid of hybrid genetic algorithm techniques, the model has been found useful for on-line speed/torque control in most field orientation control schemes as it is much easier to achieve a global minimum during the optimization process.

Chapter 7 - Performance Analysis of Unified Power Quality Controller for a Two Bus System with Different types of Loads: The chapter introduces a Unified Power Quality Conditioning (UPQC) system for a two bus system. This system can compensate for voltage and current disturbances simultaneously in two bus system.

Chapter 8 - Smart Grid - Optimized power sharing and energy storage system framework with recent trends and future ahead: In this chapter the new emerging technology, which is Smart Grid, has been emphasized on. Various related challenges, such as cyber security, power quality, interconnection, etc., have been addressed here.

Chapter 9 - STATCOM Based Solid State Voltage Regulation for Isolated Self-Excited Induction Generator: Static Compensator (Statcom) is modeled in this chapter which tries to overcome the performance disadvantages of Self-Excited Induction Generator (SEIG) in isolated environment.

Chapter 10 - Artificial Neural Network (ANN) in Network Reconfiguration for Improvement of Voltage Stability: The chapter addresses the issues concerning power system voltage levels that might result in voltage collapses and power outages. A technique for improving the stability of voltage in a power network has been suggested in this chapter that is based on system reconfiguration approach.

Chapter 11 - Global and local clustering-based regression models to forecast power consumption in buildings: This chapter proposes an effective method for efficiency in distribution of energy in building. The daily load profiles on weekdays and weekends for various buildings have also been compared.

Chapter 12 - Building Automation Systems-Recent Trends, Design and Development: In this chapter, the author presents the learning outcomes for Building Automation Systems (BASs), in terms of instrumentation engineer's perspective towards the focused area based on author's research work and experiences. The results are supported by literature review as well as practical hands-on exposure.

Chapter 13 - Contemporary Energy Management Systems and Future Prospects: The chapter explores the contemporary Energy Management Systems (EMSs), their applications, classifications, standards, and frameworks. EMSs enhance the role of users in controlling the energy management thereby shaping their energy consumption behaviors for efficient energy utilization.

Chapter 14 - Energy Management in Cloud Computing: Cloud Computing is an emerging technology that has been widely accepted world-wide to meet the growing demands of business. Cloud datacenters have thousands of servers to provide uninterrupted services to the consumers. These high-performance cloud systems consume great amount of energy thus increasing their operational costs. This chapter describes some of the methods to reduce this cost and eventually also reduce CO_2 emission and maintain Quality of Service (QoS).

Chapter 15 - Fuzzy Logic Based Approach for Power System Fault Section Analysis: Restoration of a power system after failure is one of the major characteristics that defines the system's reliability. To improve the reliability, it is important to identify and locate fault sections as quickly as possible. The chapter proposes a fuzzy logic based approach to identify faulty components in power system and diagnose the cause which lead to that particular fault.

Chapter 16 - Parameter Estimation of an High Performance I.M. Drive For Sensorless Operation: The drive motor is an important element of any motion control system. With the advancement in semiconductor technology, modern control techniques have been introduced in motion control systems. The DTC (direct torque control) algorithm with a decoupled torque and flux control with closed loop speed control has been studied in this chapter.

Chapter 17 - Restructuring Power Industry in India: With increase in population, the demand for electricity is also growing rapidly. Many nations have undergone modification in their then existing power systems to meet these demands. This chapter deals with restructuring of Power Sector with special reference to India since the developing countries, like India, faced the problem of high demand growth along with irrational tariff policies and lack of efficiency in managing the power system since the beginning of nineties decade.

Chapter 18 - Application of soft computing techniques in power engineering: In this chapter, a discharge estimator is designed and analyzed in this chapter to estimate the discharge with voltage, frequency and current without prior knowledge of head. The development of discharge estimator, PID and Fuzzy controllers are also discussed in this chapter.

Chapter 19 - Smart Grid and Cloud Computing: 'Smart' grid implies a power grid that has its 'own intelligence'. Smart grid is an integral part of cloud computing infrastructures. But, incorporating smart grids includes a number of challenges to be addressed. In this chapter, the authors discuss about the applications of cloud paradigm for smart grid.

We express our heartfelt gratitude to all the authors, reviewers and IGI-Global personnel, especially Ms. Katherine Shearer, Ms. Jan Travers, Ms. Kayla Wolfe, and Ms. Rachel Ginder, for their endless motivation and patience. Special thanks to Ms. Kashika Dubey for her constant support. We hope that this handbook will be beneficial to all the concerned readers.

Smita Shandilya
Sagar Institute of Research, Technology & Science, India

Shishir Shandilya
Bansal Institute of Research & Technology, India

Tripta Thakur
Maulana Azad National Institute of Technology, India

Atulya K. Nagar
Liverpool Hope University, UK

Chapter 1
Electrical Faults in Power Systems

Abdelkader Abdelmoumene
Boumerdes University, Algeria

Hamid Bentarzi
Boumerdes University, Algeria

ABSTRACT

The identification and classification of electrical faults have a great importance in power system analysis. They help in the dimensioning and the adequate choice of electrical equipment, especially, for protective and interrupting devices. This chapter describes the various faults undergone by the power system and removes some ambiguities causing confusions and difficulties for the correct classification of faults. Once the faults terminologies are well understood and properly assessed; they can be used efficiently to develop enhanced algorithms dedicated to fault detection, classification, isolation and diagnostics.

1. INTRODUCTION

Power system analysis usually involves the calculation of the different electrical parameters (voltages, currents, power…) under specified conditions. When the computation concerns the normal operating mode; the calculated parameters are called the nominal (or rating) values of the system, whereas, if it considers the abnormal circumstances that can effect on the normal functioning of any electrical equipment, the calculation must be regarded as fault analysis (FA).

FA aims to identify and classify faults according to:

- **Nature:** Current, voltage, frequency…,
- **Magnitude:** Over or under the magnitude of the nominal value;
- **Synchronization:** Phase shift, time;
- **Duration:** Transient or steady state mode;
- **Progression in Time:** Oscillatory, impulsive, constant….;
- **Emplacement:** Internal or external;
- **Exposition Manner:** Direct, indirect, induced…;
- **Tolerance:** Tolerable fault, dangerous fault.

In this chapter, we try to give the most comprehensive approach allowing to distinguish efficiently among the different electrical faults and remove some ambiguities causing confusions and

DOI: 10.4018/978-1-4666-9911-3.ch001

difficulties for the accurate classification of faults, while indicating the definition limits of each fault.

Once the faults terminologies, presented here, are well understood and properly assessed; they can be used efficiently to develop competitive algorithms dedicated to fault detection, classification, isolation and diagnostics.

2. THE MAIN ELECTRICAL FAULTS

2.1. Definition

A fault can be defined as an accidental change affecting the normal operation of a process (Abdelmoumene, 2010). In other words, any phenomenon that causes a modification, more or less than the nominal values of electrical quantities within their maximal tolerances, is considered as fault.

In general, we can distinguish five main families of electrical faults:

- Overcurrent (OC) faults;
- Overvoltage (OV) faults;
- Voltage dip and cut-off;
- Harmonics;
- Unbalance

2.2. Overcurrents

All currents that exceed the nominal values of service are considered as overcurrents. Generally, the OCs are caused by overload and short circuit (Abdelmoumene & Bentarzi, 2012).

2.2.1. Overloads

The overload can be due to the increase in the number of loads fed simultaneously or to the increase in the power absorbed by one or more loads. It results in an over-current which causes a rise in temperature prejudicial to the characteristics of insulators and the longevity of the transformer (Abdelmoumene, Benatrzi & Ouahdi, 2012; Abdelmoumene, Ouahdi & Benatrzi, 2011).

From the viewpoint of the duration of appearance, overloads may be classed into two categories:

- **Temporary Overloads:** Last only some moments (from some seconds to few minutes). Except certain special application; this type of overcurrent is tolerable from protective relaying.
- **Persistent Overloads:** When the overload persists and risks causing damage in the system; the protective relay must isolate the source of fault.

2.2.2. Short Circuits

A short-circuit is an accidental connection between active conductors presenting a difference of potential with a null impedance (solid short circuit) or not (impendent short circuit). It can be internal the equipment or external (Abdelmoumene & Bentarzi, 2012; Abdelmoumene, 2010).

The classification of short circuits according to the nature of the connection is shown in Figure 1.

There are also other forms of overcurrents, but, generally they are known under specific names according to the phenomenon causes. Thus, the starting current of motors and the inruch current resulting when energizing a magnetic circuit are a particular kind of overcurrents (Figure 2).

2.2.3. Inrush Currents

Inrush Current is a form of over-current results from any abrupt changes of the magnetizing voltage that leads to the core saturations (see Figure 3). The inrush current is of transient nature and, in general, it is not really dangerous for the equipment rather than undesirable event that may cause mal tripping of protective relays. The problem of

Figure 1. Types of short circuit faults on a three-phase system according to the nature of connection

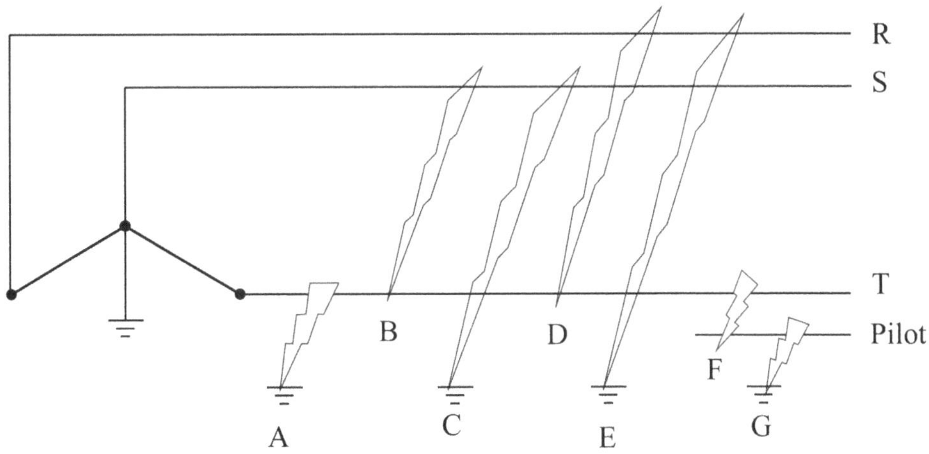

Figure 2. Classification of overcurrent faults according to their waveforms (examples of OC)

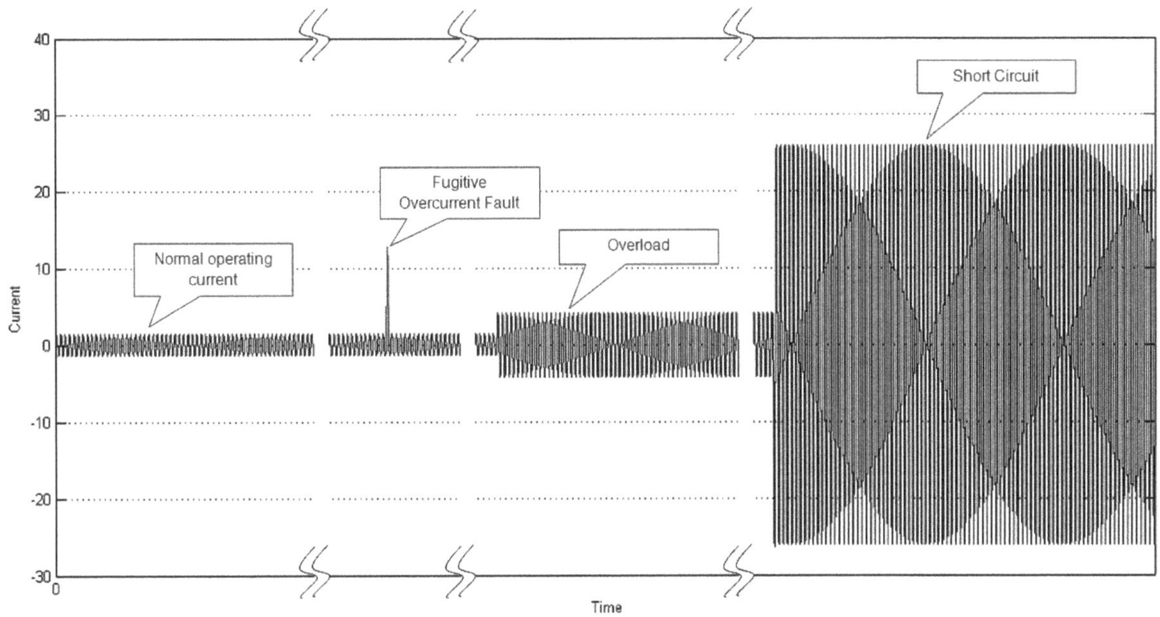

mal operation of protective relays causing by the inrush current has long considered as a challenging problem, especially for the power transformer.

2.2.4. Starting Currents

The starting current of an AC motor can vary from 3 to 9 times the nominal current as shown in Figure 4 (Schneider Electric, 2010; Medrea & Demean, 2008). This is because a large amount of energy is required to magnetize the motor enough to over-come the inertia of the system that has been at standstill. The high current drawn from the network can cause problems such as voltage drop, high transients and, in some cases, uncontrolled shutdown. High starting current also causes great

Figure 3. Typical inrush current of power transformer

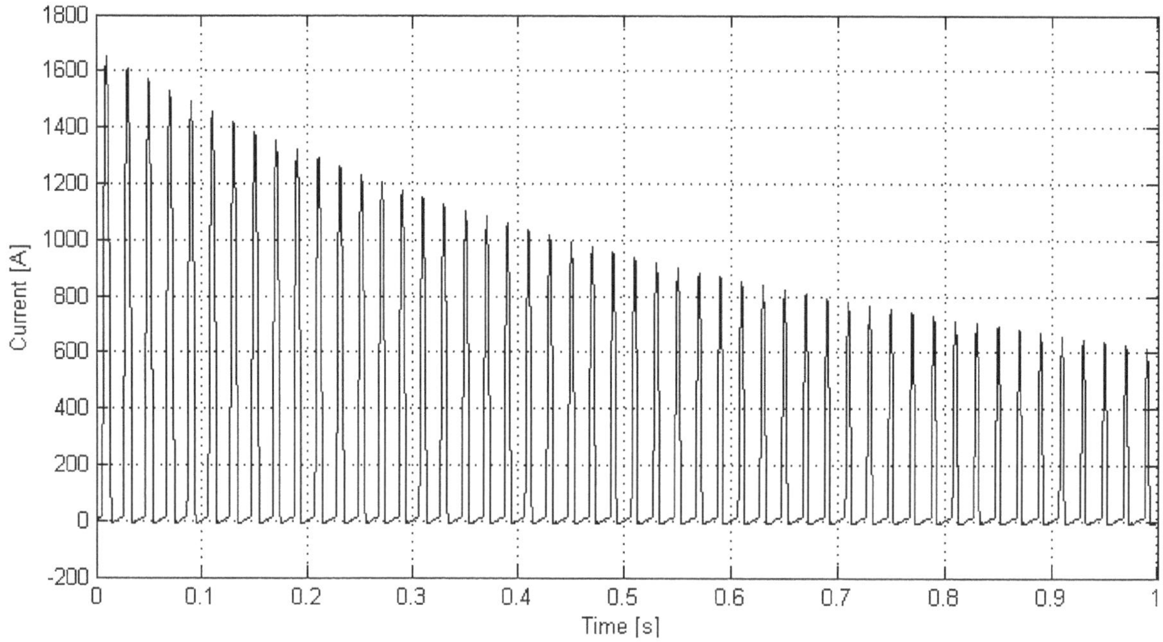

Figure 4. Starting current of induction motor

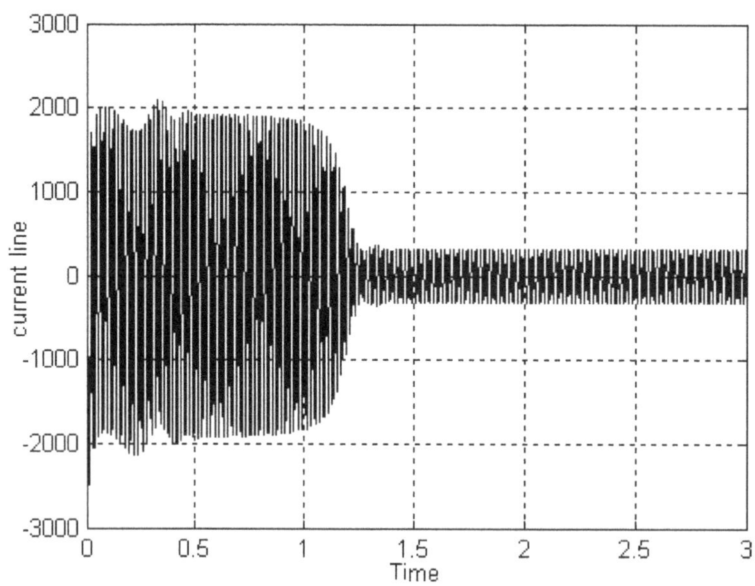

mechanical stress on the motor's rotor bars and windings, and can affect the driven equipment and the foundations. Several starting methods exist, all aiming to reduce these stresses (Havells, 2013; ABB, 2010).

The protection of induction motors should take into account this phenomenon in order to avoid the tripping during the starting period. Hence, several techniques and methods are used to handle the situation.

Figure 5 shows a flowchart that resumes the different kinds and classification of the overcurrent faults.

2.3. Overvoltage

One qualifies an overvoltage as any time function voltage that exceeds the peak voltage of the steady mode at its maximum tolerance (Abdelmoumene, & Benatrzi, 2012).

Overvoltages are caused on power systems due to external and internal influencing factors. The voltage stress caused by over voltage can damage the lines and equipment's connected to the system. Over voltages arising on a system can be generally classified into two main categories as follows:

2.3.1. External Overvoltages

This type of over voltages originates from atmospheric disturbances, mainly due to lightning. This takes the form of a surge and has no direct relationship with the operating voltage of the line. It may be due to any of the following causes (Mytech-info, 2014):

- Direct lightning stroke;
- Electromagnetically induced over voltages due to lightning discharge taking place near the line, called 'side stroke';
- Voltages induced due to atmospheric changes along the length of the line;
- Electrostatically induced voltages due to presence of charged clouds nearby;

Figure 5. Classification of overcurrents phenomena

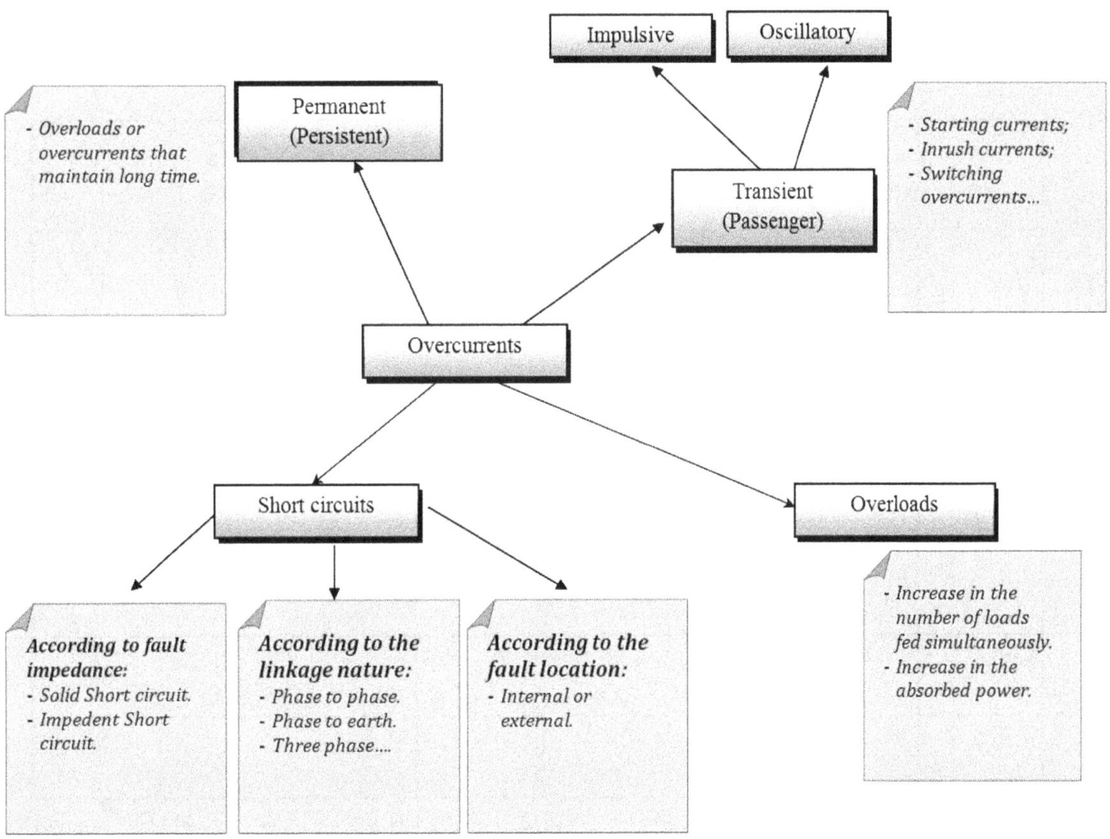

- Electrostatically induced over voltages due to the frictional effects of small particles like dust or dry snow in the atmosphere or due to change in the altitude of the line.

2.3.2. Internal Overvoltages

Switching Overvoltage is created whenever a sudden change occurs in a power circuit, especially during power switching - either closing or opening a circuit.

These over voltages are caused by changes in the operating conditions of the power system. It may be caused when switching operation is carried out under normal conditions or when fault occurs in the network. It can appear also, as shown in Figure 6, when some major load gets connected or disconnected from the long power line under normal or steady state condition (Abdelmoumene & Bentarzi, 2012).

2.4. Voltage Dip and Cut-Off

Figures 7 and 8 show transient voltages and the different types of voltage variations have been illustrated in Figure 9.

2.4.1. Voltage Dip

A voltage dip is a decrease of voltage amplitude for a period of time ranging from 10 ms to 1 s. The voltage variation is expressed in percentage of nominal current (between 10% and 100%), (Schneider Electric, 2007).

Figure 6. Example of operating overvoltage (effect of load switching on the primary of power transformer (Abdelmoumene & Bentarzi, 2012)

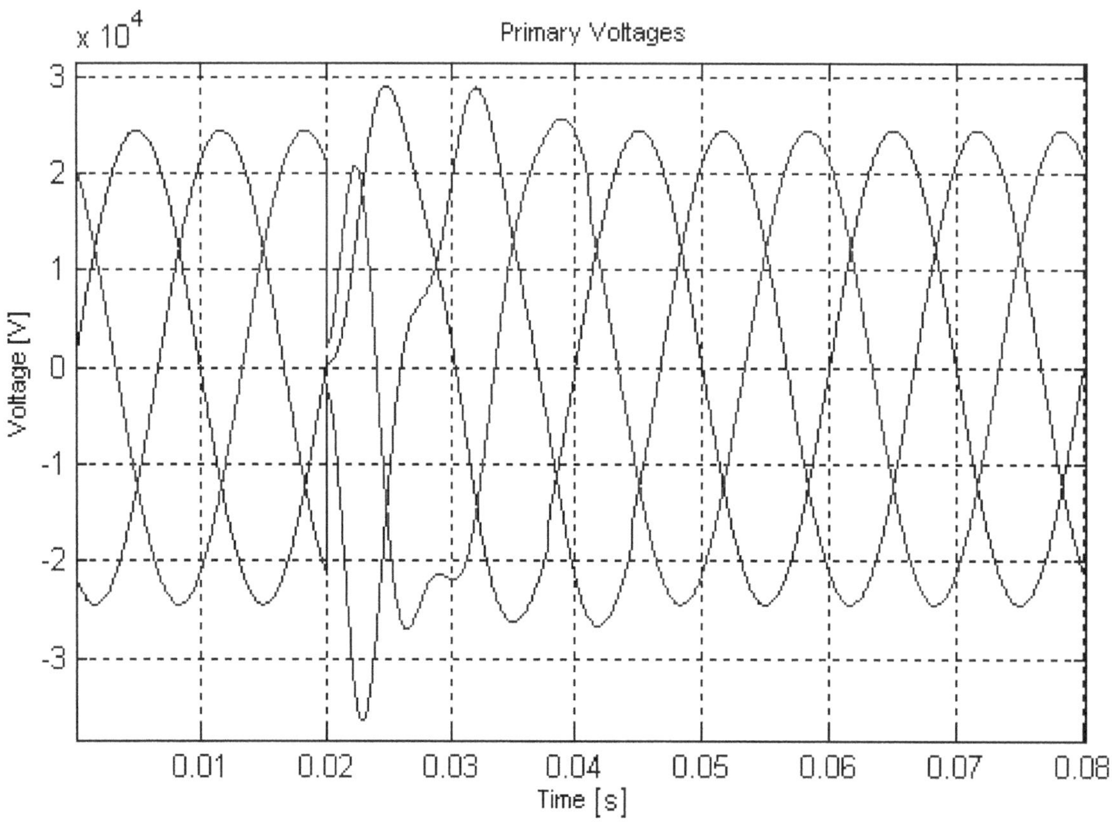

Figure 7. Examples of Voltage variations (voltage dip, cut-off)

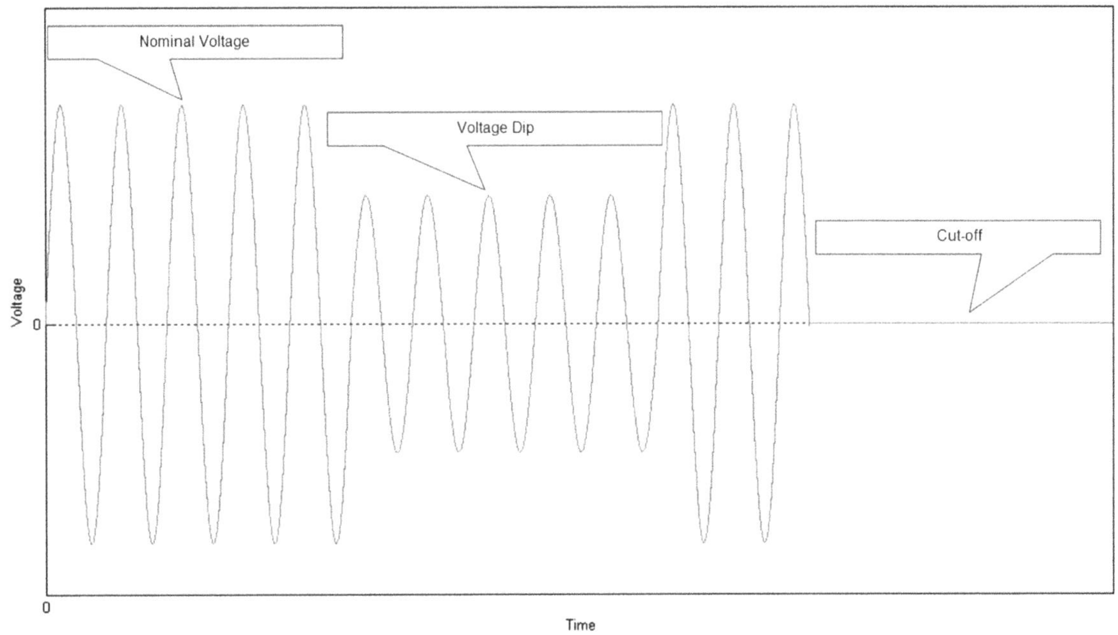

Figure 8. Transient phenomena

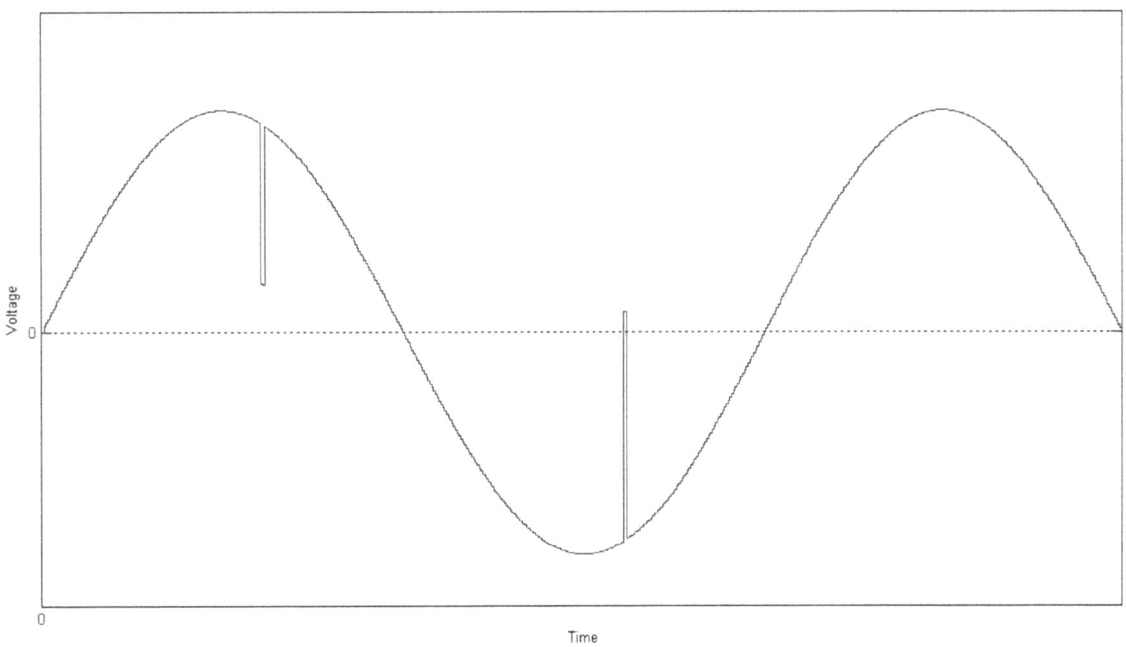

2.4.2. Cut-Off

A 100% voltage dip is termed a cut-off.

Depending on cut-off time t, the following can be distinguished (Schneider Electric, 2007):

- 10 ms < t < 1 s: micro cut-offs due, for example, to fast reset at transient faults, etc.,
- 1 s < t < 1 mn: short cut-offs due to protection device operation, switching-in of high start-up current equipment, etc.,

Figure 9. Types of voltage variations

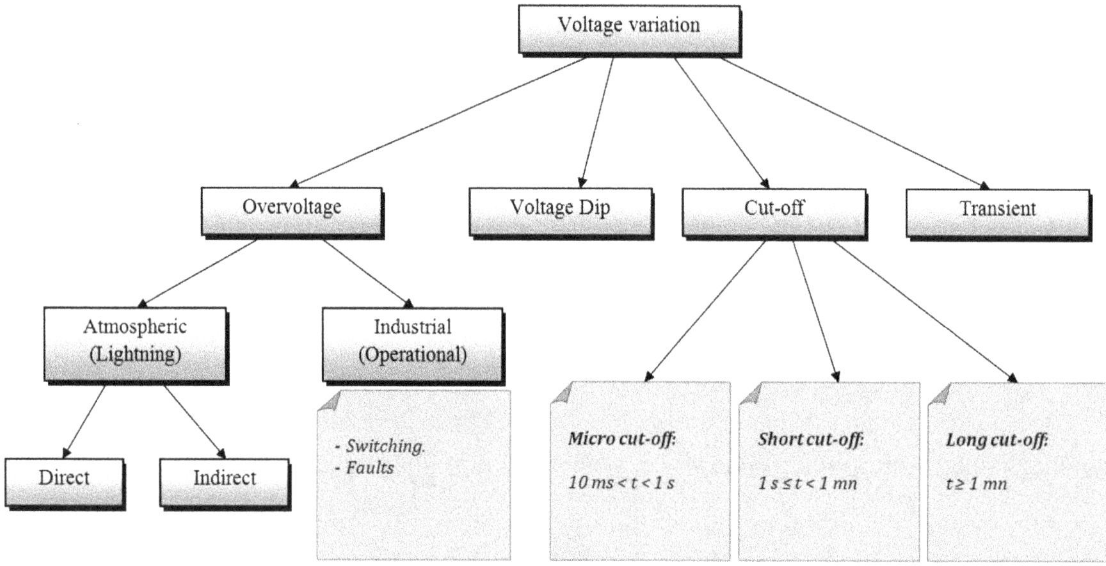

- 1 mn < t: long cut-offs generally due to HV mains.

2.4.3. *Transients Variation*

Transient phenomena are essentially fast, very high voltages, due to: lightning, operations or fault on HV or LV mains as illustrated in Figures 7 and 8.

2.5. Unbalances

In a balanced system, the three instantaneous voltages (quantities) have equal amplitudes and are separated from the other voltages by a phase angle of 120. When the quantities do not verify these terms of phase and amplitude, it is called an unbalanced three-phase system (see Figures 10 and 11).

Unbalances are generally due to single phase loads; because in this case the absorbed current on all three phases are of different amplitudes and / or phases, hence an imbalance of the three voltages. Voltage unbalance may also be due to three-phase loads, when they are not symmetrical.

An amplitude unbalance may be produced when the three voltages do not have the same RMS value, and phase unbalance when the phase shift between the three phases is not 120°.

Voltage balance is the Achilles heel of rotary and static converters, and even utility three-phase service can be unbalanced enough to harm three-phase equipment. The impact of this problem is evident by the huge industry in manufacturing of devices that monitor phase balance to protect motors.

2.6. Harmonics

It is the alteration of the sinusoidal waveform of the current and / or voltage generated by non-linear electrical systems.

The distorted signal can be decomposed into a sum of sinusoid with frequency ($n. f$) where $n \in \mathbb{N}$ and f is the frequency of the signal (Fourier series expansion).

Harmonic analysis of currents and voltages indicates that besides harmonics; there are also other components in which the frequencies are

Figure 10. Example of amplitude unbalance

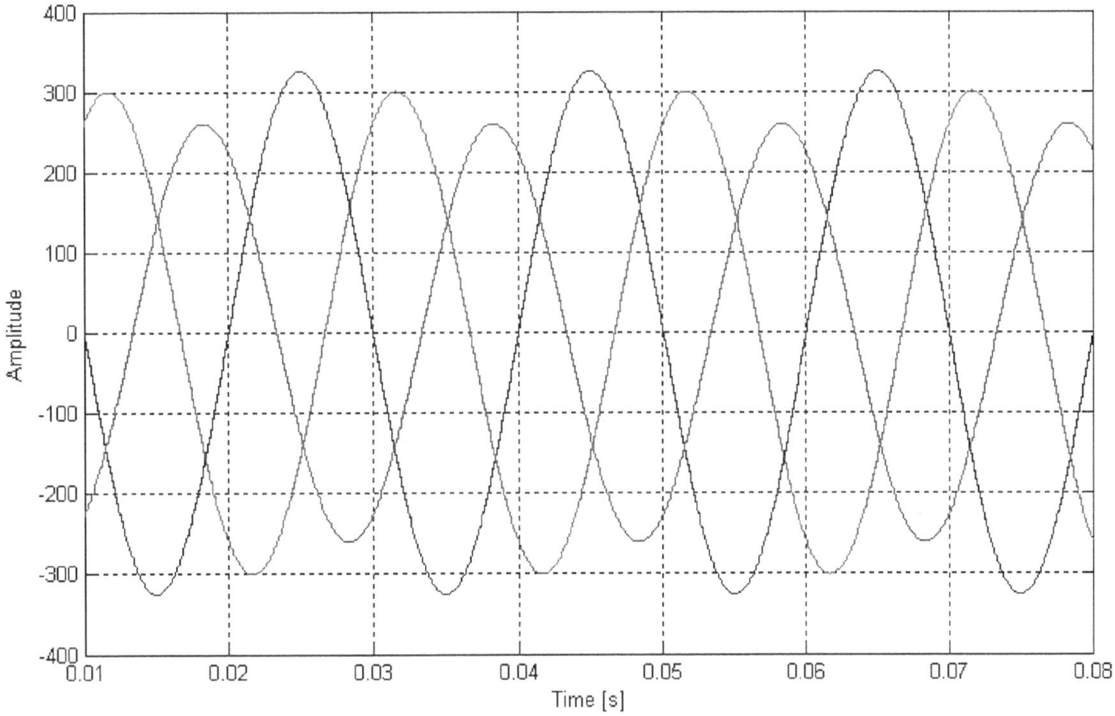

Figure 11. Example of phase unbalance

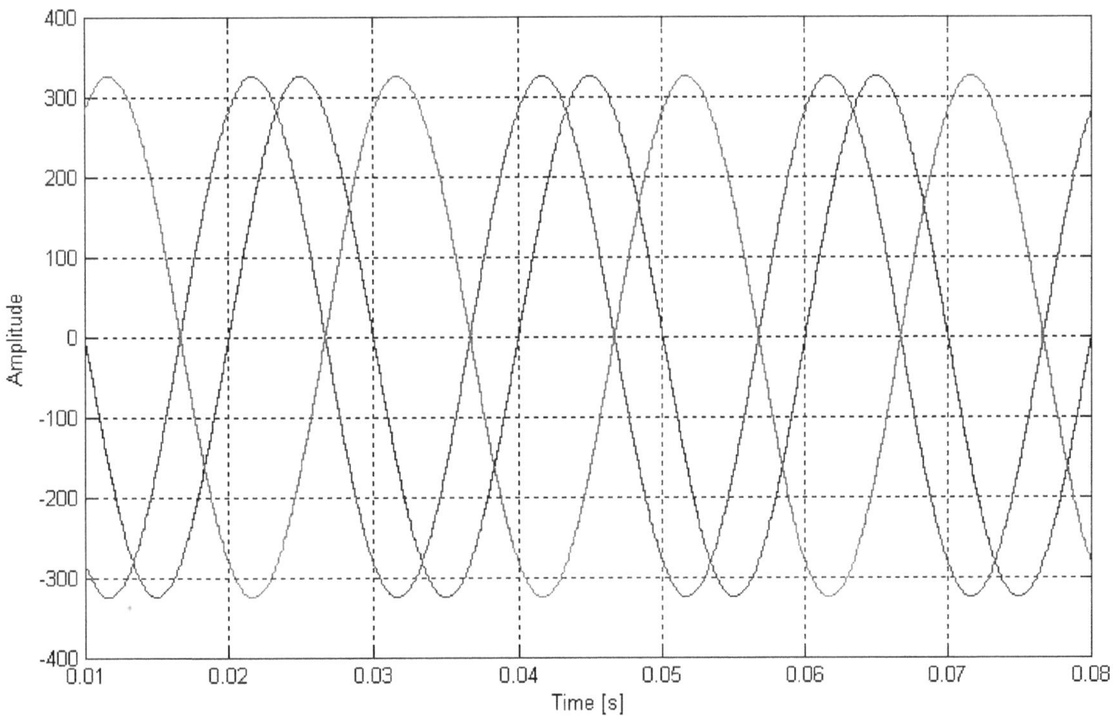

Figure 12. Types of harmonics perturbations

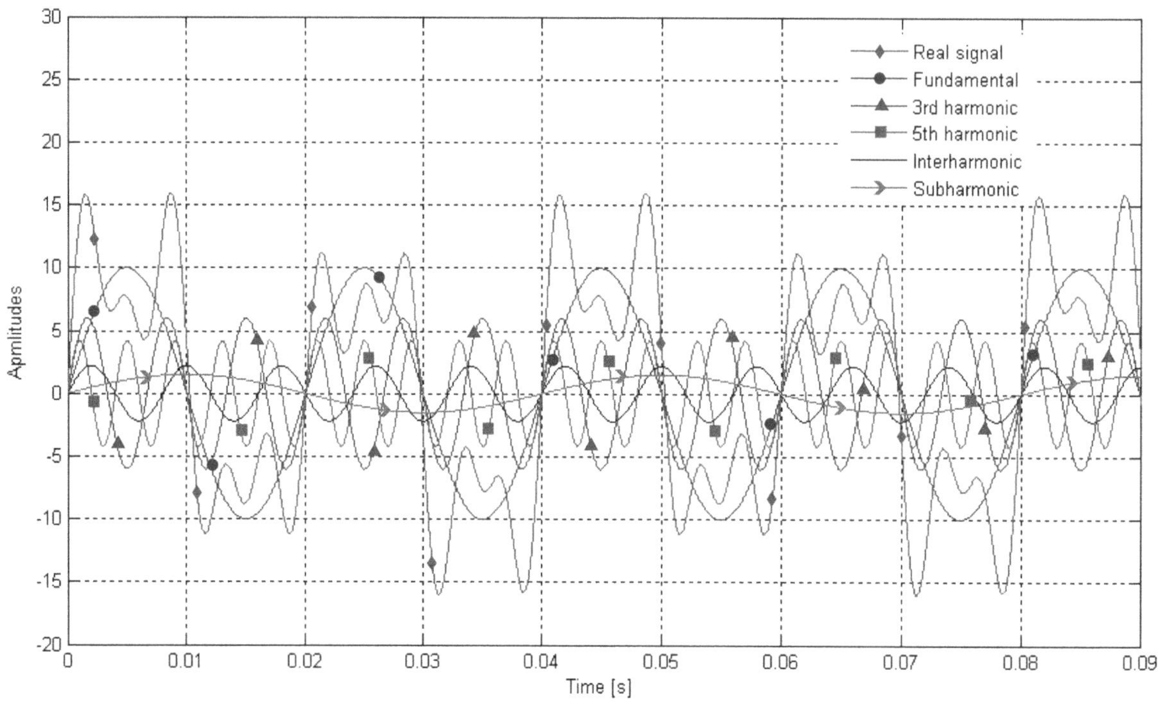

not integral multiples of the fundamental power frequency as shown in Figure 12. They are usually called interharmonics and were visualized back, for the first time, in 1971 (Farmer, Schwalb, & Eli, 1977).

- **Inter-Harmonics:** Components with frequencies between two consecutive harmonics or those components whose frequencies are not integer multiples of the fundamental power frequency.
- **Sub-Harmonics (or Infraharmonics):** A special subset of interharmonics that have frequency values that are less than that of the fundamental frequency.

3. CONCLUSION

In this chapter, we tried to give the most comprehensive approach that allows to discriminate efficiently among the different electrical faults and removes some ambiguities causing confusions and difficulties for the accurate classification of faults, while indicating the definition limits of each fault.

REFERENCES

ABB. (2010). *Technical note: Starting methods for AC motors*. ABB.

Abdelmoumene, A. (2010). Etude de la redondance sur le système de protection du transformateur de puissance [Thèse de magistère]. Boumerdes University.

Abdelmoumene, A. (2010). *Amélioration de la sureté de fonctionnement du transformateur de puissance. Journées d'Etude sur les Technologies Contrôle Commande JETCC'10*. Algeria: IAP Boumerdes.

Abdelmoumene, A., & Benatrzi, H. (2012). Study of the Influence of Power Fluctuation on Power Transformers. *International Journal of Industrial Engineering & Technology*, *2*(2), 31–37.

Abdelmoumene, A., & Benatrzi, H. (2012). Simulation du transformateur de puissance en régime de fonctionnement perturbé. *Proceedings of the deuxième Conférence Internationale sur la Maintenance, la Gestion, la Logistique et l'Electrotechnique CIMGLE'2012,* ENSET Oran, Algeria.

Abdelmoumene, A., Benatrzi, H., & Ouahdi, D. (2012, June 13-15). Power transformer protective system reliability assessment. *Proceedings of the 5th International conference on reliable engineering computing*, Brno, Czech Republic.

Abdelmoumene, A., Ouahdi, D., & Benatrzi, H. (2011). *Contribution à la réduction des risques de défaillances des transformateurs de puissance. Proceedings of the séminaire Euro-Méditerranéen sur l'Environnement et la Sécurité Industrielle SEMESI'2011.* Algeria: ENSET Oran.

AC motors starting and protection systems. (2010Schneider Electric.

Asynchronous motors. (2010). Electrical installation guide (p. 45). Schneider Electric.

BE Electra. Retrieved from http://electra.afrik-blog.com

Farmer, R. G., Schwalb, A. L., & Eli, K. (1977). Navajo project report on subsynchronous resonance analysis and solutions. *IEEE Transactions on* Power Apparatus and Systems, *96*(4), 1226–1232.

Havells. (2013). *Energy Efficient: Electric Motors*. Havells.

Medrea, N., Demean, A. (2008). Induction motor starting in a power system using edsa paladin design base software (Report No. 32). *Annals of the University of Craiova, Electrical Engineering series.*

Mytech-info. Retrieved from http://www.mytech-info.com

Technical Note: Power Quality. (2007). Schneider Electric.

Chapter 2
Stochastic Contingency Analysis Based on Voltage Stability Assessment in Islanded Power System Considering Load Uncertainty Using MCS and k-PEM

Farkhondeh Jabari
University of Tabriz, Iran

Sajad Najafi Ravadanegh
Azarbaijan Shahid Madani University, Iran

Heresh Seyedia
University of Tabriz, Iran

Behnam. Mohammadi Ivatloo
University of Tabriz, Iran

ABSTRACT

Increased electricity demands and economic operation of large power systems in a deregulated environment with a limited investment in transmission expansion planning causes interconnected power grids to be operated closer to their stability limits. Meanwhile, the loads uncertainty will affect the static and dynamic stabilities. Therefore, if there is no emergency corrective control in time, occurrence of wide area contingency may lead to the catastrophic cascading outages. Studies show that many wide area blackouts which led to massive economic losses may have been prevented by a fast feasible controlled islanding decision making. This chapter introduces a novel computationally efficient approach for separating of bulk power system into several stable sections following a severe disturbance. The splitting strategy reduces the large initial search space to an interface boundary network considering coherency of synchronous generators and network graph simplification. Then, a novel islanding scenario generator algorithm denoted as BEM (Backward Elimination Method) based on PMEAs (Primary Maximum Expansion Areas) has been applied to generate all proper islanding solutions in the simplified network graph. The PPF (Probabilistic Power Flow) based on Newton-Raphson method and Q-V modal analysis has been used to evaluate the steady-state stability of created islands in each generated scenario. BICA (Binary Imperialistic Competitive Algorithm) has then been applied to minimize total load-generation mismatch considering integrity, voltage permitted range and steady-state voltage stability constraints. The best solution has then been applied to split the entire power network. A novel stochastic contingency analysis of islands based on PSVI (Probability of Static Voltage Instability) using MCS (Monte Carlo Simulation) and k-PEM (k-Point Estimate Method) have been proposed to identify the critical PQ buses and severe contingencies. In these approaches, the ITM (Inverse Transform Method) has been used to model uncertain loads with normal probability distribution function in optimal islanded power system. The robustness, effectiveness and capability of the proposed approaches have been validated on the New England 39-bus standard power system.

DOI: 10.4018/978-1-4666-9911-3.ch002

INTRODUCTION

Power industry restructuring and competition in the deregulated electricity markets to provide increased demand causes operation of large power systems close to their stability boundaries. Although an interconnected power system may be stable against small disturbances, if there is no emergency corrective control to resynchronize all generators and prevent from fault spreading into the entire network, occurrence of large contingency may cause the system to lose stability and even lead to wide area blackout. Studies show that many wide area blackouts such as 2012 India blackout which led to massive economic losses may have been prevented by a fast feasible controlled splitting decision making (Wang, Shao, & Vittal, 2005). Hence, power system islanding is the last defense line to protect power grids from incidence of wide-area blackout. In other words, controlled system islanding also known as controlled system splitting provides the final remedial action against power system major incidence following a severe disturbance. If there is no proper remedial action in time, immediately after occurrence of large disturbance, it may lead to a catastrophic failures and power system blackout. If a system is encountered with severe instability problem, and emergency control fail to bring the faulted system back to the normal state, an islanding strategy executes by splitting the interconnected power network into several islands by disconnecting proper selected transmission lines. Achieving the proper islanding strategy which satisfies all steady-state and dynamic constraints within islands is a complicated scenario. Major efforts are needed to determine a splitting scheme with two following important characteristics (Ibrahim, 2011):

- **When to Split:** Islanding starts exactly after separating detection. There are many different techniques to detect the interconnected power system islanding.
- **Where to Split:** Many wide area blackouts may have been prevented and load-generation mismatch could have been reduced by fast, accurate, feasible controlled splitting strategies (Andersson et al, 2005; Yang, Vittal & Heydt, 2006).

Controlled intentional islanding separates a bulk power system into a number of stable islands by tripping selected transmission lines according to the minimum load-generation mismatch (L. Liu, W. Liu, Cartes & Chung, 2009). Hence, once separation is detected, the most important step is to find the optimal splitting points. In the literature, several approaches have been proposed to split a large power system into a number of stable sections following a wide area contingency. These procedures can be divided into two general categories. The first one is based on the coherent generators clustering and the second one is based on the network graph theory (Najafi, Hosseinian & Abedi, 2010).

Islanding Based on Coherent Generators Aggregating

In the literature, many approaches have been proposed to cluster coherent generators. These procedures are divided into two general categories. The first one identifies coherent machines through a time-domain simulation according to the rotor angle or speed curves. Hence, the machines with similar rotor angle or speed curves are assumed to be coherent. In the second category, slow coherency theory as a two-time-scale method is used to determine the synchronized machines. This method is based on power system oscillations. The power system oscillations can be classified into two modes; local or intra-area modes in the 1–3 Hz range and inter-area modes less than 1 Hz. The machines with the same inter-area mode swing together and they are called "coherent" in the selected inter-area modes (Liu et al, 2009).

Guowei, Junfeng, Yang, Chan and Long (2010) determines the coherent machines via EMD (Empirical Mode Decomposition) and SSI (Stochastic Subspace Identification) method. In this method, the generator rotor speed prepared from the WAMS (Wide Area Measurement System) is only used in the proposed method. Aghamohammadi and Tabandeh (2012) present a novel approach based on the correlation characteristics of the generator rotor angle oscillation. In this scheme, the correlation coefficients of generators are assessed by online measurement of rotor angle oscillations. Date and Chow (1988, 1991), Chow, Galarza, Accari and Price (1995), and Chow (1996) identify the coherent groups of synchronous generators via a slow-coherency method. Najafi et al (2010) have reported the mathematical background of Krylov subspace model reduction and coherency based aggregating. A spanning tree based DFS (Depth First Search) algorithm has then been applied to find all candidate solutions in the reduced search space. The speed of this algorithm is remarkably high and can be applied for islanding the power system in real-time. Najafi (2008, 2010 and 2013), Hosseinian and Abedi (2008, 2010) have identified the coherent machines via slow coherency algorithm. A spanning tree based BFS (Breadth First Search) algorithm has then been used to find all proper islanding scenarios. This method reduces the large initial search space of the integrated power systems islanding to the boundary branches. The proposed islanding strategy determines the best separating points according to the minimum load shedding. Liu et al (2009) have identified the desired generator groups by slow coherency theory. AMPSO (Angle Modulated Particle Swarm Optimization) is then used to find a number of candidate scenarios that provide optimal static and dynamic characteristics. You, Vittal, and Xiaoming (2004) propose an automatic islanding program based on slow coherency theory to find the optimal cutsets to form the isolated stable islands considering minimum load-generation mismatch. All generators clustering techniques reduce the large search space of the islanding studies from the entire system to the boundary interface network, significantly. The slow coherency based islanding has the following advantages (Najafi et al, 2008, 2010; You et al, 2004).

- Slow coherency based clustering is independent of the size, type and location of the disturbance. Hence, it is possible to design a defensive splitting strategy before a wide area disturbance.
- Slow coherency based clustering is independent of the generator model details. So, the classical generator model can be used.
- Slow coherency based aggregating is not sensitive to the changes of the initial condition.
- Slow coherency algorithm identifies the weakest connection by clustering the generators in the selected inter-area modes under normal operating condition.

Islanding Based on Network Graph Theory

The algorithms based on this method rise from the fact that an interconnected power system can be interpreted as a simple two-dimensional graph. The inputs of the graph are nodes and branches. In real power systems, the number of nodes and branches increase considerably. Hence, the bulk power system splitting will have a large initial search space. In some literature, scenario reduction techniques based on graph simplification have been proposed. Ibrahim (2011) presents a SNE (Source Node Expansion) algorithm based on slow coherency to split a simplified power network. The SNE initiates expansion from a source node to the connect loads until desired power balance be met. When all sources were expanded, optimal cutsets can be determined using the adjacency matrix. Kai, Da-Zhong, and

Qiang (2003) report an OBDD (Ordered Binary Decision Diagram) based three-phase method to find proper islanding solutions of interconnected power systems. A time-based layered structure has then been introduced to demonstrate the capability of the proposed strategy. The feasibility of this method has been studied by means of power system transient simulations (Sun, Zheng, & Lu, 2005). A novel approach based on the minimal cutsets with minimum net flow has been provided to split the large power system into several stable sections following a severe disturbance by Wang and Vittal (2004). Hazarika, Bhuyan, and Chowdhury (2006) describe an algorithm for determining the line outage contingency of a line taking into account of line over load effect in remaining lines and subsequent tripping of over loaded line(s) leading to possible system split or islanding of a power system. The optimally ordered sparse $[B']$, $[B'']$ matrices for the integrated system are used for load flow analysis to determine modified values of voltage phase angles $[\delta]$ and bus voltages $[V]$ to determine the over loading effect on the remaining lines due to outage of a selected line outage contingency. In case of over loading in remaining line(s), the over loaded lines are removed from the system and a topology processor is used to find the islands. A fast decoupled load flow analysis is carried out for finding out the system variables for the islanded (or single island) system by incorporating appropriate modification in the $[B']$ and $[B'']$ matrices of the integrated system. Line outage indices based on line overload, loss of load, loss of generation and static voltage stability are computed to indicate severity of a line outage of a selected line. Tian, Miao, Hongjie, and Zhou (2012) introduce a three-phase multilevel partitioning scheme for optimal splitting of power system based on Greedy growing and KL algorithm. An interesting Ant Search algorithm based on linear programming and DC power flow has been proposed by Aghamohammadi and Shahmohammadi (2012) to identify proper splitting points considering power balance and line overloading constraints. Ding, Gonzalez-Longatt, Wall and Terzija (2013) propose a two-step controlled islanding algorithm that uses spectral clustering to find a suitable islanding solution for preventing the initiation of wide area blackouts by un-damped electromechanical oscillations. The objective function used in this controlled islanding algorithm is the minimal power-flow disruption. The sole constraint applied to this solution is related to generator coherency. In the first step of the algorithm, the generator nodes are grouped using normalized spectral clustering, based on their dynamic models, to produce groups of coherent generators. In the second step of the algorithm, the islanding solution that provides the minimum power-flow disruption while satisfying the constraint of coherent generator groups is determined by grouping all nodes using constrained spectral clustering. A Mixed Integer Linear Programming (MILP) algorithm has been proposed by Trodden, Bukhsh, Grothey, and McKinnon (2013) to determine the best solution vector. The steady-state DC power flow equations and operating limits such as line losses, transmission capacity constraint and generation limits have been considered to minimize total load shedding. Then, an AC optimal load shedding problem has been solved in islanded system to provide a scenario that satisfies AC power flow. Golari, Fan, and Wang (2014) present a two-stage stochastic programming model to divide the power grids into self-sufficient islands before any multiple failures happen, and optimize islanding operations plan under severe multiple contingencies that lead to extreme situations where rolling blackouts may occur. Line switching and controlled load shedding are main tools for islanding, and the expected penalty for load shedding cost is minimized in consideration of contingencies with certain probabilities to happen. The presented model can give the system operator an efficient and optimal way to properly respond to outages. Wu, Tang, Han, and Ni (2015) apply a topology analysis and genetic algorithm combined approach for determining proper splitting strategies of large-scale power networks. Topology analysis is used to simplify the original power network into a simple equivalent network so that the splitting strategy space world be dramatically reduced; while the genetic algorithm incorporated with the breadth-first search (BFS) is employed to determine the final proper splitting strategy in the simplified power network. Two additional applications, mimicking weak connections between islands and obtaining specific pre-defined islands, of the proposed method are introduced. Pahwa, Youssef, Schumm, Scoglio, and Schulz (2013) propose a new constraint programming formulation for optimal islanding in power grid networks. This technique works efficiently for small networks but becomes expensive as size increases. To address the scalability problem, the authors propose two grid partitioning methods based on modularity, properly modified to take into account the power flow model. They are modifications of the Fast Greedy algorithm and the Bloom algorithm, and are polynomial in running time. Ding, Wall, and Terzija (2014) propose a computationally efficient controlled islanding algorithm that uses constrained spectral clustering. An undirected edge-weighted graph is constructed based on absolute values of active power flow and constraints related to transmission line availability and coherent generator groups are included by modifying the edge weights of the graph and using a subspace projection. Spectral clustering is then applied to the constrained solution subspace to find the islanding solution. To improve the clustering quality, a pre-processing procedure is used to detect and eliminate outliers in the eigenvectors of the graph before clustering. A robust k-medoids algorithm, which is less sensitive to outliers than the traditional k-means algorithm, is then used for clustering.

LARGE POWER SYSTEM CONTROLLED ISLANDING BASED ON BACKWARD ELIMINATION METHOD AND PRIMARY MAXIMUM EXPANSION AREAS

Islanding Criteria

The bulk power system optimal splitting constraints can be summarized as follows (Liu et al, 2009; Aghamohammadi & Shahmohammadi, 2012)

1. **No-Isolated Bus:** In an interconnected power system, none of buses should be removed. In other words, the islanding strategy must not eliminate any bus and all buses inside each island must be connected to each other to form integrated islands.
2. **Static Constraints:** At the steady state condition, the following constraints should be satisfied in each island.
   a. Transmission system capacity constraint (TSCC). The capacity of transmission system is limited. Hence, the splitting algorithm should be able to generate a number of scenarios that satisfy the transmission lines thermal limits.
   b. Voltage permitted range constraint. The voltage magnitude of all buses must be limited to the acceptable range in each created island. This has been considered as another islanding constraint.
   c. Static voltage stability (SVS) constraint (Jabari & Mohammadi Ivatloo, 2014).

Nowadays, large power systems are operated close to their stability margins as a result of increased consumption, power industry restructuring and competition in the electricity markets. Therefore, it is necessary to assess the power system voltage stability. The voltage stability can be evaluated using static and dynamic methods. A positive sensitivity between incremental change in bus reactive power injection and incremental change in bus voltage magnitude in all buses demonstrates the steady-state voltage stability. In this chapter, Newton-Raphson power flow and Q-V modal analysis have been used to evaluate the static voltage stability of PQ buses for each island in each generated scenario.

In Newton-Raphson technique, the linearized steady-state power flow equations can be written as follows (Kundur, 1994, pp. 990-995):

$$\begin{bmatrix} J_{P\delta} & J_{PV} \\ J_{Q\delta} & J_{QV} \end{bmatrix} \begin{bmatrix} \Delta\delta \\ \Delta V \end{bmatrix} = \begin{bmatrix} \Delta P \\ \Delta Q \end{bmatrix} \quad (1)$$

where:

ΔP: Incremental change in bus real power injection

ΔQ: Incremental change in bus reactive power injection

$\Delta\delta$: Incremental change in bus voltage angle

ΔV: Incremental change in bus voltage magnitude

$J_{P\delta}$: Sensitivity of incremental change in bus real power injection to the incremental change in bus voltage angle

J_{PV}: Sensitivity of incremental change in bus real power injection to the incremental change in bus voltage magnitude

$J_{Q\delta}$: Sensitivity of incremental change in bus reactive power injection to the incremental change in bus voltage angle

J_{QV}: Sensitivity of incremental change in bus reactive power injection to the incremental change in bus voltage magnitude

Also, Jacobian matrix is defined as follows:

$$J = \begin{bmatrix} J_{P\delta} & J_{PV} \\ J_{Q\delta} & J_{QV} \end{bmatrix} \quad (2)$$

In an interconnected power system, static voltage stability is affected by active and reactive powers. In order to reduce the computational burden, active power can be assumed to be constant at each operating point. Hence, the relation between incremental change in bus reactive power injection and incremental change in bus voltage magnitude will be represented as follows:

$$\Delta Q = J_R \Delta V \quad (3)$$

$$J_R = J_{QV} - J_{Q\delta} J_{P\delta}^{-1} J_{PV} \quad (4)$$

where, J_R is called the reduced Jacobian matrix and relates the bus voltage magnitude to the bus reactive power injection as follows:

$$\Delta V = J_R^{-1} \Delta Q \quad (5)$$

In (5), the ith diagonal element of the J_R matrix represents the Q-V sensitivity at bus i. Due to the nonlinear relation between bus voltage magnitude and bus reactive power injection, the magnitude of the sensitivity cannot evaluate the voltage security margin and stability degree in Q-V sensitivity analysis. Hence, the eigenvalues of reduced Jacobian matrix in Q-V modal analysis have been used to determine the voltage stability index. Therefore:

$$J_R = \xi \Lambda \eta \quad (6)$$

In (6), ξ is the right eigenvector matrix of J_R, Λ consists of diagonal eigenvalues of J_R, η is the left eigenvector matrix of J_R.

$$J_R^{-1} = \xi \Lambda^{-1} \eta \quad (7)$$

The relation between incremental change in bus voltage magnitude and incremental change in bus reactive power injection can be rewritten as follows:

$$\Delta V = \xi \Lambda^{-1} \eta \Delta Q \quad (8)$$

Or

$$\Delta V = \sum_i \frac{\xi_i \eta_i}{\lambda_i} \Delta Q \tag{9}$$

In (9), λ_i is the ith eigenvalue, ξ_i is the ith column of the right eigenvector matrix and η_i is the ith row of the left eigenvector matrix.

$$\eta = \xi^{-1} \tag{10}$$

Therefore:

$$\eta \Delta V = \Lambda^{-1} \eta \Delta Q \tag{11}$$

The vectors of the modal voltage variation and the modal reactive power variation are defined according to (12) and (13) respectively:

$$v = \eta \Delta V \tag{12}$$

$$q = \eta \Delta Q \tag{13}$$

Equation (14) can be written as:

$$v_i = \frac{q_i}{\lambda_i} \tag{14}$$

If $\lambda_i>0$, the voltage and the reactive power variation of the ith PQ bus are along the same direction and the system is stable. If $\lambda_i<0$, the voltage and the reactive power variation of the ith PQ bus are along opposite directions and the system is unstable. If $\lambda_i=0$, the ith modal voltage collapses.

3. **Dynamic Constraints:** At the transient condition, the dynamic stability of the formed islands should be guaranteed as follows:
   a. Coherency index. The proposed approach of this chapter considers the coherency index of New England 39-bus by a dynamic time-domain simulating in DIGSILENT environment.
   b. Dynamic voltage stability. Voltage stability is a dynamic phenomenon in a real power system. Dynamic stability analysis is more practical to study the stability characteristics in power systems. Hence, it is very important to increase the dynamic voltage stability at the transient condition of islands formation.
   c. Frequency stability. The amounts of power mismatch effects on the network frequency. If active generation increases, the frequency will be increased and if active load increases, the frequency will be decreased. Hence, it is very necessary to maintain the system frequency in an acceptable limit.
4. **Controlled Load Shedding:** In a real power system, loads can be classified into two categories; critical or essential loads and non-critical or non-essential loads. Hence, an islanding strategy considering controlled non-critical load shedding can be proposed to separate a bulk power system following a severe disturbance.
5. **Computational Burden:** In an interconnected power system with thousands of buses and branches, the size of islanding search space increases considerably. Hence, it will be an arduous challenge to find the optimal solution in a real power system. The scenario reduction techniques can be used to reduce the large initial search space into a smaller interface boundary network. In this chapter, a novel splitting scenario generator algorithm denoted as BEM (Backward Elimination Method) based on PMEAs (Primary Maximum Expansion Areas) has been used to generate all feasible islanding scenarios according to coherent groups in the simplified network graph. The proposed fast scenario generator algorithm reduces the computational burden and makes it possible to study splitting in real-time.

6. **Load Uncertainty:** In a real power system, loads are uncertain during the system operation period. Hence, it will be necessary to consider the load uncertainty under different operating conditions. In this chapter, active and reactive powers of loads have been modeled with normal probability distribution function as fallows.

Let X_i be i^{th} random variable with PDF (Probability Distribution Function) $f(x_i \mid \mu_i, \sigma_i)$ and CDF (Cumulative Distribution Function) $F(x_i \mid \mu_i, \sigma_i)$.

$$f(x_i \mid \mu_i, \sigma_i) = \frac{1}{\sigma_i \sqrt{2\pi}} \exp\left(\frac{-(x_i - \mu_i)^2}{2\sigma_i^2} \right) \quad (15)$$

$$F(x_i \mid \mu_i, \sigma_i) = \int_{-\infty}^{x_i} f(u_i \mid \mu_i, \sigma_i)\, du_i \quad (16)$$

where, μ_i and σ_i are mean and standard deviation of ith random variable. Hence, based on Inverse Transform Method (ITM) (Kroese, Taimre & Botev, 2011), active and reactive powers of loads will be random variables which obtained from following relations:

$$P_{L_i}\left(k \mid \mu_{P_{L_i}}, \sigma_{P_{L_i}}\right) = F^{-1}\left(U(k)\right) \quad (17)$$

$$Q_L\left(k \mid \mu_{Q_L}, \sigma_{Q_L}\right) = F^{-1}\left(U(k)\right) \quad (18)$$

where, $P_L(k \mid \mu_{P_L}, \sigma_{P_L})$ and $Q_L(k \mid \mu_{Q_L}, \sigma_{Q_L})$ are i^{th} active and reactive load in k^{th} scenario. $U(k)$ is an uniform random number on (0,1). We write this as $U(k) \sim U_k(0,1)$.

7. **Power Balance Criteria:** The main purpose of power systems is to maintain the balance between generation and demand. The following objective function is defined to minimize total load-generation active mismatch.

$$\text{Objective Function} = \sum_{i=1}^{n_{\text{island}}} \left| P_{G_{\text{island}}}(i) - P_{L_{\text{island}}}(i) \right| \quad (19)$$

In (19), $P_{G_{\text{island}}}(i)$ and $P_{L_{\text{island}}}(i)$ are total generation and load of the ith island respectively, n_{island} is the number of islands. In Equations (20) and (21), $P_{G_i}(\text{m})$ and $P_{L_i}(\text{n})$ are the generation of the mth machine and the consumption of the nth load in i^{th} island, respectively.

$$P_{G_{island}}(i) = \sum_{m=1}^{N_{G_i}} P_{G_i}(\text{m}) \quad (20)$$

$$P_{L_{island}}(i) = \sum_{n=1}^{N_{L_i}} P_{L_i}(\text{n}) \quad (21)$$

Scenario Reduction Techniques

In order to reduce the number of splitting scenarios, two following steps have been applied on 39-bus New England test system.

1. **Coherent Generators Clustering:** In this chapter, the rotor speed curves and speed participation factors based on modal analysis have been used to identify coherent generators of New England 39-bus power system following a severe disturbance through a time-domain simulation in the DIGSILENT environment.
2. **Proposed Network Graph Simplification**: In an interconnected power system with thousands of buses and branches, the islanding search space grows exponentially. Hence, the network graph simplification can reduce the

number of possible scenarios significantly. The proposed rules in graph simplification will be as follows.

a. Graph simplification by degree-two transfer bus: In graph theory, the degree of a bus is the number of branches connected to this bus. The bus type can be divided into three categories; load bus, generation bus and transfer bus. The bus with load and no generator is called load bus. The bus with generator and no load is called generation bus and the bus with no load and generator is called transfer bus. As shown in Figure 1, C is a degree two transfer bus. If A-C transmission line is closed, $x(i)$ will be equal to 1, else, it will be zero. Similarly, if C-B transmission line is closed, $x(j)$ will be equal to 1, else, it will be zero. The number of islanding scenarios can be reduced by considering $x(i) = x(j)$.
b. Graph simplification by parallel transmission lines: According to Figure 2, the binary states of parallel transmission lines between A and B are considered to be identical. In other words, all parallel branches between A and B are either connected (1) or disconnected (0).

DETERMINATION OF BOUNDARY NETWORK

The interface boundary network between each area with other adjacent areas is called boundary or probabilistic network which consists of boundary branches and buses. Hence, the branches or buses that are not boundary are called non-boundary or deterministic branches or buses, respectively.

Principles of Backward Elimination Method Based on Primary Maximum Expansion Areas

Determination of Primary Maximum Expansion Areas

Primary Maximum Expansion Area of island A consists of deterministic buses of island A, deterministic branches of island A and boundary network considering isolated islands constraint. As shown in Figure 3, there are three coherent groups which have been named A, B and C and colored by red, blue and green, respectively. The boundary or probabilistic interface network has been shown with purple. Considering isolated islands constraint, deterministic area of island A and boundary network will form the primary maximum expansion area of island A as a gray area.

Figure 1. Graph simplification by degree-two transfer bus

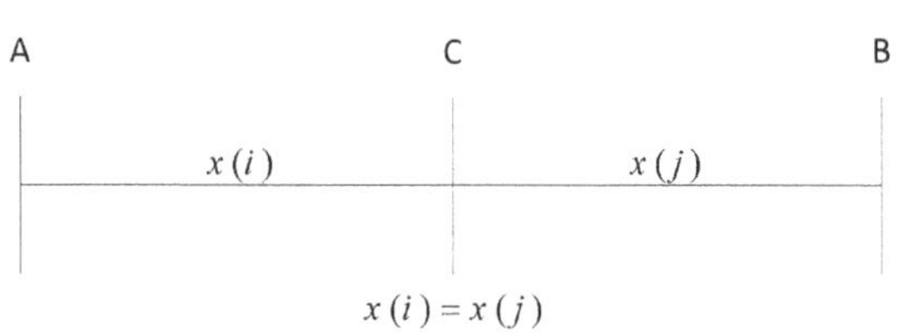

Figure 2. Graph simplification by parallel transmission lines

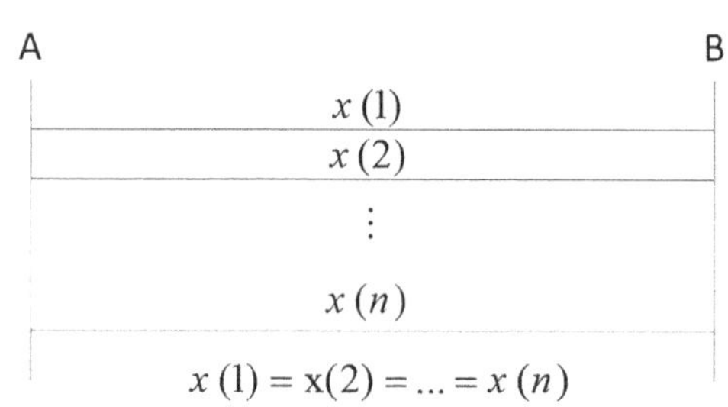

Figure 3. Concepts of primary maximum expansion areas in New England 39-bus test system

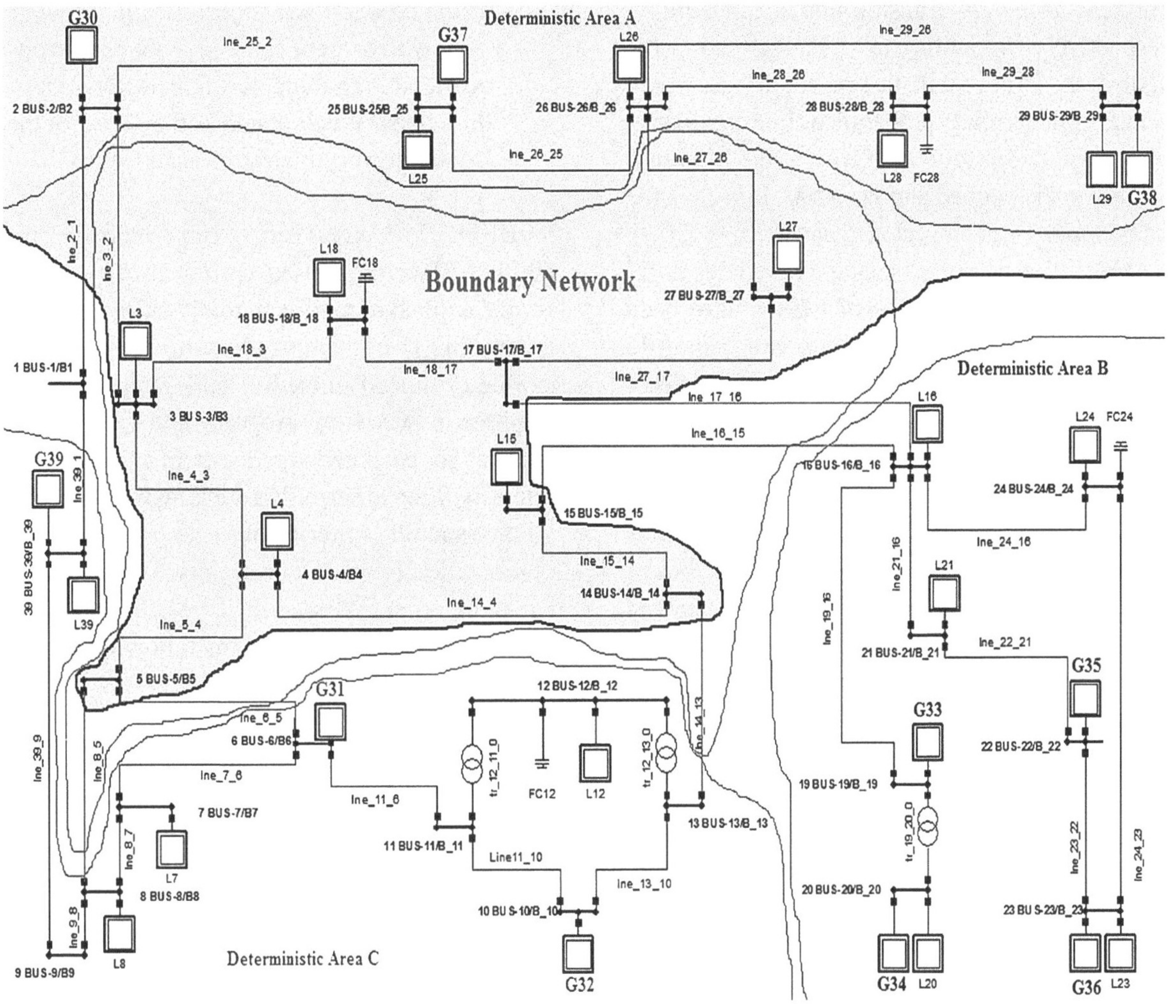

Formation of the Probabilistic Line and Bus Data Matrices Considering PMEAs

The primary line data matrices of the islands will have probabilistic nature due to the boundary transmission lines. After numbering all buses and determining the slack bus in each island (The slack bus must not belong to the set of the boundary buses) considering PMEAs of islands, the primary probabilistic line data matrix will be formed for each island as follows. For the kth boundary branch connecting i^{th} and j^{th} buses of island A, we can use:

$$Z(i,j) = (1 - x(k))M + Z_0(i,j) \tag{22}$$

$$x(k) = \begin{cases} 0 & \text{if the } k^{th} \text{ boundary branch has been opened} \\ 1 & \text{if the } k^{th} \text{ boundary branch has been closed} \end{cases} \tag{23}$$

$$Z_0(i,j) = R(i,j) + jX(i,j) \tag{24}$$

$Z(i,j)$ is the impedance of the transmission line i to j in primary probabilistic line data matrix of island A. M is a sufficiently large constant. $Z_0(i,j)$,

$R(i,j)$ and $X(i,j)$ are the impedance, resistance and reactance of i–j transmission line in per unit, respectively. According to (23), $x(k)$ shows the state of the kth probabilistic branch. In a generated scenario, the primary probabilistic line data matrix can be updated as follows. If $x(k)=0$, the kth boundary branch is opened and $Z(i,j)\cong M$. If $x(k)=1$ the kth boundary branch is closed and $Z(i,j) = Z_0(i,j)$.

Once all buses of each island have been numbered and the slack buses of islands have been determined (The slack bus must not belong to the set of the boundary buses), the probabilistic bus data matrix will be formed for each island considering PMEA of island. In order to generate proper splitting solutions, it is necessary to update the probabilistic bus data matrices of the islands in each generated scenario. Hence, if the ith bus of the island A (that is a boundary bus) is removed from island A to island B, the ith bus of island A will be eliminated from the probabilistic bus data matrix of this island through a Bus-Trajectory matrix. In other words, for each generated scenario, the probabilistic bus data matrix which is formed based on the PMEA of the island A will be updated across the Bus-Trajectory matrix of this island. The Bus-Trajectory matrix of the island A can be formed as follows. If the ith bus of the island A is a deterministic or non-boundary bus, the ith row of the Bus-Trajectory matrix of island A will be equal to 1. In other words, $BT_A(i)=1$. If the ith bus of island A is a probabilistic or boundary bus, the ith row of the Bus-Trajectory matrix of island A can be calculated via the following steps.

1. All probabilistic trajectories which connect the ith probabilistic bus of island A to the deterministic network of this island should be determined.
2. A trajectory which connects the ith bus of island A to the deterministic network of this island through boundary branches can be considered as the multiplication of the binary states of these probabilistic branches.
3. The sum of the obtained expressions for all of the probabilistic trajectories in previous step (which connect the ith probabilistic bus of island A to the deterministic network of this island) is calculated as the ith row of the Bus-Trajectory matrix of island A.

BEM, PMEAs and Binary Imperialistic Competitive Algorithm based optimization program is then applied to minimize total load-generation mismatch. The steady-state stability of islands can be evaluated in each island using Newton-Raphson power flow program and Q-V modal analysis for each generated scenario. Hence, the integrity, voltage permitted range and steady-state voltage stability criteria have been considered as optimization problem constraints. The proposed islanding strategy based on BEM, PMEAs and BICA has been summarized as following steps.

Step 1: Start
Step 2: Process power system structure
Step 3: Cluster the coherent generators, simplify network graph and determine the interface boundary network
Step 4: Determine the PMEAs
Step 5: Initialize BICA parameters
Step 6: Generate initial solution vector
Step 7: Update the line and bus data matrices of islands in the generated scenario
Step 8: Run Newton-Raphson power flow in each island considering updated line and bus data matrices
Step 9: Modal analysis to evaluate the static voltage stability and determine the stability index for all PQ buses in each island
Step 10: Are all constraints satisfied? If yes, go next; else, generate new scenario and go step 7.
Step 11: Has a satisfactory scenario been found? If yes, display optimum splitting points; else, generate new scenario and go step 7.
Step 11: End

Proposed Stochastic Contingency Analysis in Islanded Power System Based On Probability of Static Voltage Instability

In optimal islanded power system, equipment outages such as transformers, transmission lines and generation units will affect the power system stability. Hence, the stochastic contingency analysis based on static voltage instability has been proposed to predict which outages make system violations and rank the contingencies according to *PSVI* (Probability of Static Voltage Instability) under uncertain operation. It is formulated by (25) and (26) gives measure of steady-state voltage instability.

$$s_n(i,j) = \begin{cases} 1 & \mathit{If}\ \lambda_n(i,j) \le 0 \\ 0 & \mathit{Else} \end{cases} \tag{25}$$

$$PSVI_n(i) = \frac{\sum_{j=1}^{N_{Iter}} s_n(i,j)}{N_{Iter}}$$
$$i = 1,2,\ldots,N_{PQ_n},\ \ j = 1,2,\ldots,N_{Iter},\ \ n = 1,2,\ldots,N_{Islands} \tag{26}$$

N_{Iter}: Number of iterations

$\lambda_n(i,j)$: i^{th} eigenvalue of reduced Jacobian matrix of nth island in j^{th} iteration

N_{PQ_n} : Number of PQ-buses of nth island

$N_{Islands}$: Number of islands

Proposed Stochastic Contingency Analysis Based on Monte Carlo Simulation

The MCS has been applied to generate the stochastic scenarios. Proposed contingency analysis models a single transmission line outage to evaluate the voltage stability considering loads uncertainties until all credible outages have been studied. Contingencies are then ranked according to their relative severity using *PSVI*. The Flowchart of the proposed stochastic contingency analysis based on static voltage instability using MCS has been shown in Figure 4.

Proposed Stochastic Contingency Analysis Based on k-Point Estimate Method

The main drawback of the Monte Carlo method is the great number of iterations required to achieve the desired convergence. Hence, approximate methods such as k-PEM require a much lower computational burden. In literature, 2-PEM with 2m iterations are widely utilized (Su, 2005; Verbic & Cañizares, 2006). In this method, *K* points should be estimated for each random variable. Hence, for *M* input variables, *KM* scenarios will be generated. During each estimation process, the input vector is calculated as follows:

$$\left(\mu_{p_1}, \mu_{p_2}, \ldots, P_{l,k}, \ldots, \mu_{p_M}\right) \tag{27}$$

where, μ_{p_n} is the mean value of the nth input random variable and $P_{l,k}$ can be calculated as following relation:

$$P_{l,k} = \mu_{p_l} + \xi_{l,k}\sigma_{p_l} \tag{28}$$

where, μ_{p_l} and σ_{p_l} are the mean and standard deviation of the lth input random variable, respectively. The parameter $\xi_{l,k}$ is the standard location related to the number of estimated points. For *2m* point estimate method, standard locations are calculated as following relations (Hong, 1998):

$$\xi_{l,1} = \frac{\lambda_{l,3}}{2} + \sqrt{m + \frac{\lambda_{l,3}^2}{4}} \tag{29}$$

Figure 4. Flowchart of proposed stochastic contingency analysis using MCS

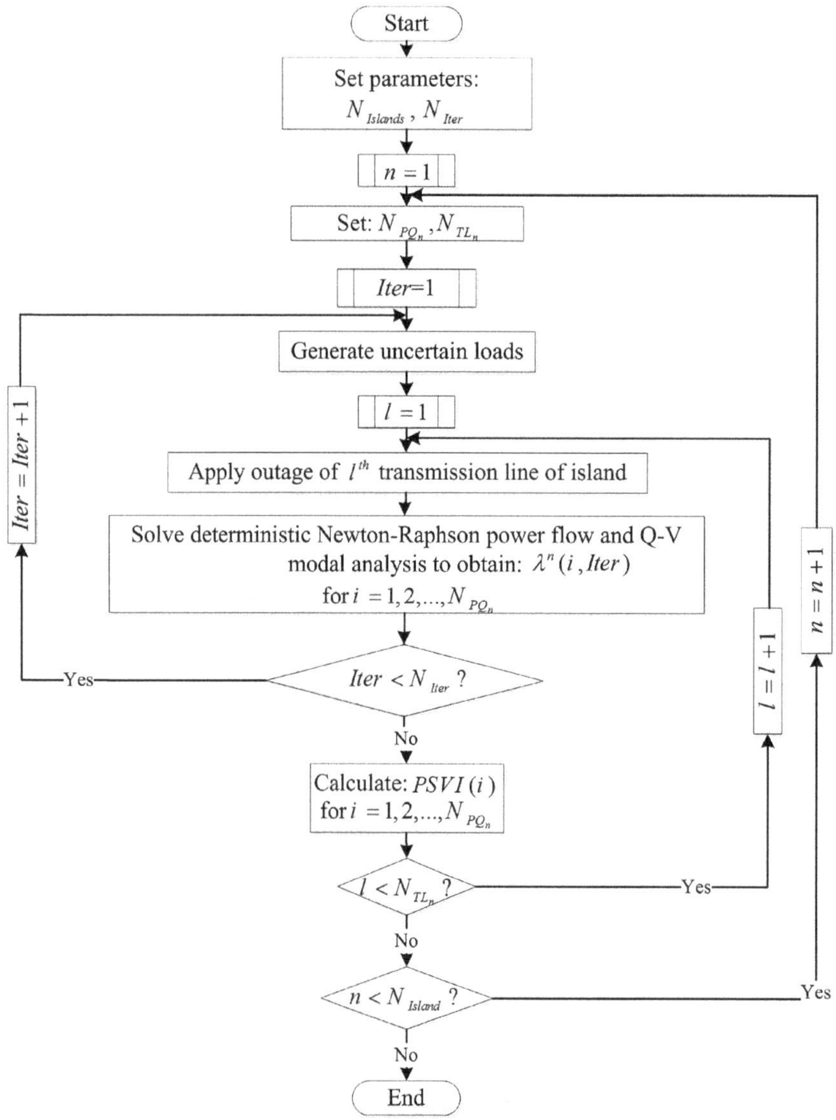

$$\xi_{l,2} = \frac{\lambda_{l,3}}{2} - \sqrt{m + \frac{\lambda_{l,3}^2}{4}} \tag{30}$$

where, $\lambda_{l,3}$ is the skewness of lth input variable and m is the number of input random variables. The moments can be calculated as (31):

$$\lambda_{l,j} = \frac{\int_{-\infty}^{+\infty} \left(p_l - \mu_{p_l}\right)^j f_{p_l} dp_l}{\sigma_{p_l}^j} \qquad j = 1,2,...,2k-1 \tag{31}$$

In this method, expected values are computed as follows:

$$E[Z] = \sum_{l=1}^{m} \sum_{k=1}^{K} \omega_{l,k} Z(l,k) \tag{32}$$

where, Z is the vector of output random variables and $E[Z]$ is the expected value of vector Z. The weighting coefficients are calculated as follows:

$$\omega_{l,1} = \frac{-\zeta_{l,2}}{m\left(\zeta_{l,1} - \zeta_{l,2}\right)} \tag{32}$$

$$\omega_{l,2} = \frac{\zeta_{l,1}}{m\left(\zeta_{l,1} - \zeta_{l,2}\right)} \tag{34}$$

Similar to previous method, the proposed stochastic contingency analysis based on static voltage instability using 2-PEM models a single transmission line outage in each island to evaluate the voltage stability considering loads uncertainties until all credible outages have been studied. Contingencies are then ranked according to their relative severity using *PSVI*. The flowchart of the proposed strategy has been shown in Figure 5.

The results of the proposed stochastic contingency analysis based on static voltage instability using 2-PEM can be compared with MCS. The probability error indices can be calculated as following relation:

$$\varepsilon_{\lambda_i|l}^{PSVI} = \left| \frac{PSVI_{MCS}\left(i \mid l\right) - PSVI_{2\text{-}PEM}\left(i \mid l\right)}{PSVI_{MCS}\left(i \mid l\right)} \right| \times 100 \tag{35}$$

where, $\varepsilon_{\lambda_i|l}^{PSVI}$ is the PSVI error index for evaluate the SVS of the ith PQ-bus considering l^{th} contingency.

SIMULATION RESULTS AND DISCUSSION

The New England 39 bus standard test system (Chow & Cheung, 1992) is simulated and utilized to confirm the efficiency of the two proposed stochastic contingency analysis approaches. At first, the Newton-Raphson power flow has been performed and complete system report under normal operating condition is shown in Table 1.

According to the above Table, total active generation, total active load and transmission losses are equal to 7001.56 (MW), 6905.76 (MW), 95.80 (MW) respectively. In order to identify synchronized machines under a severe disturbance, it is assumed that two switching events have occurred on the 3-4 transmission line at t=1sec and 17-18 transmission line at t=1.3sec. The voltage profile, rotors speed curves are shown in Figures 6 and 7, respectively. According to voltage profile, occurrence of two switching events on two transmission lines causes the catastrophic voltage collapse, system instability and wide area blackout during five seconds. Hence, the controlled islanding is necessary to avoid catastrophic system collapse. Three coherent groups {G30, G37, G38}, {G33, G34, G35, G36} and {G31, G32, G39} have been formed.

The three deterministic areas and the interface boundary network have been shown in Figure 8.

Once the coherent machines clustering and graph simplification have been applied, the splitting search space will have 2^9=512 scenarios. The 11th, 2nd and 2nd buses are assumed as the slack buses in islands A, B and C, respectively. The BEM, PMEAs and BICA based optimization programs have been run on a Lenovo with 2.10 GHz CPU, 4 GB RAM with an average calculation time less than 0.1 second for generating each islanding scenario in simplified network graph. The optimal solution

Table 1. Load flow calculation results

Total System Summary	
Generation [MW]/[Mvar]	7001.56/2419.49
Load [MW]/[Mvar]	6905.76/2886.08
Compensation [Mvar]	-657.18
Total Losses [MW]/[Mvar]	95.80/190.59
Load Losses [MW]/[Mvar]	95.80/2405.14
No Load Losses [Mvar]	-2214.55

Figure 5. Flowchart of proposed stochastic contingency analysis using k-PEM

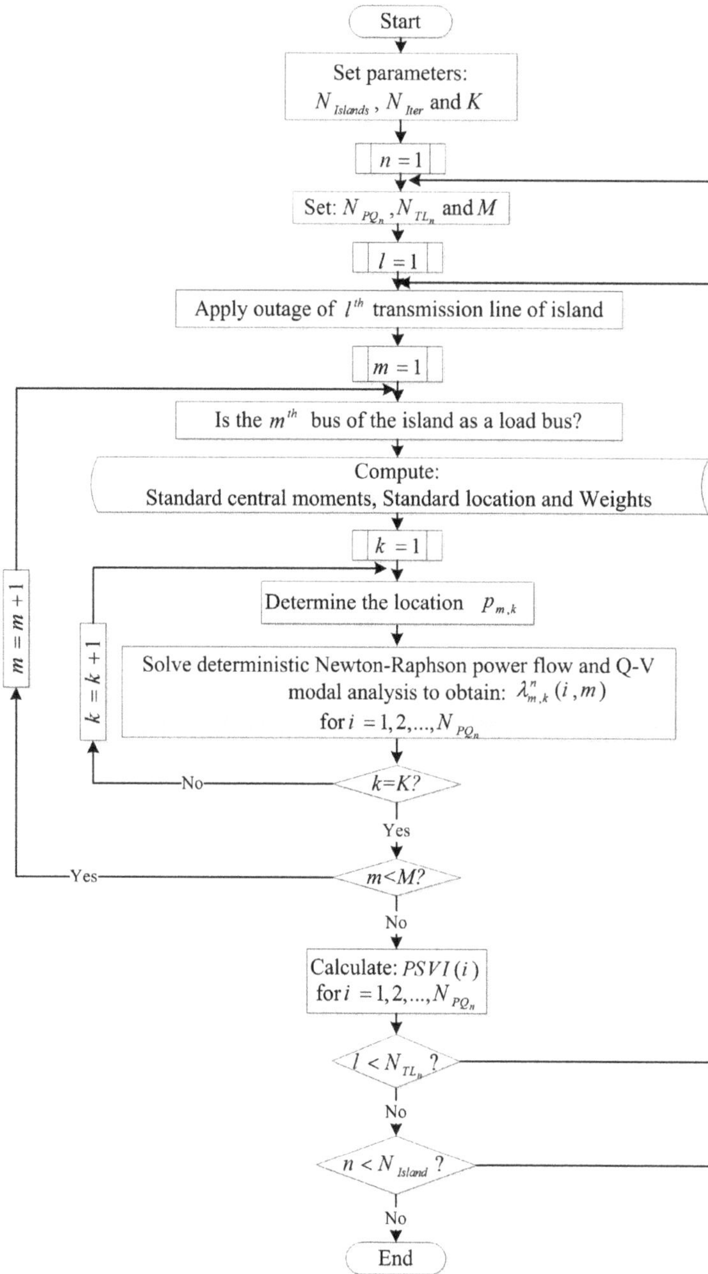

vector has been obtained from the optimization algorithm. The optimal islanded New England test system has been shown in Figure 9.

$$\begin{array}{rccccccccc} & x(1) & x(2) & x(3) & x(4) & x(5) & x(6) & x(7) & x(8) & x(9) \\ \text{Best scenario} = [& 0 & 1 & 0 & 1 & 1 & 1 & 1 & 1 & 1\,] \end{array}$$

The voltage profiles of three islands for optimal controlled and uncontrolled islanding solution have been shown in Figures 10, 11 and 12. Based on the optimal scenario, the 8th and 9th buses of island A which are similar to the 13th and 12th buses of island B, belong to the island B. The 14th, 15th,

Figure 6. Voltage profile during contingency

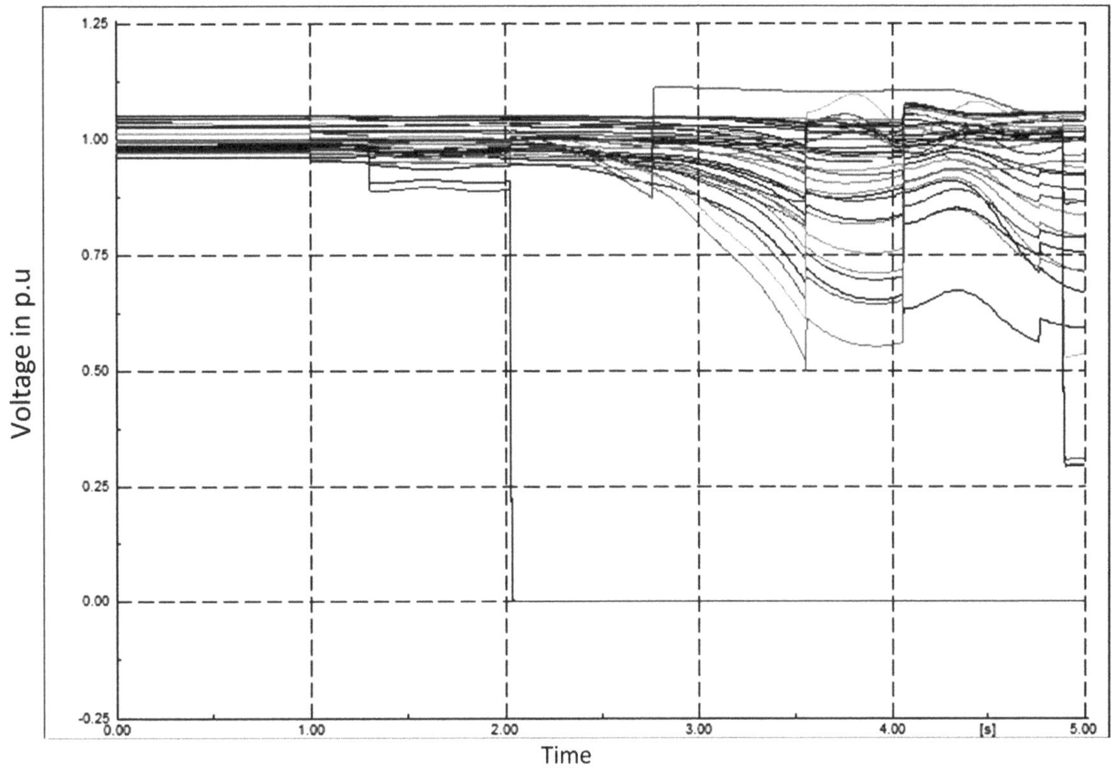

16th and 17th buses of island B which are similar to the 7th, 6th, 5th and 4th buses of island C belong to the island C.

The PQ buses in islands A, B and C are {2, 3, 4, 5, 7, 8, 9, 10}, {3, 4, 6, 7, 8, 10, 11, 12, 13, 14, 15, 16, 17} and {3, 4, 5, 6, 7, 8, 9, 11, 12, 13, 14, 15}, respectively. The static voltage stability indices of the formed islands have been shown in Figure 13. The 6th and 7th PQ buses of island A which are the 8th and 9th buses of this island will be removed from area A to island B. The 14th, 15th, 16th and 17th buses of island B which are the 9th, 10th, 11th and 12th PQ buses of this island will be removed from island B to area C. All stability indices are positive and the steady-state voltage stability has been satisfied in each island.

The load, generation and power mismatch of each island for the best scenario have been reported in Table 2. In the optimal islanded New England system, the total load-generation mismatch is equal to 45.5272 MW.

In the Inverse Transform Method, the standard deviations of the loads of islands A, B and C are assumed 15, 20 and 25, respectively. The reported data by Chow and Cheung (1992) will be used as the mean values of the uncertain loads. The results of the stochastic contingency analysis of islands A, B and C have been reported in Tables 3, 4 and 5, respectively.

The *PSVI* indices under other single contingencies are zero. Hence, these contingencies have not been reported in above Tables. It is found that there are 4, 3 and 2 single severe contingencies in islands A, B and C, respectively. According to the results, despite the great number of simulations, the proposed stochastic contingency analysis based on

Figure 7. Rotors speed curves during contingency

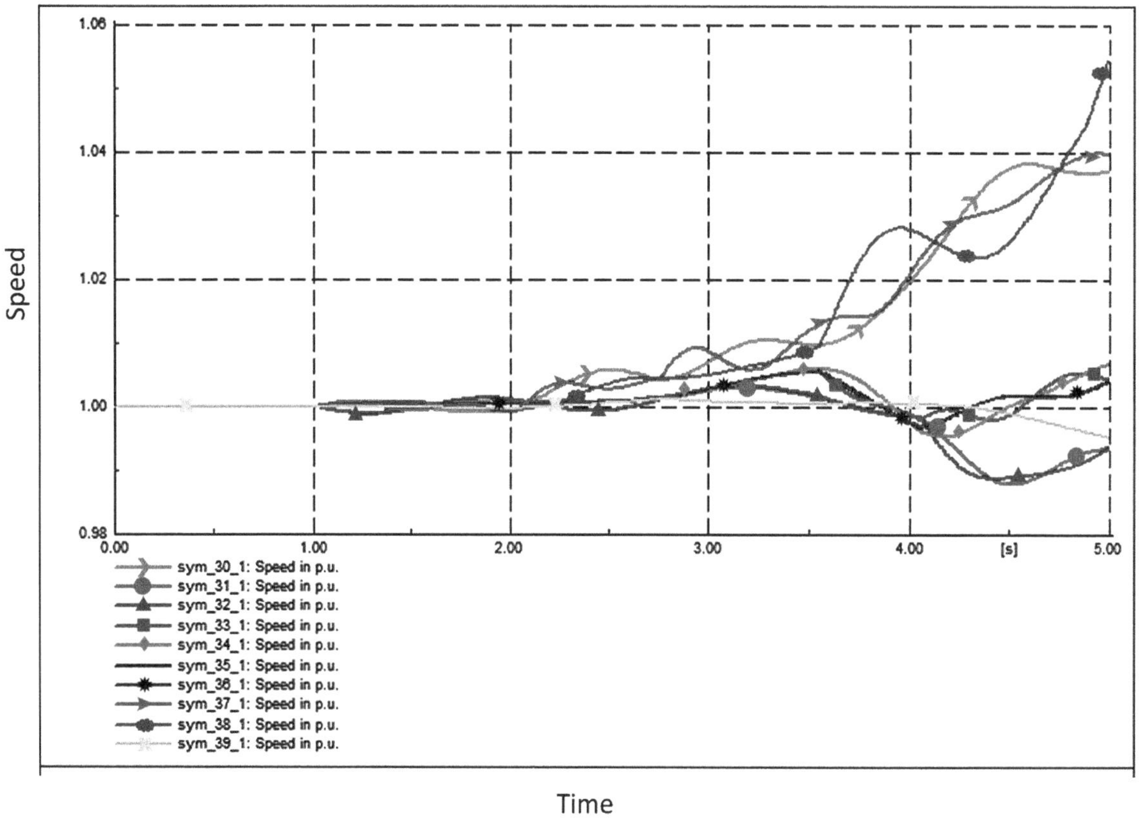

Table 2. Active load, generation and power mismatch for three formed islands

Area	Active Power in MW
Total active generation in island A	1.3506e+003
Total active load in island A	1.3325e+003
Total active imbalance in island A	18.1463
Total active generation in island B	2.1363e+003
Total active load in island B	2.1201e+003
Total active imbalance in island B	16.2336
Total active generation in island C	2.2087e+003
Total active load in island C	2.1975e+003
Total active imbalance in island C	11.1473
Total active generation	5.6957e+003
Total active load	5.6501e+003
Total active imbalance	45.5272

MCS is more accurate than 2-PEM to obtain the *PSVI* indices. Therefore, the results of the MCS have been used to rank these contingencies according to their relative severity using the *PSVI* of the critical PQ buses. In the meantime, the proposed stochastic contingency analysis based on k-PEM can be used to determine the *PSVI* of PQ buses and identify severe contingencies because of its much lower computational burden. Two proposed stochastic contingency analysis approaches based on static voltage instability of critical PQ buses using MCS and k-PEM are effective ways to determine severe contingencies, critical PQ buses and obtain the Probability of Static Voltage Instability in islanded power system.

Figure 8. Simplified graph and interface boundary network in the New England 39-bus test system

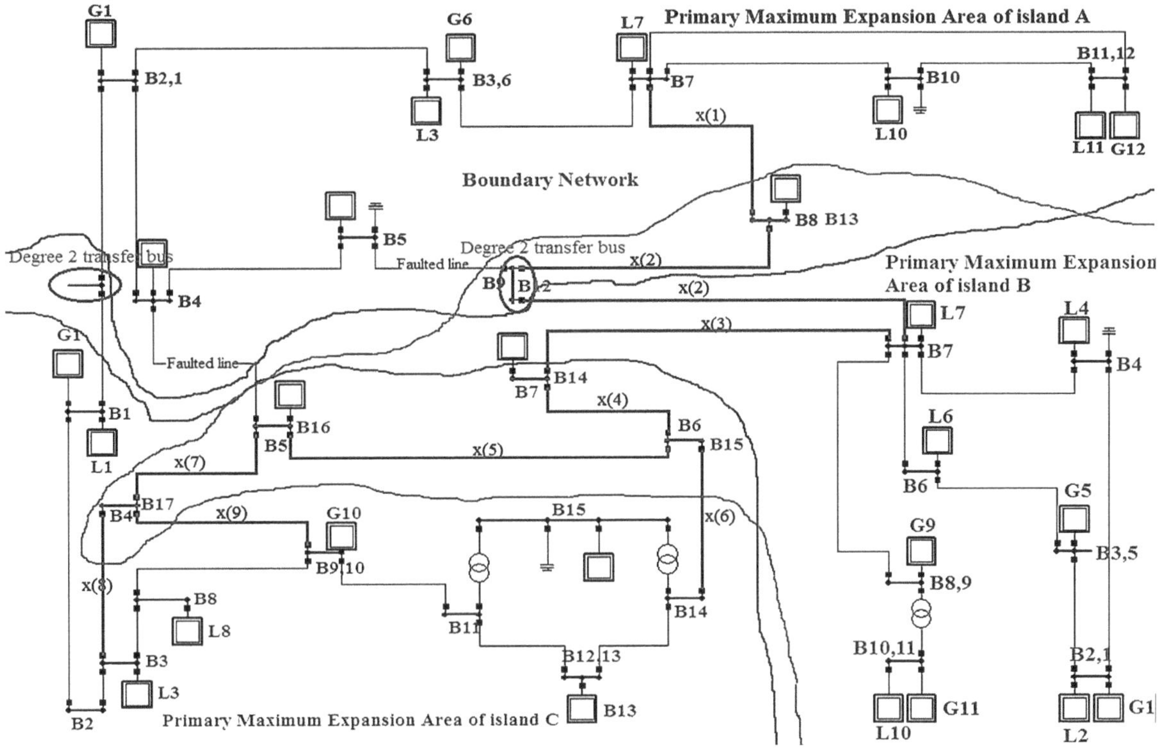

Figure 9. Optimal islanded New England 39-bus system

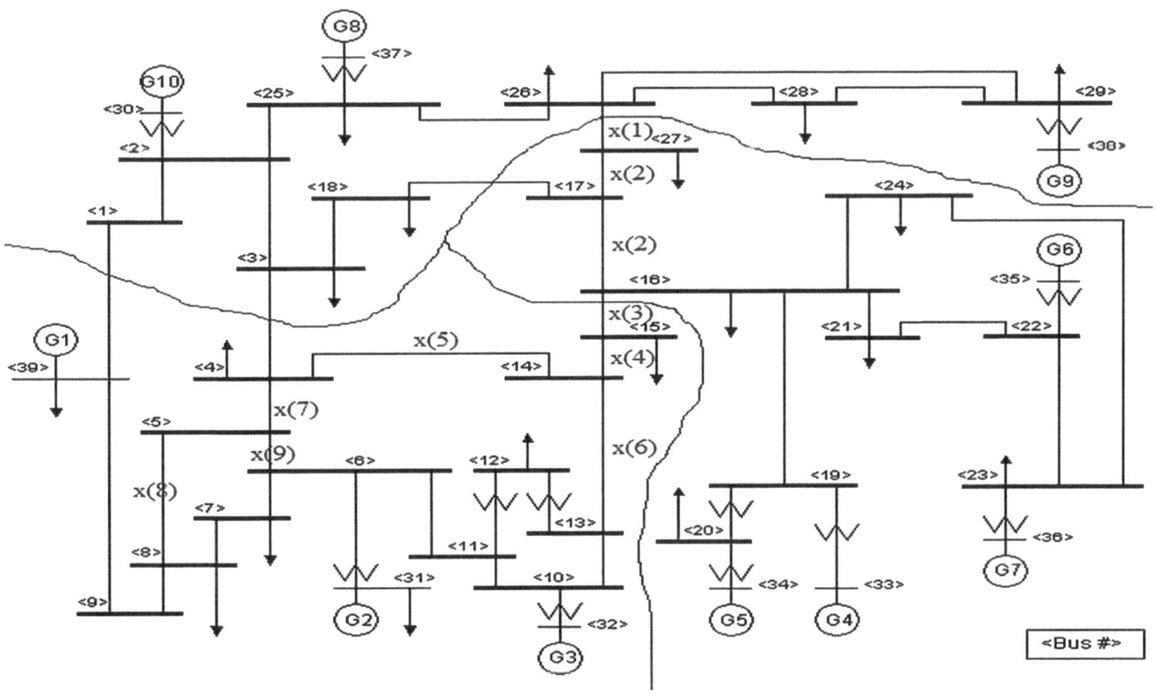

Figure 10. Voltage profile in island A

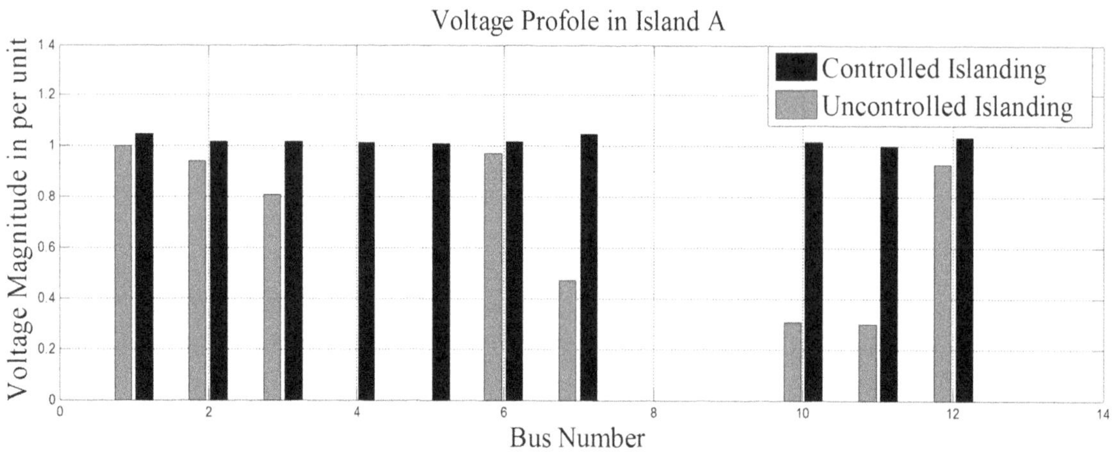

Figure 11. Voltage profile in island B

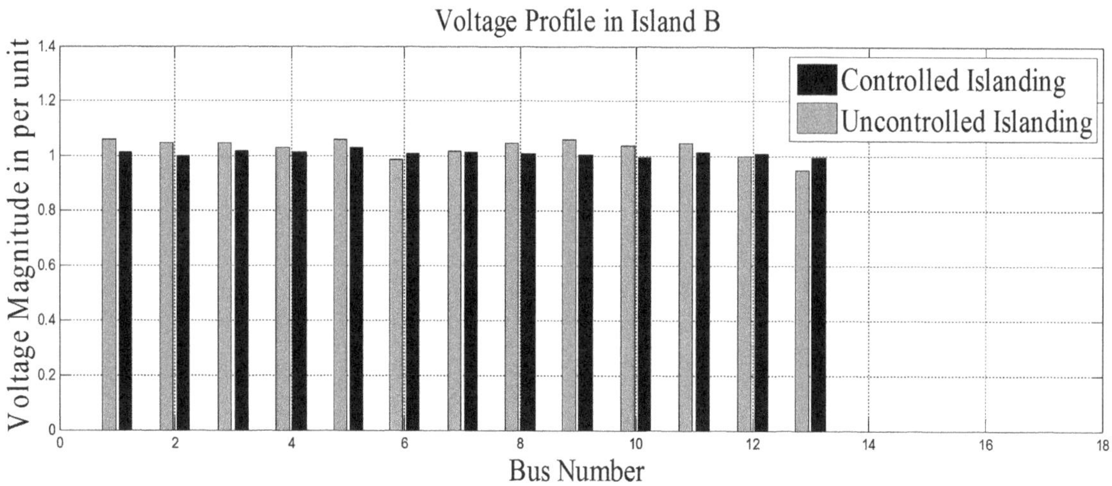

Figure 12. Voltage profile in island C

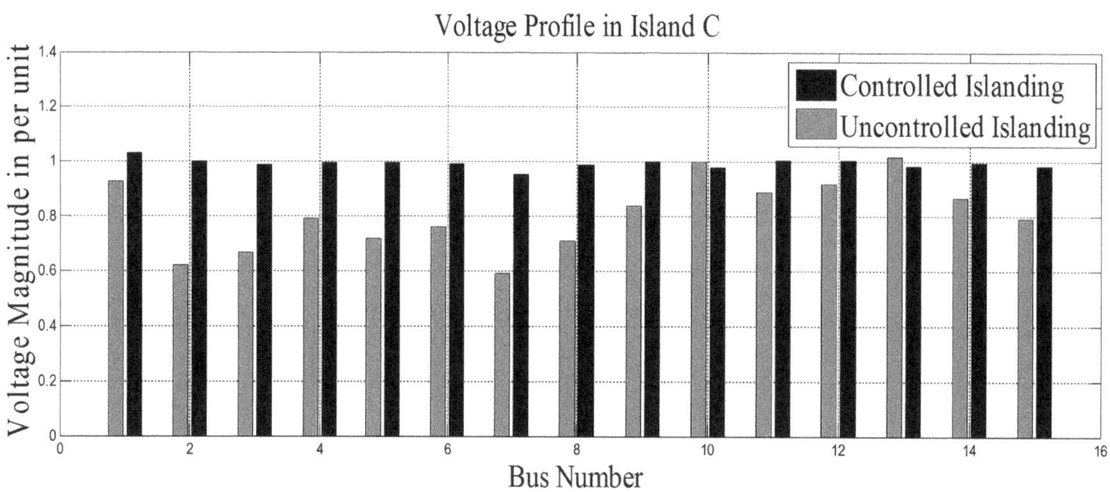

Figure 13. Static voltage stability indices in three areas

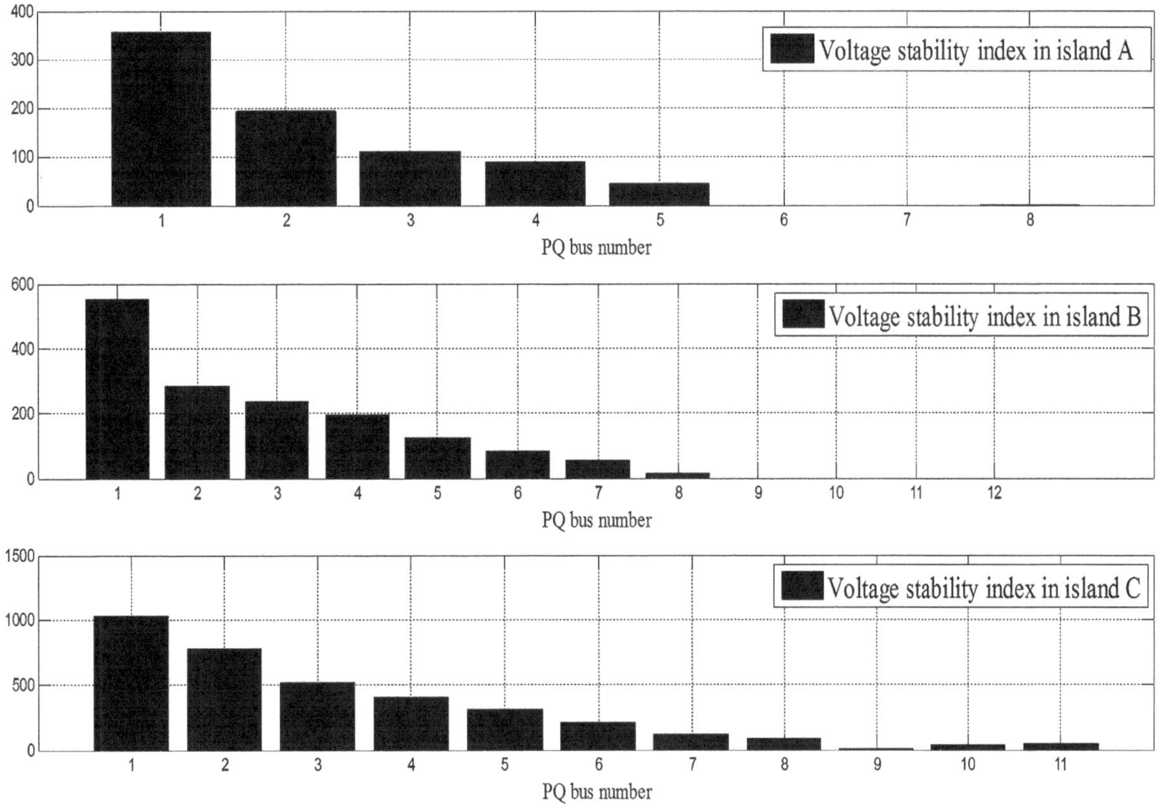

Table 3. Results of stochastic contingency analysis of island A using MCS and 2-PEM

Transmission Line Outage		$PSVI_{B_{PQ_1}}$			$PSVI_{B_{PQ_2}}$			$PSVI_{B_{PQ_3}}$		
from	**to**	**MCS**	**2-PEM**	$\varepsilon_{\lambda\|l}^{PSVI}$	**MCS**	**2-PEM**	$\varepsilon_{\lambda\|l}^{PSVI}$	**MCS**	**2-PEM**	$\varepsilon_{\lambda\|l}^{PSVI}$
2	4	0.5030	0.5000	0.59	0.5980	0.5833	2.46	0.6770	0.6666	1.54
4	5	0.3970	0.3333	16.05	0.4950	0.4166	15.83	0.5840	0.5000	14.38
2	3	0.3430	0.3333	2.83	0.4880	0.4166	14.63	0.3800	0.3333	12.29
3	7	0.1750	0.1666	4.80	0.1820	0.1666	8.46	0.3020	0.2500	17.21
Transmission Line Outage		$PSVI_{B_{PQ_4}}$			$PSVI_{B_{PQ_5}}$			$PSVI_{B_{PQ_6}}$		
from	**to**	**MCS**	**2-PEM**	$\varepsilon_{\lambda\|l}^{PSVI}$	**MCS**	**2-PEM**	$\varepsilon_{\lambda\|l}^{PSVI}$	**MCS**	**2-PEM**	$\varepsilon_{\lambda\|l}^{PSVI}$
2	4	0.5090	0.4166	16.85	0.6970	0.6666	4.36	0.5940	0.5000	15.82
4	5	0.4910	0.4166	15.15	0.5650	0.5000	11.50	0.5530	0.5000	9.58
2	3	0.3910	0.3333	14.75	0.4960	0.4166	16.01	0.4640	0.4166	10.22
3	7	0.1860	0.1666	10.43	0.2870	0.2500	12.89	0.3360	0.3333	0.80

Table 4. Results of stochastic contingency analysis of island B using MCS and 2-PEM

Transmission Line Outage		$PSVI_{B_{PQ_1}}$			$PSVI_{B_{PQ_2}}$			$PSVI_{B_{PQ_3}}$		
from	**to**	**MCS**	**2-PEM**	$\varepsilon_{\lambda\|l}^{PSVI}$	**MCS**	**2-PEM**	$\varepsilon_{\lambda\|l}^{PSVI}$	**MCS**	**2-PEM**	$\varepsilon_{\lambda\|l}^{PSVI}$
7	8	0.4580	0.4166	9.04	0.5720	0.5000	12.58	0.6920	0.6666	3.67
12	13	0.3650	0.3333	8.68	0.5130	0.5000	2.53	0.5920	0.5833	1.47
7	12	0.3470	0.3333	3.95	0.3660	0.3333	8.93	0.5620	0.5000	11.03
Transmission Line Outage		$PSVI_{B_{PQ_4}}$			$PSVI_{B_{PQ_5}}$			$PSVI_{B_{PQ_6}}$		
from	**to**	**MCS**	**2-PEM**	$\varepsilon_{\lambda\|l}^{PSVI}$	**MCS**	**2-PEM**	$\varepsilon_{\lambda\|l}^{PSVI}$	**MCS**	**2-PEM**	$\varepsilon_{\lambda\|l}^{PSVI}$
7	8	0.6050	0.5833	3.58	0.4790	0.4166	13.03	0.2940	0.2500	14.96
12	13	0.4720	0.4166	11.73	0.4030	0.3333	17.29	0.1860	0.1666	10.43
7	12	0.3900	0.3333	14.54	0.2710	0.2500	7.75	0.0950	0.0833	12.31
Transmission Line Outage		$PSVI_{B_{PQ_7}}$			$PSVI_{B_{PQ_8}}$					
from	**to**	**MCS**	**2-PEM**	$\varepsilon_{\lambda\|l}^{PSVI}$	**MCS**	**2-PEM**	$\varepsilon_{\lambda\|l}^{PSVI}$			
7	8	0.6830	0.6666	2.40	0.6210	0.5833	6.07			
12	13	0.4880	0.4166	14.63	0.5130	0.5000	2.53			
7	12	0.4560	0.4166	8.64	0.4070	0.3333	18.11			

FUTURE RESEARCH DIRECTIONS

This chapter has introduced a new approach for islanding of interconnected power system and proposed a novel strategy for stochastic contingency analyzing of optimal islanded power system based on static voltage instability using MCS and k-PEM. During each scenario, a single transmission line outage has been considered for contingency analyzing until all credible contingencies have been studied. The hierarchy proposed in this chapter provides a high-level guideline for future research such as:

- Multi-objective Optimal Placement of FACTS to Improve Short-Term Dynamic Voltage Stability and Reduce Risk of Islands Partial Blackout Considering Power System Uncertainties
- Power System Restoration Short-Term and Long-Term Scheduling to Enhance the Reliability and Security of Restored Power System: An Islands Clustering-Based Strategy
- Coherent Generators Online Clustering Based on Electrical Parameters of Synchronize Generators

Table 5. Results of stochastic contingency analysis of island C using MCS and 2-PEM

Transmission Line Outage		$PSVI_{B_{PQ_1}}$			$PSVI_{B_{PQ_2}}$			$PSVI_{B_{PQ_3}}$		
from	**to**	**MCS**	**2-PEM**	$\varepsilon_{\lambda\|l}^{PSVI}$	**MCS**	**2-PEM**	$\varepsilon_{\lambda\|l}^{PSVI}$	**MCS**	**2-PEM**	$\varepsilon_{\lambda\|l}^{PSVI}$
2	3	0.5820	0.5000	14.09	0.4660	0.4166	10.60	0.5390	0.5000	7.24
6	7	0.4740	0.4166	12.11	0.3970	0.3333	16.05	0.3430	0.3333	2.83
Transmission Line Outage		$PSVI_{B_{PQ_4}}$			$PSVI_{B_{PQ_5}}$			$PSVI_{B_{PQ_6}}$		
from	**to**	**MCS**	**2-PEM**	$\varepsilon_{\lambda\|l}^{PSVI}$	**MCS**	**2-PEM**	$\varepsilon_{\lambda\|l}^{PSVI}$	**MCS**	**2-PEM**	$\varepsilon_{\lambda\|l}^{PSVI}$
2	3	0.6980	0.6666	4.49	0.6900	0.5833	15.46	0.6980	0.6666	4.49
6	7	0.4740	0.4166	12.11	0.4550	0.4166	8.44	0.4550	0.4166	8.44
Transmission Line Outage		$PSVI_{B_{PQ_7}}$			$PSVI_{B_{PQ_8}}$			$PSVI_{B_{PQ_9}}$		
from	**to**	**MCS**	**2-PEM**	$\varepsilon_{\lambda\|l}^{PSVI}$	**MCS**	**2-PEM**	$\varepsilon_{\lambda\|l}^{PSVI}$	**MCS**	**2-PEM**	$\varepsilon_{\lambda\|l}^{PSVI}$
2	3	0.5950	0.5833	1.96	0.5630	0.5000	11.19	0.4740	0.4166	12.11
6	7	0.3020	0.2500	17.22	0.3970	0.3333	16.05	0.2020	0.1666	17.52
Transmission Line Outage		$PSVI_{B_{PQ_{10}}}$			$PSVI_{B_{PQ_{11}}}$					
from	**to**	**MCS**	**2-PEM**	$\varepsilon_{\lambda\|l}^{PSVI}$	**MCS**	**2-PEM**	$\varepsilon_{\lambda\|l}^{PSVI}$			
2	3	0.5930	0.5833	1.64	0.6780	0.5833	13.96			
6	7	0.5030	0.5000	0.59	0.6750	0.6666	1.24			

- Multi-objective Optimal Preventive Islanding Based on Modified Stochastic Backward Elimination Method to Minimize Total Load-Generation Mismatch, Maximize Static Voltage Stability Security Margin (SVSSM) and Minimize Transmission Stresses Considering Load and Large-Scale Wind Farms Uncertainties

CONCLUSION

This chapter has introduced a novel approach for splitting of large power system into several stable sections following a severe disturbance. In this chapter, an islanding scenario generator algorithm denoted as BEM (Backward Elimination Method) based on PMEAs (Primary Maximum Expansion Areas) has been proposed to generate all proper islanding solutions in the simplified network graph. The PPF program (Probabilistic Power Flow) based on Newton-Raphson method and Q-V modal analysis has been applied to evaluate the steady-state stability of created islands in each generated scenario. BICA (Binary Imperialistic Competitive Algorithm) has then been applied to minimize total load-generation mismatch considering integrity, voltage permitted range and steady-state voltage stability constraints. The best solution has then been applied to split the entire power network. A novel stochastic contingency analysis of islands based on PSVI (Probability of Static Voltage Instability) using MCS (Monte Carlo Simulation) and k-PEM (k-Point Estimate Method) have been proposed to identify the critical PQ buses and severe contingencies. In these approaches, the ITM (Inverse Transform Method) has been used to model uncertain loads with normal distribution function in optimal islanded power system. The effectiveness and capability of the proposed approaches have been validated on the New England 39 bus standard power system.

Simulation results demonstrate the computational efficiency, effectiveness and robustness of the proposed strategy to split the entire power system and stochastic contingency analysis of created islands based on static voltage instability of PQ-buses using MCS and 2-PEM. The PSVI error indices approve that the proposed stochastic contingency analysis based on static voltage instability using 2-PEM is an effective way to assess the severity of any contingency under uncertain operation. However, the MCS based stochastic contingency analysis is an accurate scheme to identify the critical PQ-buses and obtain the *PSVI* indices. The proposed Stochastic Contingency Analysis is useful and applicable in the islands short-term scheduling to detect islands weaknesses under uncertain operation.

REFERENCES

Aghamohammadi, M. R., & Shahmohammadi, A. (2012). Intentional islanding using a new algorithm based on ant search mechanism. *International Journal of Electrical Power & Energy Systems*, *35*(1), 138–147. doi:10.1016/j.ijepes.2011.10.006

Aghamohammadi, M. R., & Tabandeh, S. M. (2012, 2-5 Dec. 2012). *Online coherency identification based on correlation characteristics of generator rotor angles.* Paper presented at the 2012 IEEE International Conference on Power and Energy PECon.

Andersson, G., Donalek, P., Farmer, R., Hatziargyriou, N., Kamwa, I., Kundur, P., & Sanchez-Gasca, J. et al. (2005). Causes of the 2003 major grid blackouts in North America and Europe, and recommended means to improve system dynamic performance. *Power Systems. IEEE Transactions on*, *20*(4), 1922–1928.

Chow, J. (1996). A nonlinear model reduction formulation for power system slow coherency and aggregation. In H. Khalil, J. Chow & P. Ioannou (Eds.), *Proceedings of Workshop on Advances in Control and its Applications* (Vol. 208, pp. 282-298). Springer Berlin Heidelberg. doi:10.1007/BFb0027710

Chow, J. H., & Cheung, K. W. (1992). A toolbox for power system dynamics and control engineering education and research. *IEEE Transactions on* Power Systems, *7*(4), 1559–1564.

Chow, J. H., & Date, R. (1988, 7-9 June 1988). *A nodal aggregation algorithm for linearized two-time-scale power networks.* Paper presented at the IEEE International Symposium on Circuits and Systems.

Chow, J. H., Galarza, R., Accari, P., & Price, W. W. (1995). Inertial and slow coherency aggregation algorithms for power system dynamic model reduction. *IEEE Transactions on* Power Systems, *10*(2), 680–685. doi:10.1109/59.387903

Date, R. A., & Chow, J. H. (1991). Aggregation properties of linearized two-time-scale power networks. *IEEE Transactions on* Circuits and Systems, *38*(7), 720–730. doi:10.1109/31.135744

Ding, L., Gonzalez-Longatt, F. M., Wall, P., & Terzija, V. (2013). Two step spectral clustering controlled islanding algorithm. *IEEE Transactions on Power Systems*, *28*(1), 75–84. doi:10.1109/TPWRS.2012.2197640

Ding, L., Wall, P., & Terzija, V. (2014). Constrained spectral clustering based controlled islanding. *International Journal of Electrical Power & Energy Systems*, *63*, 687–694. doi:10.1016/j.ijepes.2014.06.016

Golari, M., Fan, N., & Wang, J. (2014). Two-stage stochastic optimal islanding operations under severe multiple contingencies in power grids. *Electric Power Systems Research*, *114*, 68–77. doi:10.1016/j.epsr.2014.04.007

Guowei, C., Junfeng, Z., Yang, D., Chan, K. W., & Long, X. (2010, 24-28 Oct. 2010). *The identification of coherent generator groups via EMD and SSI.* Paper presented at the 2010 International Conference on Power System Technology POWERCON.

Hazarika, D., Bhuyan, S., & Chowdhury, S. P. (2006). Line outage contingency analysis including the system islanding scenario. *International Journal of Electrical Power & Energy Systems*, *28*(4), 232–243. doi:10.1016/j.ijepes.2005.11.018

Hong, H. (1998). An efficient point estimate method for probabilistic analysis. *Reliability Engineering & System Safety*, *59*(3), 261–267. doi:10.1016/S0951-8320(97)00071-9

Ibrahim, I. (2011). *Source Node Expansion Algorithm for Coherency Based Islanding of Power Systems.*

Jabbari, F., & Mohammadi-ivatloo, B. (2014). Static Voltage Stability Assessment Using Probabilistic Power Flow to Determine the Critical PQ Buses. *Majlesi Journal of Electrical Engineering*, *8*(4), 17–25.

Kai, S., Da-Zhong, Z., & Qiang, L. (2003). Splitting strategies for islanding operation of large-scale power systems using OBDD-based methods. *IEEE Transactions on* Power Systems, *18*(2), 912–923. doi:10.1109/TPWRS.2003.810995

Kroese, D. P., Taimre, T., & Botev, Z. I. (2011). *Handbook of Monte Carlo Methods* (Vol. 706). John Wiley & Sons. doi:10.1002/9781118014967

Kundur, P., Balu, N. J., & Lauby, M. G. (1994). *Power system stability and control* (Vol. 7). McGraw-hill New York.

Liu, L., Liu, W., Cartes, D. A., & Chung, I.-Y. (2009). Slow coherency and Angle Modulated Particle Swarm Optimization based islanding of large-scale power systems. *Advanced Engineering Informatics*, *23*(1), 45–56. doi:10.1016/j.aei.2008.06.009

Najafi, S. (2013). An Early Warning Mechanism to Encounter Interconnected Power Systems Catastrophic Failures. *International Journal on Technical and Physical Problems of Engineering*, *5*(17), 105–112.

Najafi, S., Abedi, M., & Hosseinian, S. H. (2008, 1-3 Dec. 2008). *A strategy for frequency stability of islanded power systems.* Paper presented at the Power and Energy Conference, 2008. PECon 2008. IEEE 2nd International.

Najafi, S., Hosseinian, S. H., & Abedi, M. (2010). Proper Splitting of Interconnected Power Systems. *IEEJ Transactions on Electrical and Electronic Engineering*, *5*(2), 211–220. doi:10.1002/tee.20519

Najafi, S., Hosseinian, S. H., & Abedi, M. (2010). A Novel Strategy for Stability Evaluation of Islanded Power Systems. *IEEJ Iranian Journal of Electrical and Computer Engineering, 9*(1), 43-51.

Pahwa, S., Youssef, M., Schumm, P., Scoglio, C., & Schulz, N. (2013). Optimal intentional islanding to enhance the robustness of power grid networks. *Physica A: Statistical Mechanics and its Applications, 392*(17), 3741-3754.

Su, C.-L. (2005). A new probabilistic load flow method. *Paper presented at the Power Engineering Society General Meeting.*

Su, C.-L. (2005). Probabilistic load-flow computation using point estimate method. *IEEE Transactions on* Power Systems, *20*(4), 1843–1851.

Sun, K., Zheng, D.-Z., & Lu, Q. (2005). A simulation study of OBDD-based proper splitting strategies for power systems under consideration of transient stability. *IEEE Transactions on* Power Systems, *20*(1), 389–399.

Tian, Z., Miao, W., Hongjie, J., & Zhou, J. (2012). Multilevel partitioning algorithm for power system splitting control. *Paper presented at the Innovative Smart Grid Technologies Asia ISGT-Asia.*

Trodden, P., Bukhsh, W., Grothey, A., & McKinnon, K. (2013). MILP formulation for controlled islanding of power networks. *International Journal of Electrical Power & Energy Systems, 45*(1), 501–508. doi:10.1016/j.ijepes.2012.09.018

Verbic, G., & Cañizares, C. A. (2006). Probabilistic optimal power flow in electricity markets based on a two-point estimate method. *IEEE Transactions on* Power Systems, *21*(4), 1883–1893.

Wang, X., Shao, W., & Vittal, V. (2005). Adaptive corrective control strategies for preventing power system blackouts. *Proceedings of the 15th Power Systems Computation Conference*, Liège, Belgium.

Wang, X., & Vittal, V. (2004). System islanding using minimal cutsets with minimum net flow. *Paper presented at the Power Systems Conference and Exposition.*

Wu, Y., Tang, Y., Han, B., & Ni, M. (2015). A topology analysis and genetic algorithm combined approach for power network intentional islanding. *International Journal of Electrical Power & Energy Systems, 71*, 174–183. doi:10.1016/j.ijepes.2015.02.030

Yang, B., Vittal, V., & Heydt, G. T. (2006). Slow-Coherency-Based Controlled Islanding: A Demonstration of the Approach on the August 14, 2003 Blackout Scenario. *Power Systems. IEEE Transactions on*, *21*(4), 1840–1847. doi:10.1109/TPWRS.2006.881126

You, H., Vittal, V., & Xiaoming, W. (2004). Slow coherency-based islanding. *Power Systems. IEEE Transactions on, 19*(1), 483–491. doi:10.1109/TPWRS.2003.818729

Chapter 3
Energy-Efficient Routing Techniques for Wireless Sensors Networks

Asmaa Osamaa
Zagazig University, Egypt

Shaimaa Ahmed El-Said
Zagazig University, Egypt

Aboul Ella Hassanien
Cairo University, Egypt

ABSTRACT

Wireless sensor networks (WSNs), which normally consist of hundreds or thousands of sensor nodes each capable of sensing, processing, and transmitting environmental information, are deployed to monitor certain physical phenomena or to detect and track certain objects in an area of interests. The sensor nodes in WSN transmit data depending on local information and parameters such as signal strength, power consumption, location of data collection and accretion. Only reachable nodes are able to communicate with each other directly to collect and transmit data. The motes have limited energy resources along with constraints on its computational and storage capabilities. Thus, innovative techniques that eliminate energy inefficiencies that would shorten the lifetime of the network are highly required. Such constraints combined with a typical deployment of large number of sensor nodes pose many challenges to the design and management of WSNs and necessitate energy-awareness at all layers of the networking protocol stack. In this chapter, we present a survey of the state-of-the-art routing techniques in WSNs that take into consideration the energy issue.

1. INTRODUCTION

The popularity of WSNs has increased tremendously in recent time due to growth in advances made in wireless communication, information technologies and Micro-Electro-Mechanical Systems (MEMS) technology. Sensors networks consist of tiny, autonomous and compact devices called sensor nodes or motes deployed in a remote area to detect phenomena, collect and process data and transmit sensed information to users. The development of low-cost, low-power,

DOI: 10.4018/978-1-4666-9911-3.ch003

a multifunctional sensor has received increasing attention from various industries. Sensor nodes or motes in WSNs are small sized and are capable of sensing, gathering and processing data while communicating with other connected nodes in the network, via radio frequency (RF) channel.

Sensors Networking have profound effect on the efficiency of many military and civil applications such as target field imaging, intrusion detection, weather monitoring, security and tactical surveillance, distributed computing, detecting ambient conditions such as temperature, movement, sound, light, or the presence of certain objects, inventory control, biological detection, smart spaces, industrial diagnostics and disaster management. Deployment of sensor networks in these applications can be in random fashion (e.g., dropped from an airplane) or can be planted manually (e.g., fire alarm sensors in a facility). Initially WSN was developed for military and civilians purpose, now it is extended to wide range of applications (El-said & Hassanien, 2013) such as Disaster management and Habitat monitoring, Target tracking and Security management, Medical and health care, home automation and traffic control, machine failure diagnosis and energy management, Rescue missions, climate change, earthquake warning and monitoring the enemy territory (see Figure 1).

WSN has the potentiality to connect the physical world with the virtual world by forming a network of sensor nodes. Here, sensor nodes are usually battery-operated devices, and hence energy saving of sensor nodes is a major design issue. To prolong the network's lifetime, minimization of energy consumption should be implemented at all layers of the network protocol stack starting from the physical to the application layer including cross-layer optimization.

Routing in WSNs is very challenging due to the inherent characteristics that distinguish these networks from other wireless networks like mobile ad hoc networks or cellular networks. Presence of large number of sensor nodes and constraints in terms of energy, processing, and storage capacities requires careful management of resources. Due to such differences, many algorithms have been proposed for the routing in WSNs. These routing mechanisms have taken into consideration the inherent features of WSNs along with the application and architecture requirements. The task of finding and maintaining routes in WSNs

Figure 1. Overview of Wireless Sensor Network applications

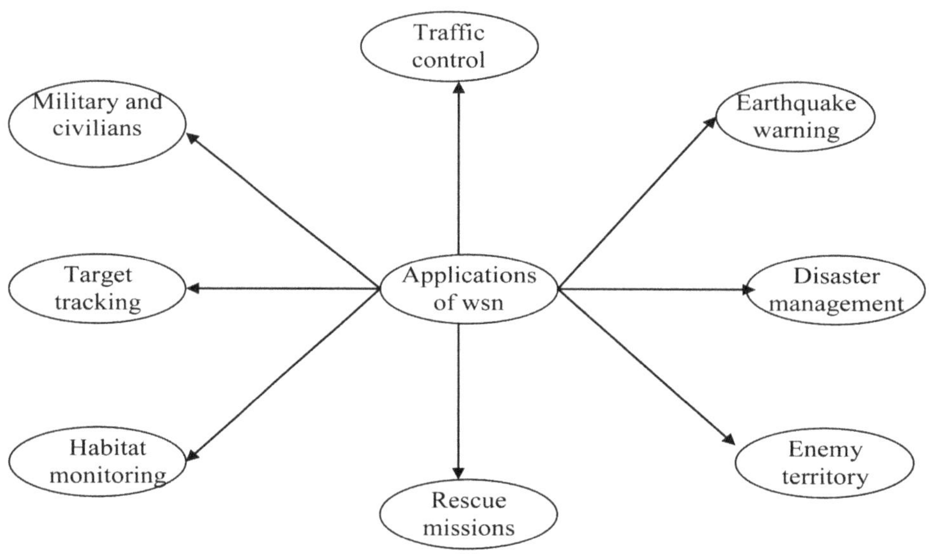

is nontrivial since energy restrictions and sudden changes in node status (e.g., failure) cause frequent and unpredictable topological changes. To evaluate the performance of different routing algorithms, we have to answer some questions about their efficiency like; how reliable is an algorithm? How efficient does it use resources? And how fast a message can be delivered? This can be quantified by the performance metrics (Misra et al., 2009) that are described in this subsection.

- *Delivery rate* is defined as the fraction of successful delivered messages over the total number of messages created at the source node.
- *Flooding rate* is used as a measure of communication overhead of multipath and flooding-based strategies. It is the ratio of the number of message transmissions needed by the algorithm and the shortest possible path between source and destination node.
- *Dilation* is the ratio of hop count for the given method and the hop count produced by the shortest path algorithm.

1.1 Wireless Sensor Network Architecture

Figure 2 shows the basic block diagram of a wireless sensor node. It consists of four basic components: a sensing unit, a processing unit, a transceiver unit and a power unit. There can be application dependent additional components such as a location finding system, a power generator and a mobilizer.

Figure 3 shows the architecture of a typical Wireless Sensor Network with the components of a sensor node. The functionalities like sensing, processing, transmission, location finding, power consumption etc. are available in each of the nodes. The decisions made by each mote rely upon its own computing capability, available energy resources and the metadata about the information it has. Each of the nodes has the capability to collect and route the data either to other nodes or to the source node. The source node (or the base station) maybe fixed or mobile and the network can be linked to some other communication network or even to the Internet so that users can access the collected data easily (Bains & Sharma, 2012).

Figure 2. Architecture of a Wireless Sensor Node

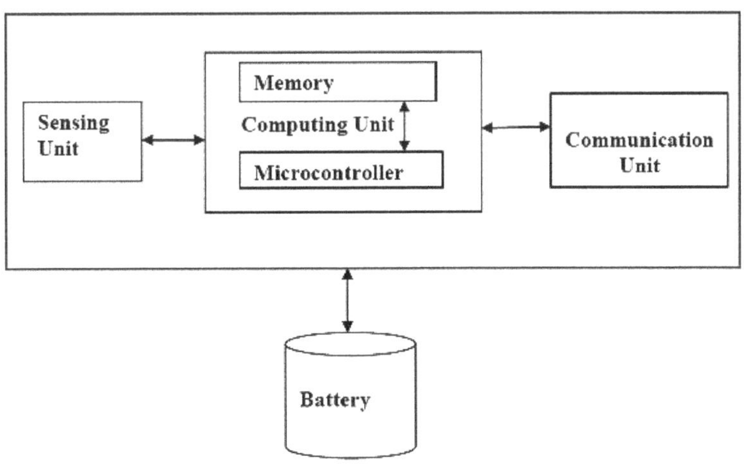

Figure 3. Architecture of Wireless Sensor Network (Bains & Sharma, 2012)

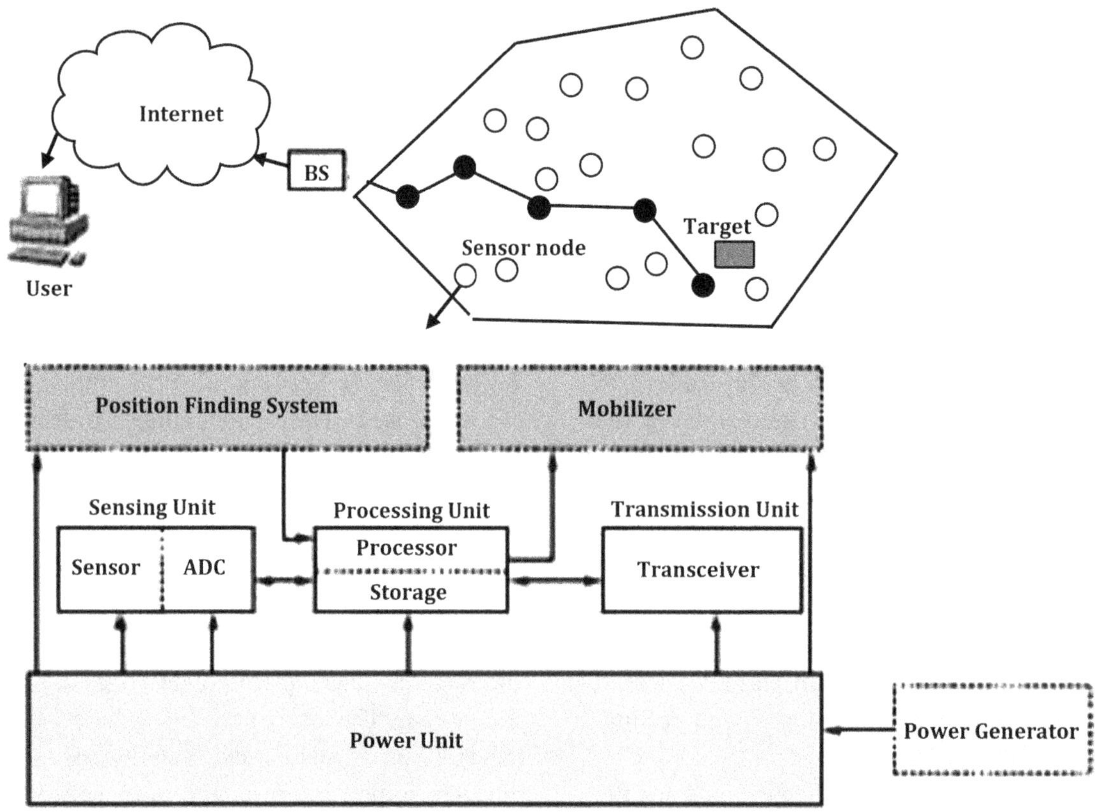

1.2 Routing Differences between WSNs and Other Wireless Systems

Routing in WSNs is very challenging due to the inherent characteristics that distinguish these networks from other wireless networks like mobile ad hoc networks or cellular networks. Routing in WSNs differs from conventional routing in fixed networks in various ways:

1. There is no infrastructure.
2. Wireless links are unreliable.
3. Routing protocols have to meet strict energy saving requirements.
4. In contrast to typical communication networks, almost all applications of sensor networks require the flow of sensed data from multiple sources to a particular BS.
5. Sensor nodes are tightly constrained in terms of energy, processing, and storage capacities. Thus, they require careful resource management.
6. In some applications, some sensor nodes may be allowed to move and change their location.
7. Design requirements of a sensor network change with application.
8. Data collected by many sensors in WSNs is typically based on common phenomena; hence there is a high probability that this data has some redundancy. Such redundancy needs to be exploited by the routing protocols to improve energy and bandwidth utilization.

1.3 Problem Statement

The battery supplies power to the complete sensor node. It plays a vital role in determining sensor node lifetime. The amount of power drawn from a battery should be carefully monitored. Sensor nodes are generally small, light and cheap, the size of the battery is limited. AA batteries normally store 2.2 to 2.5 Ah at 1.5 V. However, these numbers vary depending on the technology utilized. For example, Zinc–air-based batteries have higher capacity in Joules/cm^3 than lithium batteries. Alkaline batteries have the smallest capacity, normally around 1200 J/cm^3. Furthermore, sensors must have a lifetime of months to years, since battery replacement is not an option for networks with thousands of physically embedded nodes. This causes energy consumption to be the most important factor in determining sensor node lifetime. One of the main design goals of WSNs is to carry out data communication while trying to prolong the lifetime of the network and prevent connectivity degradation by employing aggressive energy management techniques.

The main contribution of this chapter is to provide a dedicated study of the network layer, describing and categorizing the different approaches for data routing that trying to prolong the lifetime of the network and identify some open research issues that can be further pursued. In addition, we summarize routing challenges and design issues that may affect the performance of routing protocols in WSNs.

The rest of this chapter is organized as follows. Section 2 summarizes routing challenges and design issues that may affect the performance of routing protocols in WSNs. In Section 3, a detailed survey of energy efficient routing techniques is presented. A conclusion and a summary of future research directions on routing in WSNs are provided in section 4.

2. ROUTING CHALLENGES

Routing in WSN faces many challenges (Al-Karaki & Kamal, 2004) and design issues such as node deployment, energy consumption without losing accuracy, data reporting model, node/link heterogeneity, fault tolerance, scalability, network dynamics, transmission media, connectivity, coverage, data aggregation, and quality of service (QOS). In order to design an efficient routing protocol, there are challenging factors should be addressed meticulously. In the following, we summarize some of the routing challenges and design issues that affect routing process in WSN.

2.1 Node Deployment

Node deployment in WSN affects the performance of the routing protocol. The deployment can be either deterministic or randomized. In deterministic deployment, the sensor nodes are manually placed and data is routed through pre-determined paths; but in random node deployment, the sensor nodes are scattered randomly creating an infrastructure in an ad hoc manner. Hence, random deployment raises several issues as coverage, optimal clustering etc. which need to be addressed.

2.2 Energy Consumption without Losing Accuracy

Sensor nodes can use up their limited supply of energy performing computations and transmitting information in a wireless environment. As such, energy conserving forms of communication and computation are essential. Sensor node's lifetime shows a strong dependence on the battery lifetime. The malfunctioning of some sensor nodes due to power failure can cause significant topological changes and might require re-routing of packets and re-organization of the network.

2.3 Coverage

In WSN, each sensor node obtains a certain view of the environment. A given sensor's view of the environment is limited both in range and in accuracy; it can only cover a limited physical area of the environment. Hence, area coverage is also an important design parameter in WSN.

2.4 Data Reporting Model

Data sensing and reporting in WSN is dependent on the application and the time criticality of the data reporting. In WSN data reporting can be time-driven (continuous), query-driven or event-driven. The time-driven delivery model is suitable for applications that require periodic data monitoring. As such, sensor nodes will periodically switch on their sensors and transmitters, sense the environment and transmit the data of interest at constant periodic time intervals. In event-driven and query-driven models, sensor nodes react immediately to sudden and drastic changes in the value of a sensed attribute due to the occurrence of a certain event or a query is generated by the base station (BS). As such, these are well suited for time critical applications. The routing protocol is highly influenced by the data reporting model with regard to energy consumption and route stability.

2.5 Fault Tolerance

Some sensor nodes may fail or be blocked due to lack of power, physical damage, or environmental interference. The failure of sensor nodes should not affect the overall task of the sensor network. If many nodes fail, routing protocols must accommodate formation of new links and route to the data collection BS. This may require actively adjusting transmit powers and signaling rates on the existing links to reduce energy consumption, or re-routing packets through regions of the network where more energy is available. Therefore, multiple levels of redundancy may be needed in a fault-tolerant sensor network.

2.6 Quality of Service

In some applications, data should be delivered within a certain period of time from the moment it is sensed; otherwise the data will be useless. Therefore bounded latency for data delivery is another condition for time-constrained applications. However, in many applications, conservation of energy, which is directly related to network lifetime, is considered relatively more important than the quality of data sent. As the energy gets depleted, the network may be required to reduce the quality of the results in order to reduce the energy dissipation in the nodes and hence lengthen the total network lifetime. Hence, energy-aware routing protocols are required to capture this requirement.

2.7 Data Aggregation

Data aggregation is the combination of data from different sources according to a certain aggregation function. This technique has been used to achieve energy efficiency and data transfer optimization in a number of routing protocols.

2.8 Node/Link Heterogeneity

In many studies, all sensor nodes were assumed to be homogeneous, i.e., having equal capacity in terms of computation, communication, and power. However, depending on the application a sensor node can have different roles or capabilities. The existence of heterogeneous set of sensors raises many technical issues related to data routing. For example, some applications might require a diverse mixture of sensors for monitoring temperature, pressure and humidity of the surrounding environment, detecting motion via acoustic signatures, and capturing the image or video tracking of moving objects. These special sensors can be either deployed independently or the different functionalities can be included in the same sensor nodes. Even data reading and reporting can be generated from these sensors at different rates, subject to diverse QOS constraints, and can follow multiple data reporting models.

2.9 Scalability

The number of sensor nodes deployed in the sensing area may be in the order of hundreds or thousands, or more. Any routing scheme must be able to work with this huge number of sensor nodes. In addition, sensor network routing protocols should be scalable enough to respond to events in the environment. Until an event occurs, most of the sensors can remain in the sleep state.

2.10 Network Dynamics

Most of the network architectures assume that sensor nodes are stationary. However, mobility of both BS and sensor nodes is sometimes necessary in many applications. Routing messages from or to moving nodes is more challenging since route stability becomes an important issue. Moreover, the sensed phenomenon can be either dynamic or static depending on the application, e.g., it is dynamic in a target detection/tracking application, while it is static in forest monitoring for early fire prevention. Dynamic events in most applications require periodic reporting and consequently generate significant traffic to be routed to the BS.

2.11 Transmission Media

In a multi-hop sensor network, communicating nodes are linked by a wireless medium. The traditional problems associated with a wireless channel (e.g., fading, high error rate) may also affect the operation of the sensor networks. As the transmission energy varies directly with the square of distance therefore a multi-hop network is suitable for conserving energy.

2.12 Connectivity

High node density in sensors networks precludes them from being completely isolated from each other. Therefore, sensor nodes are expected to be highly connected. This, however, may not prevent the network topology from being variable and the network size from being shrinking due to sensor node failures. In addition, connectivity depends on the, possibly random, distribution of nodes.

3. ENERGY-EFFICIENT WSN ROUTING TECHNIQUES

WSNs have inherent and unique characteristics rather than traditional networks. They have many different constraints, such as computational power, storage capacity, energy supply and etc; of course the most important issue is their energy constraints. The problem of energy consumption in WSN is not new to the research community. A lot of research activities in this field have been developed to find efficient techniques to minimize the energy consumption in WSN. To overcome this problem, many efforts have been made by the researchers across the globe for last many years.

The task of finding and maintaining routes in WSN is nontrivial since energy restrictions and sudden changes in node status (e.g., failure) cause frequent and unpredictable topological changes. The nodes gather data amongst their neighbors via their sensors process and forward the information to the user or a data sink. Being compact and wireless, they are highly energy constrained. Moreover, applications for sensor networks often require nodes to operate unattended for long periods of time on same batteries. However, it is infeasible to replace batteries of up to thousands of nodes. Thus, conserving energy so as to maximize lifetime is one of the key challenges in sensor networks (Gupta & Sharma, 2011). To minimize energy consumption, routing techniques proposed in the literature for WSN employ some well-known routing tactics as well as tactic special to WSN, e.g., data aggregation and in network processing, clustering, different node role assignment, and data-centric methods were employed. In this section, we study the design tradeoffs between energy and communication overhead savings in every routing paradigm. We also highlight the advantages and performance issues of each routing technique.

Many routing, power management, and data dissemination protocols have been specifically designed for WSN where energy awareness is an essential design issue. Overall, the routing techniques are classified as follow:

- **Network Structure:** Flat-based routing, hierarchical-based routing and adaptive-based routing.
- **Protocol Operation:** Negotiation-based routing, multipath-based routing, query-based routing and location based routing.

Figure 4 classifies the routing techniques in WSN into two categories; network structure, and protocol operation. Depending on *the network structure*, routing protocols in WSN can be classified into; flat-based (Data Centric routing), hierarchical-based (Cluster based routing) and adaptive based routing protocols. Depending on *the protocol operation*, routing protocols in WSN can be divided into; negotiation-based routing, multipath-based routing, query-based routing and location based routing.

3.1 The Routing Protocols for Protocol Operation

Negotiation Based Routing

These protocols use high-level data descriptors called *Meta data* in order to eliminate redundant data transmission through negotiations. The necessary decisions are based on available resources and local interactions. The SPIN protocol is an example of negotiation based routing protocols.

Sensor Protocols for Information via Negotiation (SPIN) (Heinzelman et al., 1999) is one of well known negotiation based routing protocols for WSN. The SPIN protocols are designed to disseminate the data of one sensor to all other sensors assuming these sensors are potential BS. Hence, the main idea of negotiation based routing in WSN is to suppress duplicate information and prevent redundant data from being sent to the next sensor

Figure 4. Classification of WSN routing protocols

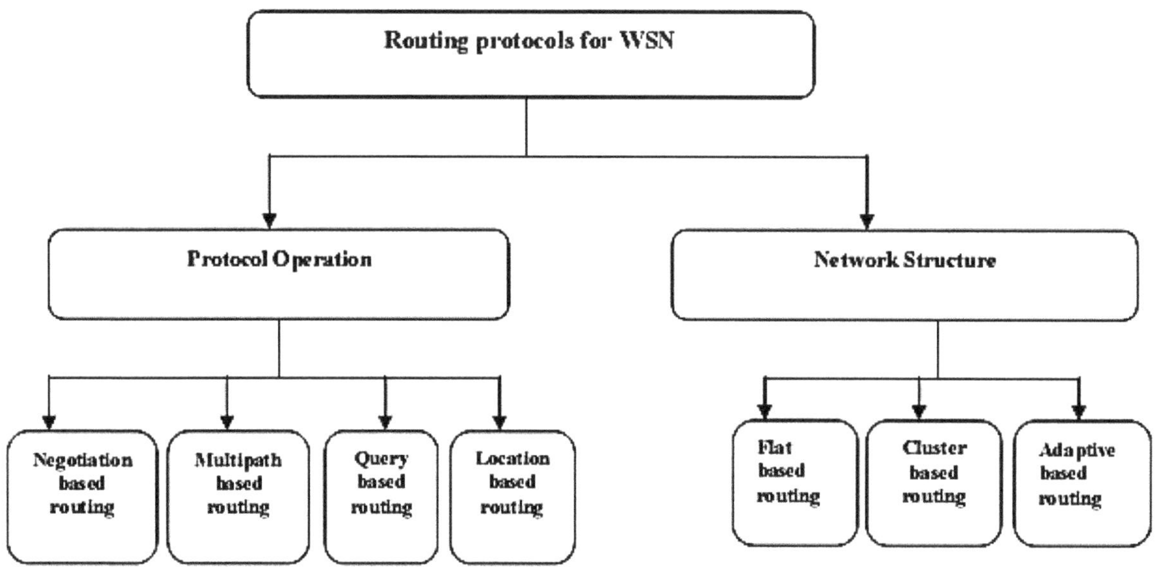

or the BS by conducting a series of negotiation messages before the real data transmission begins.

The idea behind SPIN is to name the data using high level descriptors or meta-data. Before transmission, meta-data are exchanged among sensors via a data advertisement mechanism, which is the key feature of SPIN. Each node upon receiving new data, advertises it to its neighbors and interested neighbors, i.e. those who do not have the data, retrieve the data by sending a request message. SPIN's meta-data negotiation solves the classic problems of flooding such as redundant information passing and overlapping of sensing areas thus, achieving a lot of energy efficiency. There is no standard meta-data format and it is assumed to be application specific, e.g. using an application level framing. There are three messages defined in SPIN to exchange data between nodes. Figure 5 summarizes the steps of the SPIN protocol.

One of the advantages of SPIN is that topological changes are localized since each node needs to know only its single-hop neighbors. However, SPIN's data advertisement mechanism cannot guarantee the delivery of data. For instance, if the nodes that are interested in the data are far away from the source node and the nodes between source and destination are not interested in that data, such data will not be delivered to the destination at all. Therefore, SPIN is not a good choice for applications such as intrusion detection, which require reliable delivery of data packets over regular intervals.

Multipath Based Routing

These protocols offer fault tolerance by having at least one alternate path (from source to sink) when the primary path fails. This can be achieved

Figure 5. SPIN protocol. A starts by advertising its data to B (a). Node B responds by sending a request to node A (b). After receiving the requested data (c), node B then sends out advertisements to its neighbors (d), who in turn send requests back to B (e-f). (Heinzelmanet al., 1999)

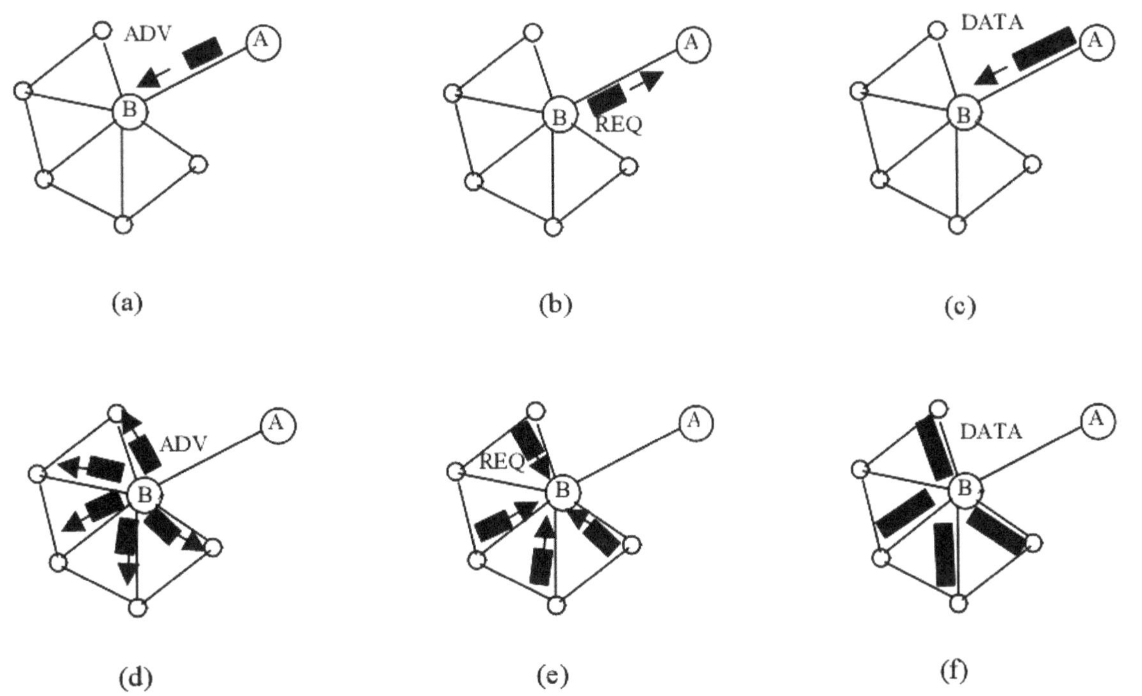

by maintaining multiple paths between the source and the destination at the expense of an increased energy consumption and traffic generation. These paths are kept alive by sending periodic messages.

Maximum Lifetime Routing in Wireless Sensors Networks (Chang & Tassiulas, 2004) is multipath based routing. It is a protocol that routes data through a path whose nodes have the largest residual energy. The path is switched whenever a better path is discovered. The primary path will be used until its energy is below the energy of the backup path. By means of this approach, the nodes in the primary path will not deplete their energy resources through continual use of the same route, thus achieving longer lifetime. A disadvantage for applications that require mobility on the nodes is that the protocol is oriented to solve routing problem in static wireless networks.

Hierarchical Power-aware Routing in Sensor Networks (Li et al., 2011) protocol enhances the reliability of WSN by using multipath routing. It is useful for delivering data in unreliable environments. The idea is to define many paths from source to sink and send through them the same sub packets (the original data packet is divided into sub packets through each path). This implies that the traffic will increase significantly (not energy ware), but increasing the reliability of the network.

Query Based Routing

In these protocols, the destination nodes propagate a query for data (sensing task or interest) from the node through the network. The nodes containing this data send it back to the node that has initiated the query.

Energy Efficient Query-based Routing Protocol (EEQR) (Ahvar & Mahdavi, 2011) is an energy aware routing protocol that consists of three major steps:

1. **Neighbors Discovery:** At the beginning, each node simply broadcasts its id and energy level by a Hello packet. When each node receives the Hello packet, it adds an entry to its "neighbor list", including "neighbor id", "energy level" and "receipt time".
2. **Agent Forwarding:** When a node witnesses an event, it creates an agent packet (a packet that is responsible for spreading rumors about the events in the network) and forwards it to the next hop.
3. **Query Forwarding:** To collect data from the network, when the geographic information is available, the best route is the shortest path and there is no need to broadcast. But if the nodes in the spatial information are not available, queries (request packet for receiving information on a particular event) are distributed across the network to discover the best route to event. When the query reaches its destination, the data can be sent to the producer of query.

Location Based Routing

Location-based protocols utilize the position information to relay the data to the desired regions rather than the whole network. Most of the routing protocols for sensor networks require location information for sensor nodes. In most cases location information is needed in order to calculate the distance between two particular nodes so that energy consumption can be estimated. Distances to next neighboring nodes can be estimated by signal strengths or by Global Positioning System (GPS) receivers.

Minimum Energy Communication Network (MECN) (Rodoplu & Ming, 1999) sets up and maintains a minimum energy network for wireless networks by utilizing low power GPS. MECN identifies a relay region for every node. The relay region consists of nodes in a surrounding area where transmitting through those nodes is more energy efficient than direct transmission. The relay region for node pair (i, r) is depicted in Figure 6. The enclosure of a node i is then created by taking the union of all relay regions that

Figure 6. Relay region of transmit-relay node pair (i, r) in MECN (Rodoplu & Ming, 1999)

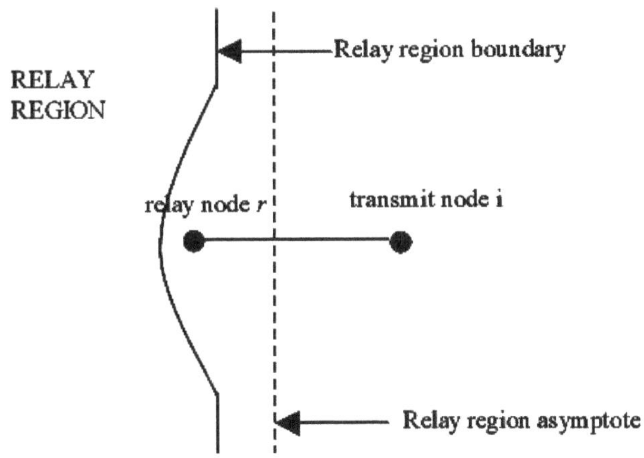

node *i* can reach. The main idea of MECN is to find a sub-network, which will have less number of nodes and require less power for transmission between any two particular nodes. In this way, global minimum power paths are found without considering all the nodes in the network. This is performed using a localized search for each node considering its relay region.

Geographic Adaptive Fidelity (GAF) (Heidemann ET AL., 2001) protocol is energy-aware location-based routing designed primarily for mobile ad hoc networks and can be applicable to sensor networks as well. GAF keeps energy by turning off unnecessary nodes in the network without affecting the level of routing fidelity. It forms a virtual grid for the covered area. Each node uses its GPS-indicated location to associate itself with a point in the virtual grid. Nodes associated with the same point on the grid are considered equivalent in terms of the cost of packet routing. Such equivalence is exploited in keeping some nodes located in a particular grid area in sleeping state in order to save energy. Thus, GAF can substantially increase the network lifetime as the number of nodes increase. A sample situation is depicted in Figure 7(a). In this figure, node 1 can reach any of 2, 3 and 4 and nodes 2, 3, and 4 can reach 5. Therefore nodes 2, 3 and 4 are equivalent and two of them can sleep. Nodes change states from sleeping to active in turn so that the load is balanced. There are three states defined in GAF. These states are discovery, for determining the

Figure 7. (a) Example of virtual grid in GAF, and (b) State transitions in GAF (Heidemann et al., 2001)

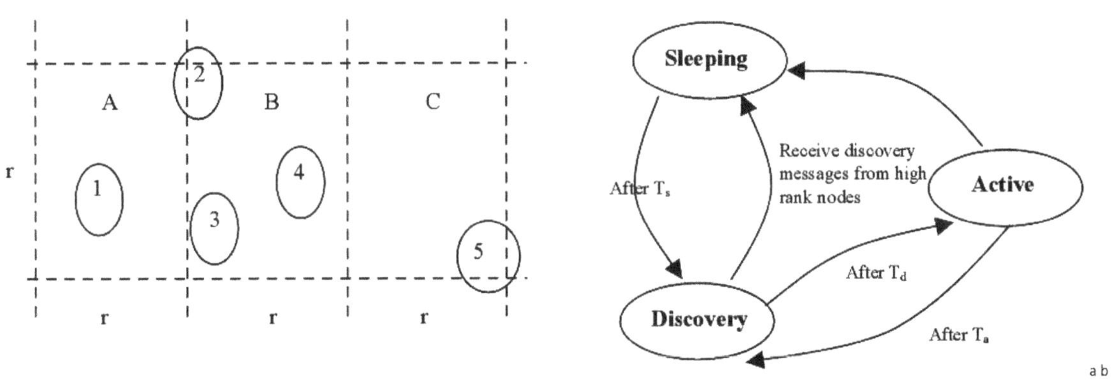

neighbors in the grid, active reflecting participation in routing and sleep when the radio is turned off. The state transitions in GAF are depicted in Figure 7(b). Which node will sleep for how long is application dependent and the related parameters are tuned accordingly during the routing process. In order to handle the mobility, each node in the grid estimates its leaving time of grid and sends this to its neighbors. The sleeping neighbors adjust their sleeping time accordingly in order to keep the routing fidelity. Before the leaving time of the active node expires, sleeping nodes wake up and one of them becomes active.

Geographic and Energy Aware Routing (GEAR) (Yu et al., 2001) is the protocol which uses geographic information while disseminating the queries to the areas of interest since data queries often includes geographic attributes. The protocol uses energy aware and geographically informed neighbor selection to route a packet towards the target area. GEAR can complement Directed diffusion (DD) by restricting the number of interests sent, and only considering a certain area rather than sending the interests to the whole network. In GEAR, each node keeps an estimated cost and a learning cost of reaching the destination through its neighbors. There are two phases in the algorithm:

1. Forwarding packets towards the target region: Upon receiving a packet, a node checks its neighbors to see if there is one neighbor, which is closer to the target region than itself. If there is more than one, the nearest neighbor to the target region is selected as the next hop. If they are all further than the node itself, this means there is a hole. In this case, one of the neighbors is picked to forward the packet based on the learning cost function. This choice can then be updated according to the convergence of the learned cost during the delivery of packets.
2. Forwarding the packets within the region: If the packet has reached the region, it can be diffused in that region by either recursive geographic forwarding or restricted flooding. Restricted flooding is good when the sensors are not densely deployed. In high-density networks, recursive geographic flooding is more energy efficient than restricted flooding. In that case, the region is divided into four sub regions and four copies of the packet are created. This splitting and forwarding process continues until the regions with only one in GEAR node are left. An example is depicted in Figure 8.

A virtual relative position based routing protocol for sensor networks that provides methods for data management is *Virtual Cord Protocol* (VCP) (Awad et al., 2008). The key characteristics of VCP are the geographical vicinity of virtual neighbors, which reduces the communication load, VCP only needs information about direct neighbors for routing, and it cannot be stuck with dead ends. The protocol is easy to be implemented on the top of a typical Media Access Control (MAC) layer. All data items are associated with numbers in a predefined range is captured by the available nodes.

Figure 8. Recursive geographic forwarding (Yu et al., 2001)

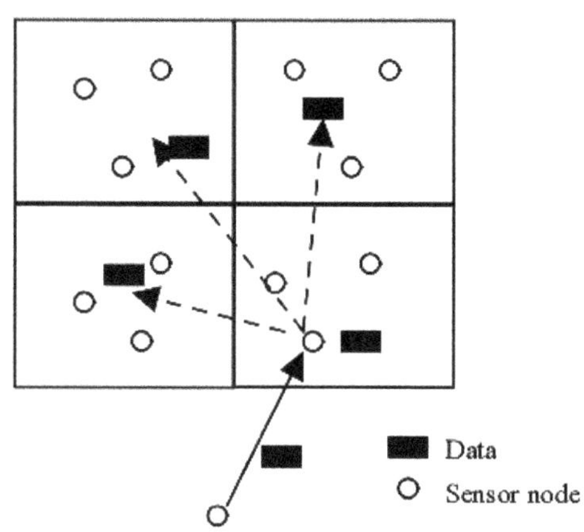

3.2 The Routing Protocols for Network Structure

Flat Based Routing

In flat-based routing algorithm, all nodes play the same role and mainly apply flood based data transferring. The drawbacks of flooding include implosion, which is caused by duplicate messages sent to the same node, overlap when two nodes sensing the same region send similar packets to the same neighbor, and resource blindness in consuming large amounts of energy without consideration for energy constraints. The well known protocols considered in flat based routing are: Sequential Assignment Routing (SAR), Directed Diffusion (DD), Energy Aware Routing (EAR)….. etc.

Sequential Assignment Routing (SAR) (Sohrabi & Pottie, 2000) was one of the first protocols for WSN that considered QoS issues for routing decisions. It is a table-driven multi-path approach striving to achieve energy efficiency and fault tolerance. SAR makes a routing decision based on three factors: energy resources, QoS planned for each path, and the packet's traffic type, which is implemented by a priority mechanism. The multipath tree is defined by avoiding nodes with low-energy while taking into account that the root tree is located in the source node and its ends in the sink nodes set. In other words, SAR creates a multipath table whose main objective is to obtain energy efficiency and fault tolerance. Although this ensures fault tolerance and easy recovery, the protocol suffers certain overhead when tables and node states must be maintained or refreshed. This problem increases especially when there are a large number of nodes.

Directed Diffusion (DD) (Intanagonwiwat et al., 2000) is a data-centric and application aware paradigm since all data generated by sensor nodes are named by attribute value pairs. The objective of the DD paradigm is to aggregate the data coming from different sources by deleting redundancy, which drastically reduces the number of transmissions so, the main consequence is that the network saves its energy and extends its life.

The idea aims at diffusing data through sensor nodes. The main reason behind using such a scheme is to get rid of unnecessary operations of network layer routing in order to save energy. DD suggests the use of attribute-value pairs for the data and queries the sensors in an on demand basis by using those pairs. In order to create a query, an interest is defined using a list of attribute-value pairs such as name of objects, interval, duration, geographical area, etc. The interest is broadcast by a sink through its neighbors. Each node receiving the interest can do caching for later use. The nodes also have the ability to do in-network data aggregation. The interests in the caches are then used to compare the received data with the values in the interests. The interest entry also contains several gradient fields. A gradient is a reply link to a neighbor from which the interest was received. It is characterized by the data rate, duration and expiration time derived from the received interest's fields. Hence, by utilizing interest and gradients, paths are established between sink and sources. Several paths can be established so that one of them is selected by reinforcement. The sink resends the original interest message through the selected path with a smaller interval hence reinforces the source node on that path to send data more frequently. Figure 9 summarizes the DD protocol.

DD has many advantages. Since it is data-centric, all communications are neighbor-to-neighbor with no need for a node addressing mechanism. Each node can do aggregation and caching, in addition to sensing. Caching is a big advantage in terms of energy efficiency and delay. In addition, DD is highly energy efficient since it is on demand and there is no need for maintaining global network topology. However, DD cannot be applied to all sensor network applications since it is based on a query-driven data delivery model. The applications that require continuous data delivery to the sink will not work efficiently with

Figure 9. Directed Diffusion protocol phases (Intanagonwiwat et al., 2000)

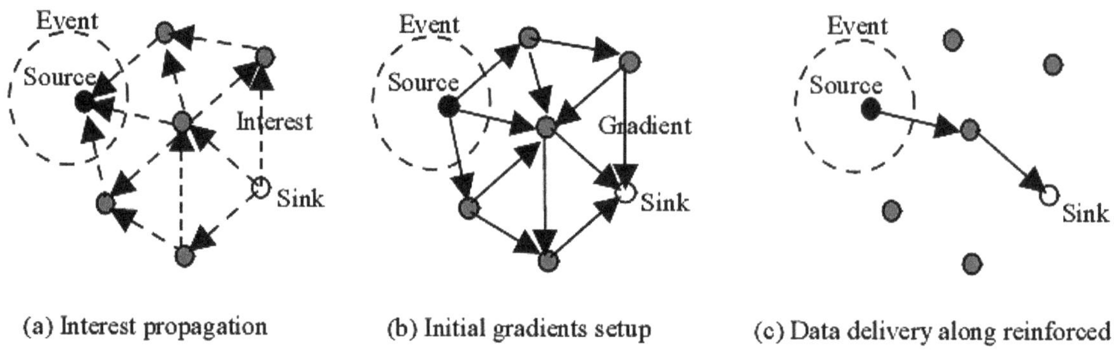

a query-driven on demand data model. Therefore, DD is not a good choice as a routing protocol for the applications such as environmental monitoring. In addition, the naming schemes used in DD are application dependent and each time should be defined a priori. Moreover, the matching process for data and queries might require some extra overhead at the sensors.

Rumor routing protocol (Braginsky & Estrin, 2002) is one of the routing protocols used in the context of event notification. It is another variation of DD. Generally DD floods the query to the entire network when there is no geographic criterion to diffuse tasks. However, in some cases there is only a little amount of data requested from the nodes and thus the use of flooding is unnecessary. An alternative approach is to flood the events if number of events is small and number of queries is large. Rumor routing is between event flooding and query flooding. The idea is to route the queries to the nodes that have observed a particular event rather than flooding the entire network to retrieve information about the occurring events. In order to flood events through the network, the rumor routing algorithm employs long-lived packets, called agents. When a node detects an event, it adds such event to its local table and generates an agent. Agents travel the network in order to propagate information about local events to distant nodes. When a node generates a query for an event, the nodes that know the route, can respond to the query by referring its event table. Hence, the cost of flooding the whole network is avoided. Rumor routing maintains only one path between source and destination as opposed to DD where data can be sent through multiple paths at low rates. However, rumor routing performs well only when the number of events is small.

Energy Aware Routing (EAR) (Rahul & Rabaey, 2002) this protocol is used to increase the lifetime of the network. This protocol maintains a set of paths instead of maintaining or reinforcing one optimal path. The maintenance and selection depends on a certain probability, which relays on how low the energy consumption of each path can be achieved. The protocol creates routing tables about the paths according to the costs. Localized flooding is performed by the destination node to maintain the paths alive. It finds a least cost and energy efficient path that meets certain end-to-end delay during the connection. The link cost used is a function that captures the nodes' energy reserve, transmission energy, error rate and other communication parameters. In order to support both best effort and real-time traffic at the same time, a class-based queuing model is employed. The queuing model allows service sharing for real-time and non-real-time traffic. The queuing model is depicted in Figure10. The protocol finds a list of least cost paths and picks a path from that list which meets the end-to-end delay requirement.

Figure 10. Queuing model in a particular sensor node (Akkaya & Younis, 2003)

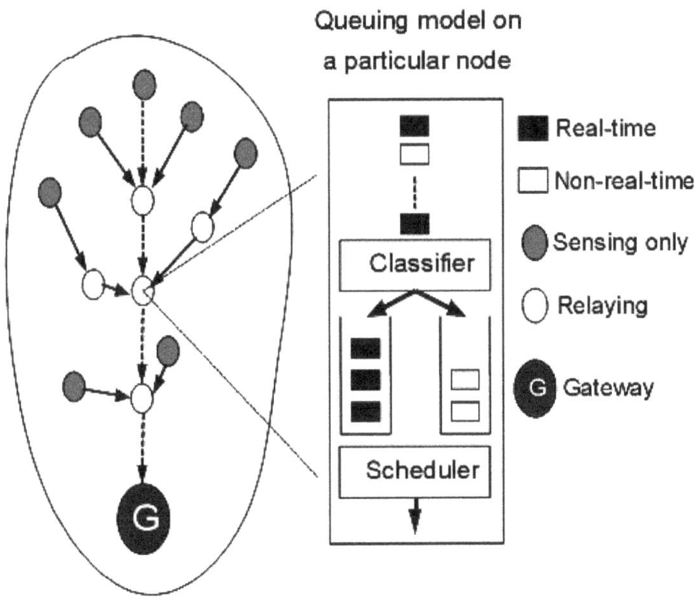

In (Zabin et al., 2008), a new energy-efficient data-centric routing protocol, named *Reliable and Energy Efficient Protocol* (REEP) is proposed. It is a fault tolerant technique. It makes sensor nodes establish more reliable and energy-efficient paths for data transmission. Table 1 summaries the REEP functional steps. The REEP protocol has been designed by taking some important points into account.

- In diffusion-based networking, all the communications are neighbor-to-neighbor.
- There are no specifically-designated routers to route the events in sensor network. Each sensor node acts as a "router" and an "end" point for communication.
- Nodes in the sensor network take routing decisions based on the local information only, because each node can only have the knowledge about their neighbors within their vicinity. Nodes cannot have the global information about the whole network because they are not uniquely identified by global addresses.
- Wireless communication is basically accomplished by broadcasting where there in no acknowledgement system like in traditional networking. So, nodes cannot get their neighbor's current status.
- Sensor nodes are capable of processing and aggregating events, as well as performing distributed sensing.

Table 1. REEP functional steps

Steps	Name
1	Sense event propagation
2	Info event propagation
3	Request event propagation

The way REEP protocol has been designed allows it to work in different scenarios; single source detects single object, single source detects multiple objects, and multiple sources detect single object

- ***Advantages***
    - Some of the important features of the REEP protocol are the introduction of the First Input First Output (FIFO) queue and the energy threshold value.

These features give a "good shape" to the REEP protocol and provide many advantages.
  - Nodes in REEP simply pick the best neighbor from its FIFO queue to request for alternate path setup without causing any flooding.
  - REEP works well with multiple objects.
  - In case of multiple objects or source nodes, REEP does not do all the works at once.
- ***Disadvantages***
  - REEP is not suitable for application where dynamic changes of the network topology need to be propagated across the network.
  - Without an appropriate design of prescheduled sensing tasks, REEP can perform poorly.

Hierarchical Based Routing

It is also known as cluster-based routing. In these protocols, the nodes can play different roles in the network and normally the protocol includes the creation of clusters.

Hierarchical routing in WSN involves the arrangement of clusters in form of hierarchy when sending information from the sensor nodes to the BS. Hierarchical routing efficiently reduces energy consumption by employing multi-hop communication for a specific cluster and thus performing aggregation of data and fusion in a way that decreases the number of data carried across the network to the sink. Cluster formation is based on residual energy in the sensor nodes and election of a cluster head (CH). A very good example of hierarchical routing protocol is *low-energy adaptive clustering hierarchy* (LEACH) (Heinzelman et al., 2000).

The LEACH approach involves formation of clusters of sensor nodes and the use of a local CH as a router to the BS (see Figure 11). Clusters are formed geographically. One cluster formation which makes use of the entire sensors area space can be obtained, other formation such as two clusters formation and three clusters formation involves equal segregation of area space. The two clusters formation and the three clusters formation are otherwise known as first level and second level hierarchy respectively. The selection of CHs within each cluster formed is carried out by electing a node that requires less transmission energy to be the CH for a particular transmission round. A completely new estimation of energy is carried out at the beginning of every transmission round to elect a new CH for the cluster.

Threshold sensitive energy efficient sensor network protocol (TEEN) (Manjeshwar & Agrawal, 2001) is a hierarchical protocol. Closer nodes form clusters and elect a CH. Each CH is responsible for directly sending the data to the sink. After the clusters are formed, the CH broadcasts two thresholds to the nodes. These are hard and soft thresholds for sensed attributes. Hard Threshold (HT) is the minimum possible value of an attribute to trigger a sensor node to switch on its transmitter and transmit to the CH. Thus, the HT allows the nodes to transmit only when the sensed attribute is in the range of interest. Hence, the number of transmissions significantly reduced. Once a node senses a value at or beyond the HT,

Figure 11. Cluster-based mechanism of LEACH in WSN (Patel et al., 2011)

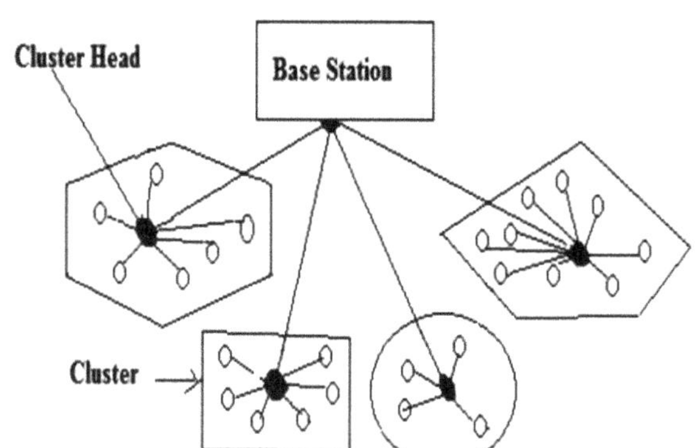

it transmits data only when the value of that attributes changes by an amount equal to or greater than the soft threshold (ST). As a result, ST will further reduce the number of transmissions if there is little or no change in the value of sensed attribute. Thus, the user can control the trade-off between energy efficiency and data accuracy. When CHs are to change (see Figure 12(a)), new values for the above parameters are broadcast. The main drawback of this scheme is that, if the thresholds are not received, the nodes will never communicate, and the user will not get any data from the network at all.

The nodes sense their environment continuously. The first time a parameter from the attribute set reaches its HT value, the node switches its transmitter on and sends the sensed data. The sensed value is stored in an internal variable, called Sensed Value (SV). The nodes will transmit data in the current cluster period only when the following conditions are true: (1) The current value of the sensed attribute is greater than the HT (2) The current value of the sensed attribute differs from SV by an amount equal to or greater than the ST. Important features of TEEN include its suitability for time critical sensing applications.

Adaptive Periodic Threshold-sensitive Energy Efficient sensor Network protocol (APTEEN) (Manjeshwar & Agrawal, 2002) is an extension to TEEN. The architecture is same as in TEEN. When the BS forms the clusters, the CHs broadcast the attributes (this is a set of physical parameters which the user is interested in obtaining information about), the threshold values (this parameter consists of the HT and the ST), and the transmission schedule to all nodes (this is a Time Division Multiple Access (TDMA)) schedule, assigning a slot to each node) and Count Time (CT) (it is the maximum time period between two successive reports sent by a node) (see Figure 12(b)). CHs also perform data aggregation in order to save energy. APTEEN supports three different query types: historical query, to analyze past data values; one-time query, to take a snapshot view of the network; and persistent query to monitor an event for a period of time.

APTEEN is best suited for both periodic sensing & reacting to time critical events such as habitat monitoring. So APTEEN is a hybrid protocol that is both proactive and reactive.

Power-Efficient Gathering in Sensor Information Systems (PEGASIS) (Lindesy & Raghavendra, 2002) protocol is a LEACH-inspired protocol. PEGASIS is an improvement of the LEACH protocol, it not exactly a cluster-based protocol, as nodes are not explicitly grouped into clusters. PEGASIS is instead a chain based approach, in which each node only communicates with a close neighbor and takes turns to transmit to the BS, thus reducing the amount of energy spent per round. This approach distributes the energy load evenly among the sensor nodes in the network. So the basic idea of PEGASIS is to have only one designated node that directly transmits to the BS in each round. This can be achieved with a linear chain-based approach, where sensor nodes form a chain, in which gathered data moves from node to node, gets fused, and eventually a designated node

Figure 12. Time line for the operation of (a) TEEN and (b) APTEEN (Manjeshwar & Agrawal, 2001)

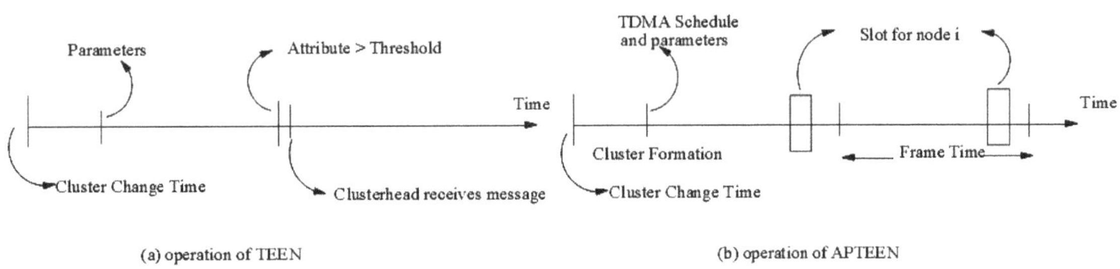

will transmit it to the BS (see Figure 13). As nodes take turns to transmit to the BS and transmissions are between close neighbors, the average energy spent by each node per round is reduced.

Virtual Grid Architecture (VGA) (Al-Karaki et al., 2004) is based on the concept of data aggregation and in-network processing. This routing paradigm considers an extremely low mobility of sensor nodes. Therefore, this protocol arranges the nodes in a fixed topology forming clusters that are fixed, equal, adjacent and non-overlapping with symmetric shapes. One node per zone is considered as CH which is in charge of aggregating and transmitting data.

Hybrid Energy Efficiency Distributed (HEED) clustering protocol (Younis & Fahmy, 2004) is an energy-efficient clustering protocol designed for WSN. In HEED, CH selection is based on two different parameters. The primary parameter is the residual energy of each node, while the second parameter measures the intracluster communication cost, i.e., the number of neighbors or node proximity or node degree so it can be called hybrid. The idea is to use the primary parameter to perform a probabilistic choice of an initial set of CHs, and the second parameter to break ties between them, e.g., when a node is within the range of multiple CHs. The HEED clustering improves network lifetime over LEACH clustering because LEACH randomly selects CHs, which may result in faster death of some nodes.

Genetic Algorithm (GA) (Hussain et al., 2007) is a hierarchical cluster-based routing (HCR) protocol. The main idea in this algorithm is after every generation, the best fit chromosome (nodes are represented as bits of a chromosome) is evaluated to assess the improvement in the fitness parameters. The fitness of a chromosome is determined by several parameters, such as node density and energy consumption. A population consists of several chromosomes and the best chromosome is used to generate the next population. In comparison with LEACH this algorithm shows that it is more efficient and the depletion of energy in sensor nodes is improved.

Figure 13. Chaining in PEGASIS

c0 → c1 → c2 → c3 → c4

↓

Base Station

Energy Balancing and Dynamic Hierarchal (EBDHR) (Rasouli et al., 2010) looks like LEACH in the way of the election of CHs and cluster formation but the difference is in the data transmission from CHs to BS. In LEACH this process is done directly but in EBDHR is done in multi hop.

Cluster Based Routing Protocol (CBRP) (Zarei et al., 2010). The operation of CBRP is divided into rounds and each round contains two phases, set-up phase and steady-state phase. In the set-up phase, clusters are generated and then in the steady-state phase, routing tree constructed and aggregated data are sent to the sink node. In CBRP protocol, each node maintains a neighborhood table to store the information about its neighbors.

Cluster Based Hierarchical Routing Protocol (CBHRP) (Golam et al., 2010) is an energy efficient cluster-based hierarchical routing protocol for WSN, where a number of clusters cover the whole region. This protocol introduces a concept of head-set instead of only one CH within a cluster. At one time, only one member of head-set is active and the remaining are in sleep mode. For a given number of data collecting nodes, the head-set members are systematically adjusted to reduce the energy consumption, which increases the network lifetime.

Hierarchical Energy Efficient Routing Protocol (HEERP) (Inam ET AL., 2013) is a protocol which aims to save energy with energy efficient routing. In this protocol the nodes are divided into different clusters which have approximately equal number of nodes. The BS then receives energy information from all nodes and a CH is selected based on maximum energy for current round. The CH receives data from every node within cluster, fused the data, and then fused data is transmitted with the help of other CHs or directly to the BS. Unlike other protocols which directly communicate with the BS, HEERP uses dependent and independent CHs for the transmission of data to the BS.

Energy-Efficient Level Based Clustering Routing Protocol (EELBCRP) (Diwakar & Kumar, 2012) is a hierarchical clustering routing protocol. Network partitioned into annular rings by using various power levels at BS and each ring having various sensor nodes. Also consider the residual energy of each node and distance from the BS of nodes as the principle of CH election.

Adaptive Based Routing Protocols

Adaptive based routing is based on the family of protocols called SPIN which is described in Negotiation based routing (Heinzelman et al., 1999). The protocols disseminate all the information at each node to every node in the network assuming that all nodes in network are potential BS. With this, the user can query any node and get the needed information immediately. The protocols use data negotiation and resource-adaptive algorithms. The nodes assign a high-level name to describe completely their collected data; this is called meta-data. Then are preformed negotiations before any data is transmitted to avoid redundant data to be transmitted. These protocols distribute the information all over the network, even when the user does not request any data.

In recent years, insect sensory systems have been inspirational to new communications and computing models like bio inspired routing. It is due to their ability to support features like autonomous, and self-organized adaptive communication systems for pervasive environments like WSN and mobile ad hoc networks. Biological synchronization phenomena have great potential to enable distributed and scalable synchronization algorithms for WSN (Atakan et al., 2009).

Minimum Ant-based Data Fusion Tree (MADFT) (Juan et al., 2007) is a sink selection heuristic routing algorithm .It is based on Ant Colony Optimization (ACO) for gathering correlated data in WSN. It first assigns ants to source nodes. Then, the route is constructed by one of the ants in which other ants search the nearest point of previous discovered route. The chosen formula is probability function composed of pheromones and costs in order to find the minimum total cost path. MADFT not only optimizes over both the transmission and fusion costs, but also adopts ant colony system to achieve the optimal solution.

Swarm Intelligence Optimization Based Routing Algorithm (Wang & Lin, 2008) is a relatively novel field. It works with the objective to balance global energy consumption and avoiding some node's premature energy exhausting. The algorithm chooses the nodes with less pheromone as next hop, taking less hop numbers into consideration. The algorithm is different from traditional ant colony algorithms. It is better than the DD routing protocol both in end-to-end delay and global energy balance. It can effectively balance the global energy consumption and prolong the network lifetime.

In Bio-inspired routing (Krings & Sam, 2009), the most popular ACO is a colony of artificial ants used to construct solutions guided by the pheromone trails and heuristic information. They are not strong or very intelligent; but they successfully make the colony a highly organized society. Swarms are useful in many optimization problems. A swarm of agents is used in a stochastic algorithm to obtain near optimum solutions to complex, non-linear optimization problems (Selvakennedy et al., 2007).

In (Batra ET AL., 2011), an *adaptation of Ant Colony Optimization* technique is demonstrated for network routing. This approach belongs to the class of routing algorithms inspired by the behavior of the ant colonies in locating and storing food.

The ant routing algorithm can be described as follows:

- A forward ant is launched from the source node toward the destination node at intervals which vary with time.
- The next hop node is selected by each forward ant according to the link probability distribution. All links have an equal probability at the initial stage.
- While on its way forward, each ant keeps a track of the list of nodes visited by it and avoids traversal of that node again
- When a forward ant reaches its destination, a backward ant is generated which moves back through all the links which were made by the forward ant during its traversal.
- When the backward ant travels, cost from the destination to each node in the path is recorded and rewards are given according to the relative goodness of the path. Probabilities of the nodes in the path are updated according to the rewards.
- Once a backward ant arrives at the source, the next launch interval is calculated according to the relative goodness of the whole path traversed.

Particle swarm optimization (PSO) (Siew et al., 2012) is one of the swarm intelligence methods designed to find optimum solution by mimicking the behavior of bird flocking and fish schooling. Adaptive cognitive and social learning factor can achieve better convergence speed and particles reselection mechanism can reduce the chances of getting trapped in local maximum.

Generally, PSO for CH selects M number of CH from network with N number of sensor nodes. Therefore, there are C_i^m number of combination. The chance for each set of combination can be evaluated by following formula which overall output describe as fitness (Siew et al., 2012). Smallest fitness, represent the best combination among others. The flowchart of PSO CHs selection can be seen in Figure 14 and described as follow:

A. Particles initialization: Randomly initialize numbers of particles, each particle represents a set of possible CH distribution.
B. Fitness evaluation: Calculate fitness for each particle.
C. Reselect particles formation: If global best did not change after a number of iterations, 25% particles from the worst set will be re-selected.
D. Update local best and global best: Each particle updates its local best and global best value.
E. Velocity and position update: Update velocity and position for each CH in particle.
F. Repeat B to E: Repeat until the maximum number of iteration reached.
G. Broadcast CHs candidates: BS broadcasts the qualify CHs candidates to the network.

4. DISCUSSIONS AND FUTURE WORKS

The sensor nodes are equipped with battery only with limited energy; energy efficient information processing is of critical importance to operate the deployed networks as long as possible. In this paper, we presented a comprehensive survey of routing techniques in wireless sensor networks which have been presented in the literature. They

Figure 14. Flowchart of PSO algorithm

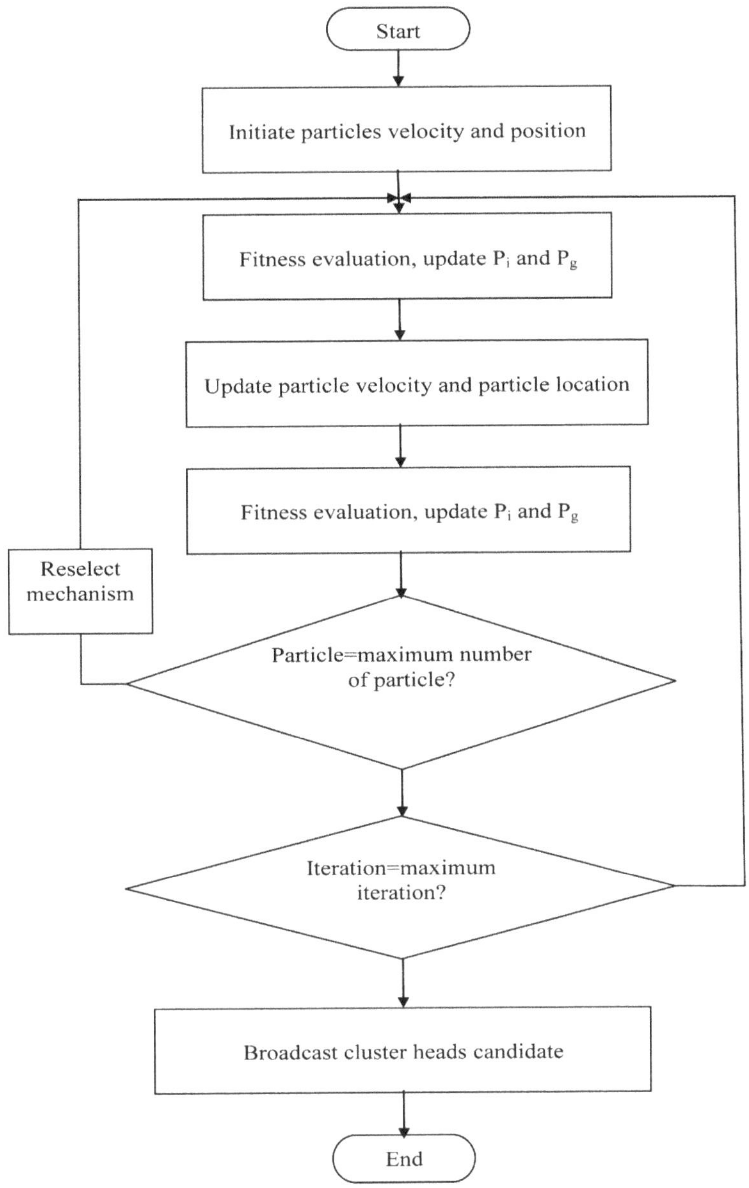

have the common objective of trying to extend the lifetime of the sensor network, while not compromising data delivery.

Table 2 compares SPIN, LEACH, and the Directed Diffusion routing techniques according to different parameters. It is noted from the table that Directed Diffusion shows a promising approach for energy-efficient routing in WSNs due to the use of in-network processing.

Table 3 shows the comparison of the different routing protocols in terms of scalability, lifetime, data diffusion and power required in WSN. From Table 4, we understand that hierarchical technique offers an approach to energy minimization and scalability features in a WSN.

We observed that there are some hybrid protocols that fit under more than one category. We summarize recent research results on data routing

Table 2. Comparison between SPIN, LEACH and Directed Diffusion

	SPIN	LEACH	DIRECTED DIFFUSION
Optimal Route	NO	NO	Yes
Network Lifetime	Good	Very Good	Good
Resource Awareness	Yes	Yes	Yes
Use of Meta-Data	Yes	NO	Yes

Table 3. Comparison of routing technique

	Data-Centric Technique	Hierarchical Technique	Location-Based Technique
Scalability	Limited	Good	No
Lifetime	Long	Long	Long
Data Diffusion	No	Yes	No
Power Required	Limited	High	Limited

Table 4. Hierarchical vs. flat topologies routing

Hierarchical Routing	Flat Routing
Reservation-based scheduling	Contention-based scheduling
Collision avoided	Collision overhead present
Reduced duty cycle due to periodic sleeping	Variable duty cycle by controlling sleep time of nodes
Data aggregation by cluster head	Node on multihop path aggregates incoming data from neighbors
Simple but non-optimal routing	Routing can be made optimal but with an added complexity
Requires global and local synchronization	Links formed on the fly without synchronization
Overhead of cluster formation throughout the network	Routes formed only in regions that have data for transmission
Lower latency as multiple hops network formed by cluster heads always available	Latency in waking up intermediate nodes and setting up the multipath
Energy dissipation is uniform	Energy dissipation depends on traffic patterns
Energy dissipation cannot be controlled	Energy dissipation adapts to traffic pattern
Fair channel allocation	Fairness not guaranteed

in WSNs in Table 5. The Table shows how different routing protocols fit under different category and also compare different routing techniques according to many metrics.

Future trends in routing techniques in WSNs focus on different directions, all share the common objective of prolonging the network lifetime. We summarize some of these directions and give some pertinent references as follows:

- **Secure Routing:** Current routing protocols optimize for the limited capabilities

Table 5. Classification and comparison of routing protocols in wireless sensor networks

	Classification	Mobility	Position Awareness	Power Usage	Negotiation Based	Data Aggregation
Spin	Flat	Possible	No	Limited	Yes	Yes
DD	Flat	Limited	No	Limited	Yes	Yes
Rumor routing	Flat	Very Limited	No	N/A	No	Yes
EAR	Flat	Limited	No	N/A	No	No
LEACH	Hierarchical	Fixed BS	No	Maximum	No	Yes
TEEN&APTEEN	Hierarchical	Fixed BS	No	Maximum	No	Yes
PEGASIS	Hierarchical	Fixed BS	No	Maximum	No	No
VGA	Hierarchical	No	No	N/A	Yes	Yes
GAF	Location	Limited	No	Limited	No	No
GEAR	Location	Limited	No	Limited	No	No
SAR	QOS	No	No	N/A	Yes	Yes
	Localization	**QOS**	**State Complexity**	**Scalability**	**Multipath**	**Query Based**
Spin	No	No	Low	Limited	Yes	Yes
DD	Yes	No	Low	Limited	Yes	Yes
Rumor routing	No	No	Low	Good	No	Yes
EAR	No	No	Low	Limited	No	Yes
LEACH	Yes	No	CHs	Good	No	No
TEEN&APTEEN	Yes	No	CHs	Good	No	No
PEGASIS	Yes	No	Low	Good	No	No
VGA	Yes	No	CHs	Good	Yes	No
GAF	No	No	Low	Good	No	No
GEAR	No	No	Low	Limited	No	No
SAR	No	Yes	Moderate	Limited	No	Yes

of the nodes and the application specific nature of the networks, but do not consider security. Although these protocols have not been designed with security as a goal, it is important to analyze their security properties. One aspect of sensor networks that complicates the design of a secure routing protocol is in-network aggregation.

- Integrating of sensor networks with wired networks (i.e. Internet).
- Developing means of establishing a coordinate system without relying on an existing infrastructure.
- Proposing routing techniques that explicitly employ fault tolerance techniques in an efficient manner

REFERENCES

Ahvar, S., & Mahdavi, M. (2011). EEQR: An Energy Efficient Query-Based Routing Protocol for Wireless Sensor Networks. *Journal of Advances in Computer Research*, *2*(3), 25–38.

Akkaya, K., & Younis, M. (2003). "An Energy-Aware QoS Routing Protocol for Wireless Sensor Networks. *Proceedings of the IEEE Workshop on Mobile and Wireless Networks MWN '03, Providence, Rhode Island.* doi:10.1109/ICDCSW.2003.1203636

Al-Karaki, J. N. Ul-Mustafa v & Kamal A. E. (2004). Data Aggregation in Wireless Sensor Networks - Exact and Approximate Algorithms. *Proceedings of IEEE Workshop on High Performance Switching and Routing (HPSR), Phoenix, Arizona, USA* (pp. 241-245).

Al-Karaki, J. N., & Kamal, A. E. (2004). *Routing Techniques in Wireless Sensor Networks: A Survey* (pp. 6–28). IEEE Wireless Communications.

B. Atakan, O. B. Akan & T. Tugcu, (2009). Bio-inspired Communications in Wireless sensor networks. In S.C. Misra, I. Woungang, & S. Misra (Eds.), *Guide to Wireless Sensor Networks* (pp. 659-687). Springer-Verlag London.

Awad, A., Sommer, C., German, R., & Dressler, F. (2008). Virtual Cord Protocol (VCP): A flexible DHT-like routing service for sensor networks. *Proceedings of the 5th IEEE International Conference on Mobile Ad Hoc and Sensor Systems* (pp. 133-142). doi:10.1109/MAHSS.2008.4660079

Bains, V., & Sharma, K. (2012). Ant colony based routing in wireless sensor networks. *International Journal of Electronics and Computer Science Engineering*, *1*(4), 2516–2524.

Batra N., Jain A. & Dhiman S., (2011). An Optimized Energy Efficient Routing Algorithm for Wireless Sensor Network. *International Journal of Innovative Technology & Creative Engineering, 1(5)*, 41-45.

Braginsky, D., & Estrin, D. (2002). Rumor routing algorithm for sensor networks. *Proceedings of the First Workshop on Sensor Networks and Applications (WSNA), Atlanta.*

Chang, J. H., & Tassiulas, L. (2004). Maximum lifetime routing in wireless sensor networks. *IEEE/ACM Transactions on Networking*, *12*(4), 609–619.

Diwakar, M., & Kumar, S. (2012). An Energy Efficient Level Based Clustering Routing Protocol For Wireless Sensor Networks. *International Journal of Advanced Smart Sensor Network Systems*, *2*(2), 55–65. doi:10.5121/ijassn.2012.2206

El-said S. A., Hassanien A. E. (2013). Artificial Eye Vision Using Wireless Sensor Networks. In *Wireless Sensor Networks: Theory and Applications*. CRC Press.

Gupta, D., & Sharma, A. K. (2011). Investigations on Energy Efficiency for WSN Routing protocols for Realistic Radio Models. *International Journal of Future Generation Communication and Networking*, *4*(3), 61–72.

Heidemann, J., Estrin, D., & Xu, Y. (2001). Geography-informed Energy Conservation for Ad-hoc Routing. *Proceedings of the Seventh Annual ACM/IEEE International Conference on Mobile Computing and Networking*, Rome, Italy (pp. 70-84).

Heikalabad, S. R., Navin, A. H., Mirnia, M., Ebadi, S., & Golesorkhtabar, M. (2010). Energy balancing and dynamic hierarchical routing algorithm for wireless Sensor networks. *IECIE Electronics Express*, *7*(15), 1112–1118. doi:10.1587/elex.7.1112

Heinzelman, W., Chandrakasan, A., & Balakrishnan, H. (2000). Energy efficient communication protocol for wireless sensor networks. *Proceeding of the Hawaii International Conference System Sciences,* Hawaii.

Heinzelman, W., Kulik, J., & Balakrishnan, H. (1999). Adaptive protocols for information dissemination in wireless sensor networks. *Proceedings of the 5th Annual ACM/IEEE International Conference on Mobile Computing and Networking*, Seattle, WA (pp. 174-185). doi:10.1145/313451.313529

Hussain, S., Matin, A. W., & Islam, O. (2007). Genetic Algorithm for Energy Efficient Clusters in Wireless Sensor Networks. *Proceedings of the Fourth International Conference on Information Technology* (pp. 147 –154). doi:10.1109/ITNG.2007.97

Inam, M., Ikram, A. A., & Ahmad, M. (2013). Hierarchical Energy Efficient Routing Protocol (HEERP) for Wireless Sensor Networks (WSN). *International Journal of Computer Science and Telecommunications*, *4*(10).

Intanagonwiwat, C., Govindan, R., & Estrin, D. (2000). Directed diffusion: a scalable and robust communication paradigm for sensor networks. *Proceedings of the 6th annual international conference on Mobile computing and networking* (pp. 56-67).

Juan, L., Chen, S., & Chao, Z. (2007). Ant System Based Any cast Routing in Wireless Sensor Networks. *Proceedings of the International Conference on Wireless Communications, Networking and Mobile Computing* (pp. 2420-2423).

Krings, A. W., & Ma, Z. S. (2009). Bio-Inspired Computing and Communication in Wireless Ad Hoc and Sensor Networks. *Ad Hoc Networks, Elsevier*, *7*(4), 742–755. doi:10.1016/j.adhoc.2008.03.003

Li, Q., Aslam, J., & Rus, D. (2011). Hierarchical Power-aware Routing in Sensor Networks. *Proceedings of the DIMACS Workshop on Pervasive Networking.*

Lindesy, S., & Raghavendra, C. (2002). PEGASIS: Power-Efficient Gathering in Sensor Information System. *Proceedings of the Aerospace Conference, Big Sky, Montana*, (Vol. 3, pp. 9-16). IEEE. doi:10.1109/AERO.2002.1035242

Manjeshwar, A., & Agrawal, D. (2001). TEEN: a Routing Protocol for Enhanced Efficient in Wireless Sensor Networks. *Proceedings of the 15th International Parallel and Distributed Processing Symposium, San Francisco* (pp. 2009-2015). doi:10.1109/IPDPS.2001.925197

Manjeshwar, A., & Agrawal, D. P. (2002). APTEEN: a hybrid protocol for efficient routing and comprehensive information retrieval in wireless sensor networks. *Proceedings of the 2nd International Workshop on Parallel and Distributed Computing Issues in Wireless Networks and Mobile computing, Ft. Lauderdale, FL.*

Misra S., Woungang I., & Misra S. C. (2009). Guide to Wireless Sensor Networks. *Springer-Verlag London.*

Patel R., Pariyani S. & Ukani V., (2011). Energy and Throughput Analysis of Hierarchical Routing Protocol (LEACH) for Wireless Sensor Network. *International Journal of Computer Applications, 20(4).*

Rahul, C., & Rabaey, J. (2002). Energy Aware Routing for Low Energy Ad Hoc Sensor Networks. *Proceedings of the Wireless Communications and Networking Conference* (pp. 350-355). *IEEE.*

Rashed Md, G., Kabir, M. H., Rahim, M. S., & Ullah, S. E. (2010). Cluster Based Hierarchical Routing Protocol For Wireless Sensor Network. *International Journal of Computer and Network Security*, *2*(5), 128–131.

Rodoplu, V., & Ming, T. H. (1999). Minimum energy mobile wireless networks. *IEEE Journal on Selected Areas in Communications*, *17*(8), 1333–1344. doi:10.1109/49.779917

Selvakennedy, S., Sinnappan, S., & Shang, Y. (2007). A biologically-inspired clustering protocol for wireless sensor networks. *Computer Communications, Elsevier*, *30*(14/15), 2786–2801. doi:10.1016/j.comcom.2007.05.010

Siew, Z.W., Wong, C.H., Chin, C.S., & Kiring, A., & KinTeo K.T. (2012). Cluster Heads Distribution of Wireless Sensor Networks via Adaptive Particle Swarm Optimization. *Proceedings of the Fourth International Conference on Computational Intelligence, Communication Systems and Network.* doi:10.1109/CICSyN.2012.25

Sohrabi, K., & Pottie, J. (2000). Protocols for self-organization of a wireless sensor network. *Personal Communications, 7*(5), 16–27. doi:10.1109/98.878532

Wang, C., & Lin, Q. (2008). Swarm intelligence optimization based routing algorithm for Wireless Sensor Networks. *Proceedings of International Conference on Neural Networks and Signal Processing* (pp. 136-141). doi:10.1109/ICNNSP.2008.4590326

Younis, O., & Fahmy, S. (2004). Distributed Clustering for Scalable, Long-Lived Sensor Networks. *Mobile Computing, 3*(4), 366–379. doi:10.1109/TMC.2004.41

Yu, Y., Estrin, D., & Govindan, R. (2001). *Geographical and Energy-Aware Routing: a recursive data dissemination protocol for Wireless Sensor Networks* (Technical Report UCLA-CSD TR-01-0023). UCLA Computer Science Department.

Zabin, F., Misra, S., Woungang, I., Rashvand, H. F., Ma, N. W., & Ali, M. A. (2008). REEP: Data-centric, energy-efficient and reliable routing protocol for wireless sensor networks. *IET Communications*, *2*(8), 995–1008. doi:10.1049/iet-com:20070424

Zarei, B., Zeynali, M., & Nezhad, V. M. (2010). Novel cluster Based Routing Protocol in Wireless Sensor Network. *International Journal of Computer Science Issues*, *7*(4), 32–36.

Chapter 4

Energy Management Strategies to Improve Electrical Networks Using Storage Systems

Juan Aurelio Montero-Sousa
University of A Coruña, Spain

José-Luis Casteleiro-Roca
University of A Coruña, Spain

Luis Alfonso Fernández-Serantes
University of A Coruña, Spain

Xosé Manuel Vilar-Martínez
University of A Coruña, Spain

Jose Luis Calvo-Rolle
University of A Coruna, Spain

ABSTRACT

The successive energy crises, usually linked to the rising prices of oil, bring about new topics of the energy systems management in general terms. Over all, the electrical system is one of these cases. In addition, a greater concern for environmental issues has introduced, to a greater or lesser extent, the generation from renewable sources in the electrical system. In this context, the possibility of developing and using electricity storage systems would manage mismatches between generation and demand at electricity networks, making them more efficiently. In this research, we propose a number of possible strategies based on technical peak shaving and valley filling. The tool is used as energy storage systems in general terms, regardless of the accumulation technique used. The classification of strategies essentially serves two criteria: optimization service and increased profitability.

1. INTRODUCTION

Independently of their nature, environmental issues are becoming increasingly important (Marsh, 2005). Especially in developed countries there are several laws, directives or guidelines that must be abided by, in order to achieve the desired sustainability (Daniels & Daniels, 2003). The international efforts in many countries to make commitments in the minimization of pollution must be highlighted, for example the Kyoto protocol (United Nations, 1998). It is based on the climate change and it is reviewed regularly to reflect the reality in terms of the facts that

DOI: 10.4018/978-1-4666-9911-3.ch004

happened (United Nations, 2005) (United Nations, 2007)(The official website of Denmark, n. d.). In this sense, power generation systems are especially significant, whose tendency, as it could not be otherwise, is that the energy input to the power system is originated in renewable resources (MacKay, 2009; Kaltschmitt, Streicher, & Wiese, 2007; Wengenmayr, Bührke, & Brewer, 2013). Thus, it achieves a reduction, as low as possible, of the pollution emissions to the atmosphere from the power generation.

It is important to add that the energy demand usually increases over the time, although in times of financial crises the demand could decrease, as it is shown in Figure 1, and the development of society (Kaltschmitt et al., 2007). In the future, it can be foreseen that this statement continues to be true or even the demand will be increased due to reasons such as the imminent introduction of the electric car (Figure 1) (Warner, 2011; Anderson & Anderson, 2010).

In addition to the current trend justified above, two important factors that greatly determine the source of energy in an electrical system must be mentioned:

- Traditional sources are exhaustible and therefore, factors such as the price are becoming less favorable (Rao, 2011).
- Certain types of energy, such as nuclear power, are very difficult to control and, in case of accidents, lead with a huge risk with catastrophic consequences (Ferguson, 2011).

Nowadays, the use of renewable energy sources for electricity generation is unavoidable. Despite the fact that these energy sources have been known for a long time, they are relatively emerging in the generation of electricity and have a long way to go. One of the most significant drawbacks of these sources is that, they are often available at the moment there is not enough demand to consume, due to the own nature of renewable energy. For example, a lot of wind can blow during the night when the power consumption drops considerably (MacKay, 2009). As an example, a specific real case occurred at a wind farm in Sotavento Galicia is shown in Figure 2.

To solve the problem of adjusting the generation and demand, there are storage methods already

Figure 1. Demand for electricity in Spain (GWh)

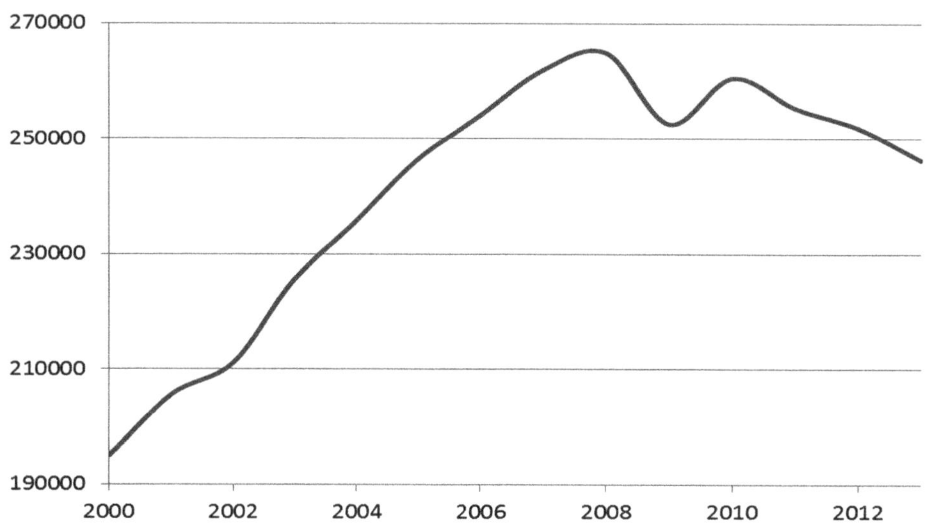

Figure 2. Comparison of wind speed (blue - m/s) and production of a wind farm (red - KWh)

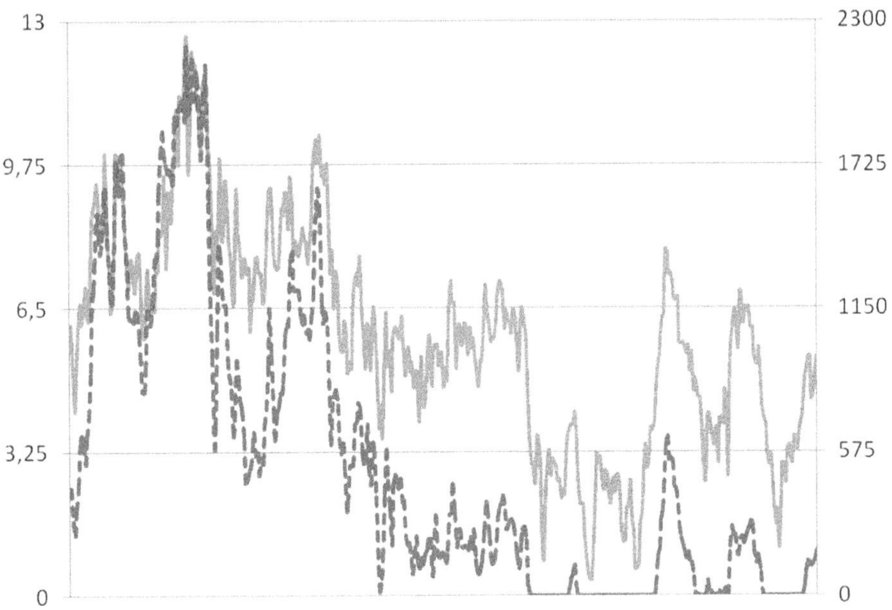

implemented industrially, such as the pumping of water into the reservoirs when there is an excess energy production. Thus, it is possible to turbine it when there is higher energy demand (MacKay, 2009). The problem of energy storage systems demonstrated and established at present is that, despite being profitable from the economic point of view, they have an extremely low performance (Huggins, 2010). For the reasons explained above, the importance of improving existing systems and developing new energy storage techniques tailored to the needs is completely justified.

It must be pointed out that the energy storage systems may have many benefits, depending on the technology used, such as (Chiang, Huang, & Liaw, 1995):

- There will be easily installed systems anywhere and of any size in terms of power, contributing to the existence of energy storage system adapted to ones' needs.
- For fast variations or specific charge, storage systems may fit in very short time intervals.
- If there is an interconnection between the storage systems, tasks transmission and distribution of energy between the different points of consumption in specific situations can be carried out.

Storage systems also contribute to carry out the management of the energy demand. Thus, the challenges already mentioned in multiple investigations arise, such as Peak Shaving (Sullivan, 1977; Gellings, 1985), and the Valley Filling (River, 2005). As well as an improvement of the quality of the electric power (Santoso, 2011) with parameters, such as maintaining synchronism, frequency regulation or ensured stability of the voltage are also challenges.

A classification proposal of the different existing techniques for managing the electricity demand is shown in this work. These systems are based on energy storage to use the energy later depending on the chosen strategy. The classification is made according to technical factors related to power distribution, and factors in terms of economic performance.

With the aim to solve the present problem it is necessary to accomplish a good model of the system performance. Due to the high nonlinearity of the problem (the generation and consumption), the traditional regression techniques are not capable to solve this challenge. To manage in a right and efficient way the energy distribution and the management approaches described on the present research, it is needed to create a model by using novel regression methods to predict the necessary variables.

The chapter will begin with this introduction. After it, a brief summary of energy economy will be discussed. Then, strategies of interconnection between renewable energy generation systems and storage systems will be analyzed in detail. The classification will be explained and a new concept to model the consumption and the generation is presented. Then, a novel generic model approach to accomplish the new strategies is presented. Finally, the reached conclusions will be shown.

2. THE NEED OF ENERGY STORAGE FROM THE ECONOMIC POINT OF VIEW

2.1. Economic Characteristics of Electricity

After the Second Industrial Revolution, electricity has become a commodity in modern industrialized societies and the access to this service, in developed societies, must be guaranteed to all citizens. Though, in recent decades, an effort has been made to reconcile these supplies with issues such as environmental conservation, quality and cost (Energía y sociedad, n. d.-b).

However, the commercialization of electric power has a number of features that differentiate it from other markets, where supply and demand can simply adjust through the balance between quantity and price.

First, electricity is a joint and two folded economic service:

- Usually, it is a private service, since supply is received if and only there is demand and it is paid for.
- And it has, to some extent, it has the same behavior as a public service in the way that it has a minimum quality and security of supply. That is the reliability of the system.

Reliability is a non-excludable service (the investments that increase the security of the supply benefit all consumers connected to the same network). Reliability is also a non-rival service (the consumption that a person makes, does not stop or reduce the amount available for consumption by others). The safety of the electrical system is not a private service, and requires the government intervention to ensure coverage rates (Utray Fabra, n. d.). Because of these characteristics of the electricity market, the markets need to be supplemented with regulatory mechanisms that are established by the different administrations (Fabra, 2007).

Moreover, it must been kept in mind that electricity is a homogeneous property, i.e., all units of the service (MWh) are indistinguishable and identical for the different agents, although the means to produces it are not homogeneous, which turns into problems to remunerate the different electricity generating systems (Utray Fabra, n. d.).

On the other hand, most of the consumers pay a sort of flat rate, a fixed price per unit consumed. This price is independent of the time of day and geographic location of the company. It does not favor the transfer demand to the time of the day in which the price should be lower, or the transfer of large industrial consumer plants to areas with high potential energy generation. Large investments often have to be made in electricity transmission networks (Benavides, n. d.).

2.2. Evolution of the Electricity Market

After the oil crisis of '73, the liberalization process of the electricity market takes place in most countries. The old natural monopolies at the state level are replaced by systems that enhance competition in one or more activities in which the electricity market can be differentiated. Following this liberalization trend, in Europe several directives have been approved, promoting the separation of regulated activities, some of which are free (transport networks) and other are competitive (generation and supply).

Theoretically, in this liberalized market, the companies make decisions of investment and bear the costs of being inefficient, both in managing the service as in making investments.

The real problem of market liberalization appears when trying to achieve a suitable design that responds to the specific characteristics of the electricity market in general terms and the ones in each country (Energía y sociedad, n.d.-a).There will be to decide between a centralized or decentralized market architecture, whether to use a remunerating system that takes the marginal costs or average costs into account, etc.

Another issue to consider is the difficulty to design and operate devices that store electricity at a reasonable price, at present. It requires that, at all times and in every point of the grid, to instantly match the generation by the producers and consumption by the users (whether these are domestic, industrial or commercial, both public and private ones), which involved a considerable effort to implement the necessary management and control mechanisms. The demand is variable over time. Throughout the different times of day, week and year the consumption patterns of users vary, and meanwhile you can use statistical models to study the behavior of users, for example, a prevision of the consumption can be made in advance, in real situations random factors always intervene (extreme cold, intense heat ...). It can be checked by studying the electricity demand curves, that there are periods of valley (low electricity demand) and peak demand (high electricity demand). The valley periods occur during the evening and on weekends and holidays. Peak periods occur mostly around midday and late afternoon (Gil Gonzalez-Madroño, 2007).

In most cases, regional imbalances occur. There are geographical areas in a country with a high power consumption, but not enough generation capacity, which makes it essential to have suitable transport and distribution lines to bring supplies wherever needed. Due to the increases in electricity demand and the peaks of that demand, the transmission lines often need to be resized, so they do not become congested or a bottleneck does not occur, which adds extra costs to the electrical system.

If we were able to make generation and demand independent by using storage systems, we would eliminate the additional costs by the need of a large generating capacity to meet peaks. By placing these storage units in strategic locations, the problems of congestion in transmission lines and distribution of electricity would also be solved.

2.3. The Influence of Renewable Energy in the Electricity Market

In recent years a great development of renewable energy, i.e., that energy which is obtained from inexhaustible sources (wind, geothermal, solar ...) has been achieved.

The implementation of wind power as an instrument of power generation poses a number of problems in relation to the management of the demand. Wind power generation is characterized by intermittency and unpredictability. The electricity is generated only when the wind blows at certain intensity (Figure 2). The demand might be high but this type of generation is unavailable due to the lack of wind, or when the wind is generating electricity, the energy is not required by consumers. With the premise that the demand and the generation have to be balanced at all times, if a sudden increase in wind generation occurs, as there is adequate wind, either we would have to reduce the production at conventional power plants or to stop production of wind farms (wind spills). In the case that wind production diminishes or disappears, it would have to be resorted to conventional power plants.

For that difficult management of wind power, due to the dependence of the existence of wind, it is necessary to have a reserve power (hot standby) to reach at least 92.5 percent of the installed wind power (Energía y sociedad, n.d.-c). Again it must be stressed that a possible solution to the problem of wind spills, when demand is very low and wind power is being produced, is the use of storage systems which allow electricity to be used when the demand reaches a peak. The use of this possibility depends on its economic and technical feasibility.

3. STRATEGIES OF INTERCONNECTION BETWEEN GENERATION SYSTEMS AND ENERGY STORAGE SYSTEMS

As dealt with in the introduction, renewable energy has the inherent characteristic of being variable systems, as far as availability is concerned (MacKay, 2009). Thus, a wind farm depends directly on the existing wind generation at all times.

It may be the case that renewable energy sources produce many moments of time when there is not enough demand. The opposite case, when demand is greater than generation, it would be necessary to supplement the power basis plants to cover the entire system demand.

The curve of energy demand of a country is not known a priori, but in general, estimates even 24 hours in advance can be "acceptable" (Egido Cortés, Rouco Rodríguez, Alonso Perez, Porras Muñoz, & Ruiz Mendoza, n. d.) (Figure 3). These estimates are used by the managers of the distribution network to engage or disengage the necessary plants at all times, in order to meet the expected demand. In case of deviations from the estimate, these could be of two types:

Figure 3. Comparison between estimated (blue) and real energy demand (red)

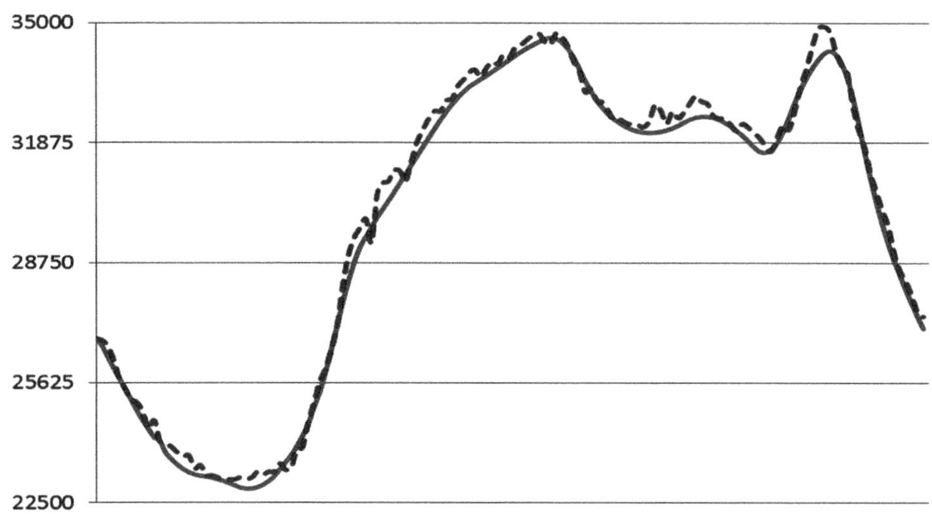

- If the available energy is greater than the consumption at any given time, certain plants will have to stop providing power to the system.
- If at some instant there is a greater demand of energy than that which is being generated, it will be needed to supply the deficiency. To deal with this case, there are several types of power plants according to the time they take to join the network (Hot Book) (Energía y sociedad, n.d.-c).

A generation of purely renewable energy would not be able to make this adjustment a priori, because if production exceeds the demand, some of the energy would be lost; but if the demand is greater than the production, it could not cope with it because of the intermittency and unpredictability of the source.

To mitigate the problem of regulation, it is possible to introduce the storage systems, and with them, carry out energy storage in the time when the generation is greater than the demand, and return it to the system when the contribution of generation plants is not enough.

Taking all this into account, three strategies for energy management within a production system are presented. The criteria on which the classification was made depends, in the first place on whom is in charge of controlling the storage facility either in load mode (accumulation), or download mode (generation). Secondly, the classification can be justified from a purely technical point view of stability and quality of the electrical system, or from a point of view of the profits.

3.1. Strategy Controlled by the System Manager

This strategy is focused exclusively to optimize the unbreakable duo between generated energy and the demand, and consequently it improves factors such as the quality of electricity. The main objective in this case, is that when more energy available than demanded, it will be accumulated in a storage system. When at a given time the electrical system demands more energy than can be generated, the storage plant would release the previously accumulated energy, thus balancing generation and demand.

In this strategy, the managers of the transmission system would control the loading and unloading of storage systems. It has to be clarified that for this strategy, the storage systems are not necessarily located at or associated with the generating plants, and would operate independently.

3.2. Strategy Controlled by Plant Generation: Avoid Penalties

In order to compete for the energy market, in many power networks, the way to proceed by the administrator of a generation plant is provide the operator of the electricity network an estimate of the energy that is going to be provided at a certain time, and with a certain deadline(Egido Cortés et al., n.d.; GaleanoGonzález & Botero Castro, n.d.). This prediction has to comply with very small margins of error. Otherwise, the system manager imposes financial penalties to the energy supplier, either for default or excess power supplies (Salmerón, 2010).

This assumption only makes sense if a storage system fully linked to a generating plant is available and, even more if it refers to renewable energy. Therefore, if a power plant has a correctly-sized storage system, it will be the storage system the one that completes the stated contribution, whenever it is not possible for the main source to provide the power indicated in the prediction. Otherwise, when you have more energy than indicated on the operator's, the left over energy will be used to recharge the storage system. Thus, in both cases, the fact matches the prediction and hence penalties are avoided.

3.3. Strategy Based on Independent Storage Plants or Related to Plant Generation

The energy market in each country sets the prices you pay at any time by the energy injected into the grid. A common case in many countries is that in addition to the price of that energy, the price also varies depending on the power demand at any given time (Lucia López & Meneu Ferrer, 2005). In peak hours of energy, or in emergency situations, the average price to suppliers for their contribution to the net is greater than in other cases, for example when the demand is very low (Figure 4).

When these price variations occur, storage systems can be used with a different aim to avoid penalties. With the margin between the cost of energy storage and the benefit that could be obtained with it, we could mention a third strategy in order to increase the profits of the power plant very significantly. Thus, the storage systems would operate in accumulation mode when the energy price in the market is very low and they would supply energy to the system when the price is high enough to pay off the expenses. Such repayment implies that the benefit from the sale of stored energy must be greater than the loss for not selling the produced energy to the market at the time, and for storing it.

4. AN INTELLIGENT ENERGY MANAGEMENT APPROACH

In this section a novel approach to create the model used to manage the above explained strategies is described. Also, due to the high nonlinearity of the system, the novel regression methods are presented to accomplish the challenge.

4.1. The Novel Approach

This subsection describes the generic procedure to create the models. The technique is represented in Figure 5. As it is shown on it, the first step is to detect and select the representative variables from the available dataset. Then, the next step is to create the model with the chosen variables. Usually, this step must be divided into other two different ones. The first one is to create clusters with the different groups of operation points. The second one is to make regression for each cluster created on the above stage.

Figure 4. Evolution of the daily cost of energy

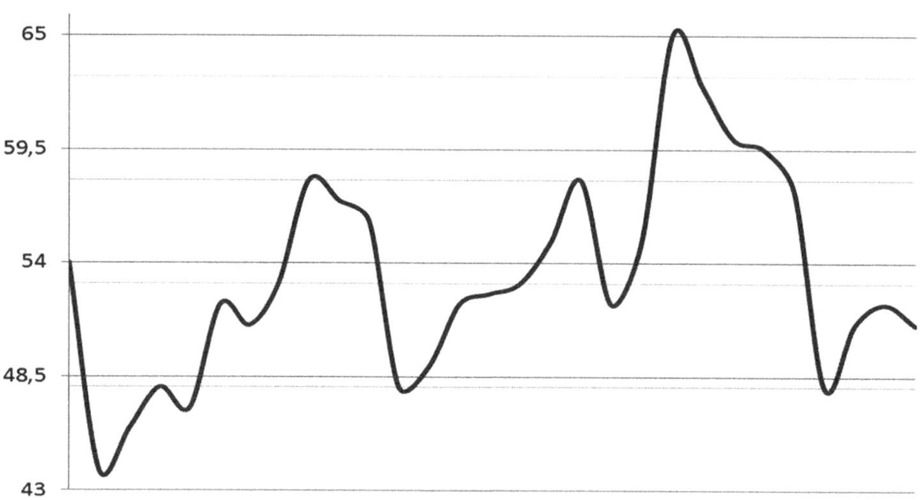

Figure 5. Generic approach for the model creation

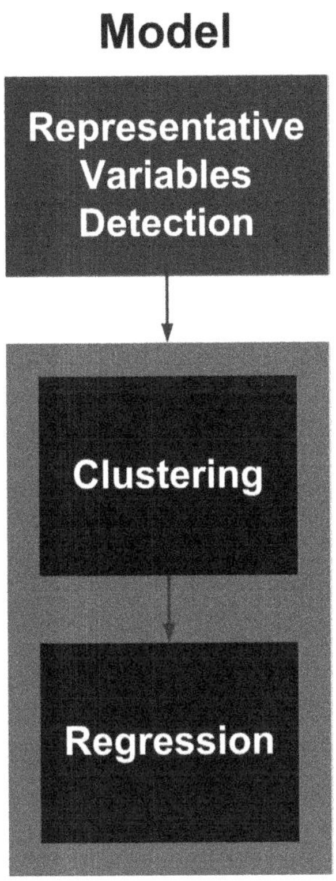

As it can be seen in Figure 6, with the created model, it is possible to predict the energy demand. Both this variable and the energy production are the inputs of a new block. This block must be a function or a rule based system with which it is possible to allow an output to control the stock management described on the section above, where the new strategies have been explained.

4.2. Obtaining the Intelligent Model

The typical procedure to obtain an intelligent model could be divided in several steps. Depending on the complexity of the system under study, it was not possible to reduce any of them. Due to the fact that the system of this research (power management) is a very complex and nonlinear system, the procedure to use must be the regression techniques with clustering.

The first step for this type of systems should be the selection of the representative variables for the system inputs. In the power management system, these variables could be different according to the season in the year, the geographical location, and so on.

For instance, in winter, in a city center, it is possible that a sport event would be a representative variable to the model, with all the lights in the stadium turned on and people watching TV at home or the bar. However, in summer, a representative variable could be a music event in a tourist region.

Figure 6. Energy stock management predicted with the model created

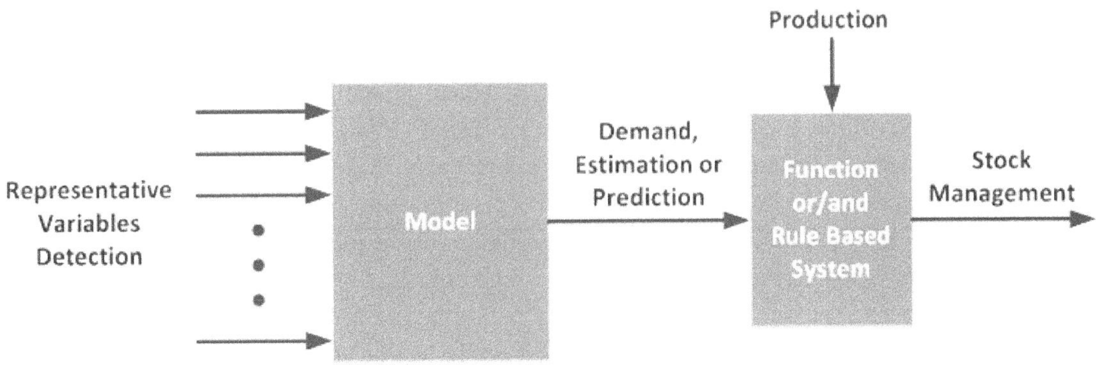

Once the representative variables are "discovered", the next step would be the clustering, by grouping input variables with influence over the system, and then, perform regression for each cluster.

4.2.1. Detection of Representative Variables

The task of finding representative variables could be done by some algorithms like PCA (Principal Component Analysis) (Quintián, Calvo-Rolle, & Corchado, 2014), or SOM (Self Organization Maps) (López García & Machón González, 2004). These algorithms allow the detection of the input variables with which the system could be modeled: the variables are most important to calculate the outputs.

The SOM algorithm also has the ability to detect the correlation between variables. This expertise aims to reduce the number of input variables to the model, thus achieving an easier model (Jove, Aláiz-Moretón, Casteleiro-Roca, Corchado, & Calvo-Rolle, 2014).

4.2.2. Clustering

When the system is a nonlinear system, one of the successful ways to perform a model is by applying the clustering techniques. This procedure groups variables with similar behavior, and divides the global model into some local models. These solutions improve the model by reducing the error between the predicted and the real outputs.

Some typical algorithms used for this task are the SOM (Corchado & Baruque, 2012) and the K-means algorithm (Fernández-Serantes, Estrada Vázquez, Casteleiro-Roca, Calvo-Rolle, & Corchado, 2014). The SOM algorithms could be used to represent the behavior of the input variables in a low dimensional representation that allows us to detect the number of clusters visually. On the other hand, the K-means algorithm does not allow us to detect the clusters, but when the user selects the right number of clusters, the K-means provides an efficient data division.

A traditional use of these two algorithms is to perform a first analysis with SOM to detect the number of clusters, and then the K-means divides the dataset. The next step should be modeling the data in each cluster.

4.2.3. Regression

Some algorithms used to perform this step are the Polynomial Regression (Casteleiro-Roca, Quintián, Calvo-Rolle, Corchado, & del Carmen Meizoso-López, 2013)(Fernández-Serantes et al., 2014), the Artificial Neural Networks (ANN) (Casteleiro-Roca et al., 2013)(Fernández-Serantes et al., 2014), the Support Vector Regression (SVR) (Casteleiro-Roca et al., 2013; Fernández-Serantes et al., 2014), and the Ensembles(Fernández-Serantes et al., 2014).

The Polynomial Regression is widely used. Despite the fact this algorithm has a lot of restrictions to model nonlinear systems, it functions properly if the clusters behavior is linear. The most important benefit of Polynomial Regression is the low computational cost.

The ANNs are very powerful algorithms, usable in nearly all cases. The typical one is the MLP (Multi-Layer Perceptron). The problem with this algorithm is that its efficiency depends on the net internal architecture; however, when the number of neurons, layers and activation functions are well selected, the achieved results are very satisfactory.

The SVR is a modification of the SVM (Support Vector Machines) classification algorithm. This algorithm is based on a nonlinear transformation of the inputs, performing regression in a high dimensional space. After this regression, the data is re-transformed again. There is an improved version of this algorithm named LS-SVR (Least Squared Support Vector Regression) that does not need any adjustment to tune the algorithm.

The Ensemble regression should only be valid when the training data covered all the situations of the system (operating range). It is based on the typical decision tree used for classification. When the system is valid, the computational cost of this algorithm is lower than the other algorithms, but it needs to use more memory to achieve good results.

4.2.4. Efficiency Measurement

The typical indicator to select the best model is to use the MSE (Mean Squared Error)(Fernández-Serantes et al., 2014; Jove et al., 2014). With this value, it is possible to compare some algorithms both for global and local models, with or without clustering. The MSE is calculated by using the model output and the real value of the system output.

This procedure could be used to compare algorithms, or to select the right number of clusters, when this number is not clear after the analysis previously explained.

5. CONCLUSION

It is well known that traditional techniques of Peak Shaving and Valley Filling allow the optimization and efficiency of the generation systems. As mentioned in the present study, when you have more energy than the required, it is stored with various kinds of techniques, in order to return it to the system later when demand is greater than the available generation. This eventual mismatch between generation and demand is currently caused to a great extent by the relatively new renewable energy farms. Such installations produce energy whenever there is the primary source, and do not take into account the energy demanded at a given time. Therefore, the technique of Peak Shaving and Valley Filling helps save the deviations between the generation and the demand with the use of storage systems.

Three strategies have been presented based on these optimization techniques of generation/ consumption.

The first strategy, coordinated by the manager or managers of the electric power system, aims at the stability of the electricity system, and it is based on a perfect match between generation and consumption, with the help of storage installation.

The next two strategies, which come from producers who attach storage facilities to generation, pursue an economic objective, in addition to the optimization of the electrical system. One of these strategies aims to deviate the least or not deviate at all, from the predictions provided to the energy market and therefore not to be penalized for these deviations. The other strategy is to store energy, which is not sold to the market, to sell it later, in times of increased profitability.

There might be the possibility of a fourth strategy consisting of a hybrid between the two associated with a generation plant, whose aim is mainly economic, in which the storage installation will be oversized in order not to deviate from the prediction and also to have the capacity to sell power at times of high profitability.

Let's point out that these techniques will have a growing and gradual progression in the future, since at present it can be economically viable to build a quick response generation plant in spite of the fact that the use throughout the year may be low in the facility when used only to cover peak demand. This tendency will be changing with the development of new and the improvement of the existing storage systems.

With the explained strategies it is possible to locate different stocks along the power network and made a right administration of them. It is very important emphasize on the novel techniques to create the model regression that allow to predict the demand. With it and with the real time energy production it is possible to make a right management of the energy storage.

REFERENCES

Anderson, C. D., & Anderson, J. (2010). Electric and Hybrid Cars. *History (Historical Association (Great Britain))*, 2010, 189.

Benavides, J. (n. d.). *Precios, inversión y economía política de la energía eléctrica*. Retrieved August 22, 2012, from http://www.scielo.org.co

Casteleiro-Roca, J., Quintián, H., Calvo-Rolle, J., Corchado, E., & del Carmen Meizoso-López, M. (2013). Intelligent Model to Obtain Initial and Final Conduction Angle of a Diode in a Half Wave Rectifier with a Capacitor Filter. In Á. Herrero, B. Baruque, F. Klett, A. Abraham, V. Snášel, A.C.P.L.F. de Carvalho, … E. Corchado (Eds.), *Proceedings of the International Joint Conference SOCO '13-CISIS'13-ICEUTE'13*. Springer International Publishing.

Chiang, S. J., Huang, S. C., & Liaw, C. M. (1995). Three-phase multifunctional battery energy storage system. *Electric Power Applications, IEE Proceedings, 142*, 275–284. doi:10.1049/ip-epa:19951856

Corchado, E., & Baruque, B. (2012). WeVoS-ViSOM: An ensemble summarization algorithm for enhanced data visualization. *Neurocomputing, 75*(1), 171–184. doi:10.1016/j.neucom.2011.01.027

Daniels, T. L., & Daniels, K. (2003). *The Environmental Planning Handbook for Sustainable Communities and Regions* (p. 524). American Planning Association.

Egido Cortés, I., Rouco Rodríguez, L., Alonso Perez, J., Porras Muñoz, E., & Ruiz Mendoza, J. L. (n.d.). Impacto en regulación frecuencia-potencia de los cambios de programa en escalón de las unidades generadoras. *Revista de La Asociación de Ingenieros Del ICAI.*

Energía y sociedad. (n. d.-a). *El proceso de liberalizacion de los sectores energéticos*. Retrieved from http://www.energiaysociedad.es

Energía y sociedad. (n. d.-b). *Energía y sociedad.*

Energía y sociedad. (n. d.-c). *Energías renovables: tecnología, economía, evolución e integración en el sistema eléctrico.*

Fabra, N. (2007). Ajustes regulatorios en el sector eléctrico español, La electricidad: mercado, inversiones y garantía de suministro. *ENEconomía Industrial*, 364.

Ferguson, C. D. (2011). *Nuclear Energy: What Everyone Needs to Know* (p. 240). USA: Oxford University Press.

Fernández-Serantes, L. A., Estrada Vázquez, R., Casteleiro-Roca, J., Calvo-Rolle, J., & Corchado, E. (2014). Hybrid Intelligent Model to Predict the SOC of a LFP Power Cell Type. In M. Polycarpou, A. P. L. F. de Carvalho, J.-S. Pan, M. Wozniak, H. Quintian, & E. Corchado (Eds.), *Hybrid Artificial Intelligence Systems*. Springer International Publishing. doi:10.1007/978-3-319-07617-1_49

GaleanoGonzález, D. A., & Botero Castro, H. A. (n. d.). Representación mediante grafcet del accionamiento de generadores sincrónicos en una central hidroeléctrica modernizada. *Dyna, Revista de La Facultad de Minas - Universidad de Colombia.*

Gellings, C. W. (1985). The concept of demand-side management for electric utilities. *Proceedings of the IEEE, 73*(10), 1468–1470. doi:10.1109/PROC.1985.13318

Gil Gonzalez-Madroño, E. (2007). *Tecnologías de acumulación de energía para el uso en la compañía eléctrica. Baterías de sodio-azufre.* Universidad Carlos III.

Huggins, R. A. (2010). *Energy Storage* (p. 406). Springer. doi:10.1007/978-1-4419-1024-0

Jove, E., Aláiz-Moretón, H., Casteleiro-Roca, J., Corchado, E., & Calvo-Rolle, J. (2014). Modeling of Bicomponent Mixing System Used in the Manufacture of Wind Generator Blades. In E. Corchado, J. Lozano, H. Quintián, & H. Yin (Eds.), *Intelligent Data Engineering and Automated Learning-IDEAL*. Springer International Publishing; doi:10.1007/978-3-319-10840-7_34

Kaltschmitt, M., Streicher, W., & Wiese, A. (2007). *Renewable Energy: Technology, Economics and Environment* (p. 596). Springer.

López García, H., & Machón González, I. (2004). Self-organizing map and clustering for wastewater treatment monitoring. *Engineering Applications of Artificial Intelligence*, *17*(3), 215–225. doi:10.1016/j.engappai.2004.03.004

Lucia López, J., & Meneu Ferrer, V. (2005). Los precios en los mercados reestructurados de electricidad: algunas lecciones básicas para la negociación derivada. *CICE Nº 69*.

MacKay, D. J. C. (2009). *Sustainable Energy: Without the Hot Air* (p. 366). Uit Cambridge Limited.

Marsh, W. M. (2005). *Landscape Planning: Environmental Applications* (I. Ed, Ed.). 4th ed.). John Wiley & Sons.

Quintián, H., Calvo-Rolle, J. L., & Corchado, E. (2014). A Hybrid Regression System Based on Local Models for Solar Energy Prediction. *Informatica*, *25*(2), 265–282. doi:10.15388/Informatica.2014.14

Rao, K. R. (2011). *Energy and Power Generation Handbook: Established and Emerging Technologies* (p. 700). American Society of Mechanical Engineers. doi:10.1115/1.859551

River, C. (2005). *Primer on Demand-Side Management with an Emphasis on Price-Responsive Programs*. Tech. Rep. prepared for The World Bank by Charles River Associates.

Salmerón, J. (2010). *Las renovables en el mercado de la electricidad: perspectivas de futuro.*

Santoso, S. (2011). *Fundamentals of Electric Power Quality* (p. 273). Surya Santoso.

Sullivan, R. L. (1977). *Power system planning* (p. 324). McGraw-Hill.

The official website of Denmark. (n. d.). *Proceedings of the Climate Conference COP '15*. Retrieved from http://denmark.dk

United Nations. (1998). *Kyoto Protocol on the Framework Convention of the United Nations about the Climatic Change.*

United Nations. (2005). *Proceedings of the Montreal Climate Change Conference.*

United Nations. (2007). *Bali Climate Change Conference*. Retrieved from http://unfccc.int

Utray Fabra, J. (n. d.). *El mercado de la electricidad: Algunas reflexiones y algunas propuestas.*

Warner, M. (2011). *The Electric Vehicle Conversion Handbook* (p. 176). Penguin Group US.

Wengenmayr, R., Bührke, T., & Brewer, W. D. (2013). *Renewable Energy: Sustainable Energy Concepts for the Energy Change* (p. 170). Wiley.

Chapter 5
Distributed Generation Capacity Planning for Distribution Networks to Minimize Energy Loss:
An Unbalanced Multi-Phase Optimal Power Flow Based Approach

Adnan Anwar
University of New South Wales, Australia

Md. Jahangir Hossain
Griffith University, Australia

Md. Apel Mahmud
Deakin University, Australia

Himanshu Roy Pota
University of New South Wales, Australia

ABSTRACT

This chapter presents an unbalanced multi-phase optimal power flow (UMOPF) based planning approach to determine the optimum capacities of multiple distributed generation units in a distribution network. An adaptive weight particle swarm optimization algorithm is used to find the global optimum solution. To increase the efficiency of the proposed scheme, a co-simulation platform is developed. Since the proposed method is mainly based on the cost optimization, variations in loads and uncertainties within DG units are also taken into account to perform the analysis. An IEEE 123 node distribution system is used as a test distribution network which is unbalanced and multi-phase in nature, for the validation of the proposed scheme. The superiority of the proposed method is investigated through the comparisons of the results obtained that of a Genetic Algorithm based OPF method. This analysis also shows that the DG capacity planning considering annual load and generation uncertainties outperform the traditional well practised peak-load planning.

DOI: 10.4018/978-1-4666-9911-3.ch005

1. INTRODUCTION

Traditionally, distribution networks are passive networks where the flow of both real and reactive power is always from the higher to lower voltage levels. The integration of small and medium sized generation into distribution networks is increasing as these types of generating units offer a number of technical, environmental and economic benefits for utilities along with consumers due to their locations close to customers (Gil, & Joos, 2008). Beside these benefits, the integration of distributed generation (DG) significantly changes the behaviour of the distribution network operations, e.g., passive distribution networks with unidirectional power flow convert into active distribution networks with bidirectional power flow. From the technical point of view, these changes create both negative and positive impacts on both distribution networks service providers (DNSPs) and customers.

The integration of DG creates a variety of well-documented negative impacts on distribution networks. The impact of integrating a single or small amount of DG may not be a significant issue but the penetration of a large fraction of DG may become problematic and affects the power flow of low and medium voltage distribution networks (Lopes et al., 2007). A wide range of conventional approaches are available for mitigating the adverse effects of integrating DG into distribution networks which generally involve the network up-gradation with considerable costs. Traditionally, DG connections are assessed based mainly on 'first come-first serve' basis as well as 'fit and forget' approach that require a DG unit to operate at fixed power factors with a limited capacity to minimize adverse impacts which conspire to limit the available capacity within distribution networks to connect DG (Vovos et al., 2007). Thus, it is essential to shift network planning and operating policies away from the conventional policy of connecting DG to electric power systems to a new and more appropriate one which is able to accommodate more DG within distribution networks.

Literature Review

There is significant ongoing research interest for evaluating the capacity of DG using a wide range of methods, objectives, and constraints within two broad approaches. The first approach is to accommodate DG with distributed and pre-specified capacities at the best locations using evolutionary computation such as genetic algorithms (GAs) and particle swarm optimization (PSO), which are able to handle distributed formulations. However, the distributed formulation of the DG capacity will not provide a truly optimal solution and the use of multiple capacities significantly extends the search space. The other approach requires the network locations of interest to be pre-specified with algorithms guiding the growth of the DG capacity within network constraints. In this case, methods should be capable of solving continuous functions of capacity. To solve these continuous functions; linear programming, gradient search, or optimal power flow (OPF) methods are employed. For example, consult the works by the authors Keane and O'Malley (2005), Rau and Wan (1994), Harrison and Wallace (2005), Vovos et al. (2005), Vovos and Bialek (2005), Ochoa et al. (2010), Dent et al. (2010a, 2010b). A major drawback of these approaches is that the perceived solution may contain a number of sites with very small available capacities when a large number of locations are searched. To overcome some of these limitations, a hybrid method (a combination of GA and OPF) is proposed by Harrison et al. (2008), where the GA is used to search a large range of combination of location and an OPF is used to define the available capacity for each combination. However, this method also requires the number of DG units to be pre-specified. Another hybrid method (a combination of PSO and OPF) is proposed by the authors Gomez-Gonzalez et al. (2012) to perform the same task as presented by Harrison et al. (2008) and it obtains better performance. A heuristic based algorithm is proposed by Anwar and Pota (2012b) which search for both DG capacities and locations.

The existing literature on DG capacity planning considers balanced distribution networks where it is assumed that the power delivery lines are transposed and the three-phase loading is balanced. But in a practical distribution system, the lines are not transposed rather than they may be single-, two- or multiphase and loads are likely unbalanced. Even when loads are balanced, because of the unequal mutual coupling between phases on the line segments, the voltages on a distribution feeder become unbalanced. Moreover, widespread use of single-phase distributed generation (DG) units, e.g., solar photovoltaics and small wind turbines have made the grid more prone to voltage imbalance (Bruno et al, 2011). Therefore, it is important to model unbalanced multi-phase distribution test system to obtain the distribution system planning and analysis results in a more accurate manner. Recently, an unbalanced three-phase optimum power flow (TOPF) has been proposed by Bruno et al. (2011) for DG capacity planning in which quasi-Newton method is used for three-phase unbalanced distribution load flow (DLF) and numerical evaluation of gradients. The quasi-Newton method as highlighted by Bruno et al. (2011) is capable of obtaining local minima very easily but it may be difficult to achieve global minima. In this case, hybrid evolutionary computation techniques are more useful.

This work aims to determine DG capacity based on an OPF flow approach for multi-phase unbalanced distribution networks. To increase the efficiency and achieve the global convergence, an adapt weight PSO (AWPSO) technique is used in this chapter where the cost of annual energy loss and unserved energy is considered as objective functions. In this chapter, the variations in load and uncertainties in solar and wind power generation is also considered and represented through time-series simulations. Finally, the accuracy and efficiency of the proposed technique is compared to that of GA based approach. This chapter shows that the DG capacity planning considering annual load and generation uncertainties outperform the traditional well practised peak-load planning.

Organization

The organization of this work is as follows- System modelling is discussed in Section 2. The formulation and operational procedure of AW-PSO is discussed in Section 3. The proposed DG capacity planning method based on OPF considering multi-phase distribution system is discussed in Section 4. Description of test system and simulation tool is presented in Section 5. The next section shows analysis structure and solving algorithm. In Section 7, results obtained from this experiment are given. Finally, the chapter concludes with some brief remarks in Section 8.

2. SYSTEM MODELING

In this section, the mathematical models used for the proposed technique are discussed. At first the models of the components used in multi-phase OPF are presented and then the models for solar and wind energy based DG are discussed briefly.

Component Modeling of Unbalanced Multi-Phase Distribution Networks for Power Flow

The components of unbalanced multi-phase distribution network are modeled using OpenDSS. This program calculates the distribution load flow based on the 'nodal admittance matrix' using Newton's method. Generally, power delivery elements (e.g., lines, transformers) in OpenDSS are defined in RMS steady-state by their impedance as shown by the following equation.

$$\hat{z} = \begin{bmatrix} \hat{z}_{ij} & \hat{z}_{in} \\ \hat{z}_{nj} & \hat{z}_{nn} \end{bmatrix} \tag{1}$$

where $\hat{z}$ is the impedance matrix and i, j represent sending and receiving node and n is the neutral point. In OpenDSS, the primitive Y_{prim} matrix for each component is calculated using the following where the primitive impedance matrix ($\hat{z}$) appears 4 times (Dugan & McDermott, 2011).

$$\hat{z} = \begin{bmatrix} \hat{z}^{-1} & -\hat{z}^{-1} \\ -\hat{z}^{-1} & \hat{z}^{-1} \end{bmatrix} \tag{2}$$

For power conversion elements (e.g., loads and generators), Norton equivalents are considered where it is modelled with a constant Y_{prim} and a 'compensation' current for the nonlinear portion (Dugan & McDermott, 2011). Once the primitive matrices of all devices are calculated, then the nodal admittance matrix (Y) of the whole network is build using a sparse matrix solver. After that, an iterative solution technique is deployed to achieve the solution of the nonlinear equations (Dugan & McDermott, 2011). During the load flow calculation, different types of load models can be used. The details of component modelling of multi-phase distribution networks for power flow are discussed by Dugan & McDermott, (2011).

Modeling of Solar DG

The PV system model of OpenDSS has been successfully used for distribution system analysis in Smith et al. (2011). In OpenDSS PV model, the PV array and the PV inverter are combined into one convenient form for impact analysis. The active power generated from the PV array, P is a function of the Irradiance, temperature (T), and the rated power, P_{mpp} at a selected temperature and an irradiance of 1.0 kW/m^2. The operating power and voltage is applied to determine the efficiency of the inverter (Smith et al., 2011). The block diagram of the PV system is given in Figure 1. According to IEEE Standard 1547 and UL 1741, typically inverters do not have reactive power capability; therefore, unity power factor is considered during the modeling. Generally, the uncertainty of the solar PV output depends on the solar irradiance and temperature. The solar irradiance data for a typical sunny day is shown in Figure 2. In this work, the annual solar irradiance and temperature data is collected from the 'Desert Knowledge Australia Solar Centre (DKASC).

Modeling of Wind DG

In this work, an induction generator model of OpenDSS is considered. The sequence equivalent circuit of a three-phase induction generator is given in Figure 3 in which where R_s is the stator resistance, X_s stator reactance, R_r rotor resistance, X_r rotor reactance, X_m magnetizing reactance, I_m magnetizing current, I_s stator current, I_r rotor current (referred to stator).

The description and validation of this model can be found in the work conducted by Dugan & Kersting (2006). In order to consider the uncertainty in wind power output, OpenDSS multiplies the rated output with time-varying normalized values. For a typical windy day, the normalized multipliers are given in Figure 4. In this work, annual normalized wind power output values are calculated based on the data obtained from the 'Australian energy market operator' (AEMO).

These models are considered during the OPF and optimization techniques which are shown in the following sections.

3. PROPOSED OPTIMIZATION TECHNIQUE

In this chapter, AWPSO is considered which is an updated version of PSO, a population-based

Figure 1. Block diagram of the PV System model reproduced from Smith et al. (2011)

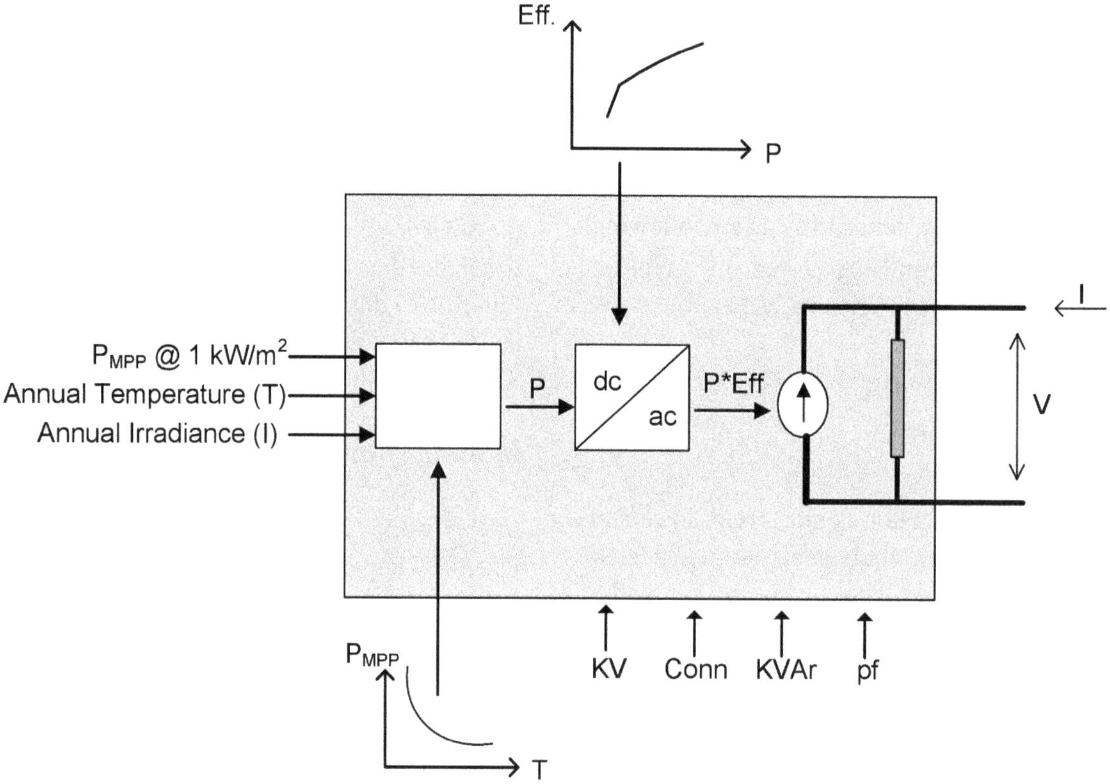

Figure 2. Solar irradiance variation of a typical day

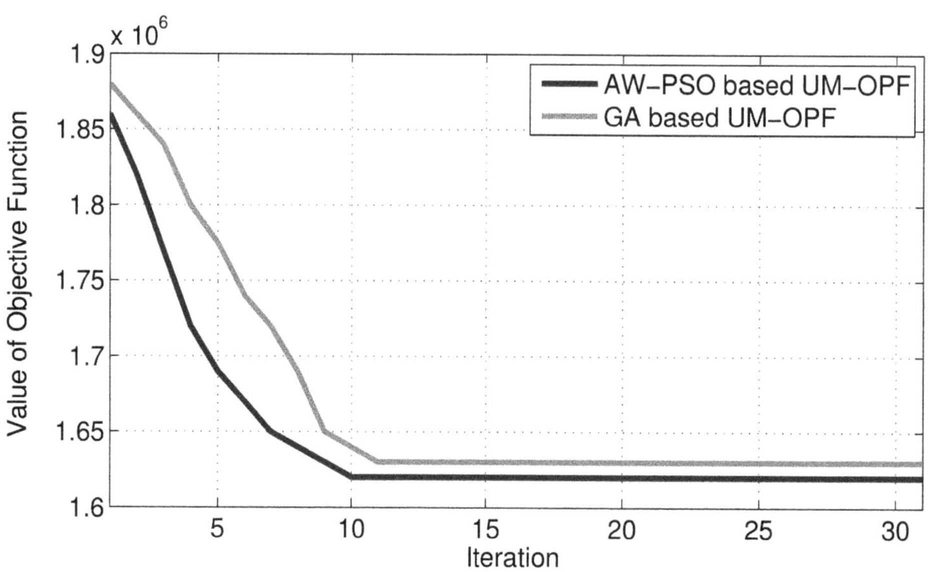

Figure 3. Induction generator model for load-flow

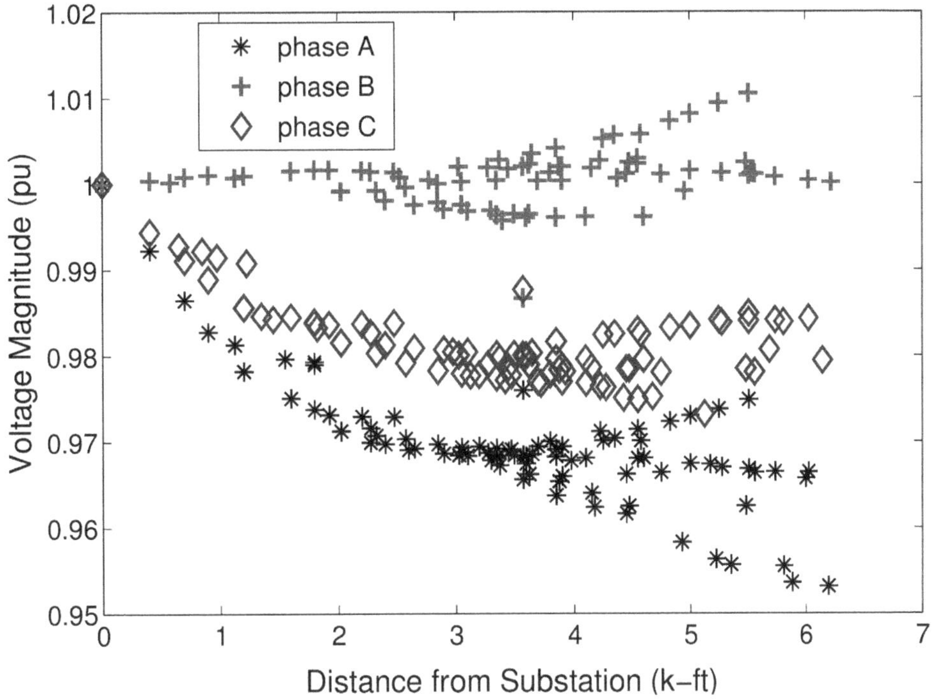

Figure 4. Wind power uncertainties with time

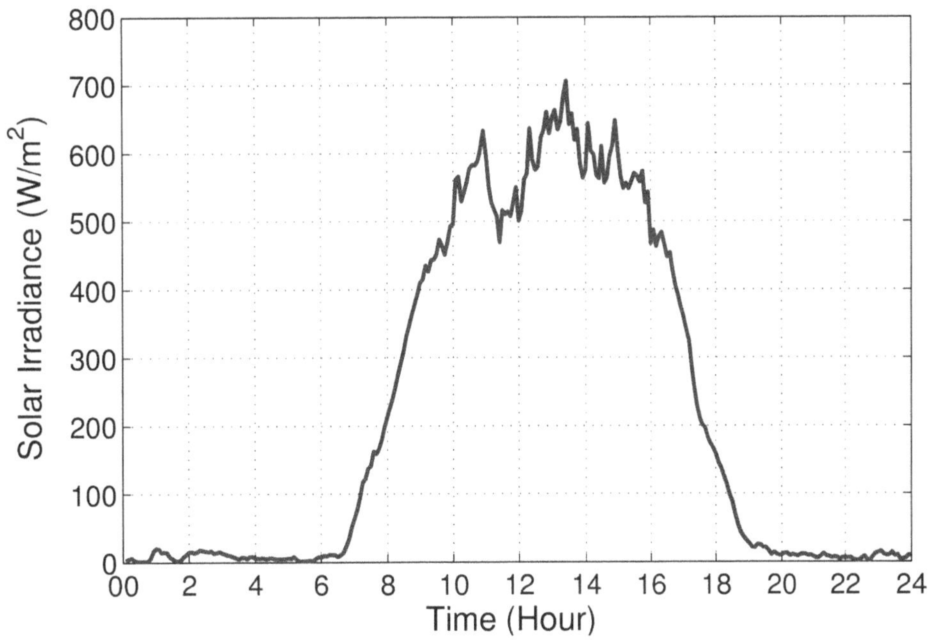

multi-objective optimization. During the optimization procedure, PSO finds the global optimum solution by searching iteratively in a large space of candidate solutions.

In PSO, each individual possible solution is represented as a 'particle'. At each iteration, the position of each particle is updated depending on the particle's own experience and by learning from other members of the swarm. This simple collective behavior helps particles to travel through a multi-dimensional search space and thus the algorithm moves towards the optimum solution gradually. The basic PSO formulation and AWPSO are discussed in the following two subsections.

Basic PSO Formulation

During the PSO formulation, objective function is defined as a minimization problem. To solve the maximization problem, the duality principle as presented in Deb (2004) can be used. Constraints also need to be defined, which can be added together with the objective function using weight vector consisting of penalty parameters. Input variables or control variables and their range have to be specified. In power system optimization, control variables of the optimization problem are different power system parameters depending on the problem which need to be solved. In this research, control variables are the generation capacities of the DG units.

As PSO is a multi-dimensional population based search technique, an *n*-dimensional initial population (swarm) and control parameters need to be initialized. The population members are referred as particles. Each particle is associated with the position vector and the velocity vector, which can be written as (Anwar, 2012, Anwar & Pota, 2012a, 2012b, and Anwar & Mahmood, 2014),

Velocity Vector: $v_i = [v_i^1, v_i^2, \cdots, v_i^n]$

and

Position Vector: $x_i = [x_i^1, x_i^2, \cdots, x_i^n]$

where n is the dimension of the search space. The initial position and velocity of each particle are initialized at the beginning of the algorithm. The initial position of the ith particle x_i can be obtained from uniformly distributed random vector $U(x_{min}, x_{max})$. Here, x_{min} and x_{max} are the lower and upper limits of the solution space respectively.

During optimization, each particle is modified by updating their position and velocity using following equations:

$$x_i(t) = x_i(t-1) + v_i(t) \tag{3}$$

where $x_i \in R^n$ and $v_i \in R^m$.

$$v_i(t) = w \bullet v_i(t-1) + r_1 \bullet c_1 \bullet \left(p_i - x_i(t-1)\right) + r_2 \bullet c_2 \bullet \left(p_g - x_i(t-1)\right) \tag{4}$$

where w is the inertia weight constant, c_1 and c_2 are the acceleration co-efficient, and r_1 and r_2 are the random numbers between 1 and 0. Acceleration co-efficients are responsible to determine the convergence behavior of the particle swarm algorithm. Typical values of these factors are 2 (Etemadi et al., 2008). In Equation (4), one major component ($p_i - x_i(t-1)$) represents self-influence of each particle and another major component ($p_g - x_i(t-1)$) represents the social cognition of particles where p_i is particle's best known position and p_g is swarm's best known position (Anwar, 2012, Anwar & Pota, 2012a, 2012b, and Anwar & Mahmood, 2014)

In the solution loop of PSO, the algorithm continues to run except one of the following stopping conditions is satisfied (Engelbrecht, 2007)

- Number of iterations reach the maximum limit
- An acceptable solution is found

- No improvement is observed over a number of iterations

Adaptive Weight PSO with Velocity Clamping

In adaptive weight PSO (AW-PSO), the value of inertia weight varies dynamically with the iteration which ensures a better convergent behaviour of the algorithm. The formulation of AW-PSO is adopted from the work proposed by Zonkoly (2011) which is shown below

$$w(iter) = w_{\max} - \frac{w_{\max} - w_{\min}}{n_t} \times n_c \quad (5)$$

where n_t is the total number of iterations and n_c is the current iteration number. Typical values of w_{max} and w_{min} are 0.9 and 0.4, respectively.

Particle's velocity is a stochastic variable and to avoid uncontrolled trajectory of particles, the velocity needs to be clamped during the optimization process. The following condition is applied if the velocity of any particle exceeds its maximum limit (Anwar, 2012),

$$v_i^n = \begin{cases} v_{\max} & if\ v_i^n > v_{\max} \\ -v_{\max} & if\ v_i^n < -v_{\max} \end{cases} \quad (6)$$

4. PROPOSED MULTI-PHASE OPF

The UM-OPF problem can be formulated as (Anwar, 2012 and Bruno et al., 2011):

$$\min f_{obj}(x, u) \quad (7)$$

subject to

$$g(x, u) = 0 \quad (8)$$

$$h(x, u) \leq 0 \quad (9)$$

with

$$x \in R^n \text{ and } u \in R^m \quad (10)$$

where f_{obj} is the objective function which needs to be minimized, x is the vector of dependent variables, u is the vector of independent variables. In Equation (8), g is the set of equality constraints which are basically load-flow equations. For power distribution systems, three-phase unbalanced load flow can be obtained using sequence or multi-phase models of power conversion and power delivery components. Therefore, different formulations for representing load flow equations are possible for unbalanced networks (Bruno et al., 2011). In Equation (9), h represents the set of inequality constraints (Anwar, 2012).

A cost minimization planning where technical terms are converted into cost can provide a valuable insight into the problem. The objective of this research is to minimize the cost of annual energy loss (C_{EL}) and the cost of unserved energy (C_{UE}). Therefore, OF can be written as

$$f_{obj}(x, u) = \min \{w_1 C_{EL} + w_2 C_{UE}\} \quad (11)$$

where w_1 and w_2 are the weights applied to the cost of annual energy loss (EL) and the cost of unserved energy (UE), respectively.

A time series simulation has been performed to determine the annual energy loss. In this research, a series of unbalanced multi-phase load flow solutions have been carried out with individual values for each time step. This type of quasi-static simulation can consider the effects of variable load and generation on distribution systems. During the annual energy loss calculation, a sequential 8760-hour simulation has been performed with 1-hr interval considering annual load and generation variation data. Total cost of the annual energy loss (CEL) can be found by multiplying per kWh cost to the total energy loss. In this analysis, energy price cost is considered 0.07$ /kWh (El-Khattam et al., 2004) and cost of unserved energy is 6$/ kWh (Dugan et al., 2001).

UE can be defined as the demanded energy over a time period projected to be above the maximum rating of power delivery components which cannot be supplied. Due to high penetration of DG, power delivery elements may surpass their maximum capacity. At that situation, load must be shed or curtailed from the network to keep the system stable. This is an interruption of electricity supply which would degrade the reliability of the distribution system. The cost of unserved energy is much higher than the normal energy price. Here, the total cost of UE has been calculated by multiplying per kWh cost of UE to the total UE.

Significant reverse power flow caused by large amount of DG increases the voltage level of the distribution network which may cause voltage rise problems (Mahmud et al., 2014). Therefore, voltage stability margin needs to be considered during DG capacity planning. In this research, voltage stability limit, which is a range ±5% of the nominal voltage, is considered as an inequality constraint to obtain a stable operation. If V_{max} and V_{min} are the minimum and maximum voltage limits and N' is the total number of nodes, the voltage stability limit can be written as

$$V_{min} \leq V_i \leq V_{max};\ i\forall N \tag{12}$$

The DG capacity limit is considered as an inequality constraint following the equation below:

$$P_{DG\text{-min}} \leq P_n \leq P_{DG\text{-max}};\ n\forall N \tag{13}$$

where $P_{DG\text{-min}}$ represents the minimum DG capacity which is 0 MW that represents 0% penetration. On the other hand, $P_{DG\text{-max}}$ represents the maximum DG capacity which is 3.53 MW that represents 100% penetration.

In this analysis, the inequality constraint is added to the objective function using penalty factors. Therefore, overall objective function can be written as

$$\min f(x, u) \tag{14}$$

where

$$f(x, u) = f_{obj}(x, u) + f_p(x, u) \tag{15}$$

subject to

$$g(x, u) = 0 \tag{16}$$

with

$$x \in R^n \text{ and } u \in R^m \tag{17}$$

With this proposed approach, the maximum DG capacity planning is shown in the subsequent sections but before going into the main results, a brief description on the test system and analysis tool is discussed in the following section.

5. TEST SYSTEM AND ANALYSIS TOOL

Distribution network working group of IEEE PES Distribution System Analysis Subcommittee has released several unbalanced multi-phase test feeders. In this work, IEEE 123 node test feeder has been chosen which has a nominal voltage of 4.16 kV. There are four step type voltage regulators, shunt capacitor banks, unbalanced loadings with all configuration of loads, both overhead and underground line segments. Single line diagram of this test feeder is given in Figure 5.

In this research, Electric Power Research Institute (EPRI)'s software 'Distribution System Simulator' (OpenDSS) is used for simulating the distribution system. It is a frequency domain tool

Figure 5. IEEE 123 node test feeder

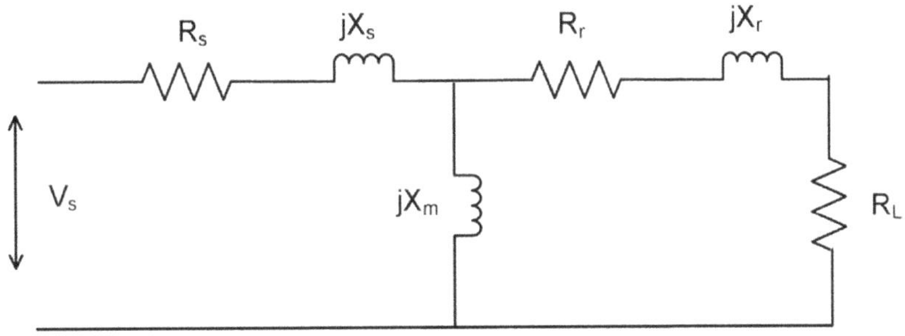

which supports all RMS steady-state analyses and some of its important attributes are discussed below:

1. It can perform sequential time simulations called 'quasi-static solutions'. This distinguished characteristic is used in many analyses of renewable resources considering load variations.
2. It can be interfaced externally using other software, such as 'Matlab.
3. It is possible to model the n-phase distribution networks.
4. It is an object-oriented program where different models can be interfaced easily.

6. ANALYSIS STRUCTURE AND SOLVING ALGORITHM

To solve the optimum DG capacity problem, a co-simulation framework has been developed which is shown in Figure 6. Unbalanced multi-phase distribution network components have been modelled using OpenDSS. For optimization purpose, an algorithm based on particle swarm intelligence has been developed using Matlab. At this stage, OpenDSS has been interfaced with Matlab through it's 'in-process Component Object Model' (COM) server 'Dynamic-link library' (DLL). At every iteration of optimization process, a time series simulation is performed to calculate annual energy loss and unserved energy considering load and generation uncertainty using OpenDSS. It is the responsibility of optimization algorithm to calculate the overall objective function considering constraints and to take intelligent decision for updating the control input to reach the global optimum solution. This type of co-simulation platform for DG capacity planning is very flexible, robust and efficient in nature. Any other distribution analysis software which can perform unbalanced multi-phase load flow and has data exchange interface can also be used for solving this research problem. As optimization algorithm is developed separately in a Matlab script, it is also possible to modify the algorithm easily or to adopt a new algorithm.

The step by step procedure of the proposed method is discussed below and the detailed analysis framework is shown in Figure 7 along with the following discussion (Anwar, 2012).

Step 1: Initialize the swarm and other PSO parameters. Dimension of the algorithm also needs to be specified, i.e., for determination of optimum capacities of three DG units, dimension will be three. As PSO is a population based search technique, therefore, initial particle positions (initial DG capacities) are randomly generated within the maximum and minimum DG capacity limits.

Step 2: Run sequential load flows to calculate annual energy loss and unserved energy using OpenDSS considering initial particle positions.

Figure 6. Analysis structure of the proposed method

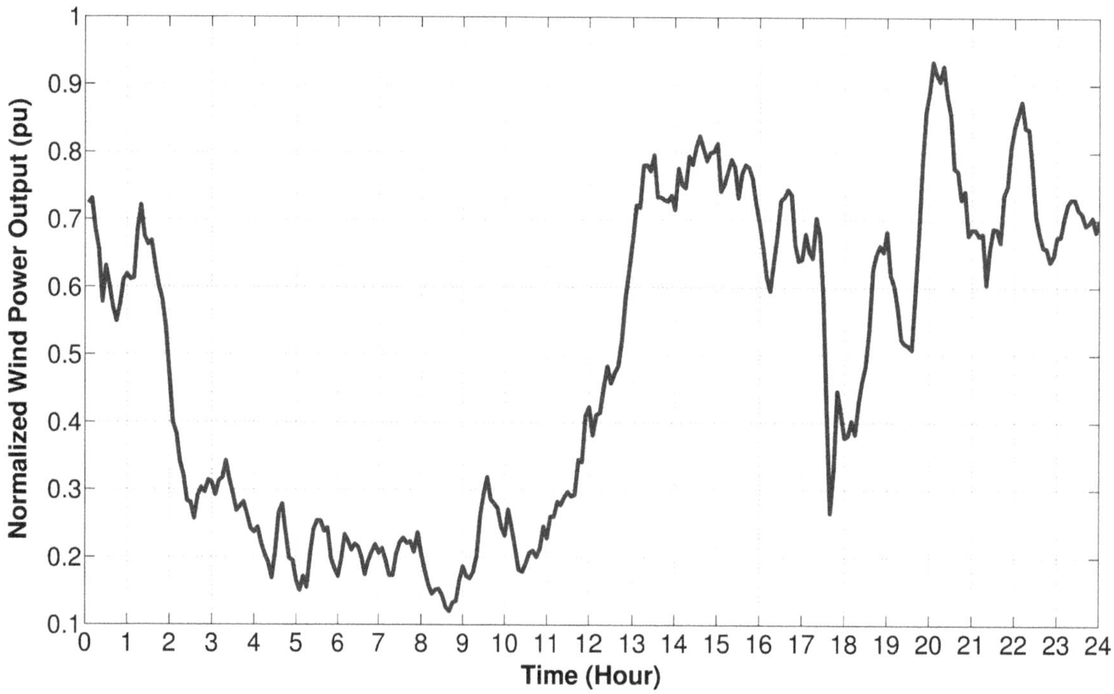

Step 3: Calculate the value of cost function considering constraints for each particle of the swarm.

Step 4: Initialize particles best known position p_i and swarms best known position p_g.

Step 5: Update velocity and position of each particle using (2) and (1).

Step 6: Run sequential load flows to calculate annual energy loss and unserved energy using OpenDSS considering new particle positions.

Step 7: Calculate the value of cost function considering constraints for each particle of the new swarm.

Step 8: Update particles best known position Pi and swarms best known position p_g if necessary.

Step 9: Check whether stopping criterion is satisfied or not. Stopping criterion is described in Section 3.1. If stopping criterion is not satisfied, go to step 5. Otherwise, print the output.

7. RESULTS AND DISCUSSIONS

The superiority of the proposed approach will be clarified from the following case studies:

Case-1: Optimum Generation Capacity of a Single DG Unit

Generally, renewable energy based DG units are installed on those locations where there is a high availability of the resources, like - high wind speed or solar irradiance. As the objective of this work is to determine the optimum capacities of multiple DG units, the detailed analysis of location selection is not the focus of this analysis. In this work, DG units are installed in the most heavily loaded buses. In Table 1, a list of most heavily loaded buses of IEEE benchmark 123 node test system is given. From the Table, bus 76 has the maximum load demand and bus 48 comes with the second highest load demand.

Figure 7. Algorithm for the proposed UM-OPF

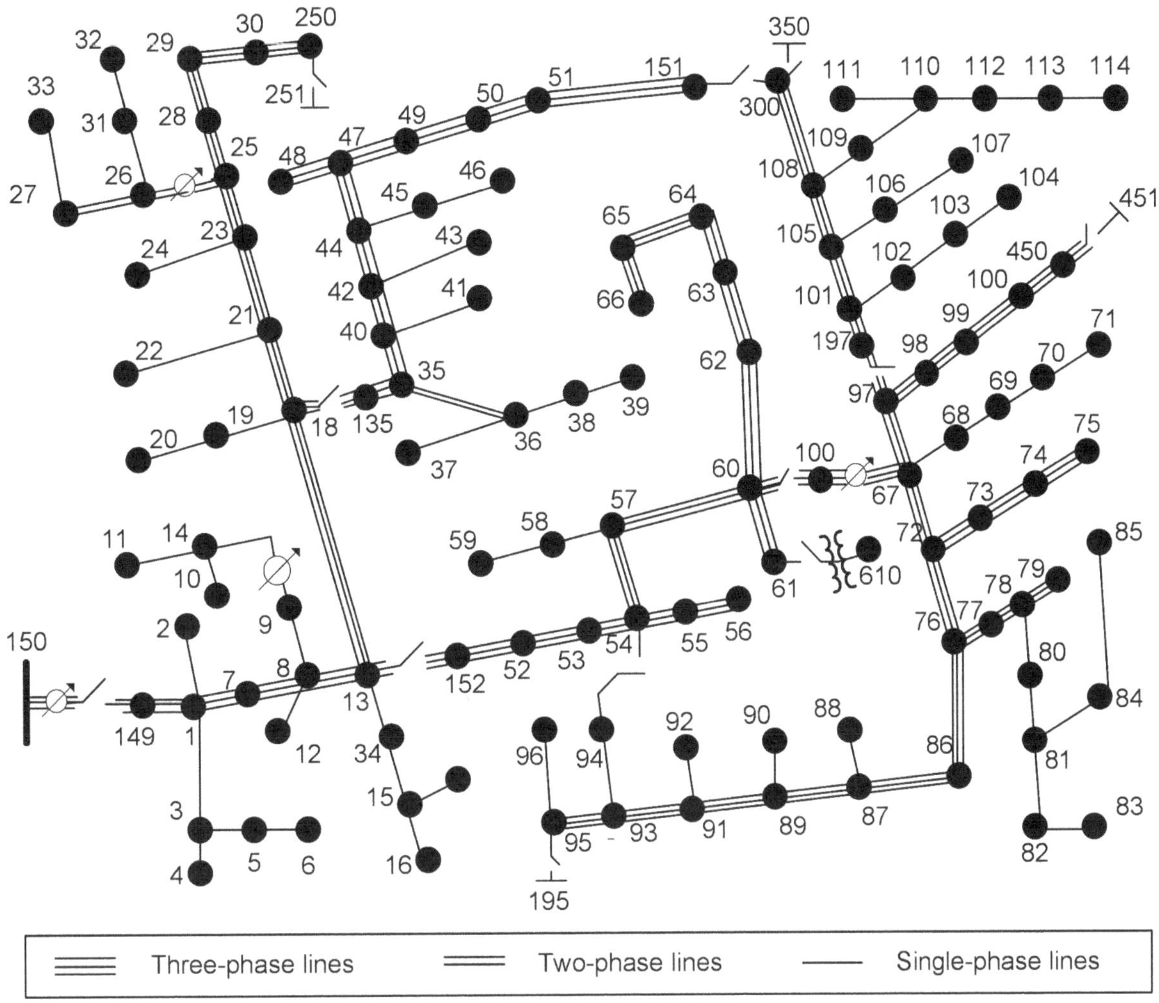

Table 1. DG sites based on bus load density

DG Site	Load Model*	Ph-1 kW	Ph-1 kVAr	Ph-2 kW	Ph-2 kVAr	Ph-3 kW	Ph-3 kVAr
Node 76	I	105	80	70	50	70	50
Node 48	Z	70	50	70	50	70	50
Node 65	Z	35	25	35	25	70	50
Node 49	PQ	35	25	70	50	35	20
Node 47	I	35	25	35	25	35	25

*where I = Constant current load, Z = Constant impedance load and PQ = Constant power load

At the first stage of the analysis, optimum DG capacities of both solar and wind type DG units are determined for different locations of the network. The trend of annual energy loss characteristic for different capacities of Solar and Wind DG at bus 76 is given in Figure 8 and Figure 9, respectively. The proposed method is employed to obtain the optimum DG capacities which are given in Table 2. The annual energy loss due to those optimum generation capacities is also given in that table. From the Table, minimum energy loss profile is obtained using a solar DG of 1.92 MW. 'Genetic Algorithm (GA)' based Unbalanced three phase OPF is used to compare the results as given in Table. 2. From the results, it can be seen that both methods find almost the same result.

Case-2: Optimum Generation Capacities of Multiple DG Units

In this case study, one wind and one solar type DG is considered. From Table. 1, two heavily loaded buses (bus 76 and bus 48) are considered for DG allocation. The aim of this analysis is to determine the optimum capacity of two different DG mix for cost minimization of the annual energy loss and unserved energy. In order to do that, the proposed AW-PSO based UM-OPF is applied considering four different alternatives as discussed below.

Alternative 1: Two wind DG units at bus 76 and bus 48.

Figure 8. Energy loss characteristics with the variation of solar DG capacities at bus 76

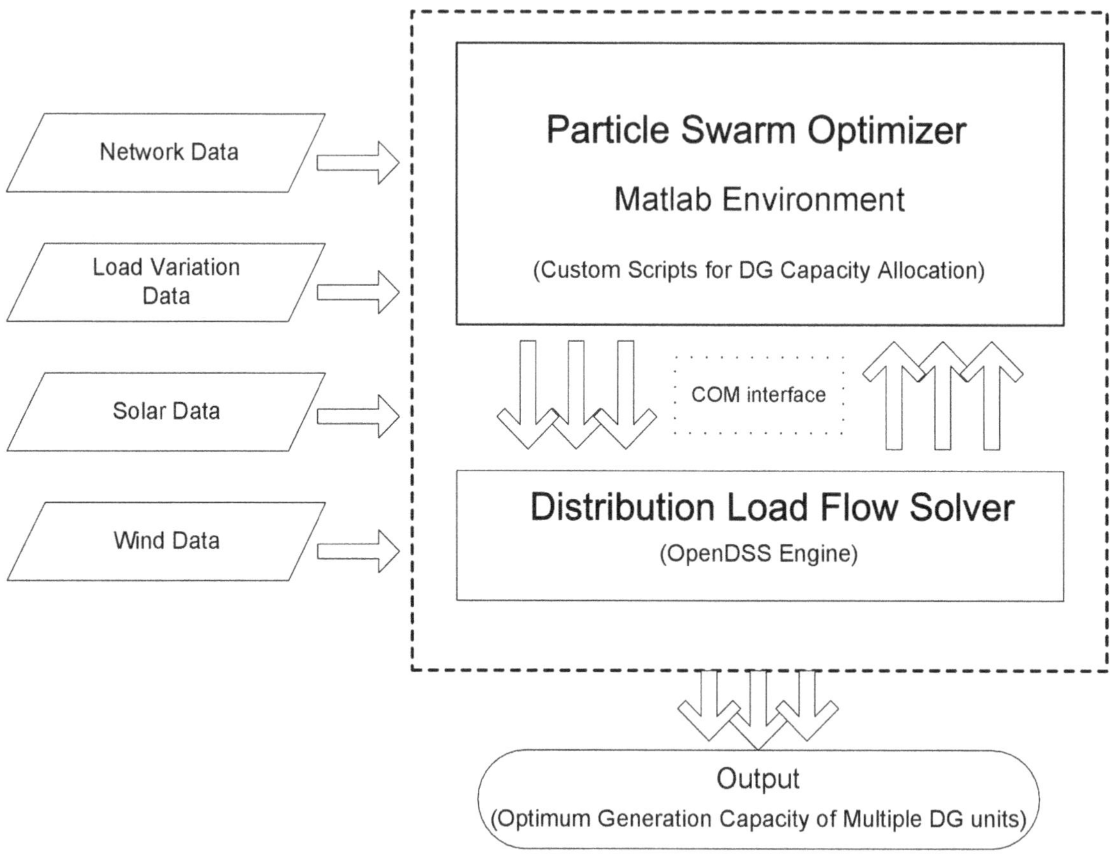

Figure 9. Energy loss characteristics with the variation of wind DG capacities at bus 76

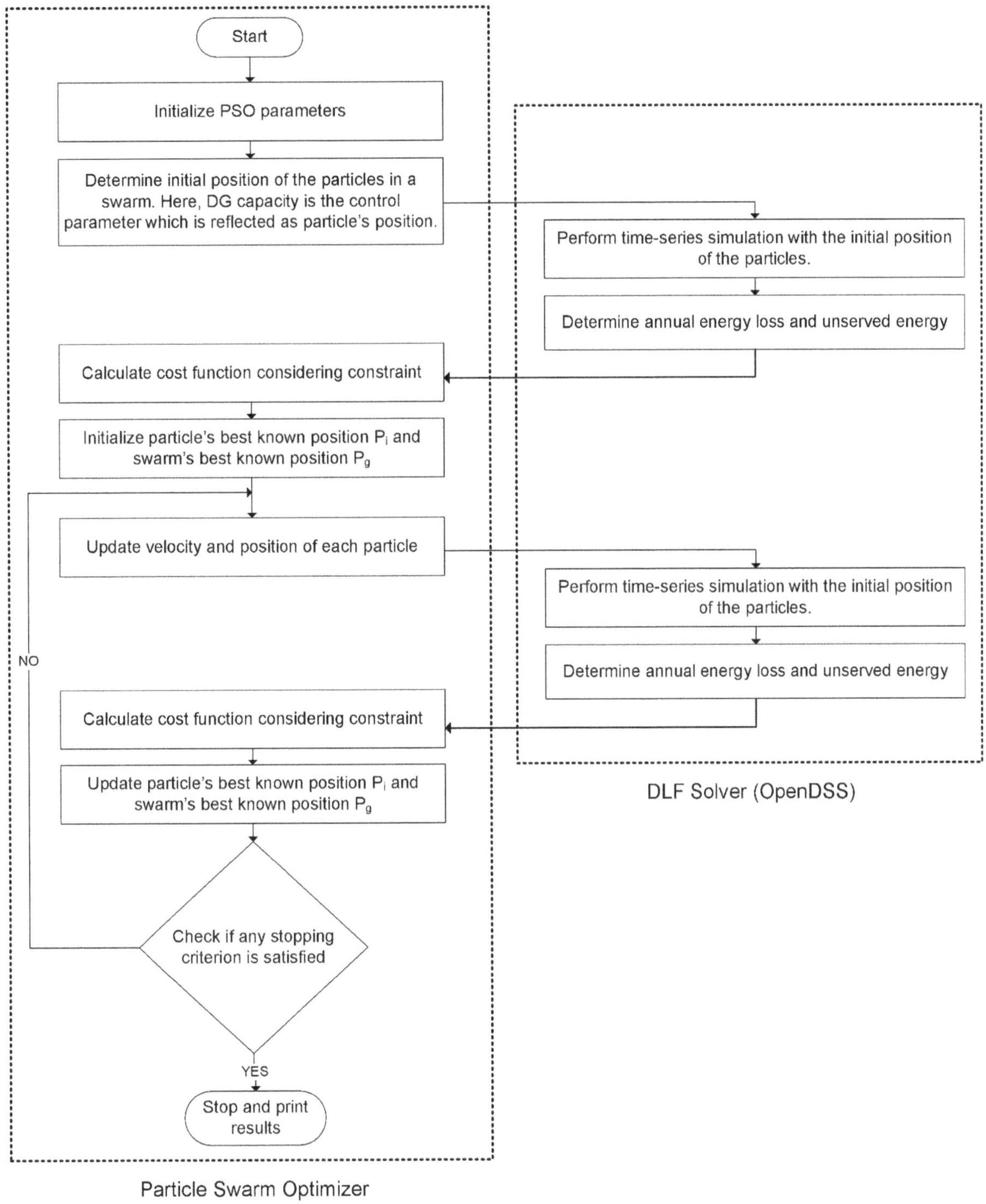

Table 2. Optimum DG Capacity for Single DG units

Location	DG Technology	Optimum Capacity (AW-PSO) (MW)	Optimum Capacity (GA) (MW)	EL (MWh)
76	Wind	1.39	1.40	267
76	Solar	1.92	1.94	237
48	Wind	0.67	0.68	318
48	Solar	1.46	1.46	282
65	Wind	0.89	0.90	308
65	Solar	1.08	1.08	276
49	Wind	0.73	0.73	318
49	Solar	1.43	1.44	285
47	Wind	1.00	0.99	319
47	Solar	1.52	1.52	280

Alternative 2: Two solar DG units at bus 76 and bus 48.

Alternative 3: One solar DG at bus 76 and one wind DG at bus 48.

Alternative 4: One wind DG at bus 76 and one solar DG at bus 48.

Results obtained from the proposed method are presented in Table 3. In this analysis, minimum cost function is obtained for 'alternative 3' in Table 3, where the capacity of solar DG is 1855 kW and the wind DG is 480 kW. At that situation, annual energy loss is 230 MWh and unserved energy is 0.28 MWh. The convergence characteristic of the proposed algorithm for 'alternative 3' is shown in Figure 10. The convergence characteristics of GA-based OPF are also compared here. Figure 10 shows that proposed method using AW-PSO has better convergence characteristics over GA-based OPF and it finds out the optimum solution more accurately and quickly.

Another case study is performed considering four DG units at four different buses. As determination of optimum capacity is the main focus, buses (Bus 76, 48, 65, and 49) are selected based on their load density as shown in Table. 1. The proposed method is implemented to obtain the optimum DG capacities which are given in Table. 4. Results show that using four solar DG units, annual energy loss is reduced to 217 MWh which is 34.36% energy loss reduction of the base case. The optimum DG units of 1320 kW, 653 kW, 363

Table 3. Comparison among different alternatives

Option	Bus	DG Type	Opt. Capacity (KW)	EL (MWh)	UE (MWh)
Alternative 1	76 48	Wind Wind	2314 363	301	0.928
Alternative 2	76 48	Solar Solar	1471 1890	330	0.590
Alternative 3	76 48	Solar Wind	1855 480	230	0.281
Alternative 4	76 48	Wind Solar	1798 1193	239	0.466

Figure 10. Convergence characteristics of the AW-PSO and GA based UM-OPF techniques

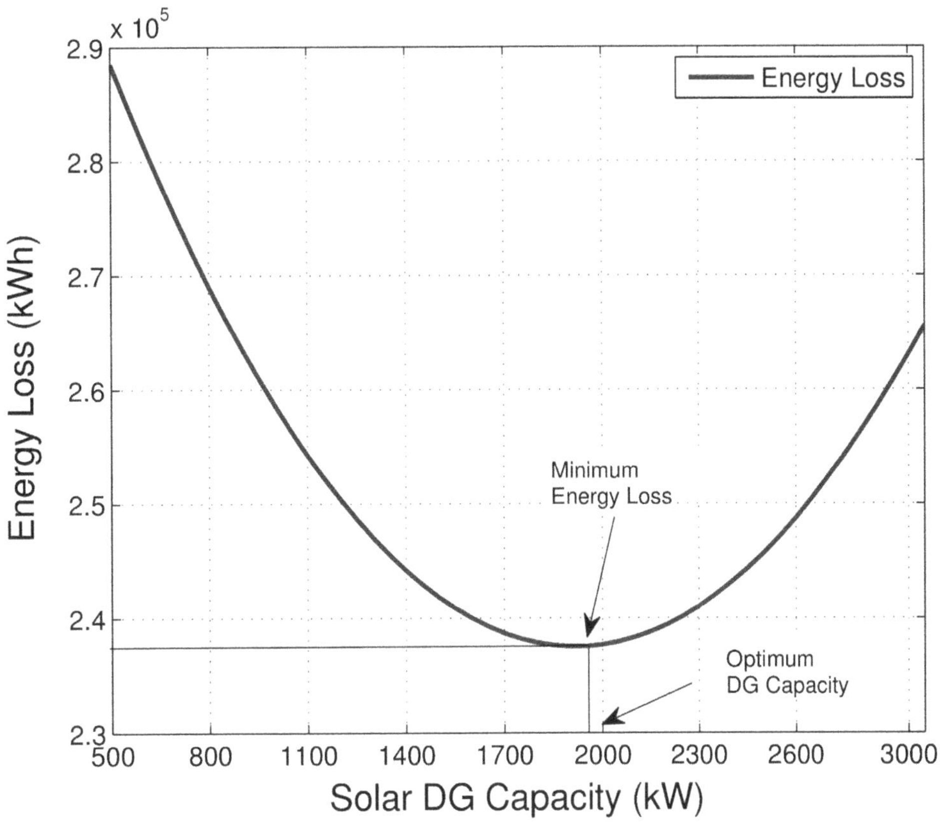

kW and 257 kW signify a total of 73.55% penetration considering total load demand of 3.52 MW. As this is a significant amount of penetration, OpenDSS is used to check the voltage profile of the network shown in Figure 11. In that figure, all node voltages of the multi-phase test system are plotted where different symbols represent different phases of the network. From the figure, it can be seen that all node voltages are within the voltage stability limit.

Case-3: Comparison with the Peak Load Planning

Peak load has been widely used for DG capacity planning (Bruno et al, 2011). If it is intended to operate DG throughout the year, DG planning considering peak load condition may not lead the optimal solution in terms of network efficiency. In this scenario, optimal planning using proposed method has been compared with the peak load planning and the base case when there is no DG unit running in parallel with the network. Annual energy loss of the system at the base case is 330.6 MWh.

At peak load condition, optimum DG capacities are determined based on power loss minimization. At that situation, optimum capacities of one wind and solar DG units at bus 76 and 48 are 1900 kW and 1380 kW, respectively. Considering these DG capacities, annual energy loss becomes 255 MWh which indicates a 22.87% loss reduction of the base case as shown in Table 5.

On the other hand, following the method based on cost minimization proposed in this work, optimum DG capacities of one wind and solar

Figure 11. Node voltages at peak load condition considering four solar DG units

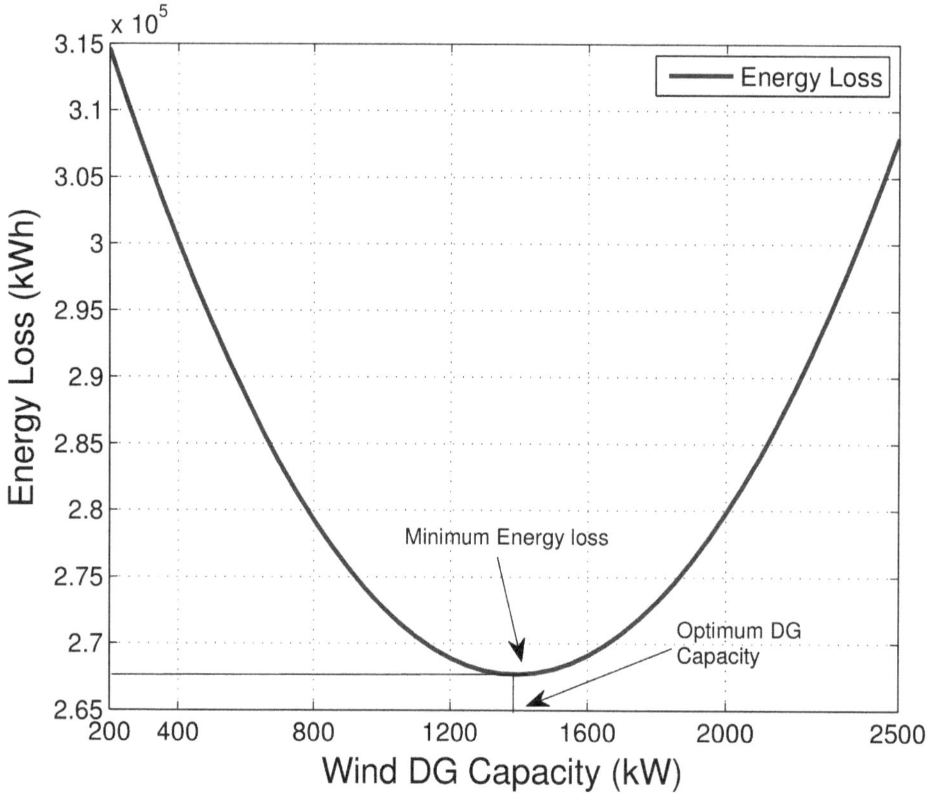

Table 4. Optimum DG Capacity for Single DG units

DG Type	Bus 76	Bus 48	Bus 65	Bus 49	EL (MWh)
Solar	1320	653	363	257	217
Wind	1256	149	82	75	265

Table 5. Comparison with the Peak Load Planning

Case	Bus	DG Type	Opt. Cap (KW)	EL (MWh)	ELR (%)
Base Case	-	-	-	330.6	0
Peak Load Planning	76 48	Solar Wind	1900 1380	255	22.87
Proposed Method	76 48	Solar Wind	1855 480	230	30.43

DG at bus 76 and 48 are 1855 kW and 480 kW, respectively. Considering these DG units, annual energy loss becomes 230MWh which reduces the base case loss by 30.43%. Therefore, the proposed method based on energy loss minimization shows better loss reduction profile compared to the peak load planning. This analysis concludes that planning based on peak load may not always lead to the optimum solution. For renewable energy based technologies, it is very important to consider all the load and generation uncertainties during planning process.

8. CONCLUSION

This research explains the procedure and effectiveness of the proposed method towards DG capacity planning. Swarm intelligence based optimization algorithm captures the global minimum solution successfully during the optimization process. Besides, intermittent nature of renewable generations and changing characteristics of load demand has been modelled using an expert distribution system simulation tool which also considers the detailed multi-phase modelling of the network. Integration of two different software platforms has made the analysis more accurate and flexible.

Although DG is a property of the end users, it is the responsibility of the utilities and the distribution planners to accommodate DG in an optimum way for safe, efficient and reliable operation. The sole purpose of this research is to determine optimum generation capacity of renewable energy based DG units in an efficient way. The proposed method in this research not only finds the global optimum solution successfully, but also is computationally very efficient. This work also shows that using optimum DG capacities, annual energy loss and unserved energy can be reduced significantly and therefore cost is minimized considerably while steady-state voltage profile of the network remains within permissible limits. DG capacity planning using proposed method also shows better results over peak load planning. It is expected that this methodology would play a vital role for capacity planning of renewable energy based DG units.

REFERENCES

Anwar, A. (2012). *Distributed Generation Planning for loss and cost minimisation in Power Distribution Systems* [Master's Thesis]. The University of New South Wales, Australia.

Anwar, A., & Mahmood, A. N. (2014). Swarm intelligence based multi-phase OPF for peak power loss reduction in a smart grid. *IEEE Power and Energy Society General Meeting,* USA.

Anwar, A., & Pota, H. R. (2012a). Optimum capacity allocation of DG units based on unbalanced three-phase optimal power flow. *IEEE Power and Energy Society General Meeting,* USA.

Anwar, A., & Pota, H. R. (2012b). Optimum allocation and sizing of DG unit for efficiency enhancement of distribution system. In *IEEE International Power Engineering and Optimization Conference (PEOCO)*, Malaysia. doi:10.1109/PEOCO.2012.6230854

Bruno, S., Lamonaca, S., Rotondo, G., Stecchi, U., & Scala, M. L. (2011). Unbalanced three-phase optimal power flow for smart grids. *IEEE Transactions on Industrial Electronics*, *58*(10), 4504–4513. doi:10.1109/TIE.2011.2106099

Deb, K. (2004). *Optimization for engineering design: algorithms and examples*. India: Prentice-Hall.

Dent, C. J., Ochoa, L. F., & Harrison, G. P. (2010b). Network distributed generation capacity analysis using OPF with voltage step constraints. *IEEE Transactions on Power Systems*, *25*(1), 296–304. doi:10.1109/TPWRS.2009.2030424

Dent, C. J., Ochoa, L. F., Harrison, G. P., & Bialek, J. W. (2010a). Efficient secure AC OPF for network generation capacity assessment. *IEEE Transactions on Power Systems*, *25*(1), 575–583. doi:10.1109/TPWRS.2009.2036809

Distribution system analysis subcommittee radial test feeders. (1991). *IEEE*. Retrieved from the http://ewh.ieee.org/soc/pes/dsacom/testfeeders/index.html

Dugan, R., & Kersting, W. (2006). Induction machine test case for the 34-bus test feeder- description. *Proceedings of the IEEE Power Engineering Society General Meeting PESGM '11*, USA.

Dugan, R., & McDermott, T. (2011). An open source platform for collaborating on smart grid research. *Proceedings of the IEEE Power and Energy Society General Meeting PESGM '11*, USA. doi:10.1109/PES.2011.6039829

Dugan, R., McDermott, T., & Ball, G. (2001). Planning for distributed generation. *IEEE Industry Applications Magazine*, *7*(2), 80–88. doi:10.1109/2943.911193

El-Khattam, W., Bhattacharya, K., Hegazy, Y., & Salama, M. (2004). Optimal investment planning for distributed generation in a competitive electricity market. *IEEE Transactions on Power Systems*, *19*(3), 1674–1684. doi:10.1109/TPWRS.2004.831699

Engelbrecht, A. (2007). *Computational intelligence: an introduction*. John Wiley & Sons. doi:10.1002/9780470512517

Etemadi, A., & Fotuhi-Firuzabad, M. (2008). Distribution system reliability enhancement using optimal capacitor placement. *IET Generation. Transmission Distribution*, *2*(5), 621–631. doi:10.1049/iet-gtd:20070515

Gil, H. A., & Joos, G. (2008). Models for quantifying the economic benefits of distributed Generation. *IEEE Transactions on Power Systems*, *23*(2), 327–335. doi:10.1109/TPWRS.2008.920718

Gomez-Gonzalez, M., Lopez, A., & Jurado, F. (2012). Optimization of distributed generation systems using a new discrete PSO and OPF. *Electric Power Systems Research*, *84*(1), 174–180.

Harrison, G. P., Piccolo, A., Siano, P., & Wallacen, A. R. (2008). Hybrid GA and OPF evaluation of network capacity for distributed generation connections. *Electric Power Systems Research*, *78*(3), 392–398. doi:10.1016/j.epsr.2007.03.008

Harrison, G. P., & Wallace, A. R. (2005). Optimal power flow evaluation of distribution network capacity for the connection of distributed generation. *IEE Proc.–Generation, Transmission and Distribution*, *152*(1), 115–122

Keane, A., & O'Malley, M. (2005). Optimal allocation of embedded generation on distribution networks. *IEEE Transactions on Power Systems*, *20*(3), 1640–1646. doi:10.1109/TPWRS.2005.852115

Lopes, J. A. P., Hatziargyriou, N., Mutale, J., Djapic, P., & Jenkins, N. (2007). Integrating distributed generation into electric power systems: A review of drivers, challenges and opportunities. *Electric Power Systems Research*, *77*(9), 1189–1203. doi:10.1016/j.epsr.2006.08.016

Mahmud, M. A., Hossain, M. J., & Pota, H. R. (2014). Voltage Variation on Distribution Networks With Distributed Generation: Worst Case Scenario. *IEEE Systems Journal*, *8*(4), 1096–1103. doi:10.1109/JSYST.2013.2265176

Ochoa, L. F., Dent, C. J., & Harrison, G. P. (2010). Distribution network capacity assessment: Variable DG and active networks. *IEEE Transactions on Power Systems*, *25*(1), 87–95. doi:10.1109/TPWRS.2009.2031223

Rau, N. S., & Wan, Y. H. (1994). Optimum location of resources in distributed planning. *IEEE Transactions on Power Systems*, *9*(4), 2014–2020. doi:10.1109/59.331463

Smith, J., Dugan, R., & Sunderman, W. (2011). Distribution modelling and analysis of high penetration PV. *Proceedings of the IEEE Power and Energy Society General Meeting PESGM '11*, USA.

Vovos, P. N., & Bialek, J. W. (2005). Direct incorporation of fault level constraints in optimal power flow as a tool for network capacity analysis. *IEEE Transactions on Power Systems*, *20*(4), 2125–2134. doi:10.1109/TPWRS.2005.856975

Vovos, P. N., Harrison, G. P., Wallace, A. R., & Bialek, J. W. (2005). Optimal power flow as a tool for fault level constrained network capacity analysis. *IEEE Transactions on Power Systems*, *20*(2), 734–741. doi:10.1109/TPWRS.2005.846070

Vovos, P. N., Kiprakis, A. E., Wallace, A. R., & Harrison, G. P. (2007). Centralized and distributed voltage control: Impact on distributed generation penetration. *IEEE Transactions on Power Systems*, *22*(1), 476–483. doi:10.1109/TPWRS.2006.888982

Chapter 6
Hybridized Genetic Algorithm Based Machine Parameters Estimation for Direct Torque Control of 3 Phase Motor for Wind Energy Systems

Vivek Venkobarao
IEEE Senior Member, India

ABSTRACT

It is generally aware that real-time parameter estimation of A.C. induction machine is very important for the efficient operation of vector-controlled drives. However, most existing methods have not taken into account of the four imperfections within the system, i.e. the existence of severe power harmonics due to the PWM inverter drive, the imbalancing of the 3-phase supply from the inverter, the variation of supply frequency under load changing and acceleration/deceleration, and changes in resistance. In this example, we have developed a new method of parameter estimation based on real time data sampling of voltages and currents, no matter they are sinusoidal or not. Here, the assumption of a synchronous rotating speed of the stator flux is not critical. With the aid of hybrid genetic algorithm techniques, the model has been found useful for on-line speed/torque control in most field orientation control schemes as it is much easier to achieve a global minimum during the optimization process.

INTRODUCTION

Basic Principle of DTC

In principle, DTC is a direct hysteresis stator flux and electromagnetic torque control which triggers one of the eight available discrete space voltage vectors generated by a Voltage Source Inverter (VSI) in order to keep stator flux and motor torque within the limits of two hysteresis bands. The correct application of this principle allows a decoupled control of flux and torque. Basically, the status of the errors of stator flux magnitude $|\psi s|$ and electromechanical torque Te are detected and digitalized by simple two- and three-level hysteresis comparators. An optimum switching

DOI: 10.4018/978-1-4666-9911-3.ch006

table is then used to determine the status of three switches S1, S2, S3 and the corresponding voltage space vector vi depending on the stator flux region (θs). The stator flux position (θs) is determined by dividing the d-q plane into six 60 degree regions. Simple three sign detectors are used to determine the sector where the stator flux exists.

Power Converter

The space vector PWM technique is used to generate the firing pulses to the power converter.

The SVPWM method is an advanced, computation intensive and the best among all PWM techniques for variable frequency drive applications. It is finding widespread applications in recent years because of the advantages it offers in its performance. The Space Vector based modulation technique is a digital technique in which the objective is to generate PWM load line voltages whose value is equal to the command voltage. This is implemented in each sampling period by selecting the appropriate switching states from the valid states of the space vector diagram. After proper calculation of the duration of the states, they are switched in proper sequence.

SPACE VECTOR TRANSFORMATION

The space vector transformation can be represented by

$$\begin{bmatrix} V_{an} \\ V_{bn} \\ V_{cn} \end{bmatrix} = \begin{bmatrix} 2/3 & -1/3 & -1/3 \\ -1/3 & 2/3 & -1/3 \\ -1/3 & -1/3 & 2/3 \end{bmatrix} \begin{bmatrix} V_{ao} \\ V_{bo} \\ V_{co} \end{bmatrix}$$

Now in order to apply the space vector modulation, the three phase voltages are transformed into a diphase frame (α,β) using *Concordia Transformation*, thereby the time varying three phase variables become time invariant vectors.

Figure 1. Block diagram of direct torque control

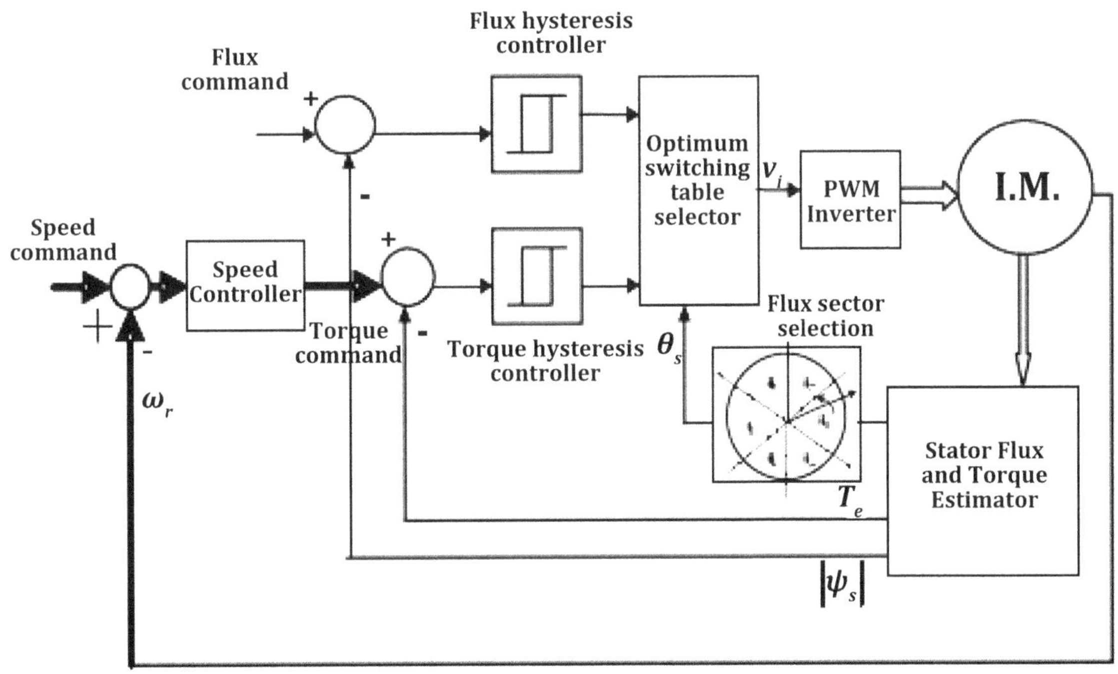

ALPHA BETA TRANSFORMATION MATRIX

$$\begin{bmatrix} V\alpha \\ V\beta \end{bmatrix} = \sqrt{\frac{2}{3}} \begin{bmatrix} 1 & -1/2 & -1/2 \\ 0 & \sqrt{3}/2 & -\sqrt{3}/3 \end{bmatrix} \begin{bmatrix} V_{an} \\ V_{bn} \\ V_{cn} \end{bmatrix}$$

Due to the presence of N states for the variable Ci (i=1,0,-1), there are Nm possible combination of the three level voltage inverter, where m is the number of phases.(here m=3). The result consists of 27 vectors grouped at the vertices of the hexagon. It is inferred that there are 3 zero vectors (V0,V7,V14) and a group of 12 vectors representing in pairs, the same co-ordinates (v11,v12), (v21,v22), (v31,v32), (v41,v42), (v51,v52) and (v61,v62). By drawing the 27 vectors, the three hexagons are distinguished from each other.

If the voltages are three balanced sinusoidal waveforms that have an amplitude Vc and an angular frequency *w*, the resulting signal in the *alpha-beta* frame becomes a vector of fixed magnitude and rotating at the same angular frequency. The objective of this technique is to approximate the line signal space with amplitude Vc with the space vectors available in each hexagon.

The small hexagon, which is defined by the six regions (I, II, III, IV, V, VI) has all the limiting vectors of same magnitude $E/\sqrt{6}$.The middle hexagon, which is also defined by six regions (a, b, c, d, e, f) has all its limiting vectors of same magnitude $E/\sqrt{2}$.The big hexagon defined by the six regions (A, B, C, D, E, F) has all its limiting vectors of magnitude $E(\sqrt{2}/\sqrt{3})$

Motor Model

The d-q induction motor model given by Boldea and Naser (1986) is considered for simulation. The continuous–time electromechanical model of an induction motor is of fifth order with stator and rotor flux linkages and speed as state variables. The model is developed without taking into account the core and hysteresis losses. The dynamic

Figure 2. Three-level inverter's hexagons vectors in the (α-β) frame

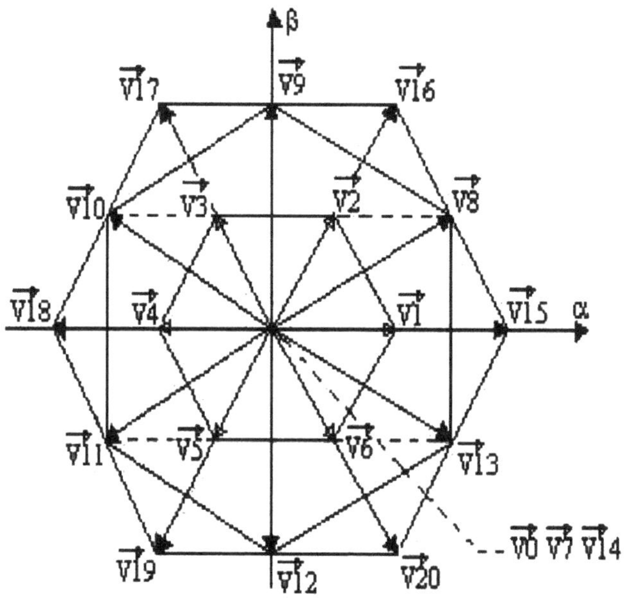

Figure 3. Three-level inverter's hexagons: Inner hexagon

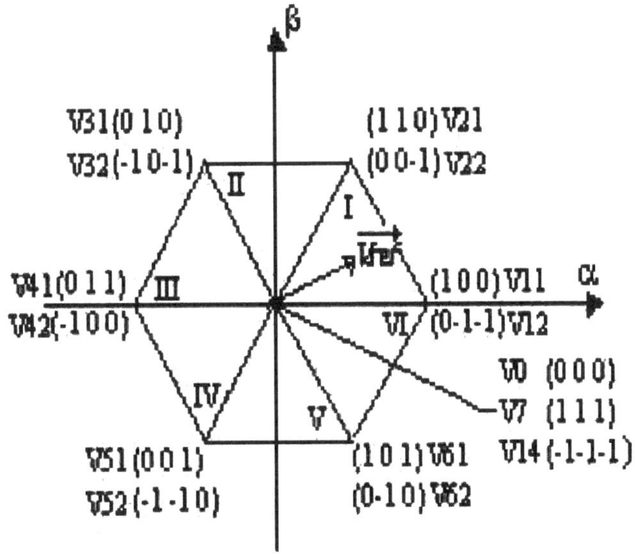

model of induction motor in synchronously rotating reference frame is described by a set of first order linear vector differential Equations

$$p[\Psi] = [A][\Psi] + [V]$$

$$T_e = \frac{n}{\left(1_1 1_2 - 1_m^2\right)} \left(\Psi_d^s \left(1_2 \Psi_q^s - 1_m \Psi_d^r\right) - \Psi_q^s \left(1_2 \Psi_d^s - 1 \Psi_d^r\right)\right)$$

$$p\acute{E}_m = -\frac{nb}{J}\acute{E}_m + \frac{n}{J}\left(T_e - T_{L(\acute{E})}\right)$$

where $[\psi]$ and $[V]$ being $[\ \psi_d^s\ \psi_d^r\ \psi_q^s\ \psi_q^r\ ]^T\ [V_d^s\ 0\ V_q^s\ 0]^T$ and $[A]$ is given as the matrix shown in Box 1.

Usage of the Soft Computing Technique for Rotor Position

Determination of 2 weighing maps for the selection of optimum switching table based on motor speed to compute rotor position.

Goal is to find rotor position such that adaptation factor Adaptation is identical to deviation

(Adaptation (actual)) - (Adaptation (predicted)) = 0

Box 1.

$$[A] = \frac{1}{\left(1_1 1_2 - 1_m^2\right)^4} \begin{bmatrix} -R_1 1_2 & R_1 1_m & \omega_1\left(1_1 1_2 - 1_m^2\right) & 0 \\ R_1 1_m & -R_2 1_1 & 0 & \left(\omega_1 - \omega_m\right)\left(1_1 1_2 - 1_m^2\right) \\ -\omega_1\left(1_1 1_2 - 1_m^2\right) & 0 & -R_1 1_2 & R_1 1_m \\ 0 & -\left(\omega_1 - \omega_m\right)\left(1_1 1_2 - 1_m^2\right) & R_2 1_m & -R_2 1_1 \end{bmatrix}$$

Figure 4. Three-level inverter's hexagons: Middle hexagon

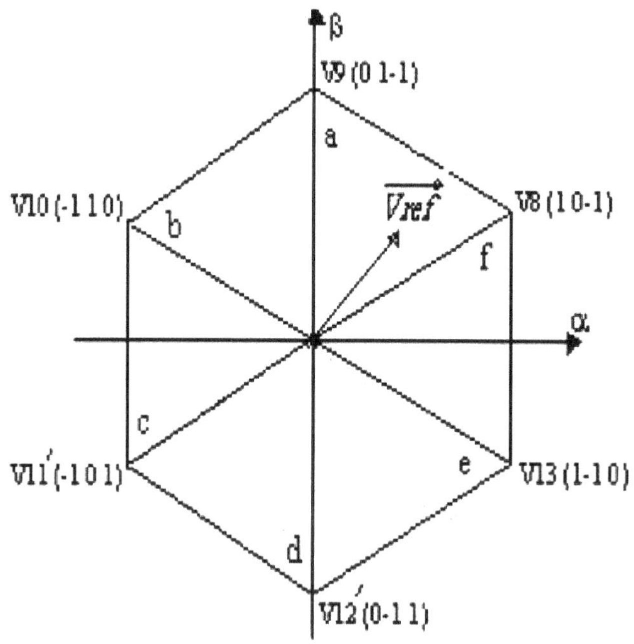

Figure 5. Three-level inverter's hexagons: Outer hexagon

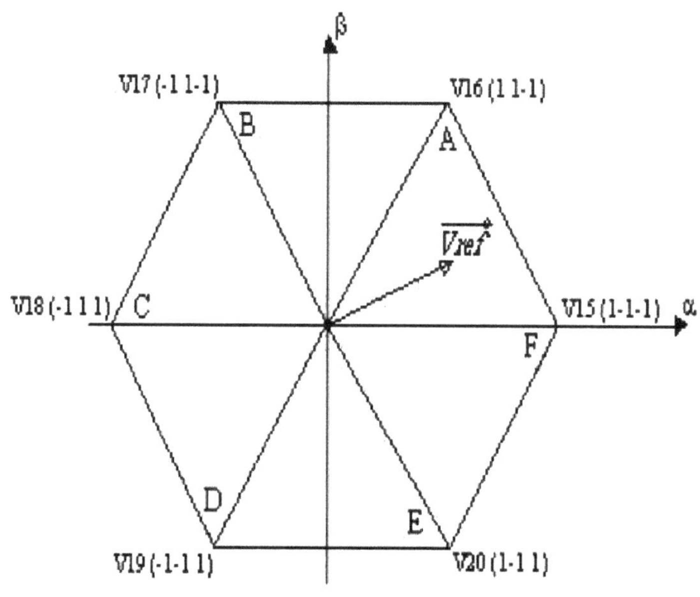

The fitness equation is given as

Error= Sigma(Adaptation (i,j) (actual)- Adaptation (i,j) (predicted))

Fitness= Error+PEN1+PEN2+------

where PEN can be

1. Penalty due to disconnected maps
2. Penalty due to variation in Adaptation learning factors
3. Penalty due to islanding

The proposed algorithm is presented as a flow chart (Figure 6) for the above described problem.

CROSSOVER LOGIC

Mutation Logic

1. The chromosome will undergo mutation if the error is greater than the permissible window (Figure 7).
2. The probability of an element to undergo mutation is given by the exponentially decaying curve which is a function of exp(-0.5*Fitness(i)).
3. When the error is greater than 5% and mutation probability permits the chromosome to undergo mutation. The selection of parents are generated randomly

IMPLEMENTATION OF PENALTY

Multiple Dependencies

The learning of torque and other parameters cannot happen in multiple locations of the adaptation

Figure 6. Application of Genetic Algorithm
Total Population chosen: 25000

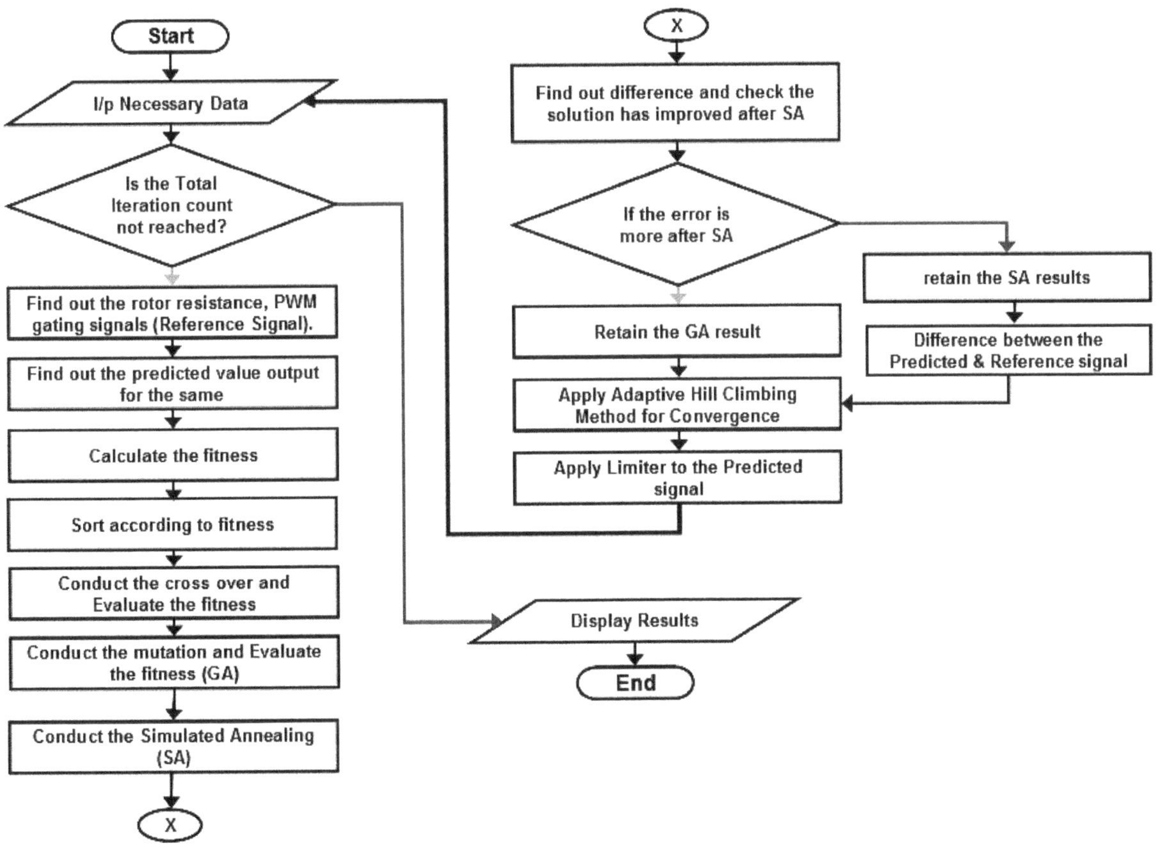

Figure 7. Ratio of Gender of chromosomes 60%(Male)

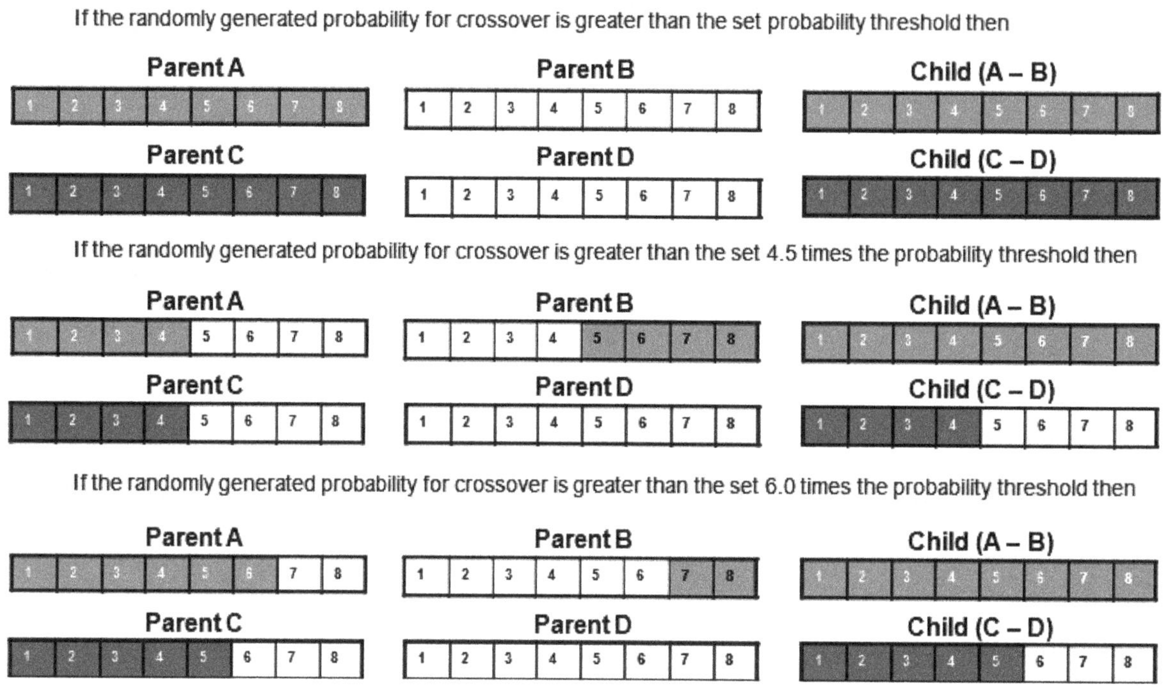

maps. The learning happens only in the selected zones. Any point outside the zone reduces the stability of the system.

In the below example matrix element a(4,4)is not in learning zone (Figures 8 and 9).

Non Plausible Value

The learning of torque and other parameters cannot have non plausible values which may be too high or low for the given operating point.

Figure 8.

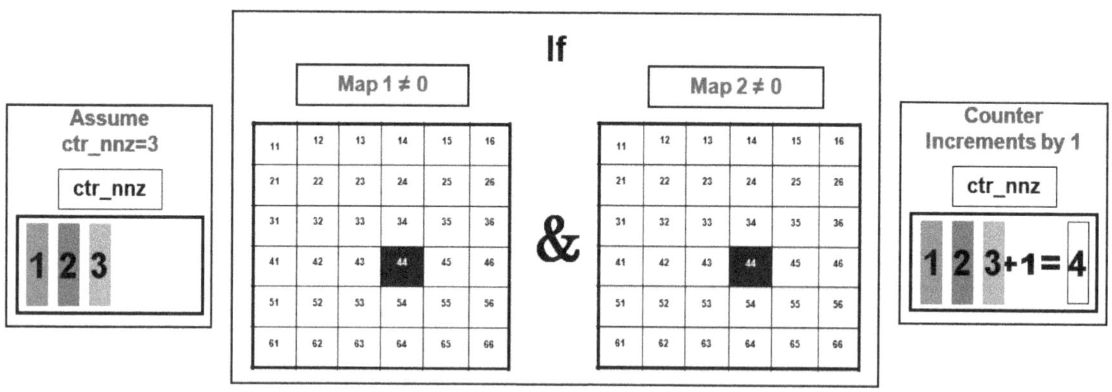

Figure 9.

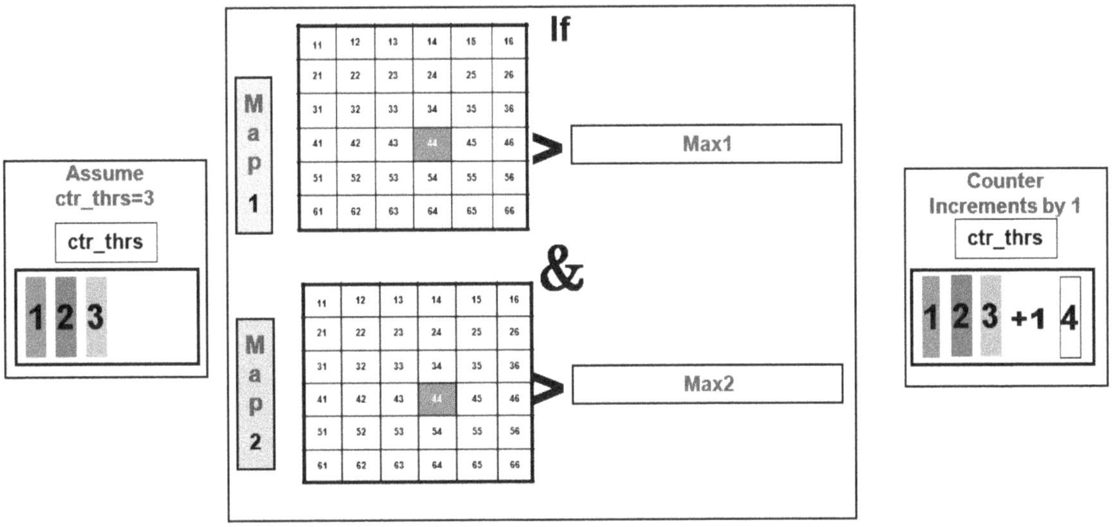

Improper Learning Zones

Penalty is added if there are fluctuating zones in both the maps. If there are different zones system becomes highly unstable as for each operating point the algorithms gives multiple learning values (Figures 10, 11, and 12).

It seen that there is no improvement in solution after 50 iteration. This is because the chromosomes have not dissipated the extra energy they possessed. This energy is made to reduce substantially in the chromosome by making it climb a Hill. This technique is called "Adaptive Hill Climbing".

Figure 10. Simulation of algorithm for rotor position estimation

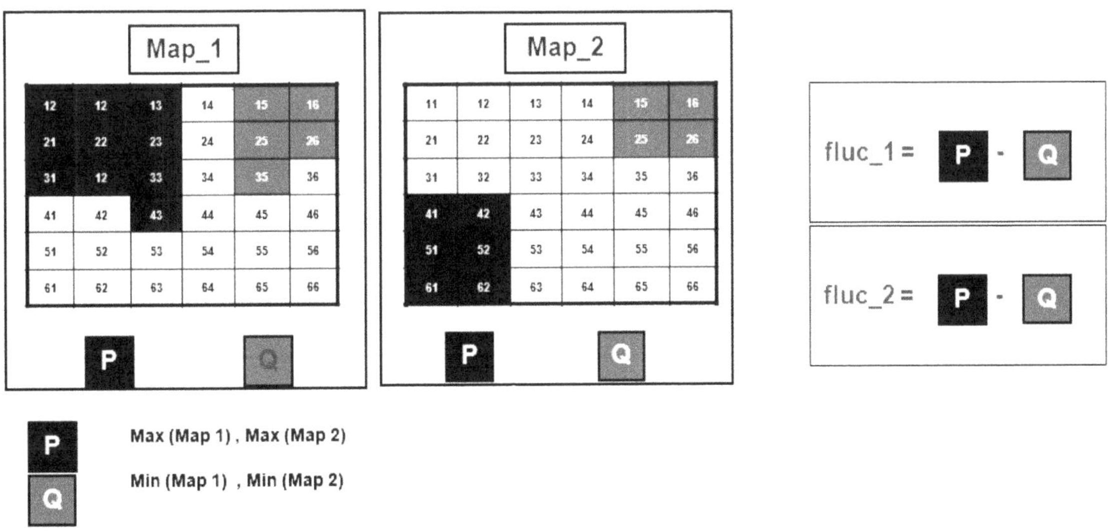

Figure 11. Ideal case in case of linear models

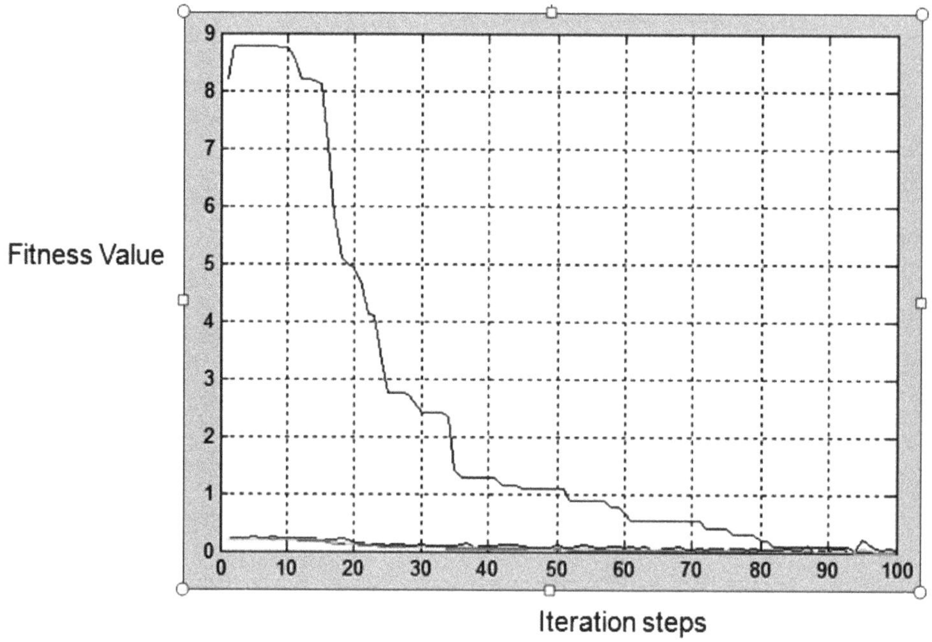

Figure 12. Actual case in case of non linear models

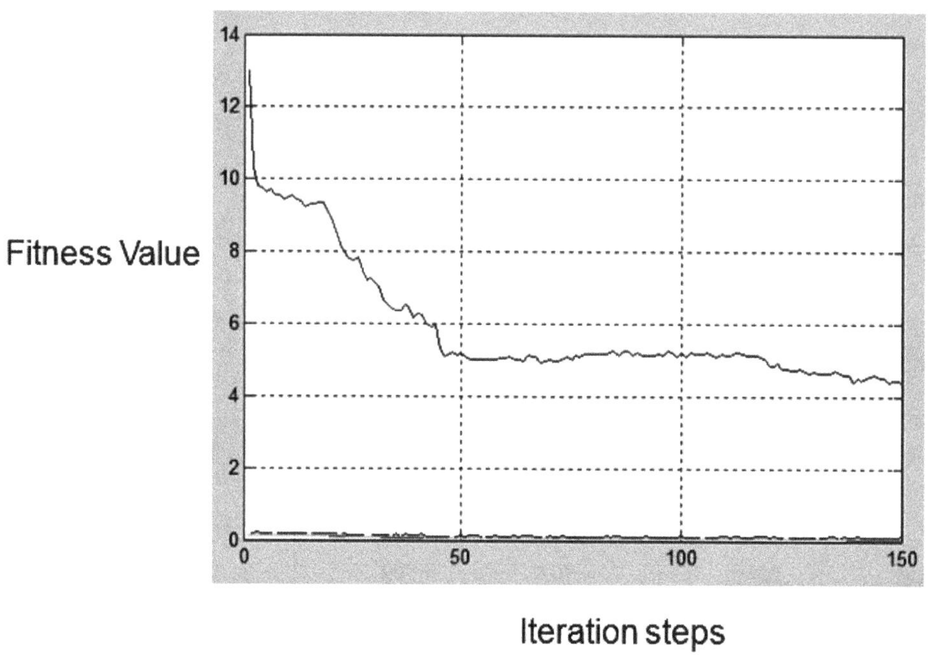

Adaptive Hill Climbing

Hill climbing is a technique for certain classes of optimization problems. The idea is to start with a sub-optimal solution to a problem (i.e., start at the base of a hill) and then repeatedly improve the solution (walk up the hill) until some condition is maximized (the top of the hill is reached) (Figures 13 and 14).

In certain cases the solution does not improve even after applying Adaptive Hill Climbing. In such cases the hybridization is achieved by simulated annealing

Simulated Annealing

Simulated annealing algorithm of the original idea was proposed in 1953, in the Metropolis, Kirkpatrick put it successful application in the combinatorial optimization problems in 1983.

Simulated annealing algorithm from the solid annealing principle, the solid heating to fully high, let its slowly cooling, heating, solid internal particle with temperature rise into disordered shape, internal energy increases, and slowly cooling particles gradually orderly, and in each temperature are to reach an equilibrium state, and the last in the room temperature reaches ground state, internal energy is reduced to the minimum.

Modified Simulated Annealing

The slow convergence of standard SA algorithm is a common drawback in its application. A similar problem existed in using standard SA for a simple one-dimensional dispersion problem. It is therefore necessary to modify the standard SA algorithm to speed up the convergence and reduce the computational cost. In this study we proposed following three strategies to improve the SA optimization speed.

The initial temperature using the following formulae

$$\chi_0 = \exp\left(\frac{-\bar{\Delta} f^+}{T_0}\right)$$

where X0 is the initial probability for acceptance, and was set as 0.8 in current work; Delta f is the averaged increment of objective function over the averaged value of objective function after multiple perturbations. Here we used 10000 times of perturbations to determine Delta f and calculate T0.

$$tanhx = \frac{sinhx}{coshx}$$

Figure 13. Genetic algorithm and adaptive hill climbing in case of non linear models

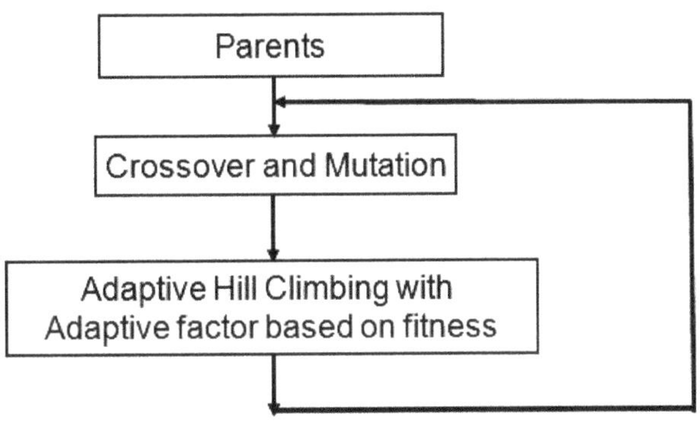

Figure 14.

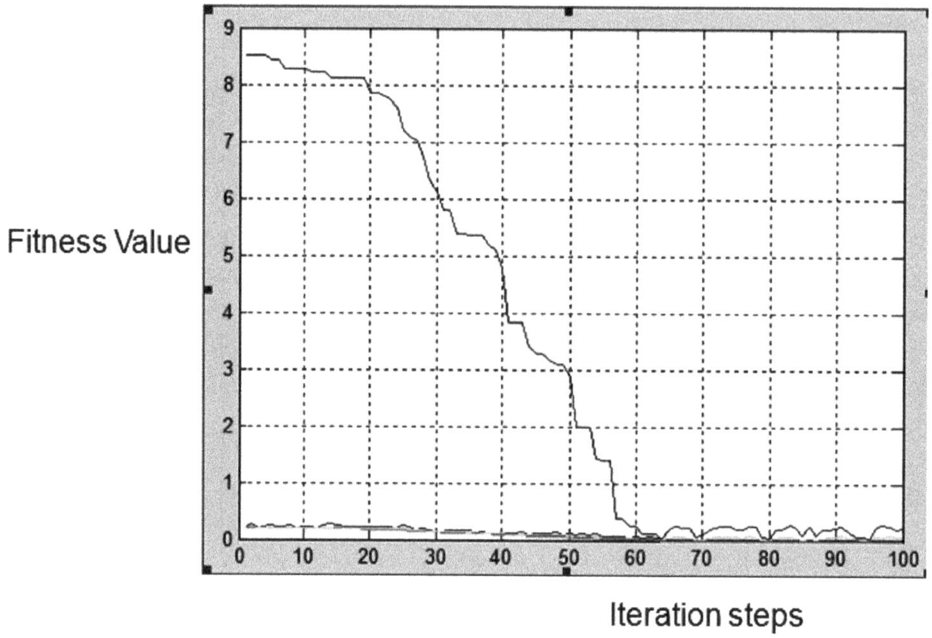

$$tanhx = \frac{e^{2x} - 1}{e^{2x} + 1}$$

$$\tanh x = \frac{\sinh x}{\cosh x}$$

- **Perturbation Generating Function:** The uniform random number is used to produce parameter set perturbation. rand(0,1) is an uniform random number between 0 and 1.
- **Cooling Schedule:** we used a power cooling schedule for temperature reduction. The cooling is varied between 0.4 and 0.95 in the current study. This is done by simulating a high pass filter

$$T = \left(1 - e^{-(2*pi*f_c*dt)}\right) * dT$$

$$f_c = -\frac{\ln(T/dT) - 1}{2*pi*dT}$$

- **Parameter Space Constraints:** Theoretically any nonnegative real numbers could be the transport parameter values. In a specific application, however, practitioners can often specify a much narrower range for each parameter to be estimated. Obviously, the narrower the range is, the faster it will be for the computation to converge.
- **A Valve for Inner Loop Break:** Normally a maximum number of perturbations within each temperature are specified as the inner loop break condition.
- **Inner Loop Memory:** Normally the end of inner loop parameter estimates will be passed onto the next loop after the temperature is reduced. We added a memory function to the inner loop which would remember the parameter set of the minimum objective function occurred during current inner loop, and used it as the beginning parameter estimates for the next inner loop.

SIMULATION PARAMETERS

Results and Discussion

The matrix shown below shows the performance of the system with different combinations of algorithms. The performance index or the performance factor of the system is a factor corresponding to the product of error and the system time taken for the solution to converge. Lower the values better the prediction (Table 1).

From the Figure 15 results we see that the performance of the system is improved considerably by the usage of the proposed hybridized

Table 1. The performance of the system is the function of fitness and the computational time

	Performance Factor					
DTC Performance	Rating of Motor	GA	SA	AHC	GA+ SA	GA+SA+AHC
	1	2	2.3	2.2	1.96	1.8
	2	3	4.3	4	2.5	2.1
	3	3.5	4.4	4	3	2.5
	4	4.5	6	5.5	4	3.2
	5	6	7	5.5	4	3.2
	10	10	9.5	10	9	8
	50	15	16	16	14.5	12
	100	18	19	18.5	17.5	15
	200	22	24	24.3	21	18

Figure 15.

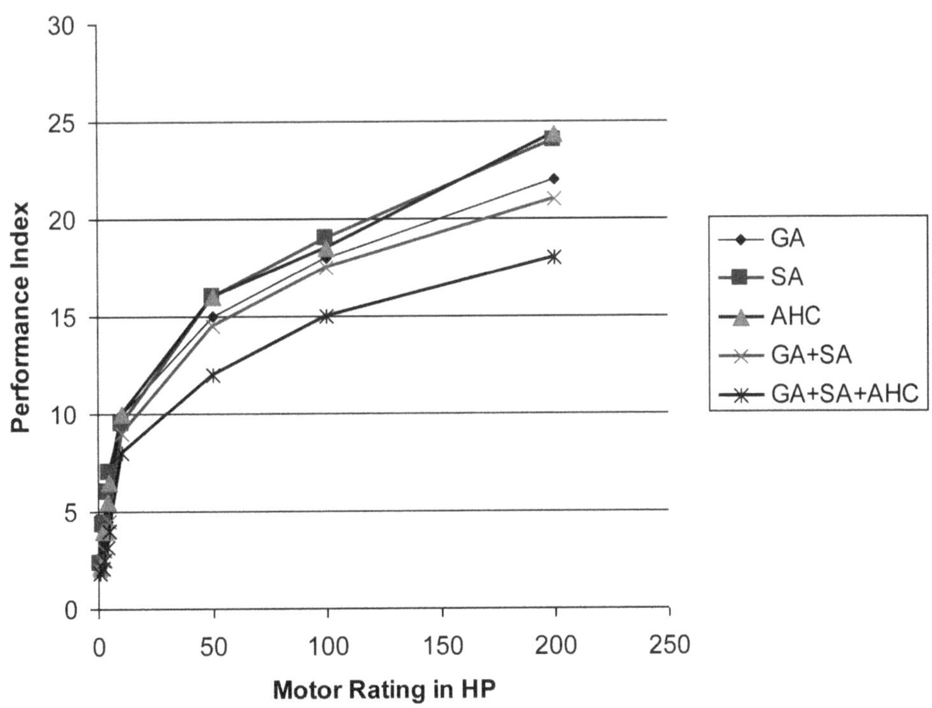

genetic algorithm which is combination of Genetic Algorithm (GA), Simulated Annealing (SA) and Adaptive Hill Climbing (AHC).Hence we can use the proposed algorithm for these applications where the system non linearity is high.

Data of Induction Motor Used for Study

$R_1 = 4.85\ \Omega$;
$R_2 = 5.90\ \Omega$;
$l_1 = 0.831$ H;
$l_2 = 0.833$ H;
J = 0.01;
b = 0.06;
n = 2.

REFERENCES

Boldea, I., & Naser, S.A. (1986). Electrical machine dynamics. New York: McMillan Publishing.

Brogliato, P., & Orhant, P. (1998). Contact Stability analysis of a One degree of freedom Robot. *Journal of Dynamics and Control*, 1998, 37-53.

Khaligh, A., & Emadi, A.Alireza Khaligh Ali Emadi. (2007). Suitability of the pulse adjustment technique to control single DC/DC choppers feeding vehicular constant power loads in parallel with conventional loads. *Int. J. Electric and Hybrid Vehicles*, *1*(1), 20–45. doi:10.1504/IJEHV.2007.014445

Venkobarao, V., & Chinnagouder, C. (2010). Sliding mode controller for internal combustion engine for energy-efficient operation of hybrid electric vehicles. *Int. J. Electric and Hybrid Vehicles*, *2*(4), 282–297. doi:10.1504/IJEHV.2010.034981

Vivek, V., Kumudini Devi, R.P., & Chellamuthu, C. (n. d.). Design and development of energy efficient discharge controllers for Induction motor driven pump for Pulp and Paper Industry. *International Journal of Modeling and Simulation*, 45-51.

Chapter 7
Performance Analysis of Unified Power Quality Controller for a Two Bus System with Different types of Loads

Jaideep Srikakolapu
Maulana Azad National Institute of Technology, India

Toshi Sharma
Maulana Azad National Institute of Technology, India

ABSTRACT

In this chapter, a Unified Power Quality Conditioning (UPQC) system has been presented for a two bus system. This is capable of compensating voltage and current disturbances simultaneously in a two bus system. In this topology one shunt voltage source converter (VSC) and two series VSC are present. All the converters are connected back to back on the dc side with a common dc link capacitor. Power transfer from one feeder to other feeder is made through this dc link during sag/swell and interruption. The performance of the UPQC and the proposed control algorithm has been validated in MATLAB/SIMULINK environment on a two bus system.

1. INTRODUCTION

The purpose of a power transmission line is to group generating stations and Load centres. It is generally constrained by line impedance, operating variables voltages and currents. Therefore generating stations may not be able to supply power demand. In addition now a days deformation of Power Quality which is effected by non linear loads and electronically switched devices is also a constraint. This affects the cost and reliability of the supply of the electricity. This may lead to large power flows with inadequate control, excessive reactive power in various parts of the power system, large dynamic swings between different parts of the system, thus the full potential of transmission line can not be utilized.

This need emerged new technology Flexible AC Transmission systems (FACTS). FACTS technology uses switching power electronics to

DOI: 10.4018/978-1-4666-9911-3.ch007

control power flow. The FACTS technology is a collection of controllers, which can be applied individually with others to control one or more of the interrelated system parameters (Padiyar, 2007). FACTS controllers can enable a transmission line to carry power closer to its thermal rating. The idea of FACTS is to use power electronics for controlling power flow in a transmission line and allowing it to fully load up to its capability (Hingorani & Gyugyi, 2011).

The over view of the FACTS devices is shown in the Table 1. The thyristor based FACTS devices are Static VAR compensator (SVC), Thyristor Controlled Series Compensator (TCSC) and Dynamic Flow Controller (DFC). These devices have low losses due to their low switching frequency.

The other type of FACT devices are voltage source converters based uses mainly Insulated Gate Bipolar Transistors (IGBT) or Insulated Gate Commutated Thyristor (IGCT) as devices. The FACT devices are namely Static Synchronous Compensator (SSC or STATCOM), Static Synchronous Series Compensator (SSSC) and Unified Power Flow Controller (UPFC). These types of FACTS controllers are more advanced Technology. The use of IGBT with Pulse Width Modulation (PWM) allows high modulation frequencies to get low harmonics in the output voltage in magnitude and phase. With the increase in switching frequency, the switching losses also increase.

Table 1. FACTS devices

	Thyristor Based	VSC Based
Shunt Device	Static VAR compensator (SVC)	Static Synchronous Compensator (SSC or STATCOM)
Series Device	Thyristor Controlled Series Compensator (TCSC)	Synchronous Series Compensator (SSSC)
Shunt and Series Device	Dynamic Flow Controller (DFC)	Unified Power Flow Controller (UPFC)

The shunt compensating FACT devices SVC and STATCOM are primarily for reactive power compensation, a current is injected in to the system at the point of common coupling and it's a good way to control the voltage. The SVC provides smother and precise control in comparison with mechanically switched shunt compensation and improves stability. The STATCOM in comparison with SVC provides power quality in against even dips and flickers. Shunt compensator can not provide control of the power flow in the lines.

The series compensating FACT devices TCSC and SSSC are also for reactive power compensation. These devices drive the voltage in to the line and impact the current and power flow directly. A conclusion can be made that if the current, power flow or damp oscillations are to be controlled then series compensator is used.

Power flow devices shift power flows from overloaded parts of the power system to areas with free transmission capability. Phase Shifting Transformer (PST) are the most common devices in this area but they have limitations like low control speed together with a high wearing and maintenance. Unified Power Flow controller (UPFC) provides power flow control together with independent voltage control. The Dynamic Power Flow Controller (DFC) is introduced to fill the gap between UPFC and PST. Thyristor switched capacitor with a PST will provide dynamic controllability over parts of the control range.

The application of FACTS in distribution systems has resulted new type of compensating devices. A unified power-quality conditioner (UPQC) is the extension of the unified power-flow controller (UPFC) concept at the distribution level. It consists of combined series and shunt converters for simultaneous compensation of voltage and current imperfections in a supply feeder.

In this chapter, a new configuration of a UPQC called the multi converter unified power quality conditioner (MC-UPQC) is presented. The proposed topology can be used for simultaneous compensation of voltage and current imperfections in both feeders by sharing power compensation capabilities between two adjacent feeders which are not connected. The system is also capable of compensating for interruptions without the need for a battery storage system. The performance of the UPQC and the proposed control algorithm has been validated in MATLAB/SIMULINK environment on a two bus system.

2. SYSTEM CONFIGURATION AND PRINCIPLE OF OPERATION

The proposed system configuration of a UPQC is shown in above Figure 1. The system configuration has been proposed to a two bus system. The system consists of two bus bars connected to a non linear sensitive load and a sensitive critical load. The harmonics and disturbances caused by these loads to the both feeders are simultaneously cleared

Figure 1. UPQC System configuration

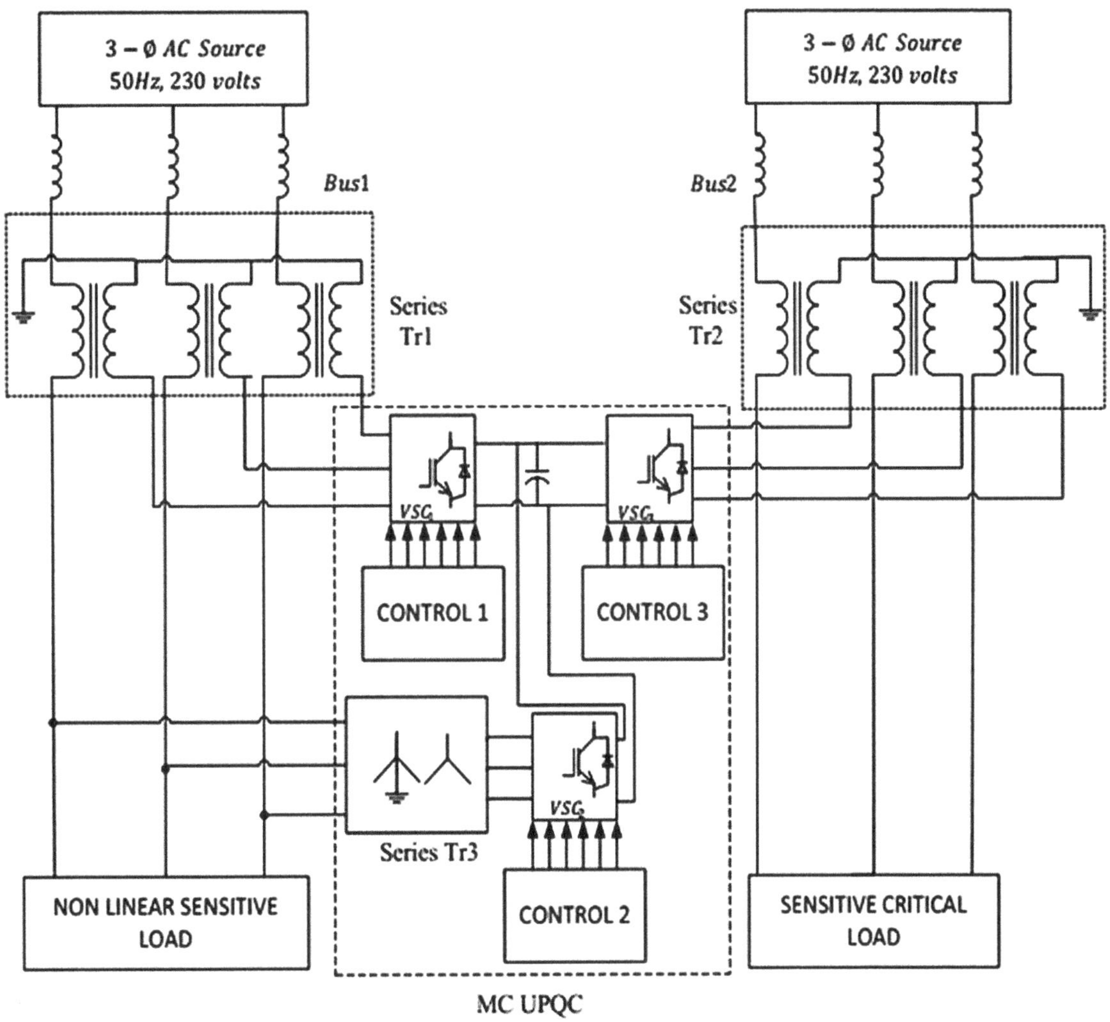

using three VSCs. System consists of three VSCs (VSC1, VSC2, and VSC3) which are connected back to back through a common dc-link capacitor. In the proposed configuration, VSC1 is connected in series with BUS1 and VSC2 is connected in parallel with load L1 at the end of Feeder1. VSC3 is connected in series with BUS2 at the Feeder2 end. Each of the three VSCs in Figure 2 is realized by a three-phase converter with a commutation reactor and high-pass output filter. As shown in Fig, all converters are supplied from a common dc-link capacitor and connected to the distribution system through a transformer. Secondary (distribution) sides of the series-connected transformers are directly connected in series with BUS1 and BUS2, and the secondary (distribution) side of the shunt-connected transformer is connected in parallel with load L1. The aims of the UPQC shown in Figure 1 are:

1. To regulate the load voltage against sag/swell and disturbances in the system to protect the nonlinear/sensitive load L1;
2. To regulate the load voltage against sag/ swell, interruption, and disturbances in the system to protect the sensitive/ critical load L2;

Figure 2. Control algorithm for shunt and series VSCs

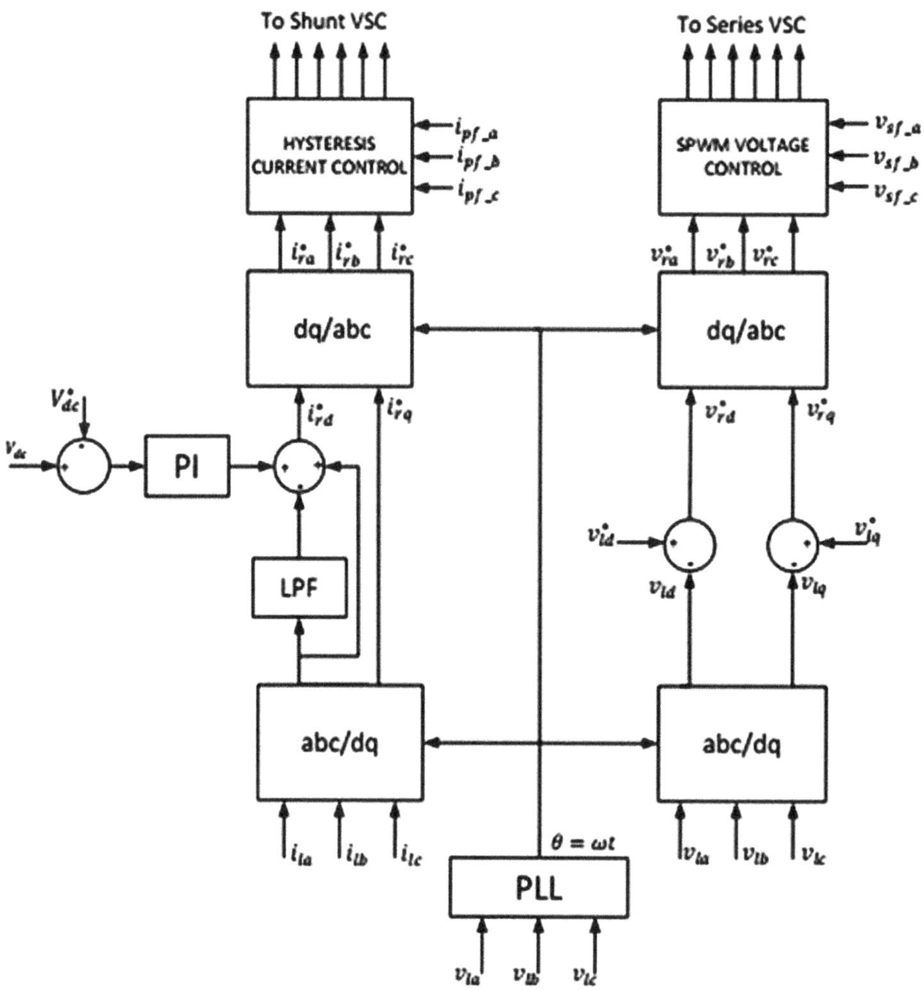

3. To compensate for the reactive and harmonic components of nonlinear load current at Load 1.

3. CONTROL ALGORITHM

The UPQC consists of two series VSCs and one shunt VSC as shown in Figure 2 which are controlled independently. The switching control strategy for series VSCs and the shunt VSC are selected to be sinusoidal pulse width-modulation (SPWM) voltage control and hysteresis current control, respectively. Shunt-VSC: Functions of the shunt-VSC are:

1. To compensate for the reactive component of load L1 current;
2. To compensate for the harmonic components of load L1 current;
3. To regulate the voltage of the common dc-link capacitor.

3.1 Shunt VSC Control

The shunt VSC is controlled in rotor reference frame. The pictorial view of the control of shunt VSC is given in Figure 2. The Load current i_{abc} is measured and transformed in to the dq synchronous frame by using Equation 1.

The Load currents are transferred to dq-axis currents as,

$$\begin{bmatrix} i_{ld} \\ i_{lq} \end{bmatrix} = \frac{2}{3}\begin{bmatrix} \cos\theta & \cos\left(\theta-\frac{2\pi}{3}\right) & \cos\left(\theta-\frac{4\pi}{4}\right) \\ -\sin\theta & -\sin\left(\theta-\frac{2\pi}{3}\right) & -\sin\left(\theta-\frac{4\pi}{3}\right) \end{bmatrix}\begin{bmatrix} i_{la} \\ i_{lb} \\ i_{lc} \end{bmatrix} \quad (1)$$

These sinusoidal voltages are transformed to stationary αβ reference frame for calculation of angle θ as,

$$\begin{bmatrix} v_{l\alpha} \\ v_{l\beta} \end{bmatrix} = \begin{bmatrix} 1 & \frac{-1}{2} & \frac{-1}{2} \\ 0 & \frac{\sqrt{3}}{2} & \frac{-\sqrt{3}}{2} \end{bmatrix}\begin{bmatrix} v_{la} \\ v_{lb} \\ v_{lc} \end{bmatrix} \quad (2)$$

$$\theta = \tan^{-1}\left(\frac{v_{l\beta}}{v_{l\alpha}}\right) \quad (3)$$

By transforming the fundamental positive sequence component from Equation (1), the AC quantities have been extracted using low pass filters (LPFs) and they have been removed from the d component of line current.

$$i_{ld} = \bar{i}_{ld} + \bar{I}_{ld} \quad (4)$$

where, $\bar{I}_{ld}$ is the DC component of the line current and $\bar{i}_{ld}$ is the AC component of the line current

$$\bar{i}_{ld} = i_{ld} - \bar{I}_{ld} \quad (5)$$

The DC link voltage is selected such that it should be more than peak of the line voltage for satisfactory operation of VSC.

$$V_{dc}^{*} = \frac{\sqrt{6}v_{la1}}{m_a} \quad (6)$$

where, v_{lal} is the fundamental RMS value of the base phase load voltage, m_a is the modulation index.

By comparing the reference DC link voltage V_{dc}^{*} with the sensed DC link voltage (V_{dc}) an error is obtained as:

$$V_{dce}(n) = V_{dc}^{*}(n) - V_{dc}(n) \quad (7)$$

The DC voltage error is passed through a PI controller (PI) which provides a reference d-axis component of current Δi_e:

$$\Delta i_e(n) = \Delta i_e(n-1) + [k_p\{V_{dce}(n) - V_{dce}(n-1)\} + k_i V_{dce}(n) \tag{8}$$

The reference dq axis currents for hysteresis current control are given by

$$i^*_{rd} = \Delta i_e + i_{ld} - \bar{i}_{ld} \tag{9}$$

$$i^*_{rq} = i_{lq} \tag{10}$$

The above reference currents are transformed in to abc parameters using the following Equation.

$$\begin{bmatrix} i^*_{ra}(n) \\ i^*_{rb}(n) \\ i^*_{rc}(n) \end{bmatrix} = \begin{bmatrix} \cos\theta & \sin\theta \\ \cos\left(\theta - \frac{2\pi}{3}\right) & -\sin\left(\theta - \frac{2\pi}{3}\right) \\ \cos\left(\theta - \frac{4\pi}{3}\right) & -\sin\left(\theta - \frac{4\pi}{3}\right) \end{bmatrix} \begin{bmatrix} i^*_{rd}(n) \\ i^*_{rd}(n) \end{bmatrix} \tag{11}$$

The θ is obtained using Equation (3). The obtained currents are processed through a hysteresis current controller for obtaining the gate pulses for shunt VSC.

3.2 Series VSC Control

Series-VSC: Functions of the series VSCs in each feeder are:

1. To mitigate voltage sag and swell;
2. To compensate for voltage distortions, such as harmonics;
3. To compensate for interruptions (in Feeder2 only).

The control block diagram of each series VSC is shown in Figure 3. The bus voltage (v_{labc}) is detected and then transformed into the synchronous dq0 reference frame using

$$\begin{bmatrix} v_{ld} \\ v_{lq} \end{bmatrix} = \frac{2}{2}\begin{bmatrix} \cos\theta & \cos\left(\theta - \frac{2\pi}{3}\right) & \cos\left(\theta - \frac{4\pi}{3}\right) \\ -\sin\theta & -\sin\left(\theta - \frac{2\pi}{3}\right) & -\sin\left(\theta - \frac{4\pi}{3}\right) \end{bmatrix} \begin{bmatrix} v_{la} \\ v_{lb} \\ v_{lc} \end{bmatrix} \tag{12}$$

The θ is obtained using Equation (3). The dq voltages are compared with the reference voltages and the respective error voltages are given by

The error voltages $\left(v^*_{lde}\right)$ and $\left(v^*_{lqe}\right)$ are obtained as:

$$v^*_{lde}(n) = v^*_{ld}(n) + v_{ld}(n) \tag{13}$$

$$v^*_{rd} = v^*_{lde}(n) \tag{14}$$

$$v^*_{lqe}(n) = v^*_{lq}(n) + v_{lq}(n) \tag{15}$$

$$v^*_{rq} = v^*_{lqe}(n) \tag{16}$$

From the above dq voltages three phase reference voltages are obtained as follows. Three phase sinusoidal reference voltages are obtained as,

$$\begin{bmatrix} v^*_{ra}(n) \\ v^*_{lbe}(n) \\ v^*_{lce}(n) \end{bmatrix} = \begin{bmatrix} \cos\theta & \sin\theta \\ \cos\left(\theta - \frac{2\pi}{3}\right) & -\sin\left(\theta - \frac{2\pi}{3}\right) \\ \cos\left(\theta - \frac{4\pi}{3}\right) & -\sin\left(\theta - \frac{4\pi}{3}\right) \end{bmatrix} \begin{bmatrix} v^*_{rd}(n) \\ v^*_{lqe}(n) \end{bmatrix} \tag{17}$$

The θ is obtained using Equation (3). Three phase reference error voltages are used to obtain

Figure 3. Feeder bus1 voltage component, series compensating voltage component, and load voltage component

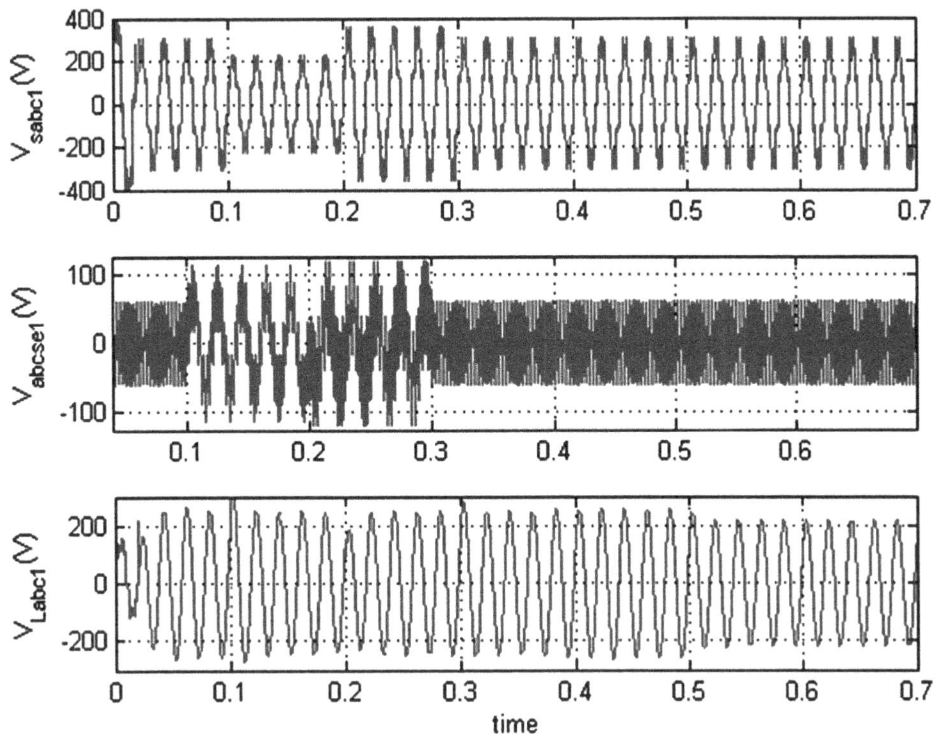

switching pulses using three phase PWM controller. The error signal is for each phase is compared with triangular carrier signal to generate switching signals for each leg of a VSC.

4. SIMULATION RESULTS

The proposed UPQC and its control schemes have been simulated in MATLAB/SIMULINK environment. In this section, the simulated results have been presented, and the performance of the proposed UPQC system has been validated.

4.1 Distortion and Sag/ Swell on the Bus Voltage

The power system in Figure 2 consists of two three-phase three-wire 380(v) (rms, L-L), 50-Hz utilities. Feeder-1 bus voltage V_{abc1} contains the seventh-order harmonic with a value of 22%, and the Feeder-2 bus voltage V_{abc2} contains the fifth order harmonic with a value of 35%. The voltage, V_{abc1} contains 25% sag between $0.1s < t < 0.2s$ and 20% swell between $0.2s < t < 0.3s$.

V_{abc2}contains 35% sag between $0.15s < t < 0.25s$ and 30% swell between $0.25s < t < 0.3s$. A three-phase diode bridge rectifier has been considered as nonlinear/sensitive load

L_1. The bridge has an RC load of 10Ω and 30 F. A balanced RL load of 10Ω and 100mH has been considered as critical load L_2. The UPQC is switched on at t=0.02sec. V_{abc1}, the corresponding compensation voltage injected by VSC_1 is V_{abcse1}, and voltage at load L_1 is V_{Labc1}, are shown in Figure 3.

In Figure 3 among three phase voltages, waveforms of phase A has been shown for simplicity.

Similarly, Feeder-2 bus voltage (V_{sabc2}), the corresponding compensation voltage injected by VSC_3 (V_{abcse3}) and load voltage (V_{Labc2}) are shown in Figure 4. The distorted voltages of Feeder bus-1 and Feeder bus-2 have been satisfactorily compensated for different loads at feeder bus-1 and feeder bus 2 with good dynamic response.

The nonlinear load currents (I_{sabc1}) and its corresponding compensation current (I_{abcsh1}) which has been injected by VCS_2, compensated Feeder 1 Load current (I_{sabc1}), and the dc-link capacitor voltage (V_{DC})have been shown in Figure 5.

When a fault occurs in Feeder bus -2 (in any form of L-G, L-L-G, and L-L-L-G faults), the voltage across the sensitive/critical load L_2 is interrupted. This voltage imperfection can be compensated by VSC_2.

In this case, the power required by load L_2 is supplied through VSC_2 and VSC_3. This implies that the power semiconductor switches of VSC_2 and VSC_3 must be rated such that total power transfer is possible. This may increase the cost of the device, but the benefit that may be obtained can offset the expense. In this configuration, the critical load on Feeder 2 has been fully protected against distortion, sag/swell, and interruption.

The performance of the UPQC under a fault condition on Feeder bus 2 has been tested by applying a three-phase fault to ground on Feeder 2 between $0.3s < t < 0.4s$. Simulation results have been shown in Figure 6.

To verify the system behavior during a load change the non linear load has been doubled at t = 0.5s. The other loads i.e. load at feeder bus -2

Figure 4. Feeder bus2 voltage component, series compensating voltage component, and load voltage component

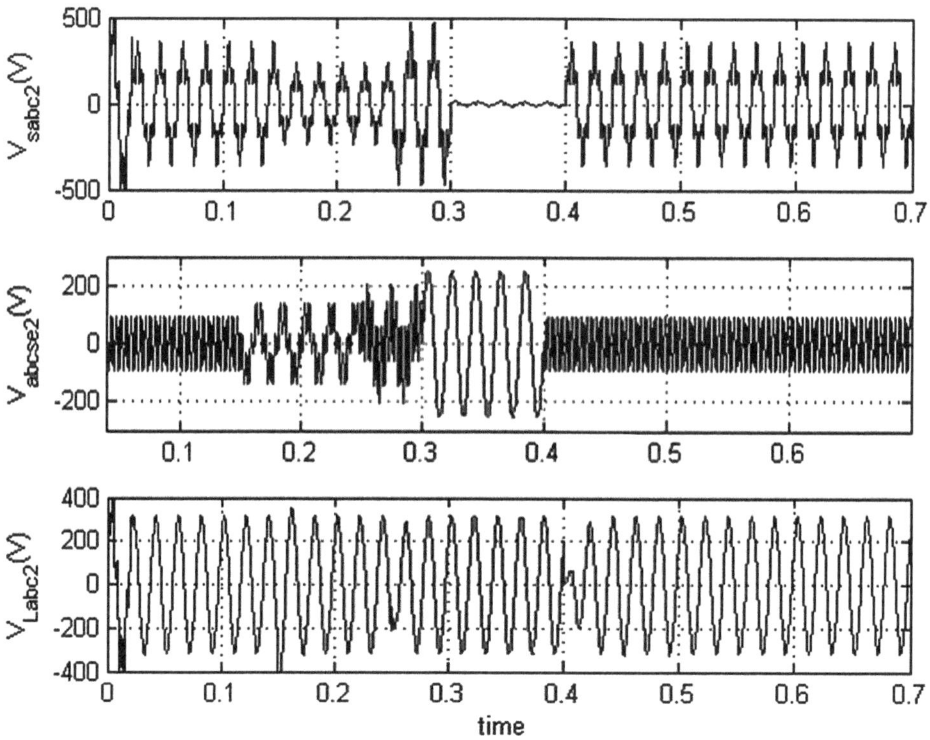

Figure 5. Load current, compensating current, and capacitor voltage for a non linear load

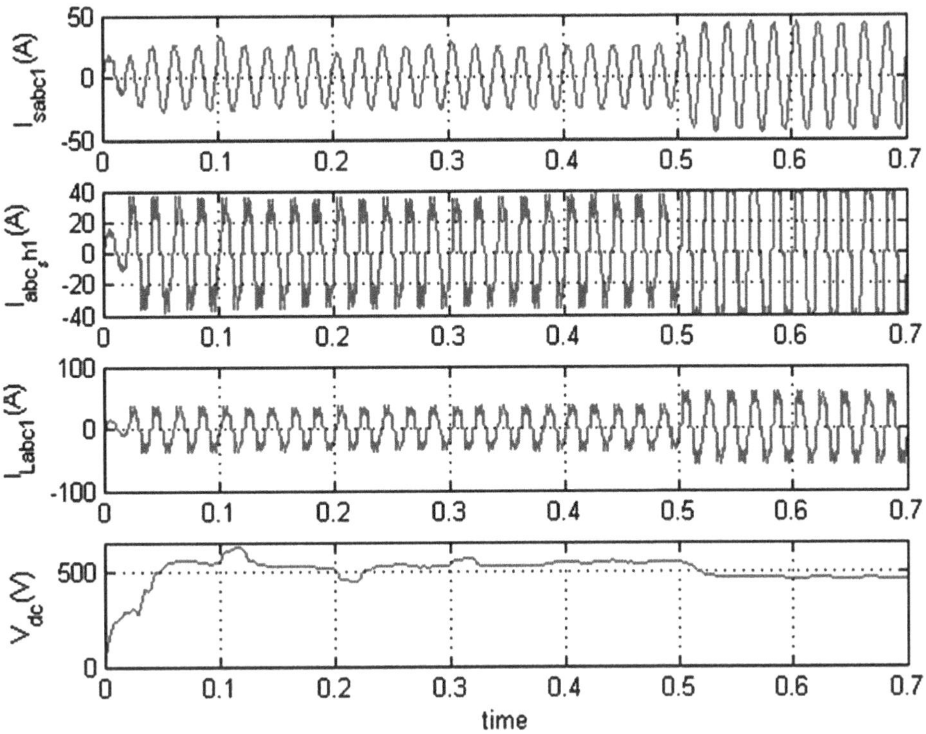

Figure 6. Feeder bus2 voltage, compensating voltage from VSC2, load 1 voltage, load 2 voltage for a fault occurred at feeder bus bar- 2

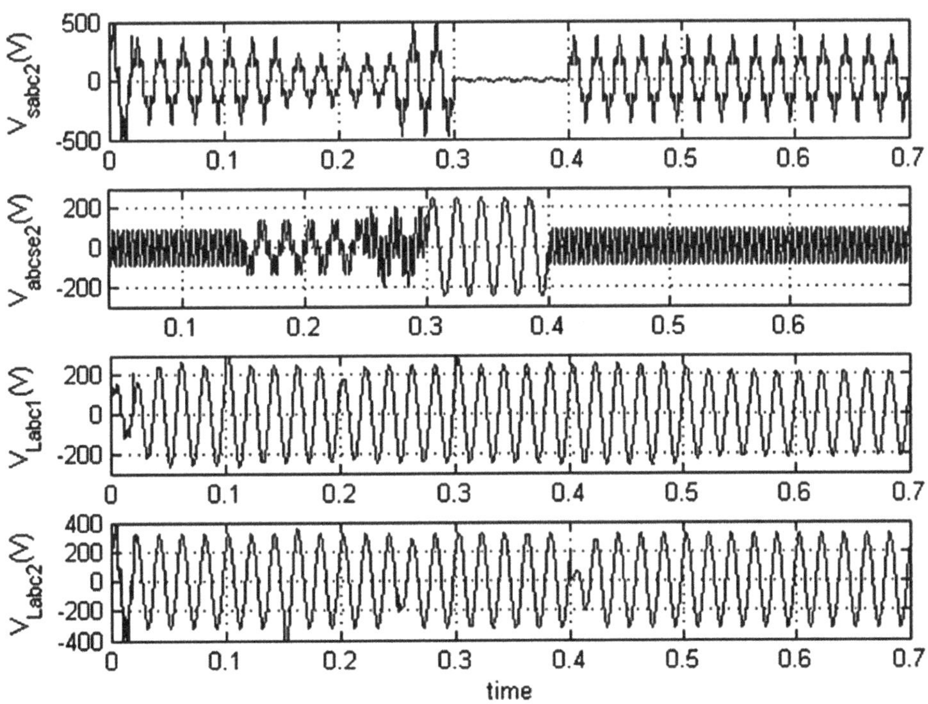

Figure 7. Load current at feeder 1, load 1 voltage, load 2 voltage and DC- link capacitor voltage for a nonliner load with change in load

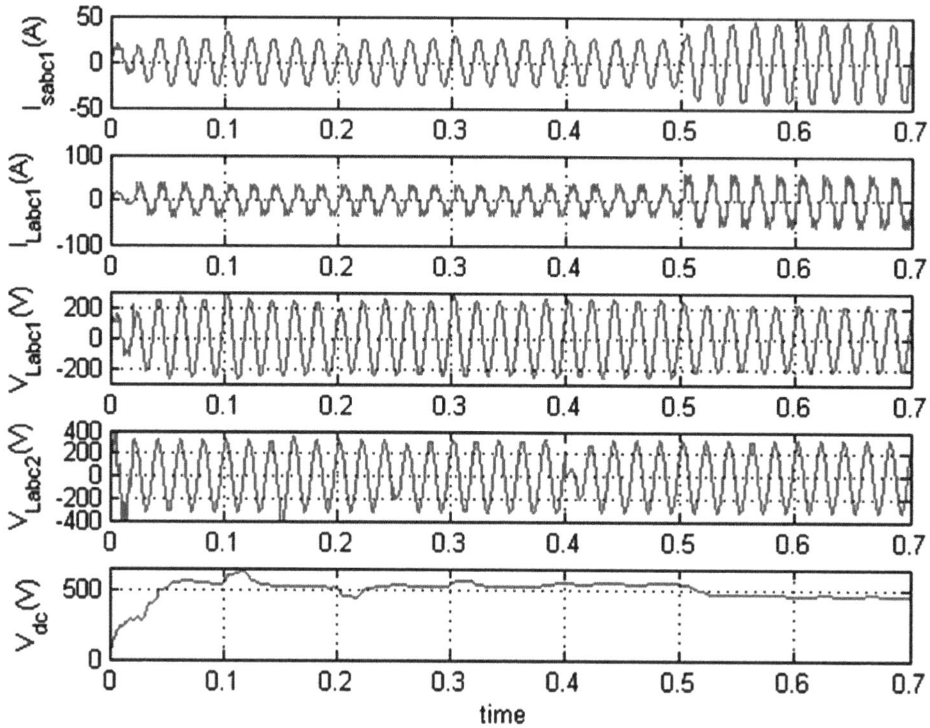

Figure 8. Feeder bus 1 voltage, series compensating voltage, load 1 voltage for unbalanced source voltage

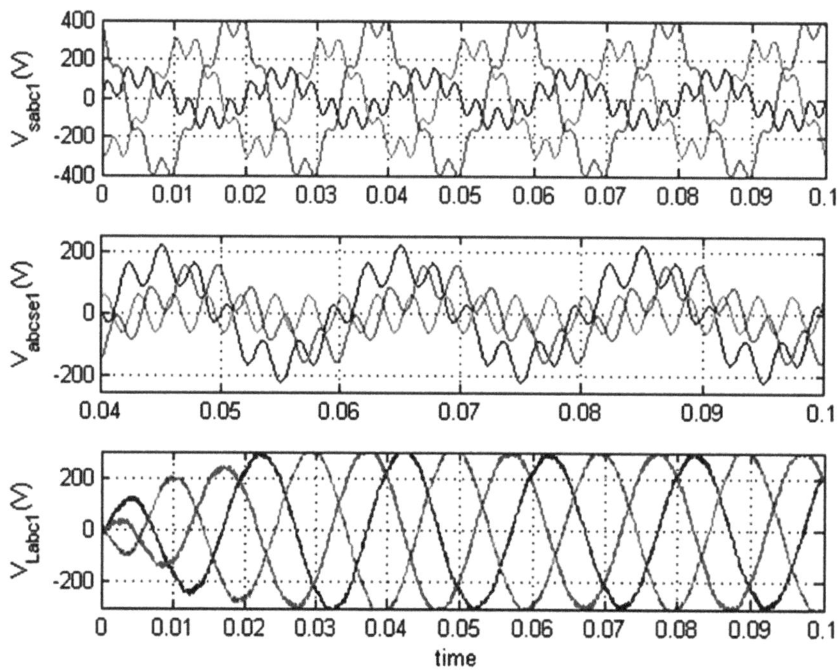

have been kept unchanged. The system response has been shown in Figure 7.

To verify the control algorithm capability for unbalanced voltage compensation, a new simulation has been performed and their results have been shown in Figure 8. Three phase Feeder-1 bus voltages, series compensated voltages and Load voltages have been shown in Figure 8.

The simulation results evaluates that the harmonic components and unbalance in feeder bus voltages have been compensated by injecting the proper series voltages in the Feeder bus bar-1. The load voltage after this compensation has been balanced with regulation in amplitude.

5. CONCLUSION

In this chapter, compensation of voltage and current in adjacent feeders has been proposed. The unified power-quality conditioner in this configuration consists of three VSCs, which can be increased according number of feeders. Compared to a conventional configuration of UPQC, the proposed configuration is capable of fully protecting critical and sensitive loads against distortions, sags/swell, and interruption in two-feeder systems. The idea can be theoretically extended to multibus/multifeeder systems by adding more series VSCs. The performance of the UPQC has been evaluated under various faulty conditions and results have been verified in MATLAB environment.

REFERENCES

Akagi, H. (1996). New trends in active filters for power conditioning. *IEEE Transactions on Industry Applications*, *32*(6), 1312–1322. doi:10.1109/28.556633

Akagi, H., Kanazawa, Y., & Nabae, A. (1984). Instantaneous reactive power compensation comprising switching devices without energy components. *IEEE Transactions on Industry Applications*, *20*(3), 625–630. doi:10.1109/TIA.1984.4504460

Buso, S., Fasolo, S., Malesani, L., & Mattavelli, P. (2000). A Dead Beat Adaptive Hysteresis Current Control. *IEEE Transactions on Industry Applications*, *36*(4), 1174–1180. doi:10.1109/28.855976

Fardanesh, B., Shperling, B., Uzunovic, E., & Zelingher, S. (2000). Multi-converter FACTS Devices: The generalized unified power flow controller (GUPFC). *Proc. IEEE Power Eng. Soc. Summer Meeting* (Vol. 4, pp. 2511–2517). doi:10.1109/PESS.2000.867513

Geddada, N., Karanki, S. B., & Mishra, M. K. (2012). Synchronous reference frame based current controller with SPWM switching strategy for DSTATCOM applications. *Proc. of IEEE Conf. PEDES* (pp. 1-6). doi:10.1109/PEDES.2012.6484393

Hingorani, N. G., & Gyugyi, L. (2011). *Understanding FACTS*. Delhi: IEEE Press.

Jindal, A. K., Ghosh, A., & Joshi, A. (2007). Interline unified power quality conditioner. *IEEE Transactions on Power Delivery*, *22*(1), 364–372. doi:10.1109/TPWRD.2006.881581

Kundur, P. (1994). *Power System Stability and Control* (1st ed.). New York: MacGraw-Hill.

Padiyar, K. R. (2007). *FACTS Controllers in Transmission and Distribution*. New Delhi: New Age International.

Rashid, M. H. (2007). *Power Electronics Handbook: Devices, Circuits and Applications* (2nd ed.). Academic Press.

Saeedifard, M., Nikkhajoei, H., & Iravani, R. (2007). A space vector modulated STATCOM based on a three level neutral point clamped converter. *IEEE Transactions on Power Delivery*, *22*(2), 1029–1039. doi:10.1109/TPWRD.2007.893448

Xia, L., & Du Taihang, T. S. (2011). A fuzzy logic variable hysteresis band current control technique for three phase shunt active filter. *Proc. IEEE Conf. CASE* (pp. 1-4).

Chapter 8
Smart Grid:
Optimized Power Sharing and Energy Storage System Framework with Recent Trends and Future Ahead

Satya Prakash
SKN Sinhgad Institute of Technology & Science, India

Vivek Saxena
SKN Sinhgad Institute of Technology & Science, India

Swati Jaiswal
SKN Sinhgad Institute of Technology & Science, India

ABSTRACT

The authors in this chapter have put in efforts to throw light on optimization of power through the emerging technology- Smart Grid. The chapter has been divided in part which cover each section associated with the smart grid. The chapter starts with the knowhow of the system and goes till the implementation challenges. The authors have focused on the challenges associated, like cyber security, interconnection, power quality and many more. The challenges have been discussed with its significance. The most emerging technology to be harnessed to maximum, authors have also dealt with the possible solutions to the problems prevalent in the implementation of the smart grid technology. Generally the problem with the frame work has been with the integration and data handling. The chapter will help the researchers to further investigate the associated area of smart grid and find solutions to problems in smart grid.

INTRODUCTION

The global demand for the electrical energy has increased manifold since its invention with the development of modern society. The extensive interconnection between the grids to meet increasing demand & to improve the efficiency has resulted into complex structure of power system and its highly dynamic & unpredictable operation. The dependency on the depleting non renewable resources of the energy for generating electricity has raised the concern for security & reliability

DOI: 10.4018/978-1-4666-9911-3.ch008

of supply in future. The existing power grid is operating under stressed condition with minimal safety margins as the amount of transportation of electrical power is increasing at rapid rate without significant addition in the grid infrastructure. The existing power system is designed to support centralized generation while its operation is limited to one way flow of energy & information and it hasn't changed much in decades.

Growing complexities due to rapid connection of ever increasing loads & its variety has made the system to operate at poor efficiency & vulnerable to frequent failures & blackout. The connection of nonlinear loads to the grid is deteriorating the quality of power by inducing the harmonics in the supply. The pilferage of power is resulting into huge losses to utilities. The usage of non renewable energy resources for production of electricity has increased the carbon emission which is major cause for global warming. The limitations of conventional power system, increased competition due to deregulation & informed customers are posing numerous challenges to utilities & forcing them to change the mode of operation of existing electrical grid.

The Smart grid is the future of electrical power system as it offers solutions to the problems of conventional grid structure. The smart grid integrates the new technologies, extensive automation & harnesses the potential of information technology enabling the system to operate more efficiently & reliably. It allows the two way flow of information, data & energy between utility & customer and vice-versa. With the emphasis on the integration of renewable energy generating unit, electrical vehicles, smart homes, the necessity of smart grid has become indispensible need of the hour. The smart grid enables the effective usage of improved grid, enhances operating efficiency, security, resilience & make the grid self healing. This reduces the chances of the failure of the grid & hence reliability increases. Also it motivates the customers to participate actively in managing their load by providing them relevant data on real times basis. This offers overall economy on the part of the customer & ultimately improves the overall efficiency by matching the demand & supply.

The backbone of the smart grid is the intensive utilization of communication infrastructure in sharing the real time critical data among various utilities & customers for optimized usage of the grid. It automates the control & protection operation of the power system by replacing the existing relays & electromechanical protective devices with the smart sensors & intelligent electronics devices (IEDs) which are capable of communicating, fast, accurate & reliable in operation.

This chapter elaborates the basics of smart grid & its importance in modern time. It underlines the latest technologies included & challenges in the implementation of smart grid. It also briefly discusses the probable solutions for these problems. All the shortcomings in the traditional grid have been worked out and evolved a better grid system for fighting the challenges of present, which can be noticed in the *Table 1*(Joy Jeena, Jasmin E A, Rajan John Viju, 2013).

Table 1. Traditional vs Smart Grid

TRADITIONAL GRID	SMART GRID
Centralized Generation	Distributed Generation
Manual Restoration	Self-healing
Lack of real time monitoring	Extensive real time monitoring
Slow Reaction time	Extremely quick reaction time
Failures and Blackouts	Adaptive and Islanding
Electromechanical	Digital
One way Communication	Two way communication
No energy Storage	Energy Storage
Total control by Utility	Increased customer participation

BACKGROUND: SMART GRID

The smart grid is the evolution of the conventional power grid that employs the advance infrastructure of digital & information technology to manage the demand and regulates the working of the power system so as to achieve the sustainable, reliable & economical operation. The smart grid focuses on automating various operations for bringing down the overall AT&C losses, power theft, demand response & demand side management, power quality management & outage management as visually seen in Figure 1. It also facilitates the integration of renewable generating units to smart transmission & distribution network which incorporates numerous sensors connected through advance communication network & data acquisition system.

Definition

According to the U.S. Department of Energy (DoE):

"Smart grid" generally refers to a class of technologies that people are using to bring utility electricity delivery systems into the 21st century, using computer-based "remote control and automation." These systems are made possible by "two way digital communications technologies" and computer processing that has been used for decades in other industries. They are beginning to be used on electricity networks, from the power plants and wind farms all the way to the consumers of electricity in homes and businesses. They offer many benefits to utilities and consumers – mostly seen in big improvements in "energy efficiency and reliability" on the electricity grid and energy user's homes and offices.

According to IEEE:

An "automated", widely distributed energy delivery network characterized by a "two-way flow of electricity and information", capable of monitoring and responding to changes in everything from power plants to customer preferences to individual appliances.

Figure 1. Overview of a Smart Grid Technology

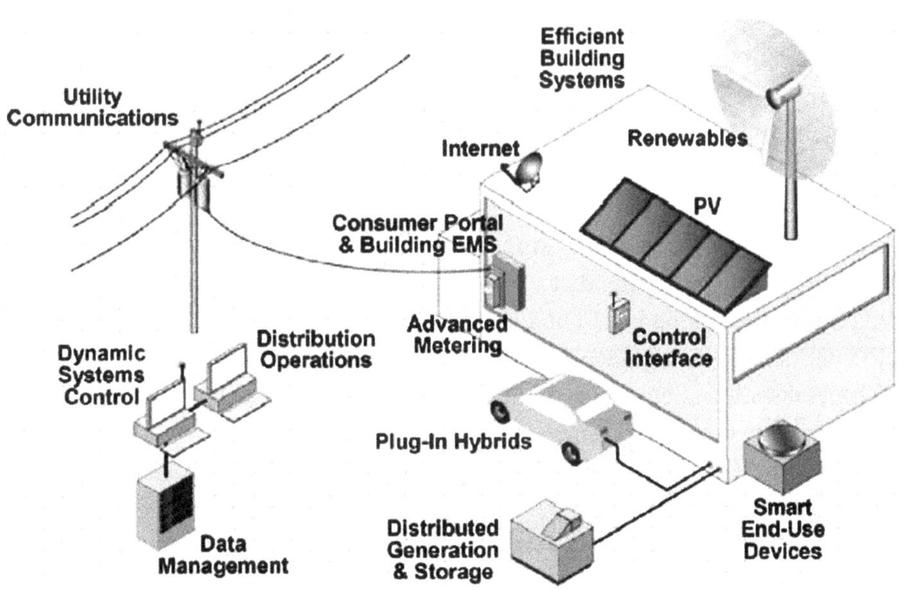

A smart grid is the electricity delivery system (from point of generation to point of consumption) "integrated with communications and information technology."

Driving Factors for Micro Grid

- It is expensive affair to expand or fix Transmission & Distribution infrastructure. The US businesses alone lose $80B to $150B per year in revenues due to power outages (NETL, 2011).
- The centralized generation facilities largely depend on the non renewable energy resources for the production of the electricity. Hence per unit cost of electricity varies as the prices of gas, coal fluctuates in international market.
- The Power quality, efficiency & reliability enhances with implementation of distributed generation scheme.
- The microgrid offers Cyber and physical security to the grid information & infrastructure respectively.
- The recent innovation in technologies also made it simple to install & operate small power system at the premises of the consumer.

Features of the Smart Grid

The various salient features of the smart grid can be listed as follows:

- **Accommodates all generating & storage options:** It not only accommodates the centralized power generating units but also enables to integrate the distributed generating (DG) units installed at customer premises. The DG mostly consists of renewable resources for production of electricity such as photovoltaic cell, wind mills etc.
- **Optimized utilization of assets & improved operating efficiency:** The application of modern technologies enables to optimize the use of its assets. The protection & control devices can be adjusted to achieve reduction in losses & eliminates the congestion.
- **Motivates customers to participate actively in demand management:** The consumer can manage their pattern of using electricity & can gain economical benefits by scheduling their demand according to cheap real time pricing available at the time of purchase.
- **It promotes new products, services and markets:** The deregulation of utilities has induced the competition among them. This offers variety of choices to customers to select from competing services as per their needs. It has improved the services, products & market
- **Manages the quality of power for range of needs:** The need of quality of power varies from customer to customer, hence smart grid offers the supply of varying grades of electricity at different tariffs.
- **Resilient and self healing:** Smart grid makes the system resilient i.e. it enables the system to recognize the impending problems & then take the appropriate actions accordingly. Once the fault is recognized it quickly isolates the faulty region before it spreads into system and results into cascaded failure. Self-healing is the property that enables a system to perceive that it is not operating correctly and, without human intervention, make the necessary adjustments to restore itself to normality
- **Extensive automation & advance data acquisition system:** Smart grid is replacing the existing slow acting relays and electromechanical protection devices with static & digital smart sensors and intelligent devices contributing the automation of transmission & distribution network. It utilizes the advance infrastructure for data

acquisition system to capture the real time data & share among the various stakeholders to improve the working & efficiency of the system.

- **Advance support for demand response:** The smart grid allows the interaction between generators & loads in real time. Their coordinated response flattens the demand response curve. This reduces the requirements of the reserve generators to put into action thereby resulting into economical operation.

Functions of the Smart Grid

As enumerated by the smartgrid.gov, (US Department of Energy 2003), various functions of Smart Grid are as follows:

- **Limiting fault & disturbances:** The severity faults & disturbances can be limited by using advance sensors & SCADA. Their fast & coordinated response enables the utilities to prevent the faults exceeding beyond safety margins.
- **Wide area monitoring & control:** The smart grid enables to monitor the power system spreading over huge geographical area on real time basis. It also helps in taking the preventive, protective & corrective measures well in advance.
- **Dynamic capability rating:** It is achieved by determining the ability of various elements to carry load in real time.
- **Control of power flow:** It is enabled by using tools such as FACTS device, phase angle regulating transformer, series capacitors, superconductors etc.
- **Adaptive protection:** This is achieved by flexible settings of the protective relays that can be changed in real time on the basis of the signals received from sensors or central control station.
- **Automated switching of feeders and lines:** The faulty sections of the feeders & the transmission lines can be isolated & reconfigured automatically using digital sensors, relays, circuit breakers & advance communication system.
- **Automated islanding & reconnection:** Smart grid consist number of microgrids (Distributed generating units) that are connected to main grid. These Micro grids are capable of operating in island mode i.e. without connecting to main grid. The smart grid features the automated separation & reconnection of the microgrids according to need & requirement.
- **Automated control of reactive power & voltage profile:** The reactive power flow can be controlled using synchronized operation of various reactive power sources such as capacitor banks, voltage regulators, OLTCs etc. The smart automatically switches these devices in response to local events or the command signals received from the central control station.
- **Diagnosis & reporting of equipment condition:** It enables the on line monitoring of performance, condition & working of various equipment connected to the system. It reports the any abnormal behaviour or condition to the central station so that preventive actions can be taken well in advance.
- **Real time measurement & management of the load:** The usage of advance metering infrastructure helps in determining the load connected to the grid in real time. Thus it offers the efficient management of the load.
- **Real time transfer of load:** It is achieved through real time reconfiguration and optimization of feeder to relieve load on equipment. This improves asset utilization, distribution system efficiency, and enhances system performance.

- **Optimization of customer electricity usage:** The advance metering infrastructure provides reliable information to customer which helps them in taking decisions about the usage of electricity. This optimizes the use & helps them in achieving the overall economy.

Components of the Smart Grid

The major components of the smart grid can be broadly classified in three areas:

- **Hardware:** The hardware consists of smart sensors & digital equipment which are referred as the intelligent electronics devices (IEDs). These devices are capable of fetching data & communicate using SCADA. It covers the equipment ranging from automated capacitor banks, switches, reclosers, sensors, advance meters etc.
- **Communication Network:** The communication network links the hardware installed at the field to master control station. It transmits the data captured by sensors using network to the control station. The feasibility of type of communication network depends upon the specific usage & the area to be covered. It may range from simple LAN to GPRS (Satellite) communication. The area to be covered may range from small home (Square meters) to Interstate grids (square kilometres).
- **Computing Platform:** The computing platforms are the different software that receives the data & perform analysis. The result of analysis is used to decide the appropriate control action or update the information on various portals for reference. The utility of different computing platform depends upon the specific need, for e.g. customer information system (CIS) is used to maintain the profile of customers, their billing records etc., however meter data management system (MDMS) is used to main the data send by advance smart meters.

Further the components of Micro Grid have following detailing (Figure 2):

- Distributed Energy Resources (DER)
- Micro grid Master Controller
- Smart Energy Storage
- Smart Sensors & Relays
- Power Electronics interfaces
- SCADA
- Communication Network

Figure 2. Smart Grid component providers

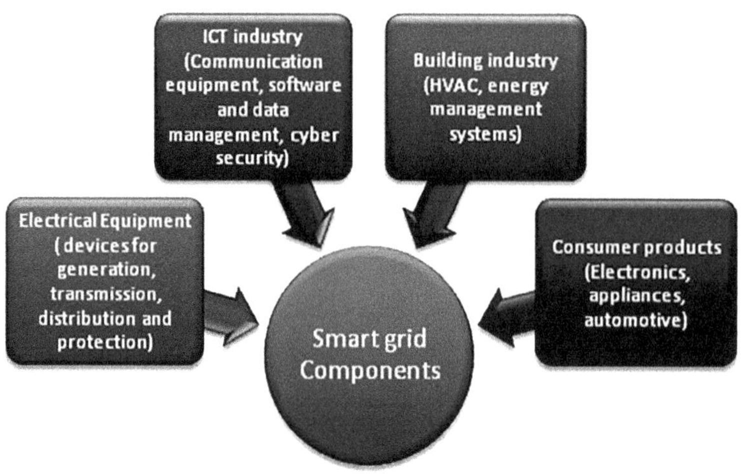

Distributed Energy Resources (DER): These are the standalone or hybrid generating units which usually converts the renewable energy resources such as solar energy, wind energy into electrical energy. For example the photovoltaic cells, wind turbines, fuel cell etc.

Microgrid master controller: The microgrid master controller matches the load with the generation in both modes of the operation of the micro grid i.e. island mode or grid connected mode. It optimizes the integration, dispatch & control of DER and loads. It ensures the combination of DER to be used so as to achieve economical generation. It also efficiently maintains the reliability and manages frequency and voltage.

Smart Energy storage options: Smart energy storage options are needed to store the surplus electricity produced & can be utilized in the situation of generation outage. This contributes in the stability of microgrid & enhances the reliability of the supply.

Smart Sensors & Relays: The microprocessor based relays & sensors are employed for the protection of microgrid. These relays have capability to communicate with the control station & also have ability to execute the control command received from the control station

Power Electronics Interfaces: The electricity has to be converted from A.C to D.C & vice versa as per the need & requirement. This conversion is done with the help of Power electronics interfaces which are usually the Inverter, Chopper etc. Apart from this they are also used in integrating the micro grid with Main grid

SCADA: SCADA is used to supervise & control the microgrid.

Communication Network:

- Fast and secure communication to monitor real-time network status
- Optimize operations and control of DERs & loads
- Continuous monitor and trend microgrid components health
- Smart metering to obtain load and DERs profile
- Electricity pricing and demand response capabilities
- Continuous communication between utilities and energy markets in Figure 3.

Figure 3. Anatomy of Microgrid

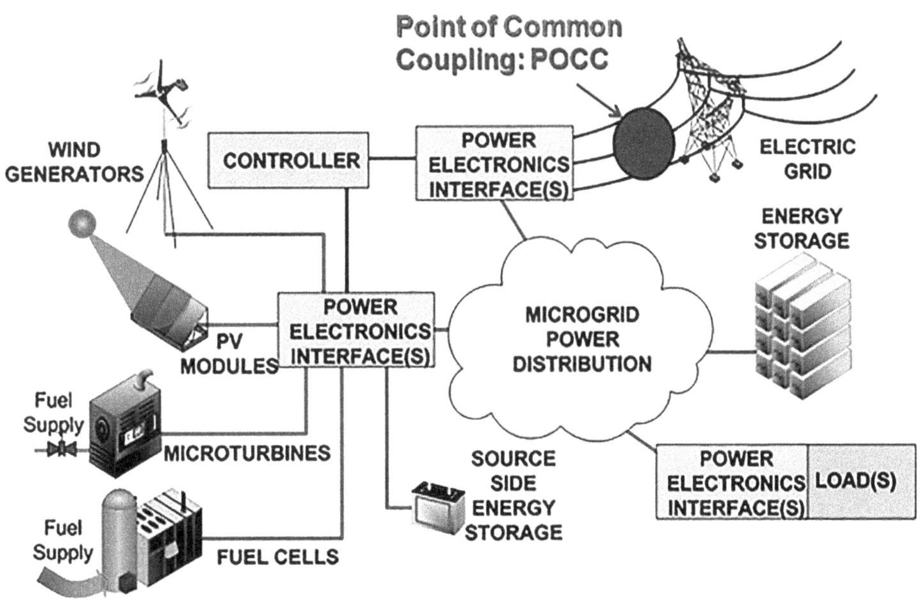

The various components & relation between them is shown in Table 2.

Smart Storage System for Micro Grid

(Haisheng Chen, Thang Ngoc Cong, Wei Yang, Chunqing Tan, Yongliang Li & Yulong Ding, 2009)

The smart storage system refers to the system that converts the excess electrical energy into storable form which can be converted into usable form at peak demand period. This flattens the load on generating stations & results into economical production of electricity. With the advent of distributed generating sources the smart storage became more popular & viable option. The various storage systems available for smart grid are as follows:

- Battery
- Flow Battery
- Fuel Cell
- Super capacitor
- SMES (Super conducting magnetic energy storage system)
- Pumped Hydro
- Compressed Air Energy Storage

Battery

It is a device that converts the chemical energy into electrical energy. The smart grid employs rechargeable batteries or secondary batteries for storing the electrical energy. A battery comprised of one or more electrochemical cells and each cell consists of a liquid, paste, or solid electrolyte together with a positive electrode (anode) and a negative electrode (cathode).

During discharge, electrochemical reactions occur at the two electrodes generating a flow of electrons through an external circuit. The reactions are reversible, allowing the battery to be recharged by applying an external voltage across the electrodes

The batteries used in power system applications so far are deep cycle batteries (similar to the ones used in Electric vehicles) with energy capacity ranging from 17 to 40MWh and having efficiencies of about70–80% (Anderson, Boulanger, Powell, & Warren, 2010).

The different types of Battery are:

- Lead acid battery
- Nickel Cadmium battery
- Sodium Sulphur
- Sodium nickel chloride
- Lithium ion batteries

Flow Battery

This type of battery consists of two electrolyte reservoirs from which the electrolytes are circulated (by pumps) through an electrochemical cell

Table 2. Components of Smart Grid

Business Applications and Projects	**Customer**	**Smart meter**		**Grid**		**Interconnection of DG**		
Computing Platform	Advance metering infrastructure (AMI)	Distribution management system (DMS)	Outage information system (OIS)	Geographic information system (GIS)	Meter data management system (MDMS)	Asset management (AM)	Customer information system (CIS)	**Security**
Communication Network	WAM WiMax Cellular		LAN RF Mesh PLC	Home Area Network (HAN) Zigbee Home Plug				
Energy T&D infrastructure (Hardware)	CAP Banks	Reclosers	Switches	Sensors	Transformer	Meters	Smart Storage	
	Substation		Transmission lines			Customers		

comprising a cathode, an anode and a membrane separator. This membrane permits ion flow, but prevents mixing of the liquids. Electrical contact is made through inert conductors in the liquids. As the ions flow across the membrane, an electrical current is induced in the conductors. The chemical energy is converted to electricity in the electrochemical cell, when the two electrolytes flow through. Both the electrolytes are stored separately in large storage tanks outside the electrochemical cell. The size of the tanks and the amount of electrolytes determines the energy density of these batteries which is depicted in Figure 4. Flow batteries are often called redox flow batteries, based on the redox (reduction–oxidation) reaction between the two electrolytes in the system.

The full-cycle efficiency of the battery is relatively low (75% - 80%), due to the energy needed to circulate the electrolyte and losses due to chemical reactions, e.g. forming of hydrogen.

Fuel Cell

A fuel cell is an electrochemical energy conversion device. It produces electricity from external supplies of fuel (anode side) and oxidant (cathode side). These react in the presence of an electrolyte. Generally, the reactants flow in and reaction products flow out, while the electrolyte remains in the cell. When the fuel is hydrogen the only by products are pure water and heat. The overall process is the reverse of water electrolysis. In electrolysis, an electric current applied to water produces hydrogen and oxygen; by reversing the process, hydrogen and oxygen are combined to produce electricity and water. A fuel cell can be seen with profit as a "chemical factory" that continuously transforms fuel energy into electricity as long as fuel is supplied. Unlike batteries, it can generate power without recharging, as long as it is being fed with fuel.

The unit fuel cell structure called the membrane electrode assembly (MEA) typically consists of an electrolyte in contact on its both sides with two electrodes, one negative electrode (anode) and one positive electrode (cathode) as shown in Figure 5. Fuel is continuously fed to the anode side and oxidant is continuously fed to the cathode side. Fuel cell reactants are classified as fuels and oxidants on the basis on their electron donor and

Figure 4. Flow Battery

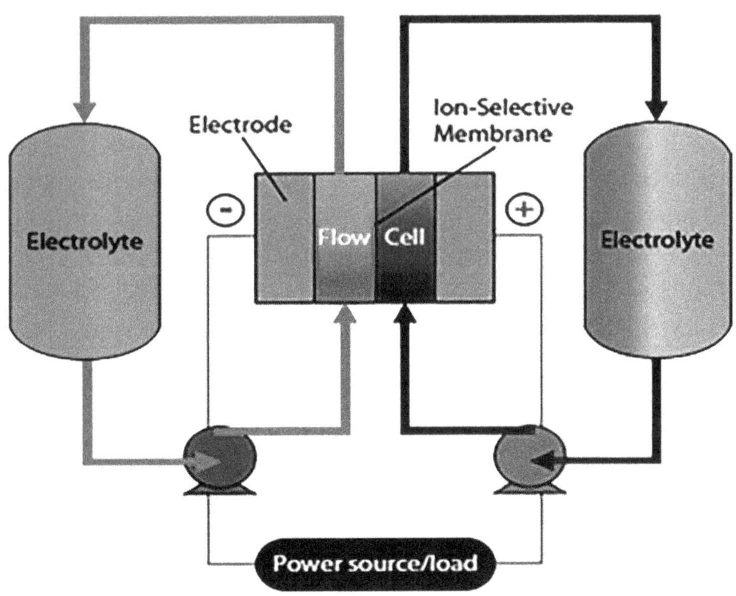

Figure 5. Fuel Cell

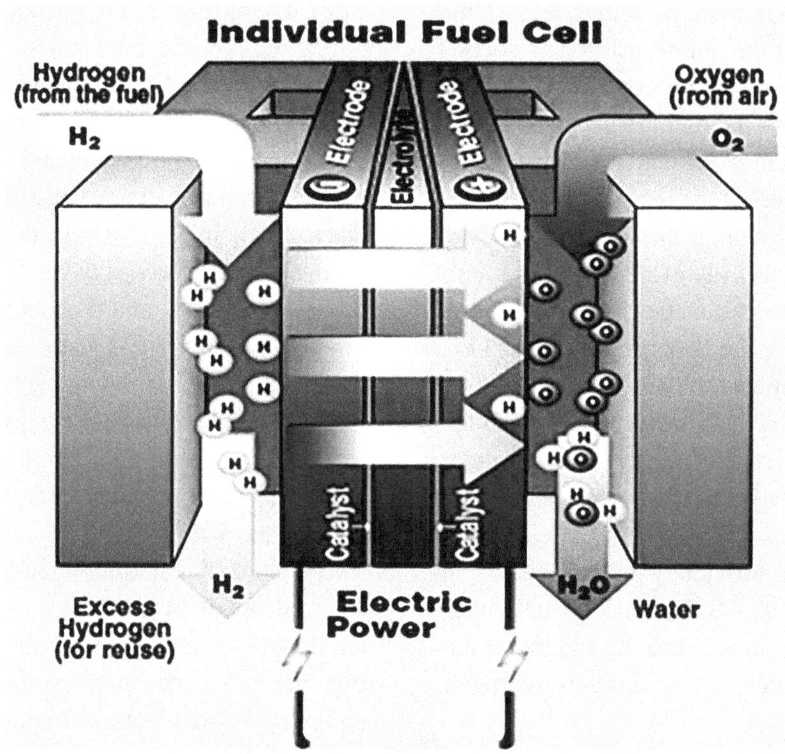

electron acceptor properties. Oxidants mainly include pure oxygen and oxygen containing gases e.g. air, or halogens e.g. chlorine. Fuels include pure hydrogen and hydrogen-containing gases, e.g. methanol, ethanol, natural gas, gasoline, biogas, diesel, etc.

Super Capacitor

Super capacitors (SC's) are energy storages having similarities with both batteries and conventional capacitors. Unlike batteries, SC's store electrical energy, not chemical energy. Unlike capacitors SC's contain moving ions. Super capacitors have many names: electrochemical capacitor (EC Capacitor), Electric double-layer capacitor (EDLC), ultra capacitor, All meaning the same thing.

Super capacitors can be fully charged and discharged in seconds, even in several degrees below zero. SC's can be charged and discharged even up to a million times.

SC's have very low energy densities & high self discharge in contrast to conventional batteries. In discharged state all the ions are distributed randomly within the cell. In charged state all the positive ions travel to the negative terminal and vice versa. The higher the carbon electrode surface area is, the higher the cell capacitance is. Conventional capacitor has solid dielectric and current collectors' smooth surfaces.

Super capacitor has rough electrode surfaces and a liquid dielectric. In contrast with traditional capacitors, electric double layer capacitors do not have a conventional dielectric shown in Figure 6. Rather than two separate plates separated by an intervening substance, these capacitors use "plates" that are in fact two layers of the same substrate, and their electrical properties, the so-called "electrical double layer", result in the effective separation of charge despite the vanishingly thin (on the order of nanometers) physical separation of the layers.

Figure 6. Super Capacitor

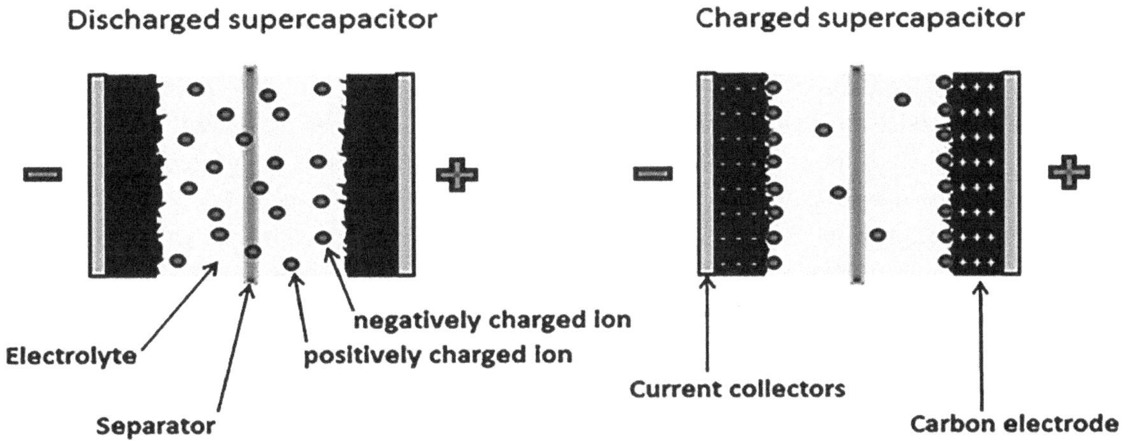

In general, electric double-layer capacitors improve storage density through the use of a nonporous material, typically activated charcoal, in place of the conventional insulating barrier. Activated charcoal is a powder made up of extremely small and very "rough" particles, which in bulk form a low-density volume of particles with holes between them that resembles a sponge. The overall surface area of even a thin layer of such a material is many times greater than a traditional material like aluminium, allowing many more charge carriers (ions or radicals from the electrolyte) to be stored in any given volume. The amount of energy stored per unit weight is considerably lower than that of an electrochemical battery. To effectively store and recover energy requires sophisticated electronic control and switching equipment. Has the highest dielectric absorption of all types of capacitors.

Super Conducting Magnetic Energy Storage System

SMES exhibits a very high energy storage efficiency (typically >_97%) and a rapid response (within a few milliseconds) in comparison with other energy storage systems, but only for short periods of time. SMES also has a high cycle life and, as a result, is suitable for applications that require constant, full cycling and a continuous mode of operation. These features make SMES suitable for use in solving voltage stability and power quality problems for large industrial customers. The major problems confronting the implementation of SMES units are the high cost and environmental issues associated with strong magnetic field (Bollen., Zhong, Zavoda Francisc, Meyer, McEachern, & López, 2010).

Pumped Hydro Energy Storage System

Pumped-storage hydroelectricity (PSH) is a type of Hydro electric energy storage used by electric power system for load balancing. The method stores energy in the form of gravitational potential energy of water, pumped from a lower elevation reservoir to a higher elevation

It consists of two reservoirs located at different elevations:

1. A unit to pump water to the high elevation (to store electricity in the form of hydraulic potential energy during offpeak hours).
2. A turbine to generate electricity with the water returning to the low elevation (converting the potential energy to electricity during peak hours).

The amount of stored energy is proportional to the height difference between the two reservoirs and the volume of water stored. PHS is a mature technology with large volume, long storage period, high efficiency and relatively low capital cost per unit of energy. Owing to the small evaporation and penetration, the storage period of PHS can be varied from typically hours to days and even years. The typical rating of PHS is about 1000 MW (100 MW–3000 MW) and facilities continue to be installed worldwide at a rate of up to 5 GW per year. It is generally applied for energy management, frequency control and provision of reserve.

Compressed Air Energy Storage System

CAES is the only other commercially available technology (besides the PHS) capable of providing very large energy storage deliverability (above 100 MW with single unit). It consists of five major components as shown in Figure 7

- A motor/generator that employs clutches to provide alternate engagement to the compressor or turbine trains.
- An air compressor of two or more stages with intercoolers and after-coolers, to achieve economy of compression and reduce the moisture content of the compressed air.
- A turbine train, containing both high- and low pressure turbines.
- A cavity/container for storing compressed air, which can be underground rock caverns created by excavating comparatively hard and impervious rock formations, salt caverns created by solution- or dry-mining

Figure 7. Compressed air storage system

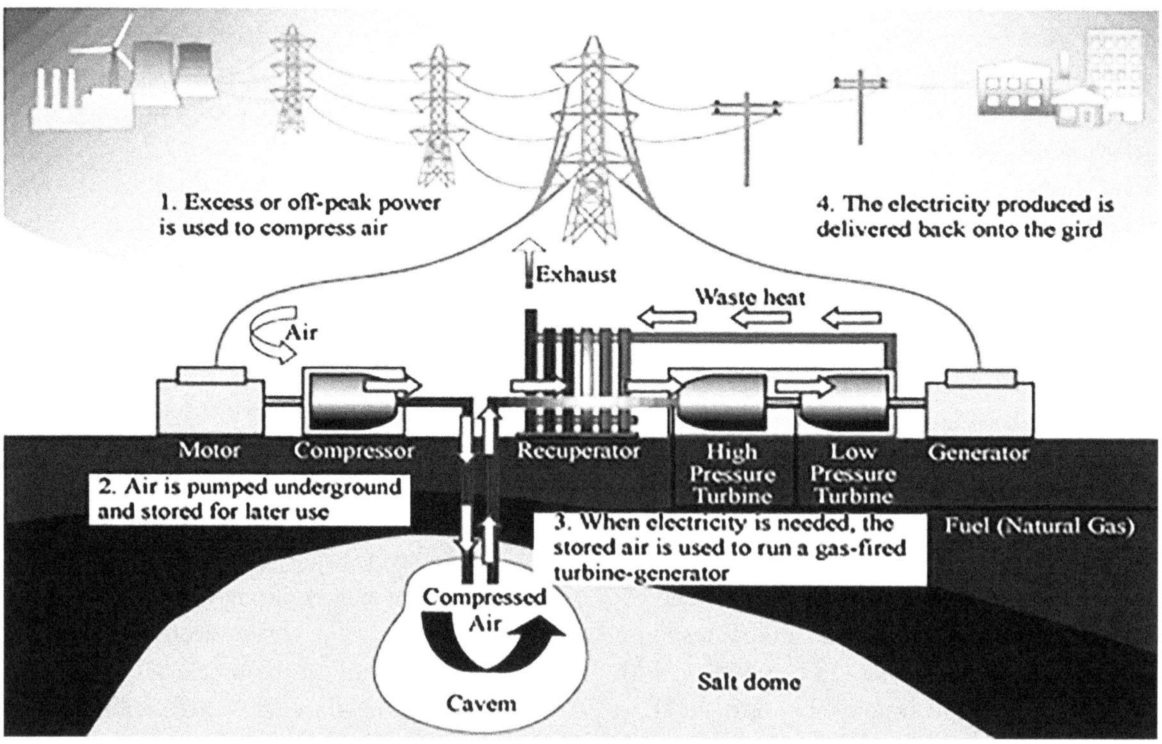

of salt formations, and porous media reservoirs made by water-bearing aquifers or depleted gas or oil fields, e.g. sandstone and fissured lime.

- Equipment controls and auxiliaries such as fuel storage and heat exchanger units.

CAES works on the basis of conventional gas turbine generation. It decouples the compression and expansion cycles of a conventional gas turbine into two separated processes and stores the energy in the form of elastic potential energy of compressed air.

During low demand, energy is stored by compressing air into an air tight space, typically 4.0–8.0 MPa. To extract the stored energy, compressed air is drawn from the storage vessel, heated and then expanded through a high pressure turbine, which captures some of the energy in the compressed air. The air is then mixed with fuel and combusted with the exhaust expanded through a low pressure turbine. Both the high and low pressure turbines are connected to a generator to produce electricity. The waste heat of the exhaust is potentially captured via a recuperator before being released. CAES systems are designed to cycle on a daily basis and to operate efficiently during partial load conditions.

This design approach allows CAES units to swing quickly from generation to compression modes. Utility systems that benefit from the CAES include those with load varying significantly during the daily cycle and with costs varying significantly with the generation level or time of day. In addition, CAES plants can respond to load changes to provide load following because they are designed to sustain frequent start-up/shutdown cycles. CAES systems also have improved environmental characteristics in comparison with conventional intermediate generating units. CAES has a relatively long storage period, low capital costs and high efficiency. The typical rating for a CAES system range of 50–300 MW.

Key Benefits of Micro Grid

Increased Reliability

Smart microgrids increase reliability locally through the establishment of a specific reliability improvement plan that integrates redundant distribution, smart switches, automation, power generation, power storage and other smart technologies. Local power generation and storage allow portions of the grid and critical facilities to operate independent of the larger grid when necessary and thus eliminate blackouts.

Helps Consumer to Save Money

The economical losses to business due to power outages can be reduced significantly using smart microgrid as consumer can produce power in real time at lower costs.

Generates Revenue

The surplus energy generated at consumer premises can be transmitted back to the main grid in return for payments.

Encourages Economic Growth

The microgrids can boost the economies through new job creation at the local level and new business opportunities for stakeholders

Makes the Grid Future Proof

The smart microgrids positioned strategically than the centralized grid to meet the known and unknown needs of the future. They allow local communities and commercial campuses to increase the overall electricity supply quickly and efficiently through relatively small local generators, solar cells, wind turbines and other means.

Reduces Carbon Footprint

The smart microgrid reduces the carbon emission. It has ability to use recover "waste" heat to displace coal-fired generation. A local power generator can be renewable- or natural gas-fuelled. The smart microgrid can effectively reuse the energy that is produced during electricity generation for heating buildings, hot water, sterilization, cooling and even refrigeration

EMERGING TECHNOLOGIES IN SMART GRID & PROBLEMS ASSOCIATED WITH THEM

Distributed Generating Units or Micro Grid

Microgrid is the small independent power system that supports the distributed generation system. As the transmission line connecting the generating unit & the load decreases, there is significant reduction in the losses. It enables the easy integration of renewable energy resources to the main grid and the pathways have been covered in Figure 8.

Definition

According to the U.S. Department of Energy (DoE):

A group of "interconnected loads and distributed energy resources" (DER) with clearly defined electrical boundaries that "acts as a single controllable entity" with respect to the grid [and can] "connect and disconnect from the grid" to enable it to "operate in both grids connected or island mode."

Value Proposition of Micro Grid

- ***Efficiency***
  - Reduce fuel consumption
  - Supply close to demand minimize distribution losses
- ***Reliability***
  - Optimally manage on-site energy resources 24/7
  - Power quality and reliability at the local level
- ***Energy Security***
  - Ensure energy supply for critical loads utilizing on-site generation

Figure 8. Developing smart grid pathways

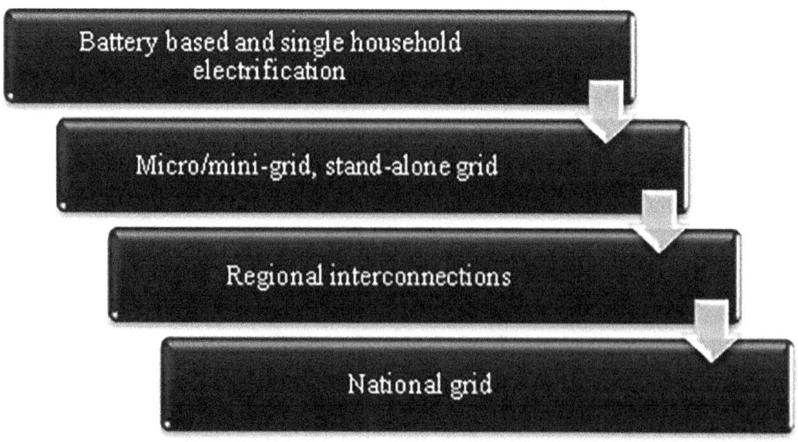

- Grid independence capability
- ***Economic Savings***
  - Peak Shaving/Load shifting and supply management with demand response
  - Enables hedging against energy cost fluctuation
  - Reduction of cost of electricity with on-site generation and effective energy management
- ***Sustainability***
  - Reduction of carbon footprint by integrating cleaner fuel resources

Concept of Vehicle to Grid (V2G)

The conventional view expects battery vehicles to be plugged in to charge their batteries. Hybrid and fuel cell vehicles generate electricity from the fuel in their tanks.

Plug-in hybrids can either run from fuel or can charge from the grid. In the conventional view, electricity never flows from vehicle to the grid.

The V2G concept is that battery, hybrid, and fuel cell vehicles all can send power to the electric grid. For battery and plug-in hybrid vehicles, the power connection is already there. For fuel cell and fuel-only hybrids, an electrical connection must be added as shown in Figure 9. Aims at using the energy stored in electric vehicles to compensate for the peak load of the grid. From the vehicle owners' point of view, since electricity price is determined by demand, the transport cost can be relatively reduced by drawing "cheap" energy from the grid, and vice versa. Vehicle-to-grid (V2G) describes a system in which plug-in electric vehicles, such as electric cars (BEVs) and plug-in hybrids (PHEVs), communicate with the power grid to sell demand response services by either delivering electricity into the grid or by throttling their charging rate.

The concept allows V2G vehicles to provide power to help balance loads by "valley filling" (charging at night when demand is low) and "peak shaving" (sending power back to the grid when demand is high). It can enable utilities new ways

Figure 9. Vehicle to grid

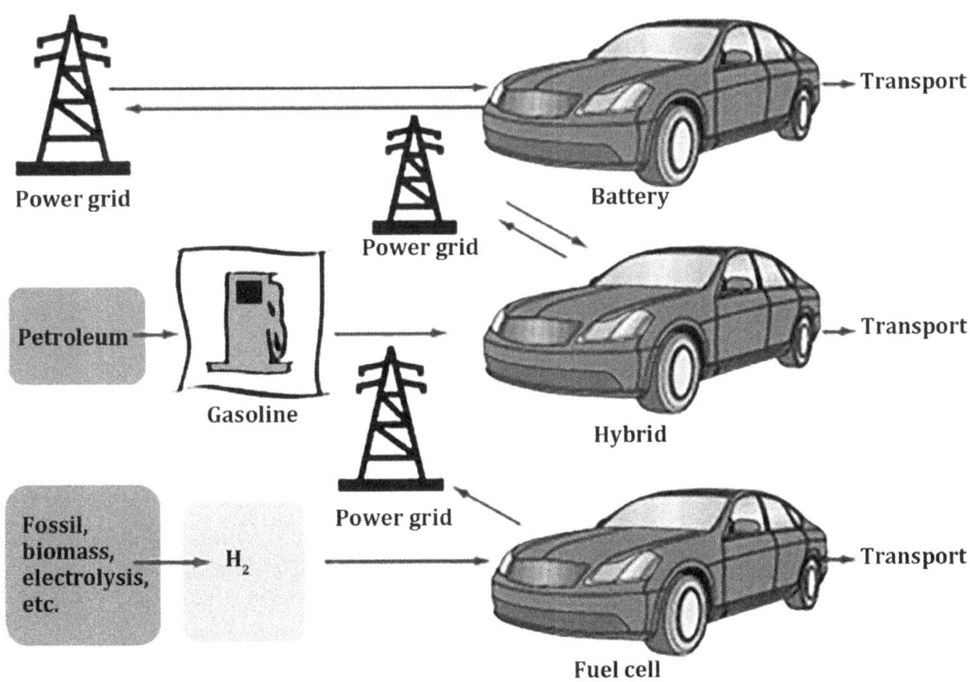

to provide regulation services (keeping voltage and frequency stable). (Letendre & Kempton, 2002)

SMART GRID: CHALLENGES TO BE REVIEWED

Issues and concerns are similar to many economics. Issues related to smart grid covers various areas including non-technical areas such as business scenario, policy, regulations, sustainable investment &development. The technical side mainly deals with the issues of interconnection, low carbon emission, quality of power, security of energy& control of smart grid.

Integration between Energies and Devices

Integration has been a major issues of concern for successful functioning of a smart grid. It is difficult to design a system for interconnection of all renewable sources, which is again not a very reliable source. Connectivity always brings challenges with security of the system. Data lost and data theft becomes a major area of concern. Base of the system lies in the interconnection of the entire network as per the defined topology. The system adopted for the integration must be user friendly and easy to handle. The area is vast and is the back bone for proper functioning of the smart grid. It starts with bringing the utilities to a common platform based on similar attributes, all the power levels have to be synchronised. Every part of the system has to be made functional on a common methodology which will be understood throughout the network. Integration will mainly be done by the software and the communication techniques. Every device in the network has to receive and send information for analysis of data and its implementation. Control on the smart grid is possible only if the grid is integrated as per the attributes decided for communication. Network efficiency has to on top notch to exploit the system, power quality will be adversely affected in absence of an efficient network.

Interference between Devices and Power Line Communication

Smart grid needs to communicate between various devices, customers, generators, distributers and grid operators for its effective operation. For establishing reliable communication system among them power line communication comes out to be the best option. Hurdle with the use of this system is inclusion of other disturbances involved with communication system. This way frequency of operation becomes a major concern. A lot has to be done in this respect to have flawless communication with minimum disturbance level.

Improving Voltage Quality

Power quality has been more talked off when it comes to power system. In order to improve the quality of power basic law states customer satisfaction, this covers reliability, affordable price and voltage quality. Important is these all have to be done in consideration with the available budget. (McBee & Simões, 2012)

Sensitivity of Devices (Continuity)

Tripping of interconnected generator station will occur unless we have balanced production and consumption. Our system must be designed in such a manner that it could be able to withstand fractional sag and swells in the system. The challenge persists when it comes to optimization of disturbances in the interconnected system and to maintain the continuity of supply to customers. Management of peak demand can be considered, around which solution for sensitivity could be achieved.

Cyber-Security and Data Privacy

Security is no more limited to solution to electricity theft. With advancement in technology being used in interconnection of generation, transmission and distributed system security issues have taken a entirely different form. As these existing systems need to be operated and controlled with communication among them for efficient operation of smart grid, so complex computer network came into existence. With complex network there came network attacks in form of cyber security vulnerabilities. These vulnerabilities can be thrown into the system from public network from across the globe. Moreover the vulnerabilities can be transmitted to any section of the system and can simultaneously affect many systems. These threats can send customer off power for uncertain time period depending on the threat spread into the network. The issues are reliable access to useful information, protection from improper information i.e. harmful information, unauthorized modification or destruction of valid information and restriction to accessibility of information to unauthorized individuals. Data management has also to be looked upon and malicious data have to be validated (Massoud, 2011).

Standardization of Technology

To shift to a new technology work force is being required with additional skill set to analyze, interpret, decision making and control of smart grid. Shifting to a new space requires set of skilled engineers and managers to take over. Needs a plan to provide the required skill set to adequate work force. Standardization of technology has been considered as one of the major obstacle in the road map to implementation. The technologies being used in transmission and distribution system is ageing. To meet the demand and maintain power quality we need to reconsider electricity system.

Policy and Regulations

Policies need to be redesigned for balance between the consumers and shareholders. Trends of liberal market are prevalent and have to be encouraged in a positive manner for healthy competition. The regulatory model has be reframed and modeled according to the needs of both the customer and the shareholders. To succeed in the implementation the reforms must attract the private financers. Layout prepared must consider balancing between the consumers and utilities. Existing regulations have been a big hindrance in the investment of new technologies.

Lack of Awareness

The concept of smart grid must be made clear to both the consumers and utilities. They must be aware of the basic know how of the technology. The advantages of the system must be clearly stated to them. Prior to formation of policies and the framework, policy makers and the regulators have to be aware of every single and minute point, considering the future aspects as well. Focus must be laid on research of the technology, with all ifs and buts before actually going into the implementation phase.

Power Quality and EMC in Smart Grid

We need to develop electromagnetic compatibility for communication and rest of the system developed in the smart grid technology. All the devices to be incorporated in the system have to be tested in well equipped laboratories specially dedicated to smart grid research. Electromagnetic interference will also be a major area of concern. Proper communication will take place between the consumers and the utilities if EMC and EMI have properly studied and devices used are checked in labs thoroughly. While considering the compatibility issues with smart grid the power quality also is a major area of concern. Effective delivery of power to consumers is incomplete without power quality. With harmonics in the grid the system will be more vulnerable to the faults. This distortion occurs with variation of load on the grid, so the load has to be monitored and analyzed for effective functioning of the grid (Vaimann, Niitsoo, Kivipold & Lehtla 2012).

Major Issues with Implementation of Micro Grid

Even though the microgrid offer inherent advantages over the centralized generation system as it creates two major problems when integrated with main grid:

- The output of the distributed generating units are usually D.C or A.C of variable frequency which is needed to modified before synchronizing to the main grid. This requires power electronic interfaces at the point of common coupling (POCC) which in turn induces harmonics in main supply.
- It requires specially designed protection scheme as the integration usually modifies the fault current values which results into false tripping of relays & other protective devices and malfunctioning of the system.

Protection Strategy of the Micro Grid

Typical microgrid system comprises of distributed generation units with inverters and incorporate control systems that enable flexible operations. Generally, it connected to the power delivery system at a point of common coupling, thus appearing as a controllable single subsystem to the utility grid. In grid connected mode, the micro sources act as constant power sources, which are controlled to inject the demanded power into the network. In autonomous mode, micro sources are controlled to supply all the power needed by the local loads while maintaining the voltage and frequency within the acceptable operating limits.

One of the major challenges of micro grid protection system is that it must respond to both island and grid connected faults. In the first case the protection system should isolate the smallest part of the micro grid when clears the fault. In the second case the protection system should isolate the micro grid from the main grid as rapidly as necessary to protect the micro grid loads. Modern micro grid consist of many different distributed energy resources like solar PV, wind turbines, fuel cells, small-scale hydro, tidal and wave generators, micro- turbines, combined heat power (CHP) systems, energy storage, etc. When different type of distributed resource connected to micro grid and Utility grid, the DR contribute fault current to the system and the contribution level depends on distribution resource type

To ensure safe operation of micro grid the protection equipment should be updated accordingly. Central control unit communicate with all relays and distributed generators in the micro grid to record their status as ON/OFF, their rated current and their fault current contribution. Communicate with relay is required to update the operating current and to detect the direction of the fault currents and thus mitigate the fault properly. The control unit also records the status of utility grid as connected or micro grid is islanded for adaptive protection.

Protection Issues of the Micro Grid

Threshold value of fault currents for grid connected and islanded operation of microgrid are different. The short circuit power varies significantly. Faults also causes loss of sensitivity, over current, earth leakage, disconnection of generators, islanding, reducing reach of over current relays, single phase connections and loss of stability. Depending upon location of faults with respect to distributed generators and existing protection equipment, problems like bidirectional power flow and change in voltage profile occurs

Various major issues associated with the protection of Micro grid are as follows:

- Modification in fault current level
- Device discrimination
- Reduction in reach of Impedance relay
- Reverse power flow
- Sympathetic tripping
- Islanding
- Single phase connection
- Selectivity

Modification in Fault Current Level

When large number of small distributed generation units that uses synchronous or induction generator units are connected to distribution network or grid it changes fault current level as both types of generators contribute towards fault currents.

When inverter interfaced DG units are used, fault current is limited to a lower value. As fault current is not high as compared to load current, some of the relays do not trip, others that respond to fault operate with the time delay. The undetected fault spreads out in the system and can damage the equipment

Device Discrimination

In the power system network that has generation sources at the end of network, fault current decreases with increase in distance as the impedance increases. The variation in magnitude of fault current is used for discrimination.

In case of islanded microgrid with inverter interfaced distributed generation units, fault is limited to a lower value, fault level at the locations of feeder will be almost constant.

The traditional current protection scheme which uses the variation in magnitude of fault current for discrimination does not work properly.

Reduction in reach of Impedance Relay

The reach of impedance relay depends upon the distance between the relay location and fault point, maximum distance means minimum fault current that is detected.

When DG is present in the system, distance relay may not operate according to defined zone settings. When faults occurs downstream of the bus DG connected to utility network, impedance measured by relay located in upstream is higher than real fault impedance. This affects grading of relays and causes delayed operation or sometimes relay does not operate at all.

Reverse Power Flow

Main challenge for protecting the microgrid arises because power can flow in both the directions in each feeder of microgrid. Sources are located in both sides of load due to which power flows in opposite direction from two sources towards the load.

Power flow also changes its direction in case of distribution network with embedded generation when local generation exceeds local consumption. The reverse power flow can also cause power quality problems resulting in variation of voltage.

Sympathetic Tripping

This occurs when protective device operates for faults in an outside protective zone. DG contributes towards the fault; relay operates along with another relay which actually sees the fault resulting in malfunctioning of protective scheme.

Islanding

The DG creates a problem when part of distributed network with DG unit is islanded. Islanding is due to fault in the network. If generator continues to supply power despite the disconnection of utility, fault might persist as fault is fed by DG. If the control for the voltage is not provided, it results in unexpected rise in voltage levels in case of islanded operation

Single Phase Connection

Some DG sources inject single phase power into the distribution grid, for example PV systems. This affects balance of three phase currents, due to unbalance current in the neutral conductor increases which also results in flow of stray currents to earth. This current should be limited to prevent overloading.

Selectivity

Protection system is said to be selective if the protection device closest to the fault operates to remove the faulty section. Without DG, there is power flow in only one direction, during normal operation as well as when there is fault, by using time graded over current relays selectivity can be obtained .

When DG is integrated with the grid, this system becomes inadequate. There is possibility of disconnection of healthy feeder by its own protective relay because it contributes to the short circuit current flowing through fault in the neighbouring feeder.

The tripping current for electrical protective device is between maximum load current and minimum fault current. Fault current and load current depends upon the state of grid, state of distributed generators and whether microgrid is operating in islanded mode. (Kaushal, 2011)

TECHNOLOGIES TO OVERCOME CHALLENGES

Solution for Protection Issues of the Micro Grid

Protection of Inverter Interfaced DG Units

Conventional protection systems cannot give reliable protection for inverter interfaced units, because there is limited fault current. The solution can be achieved by using inverters which have high fault current capability that is up rating the inverter, using faster communication system between Inverter and protective relays, and introduction of energy storage devices that are capable of supplying large current in case of faults

Differential Protection Scheme

The protection scheme for microgrid with Inverter interfaced DG units cannot differentiate between fault current and an overload current, which results in nuisance tripping when system is overloaded, in some instances traditional protection scheme.

For proper clearing of fault in an islanded microgrid and to ensure selectivity, it is important that different distributed generators should effectively communicate with each other. Use of evolving distribution system version of pilot wire line differential scheme is required for protection

Balanced Combination of Different DG System

To obtain the protection of an isolated microgrid is to use DG units with synchronous generators or to use inverters having high fault current capability or to use combination of both types of DG units so the conventional protection schemes can be properly used.

Inverter Controller Design

Protection scheme for islanded microgrid is dependent on type of inverter controller, controller can actively limit the available fault current from inverter interfaced distributed generator units.

Adaptive Protection for Microgrid

Adaptive protection scheme solves the problem both in grid connected as well as islanded mode. In adaptive protection system there is automatic readjustment of relay settings when microgrid changes from grid connected mode to islanded mode and from islanded mode to grid connected mode. Adaptive protection is an online system that modifies preferred protective response to change in system conditions or requirements in timely manner by means of external generated signals or control actions.

Solution on Broad Prospective for Smart Grid

Outage Management System

Integrating the (OMS) outage management system with smart grid will allow the real time identification of trouble locations. This can play a vital role in optimization of the system. With such a system utilities have not to wait for any response from the consumer and so will be able to get the information in real time Figure 10. This will help in reliable supply, ultimately improving the power quality. The delays due outages can be averted with the application of OMS. The hindrance lies in the commissioning of the OMS system, in other words integration of OMS with network of smart grid is a challenge.

Figure 10. Components in Information Management System

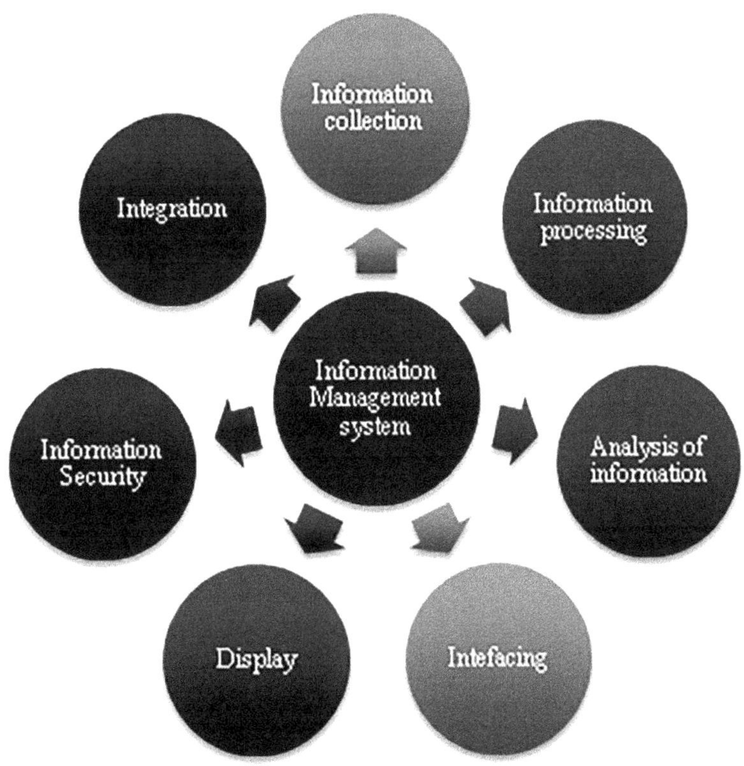

Wide-Area Monitoring and Control

Monitoring, analysis and control of real time power system attributes over a large area can be helpful in understanding the systems' functionality, requirement and performance. This includes control and automation (WAAPCA), wide area situational awareness (WASA), wide-area monitoring systems (WAMS), and wide- area adaptive protection. The system will also help us to overcome blackouts and facilitate the integration of various resources together.

Advanced Metering Infrastructure

Advanced metering infrastructure (AMI) groups various technologies getting involved in metering system along with smart meters. Instead of having one way information sharing AMI will inculcate two way communications in Table 3, between the consumer and utilities. With sharing of data the consumption of energy is available with price as well as duration. AMI also supports remote access, which means connection and disconnection can be handled from remote location.

Automated Demand Response

A significant amount of research is being carried out data processing and enabling automation technology. Energy has to be regulated through automation based on parameters such as availability of power, demand, consumption, and duration of demand, tariff or any other factor. Smart grid's maximum potential can be harnessed through well connected communication system, for receiving the valuable data of information, which can further be processed for perfect automation on

Table 3. Hardware and the software

Technology Area	Hardware	Systems and Software
Wide-area monitoring and control	Phasor measurement units (PMU) and other sensor equipment	Supervisory control and data acquisition (SCADA), wide-area monitoring systems (WAMS), wide-area adaptive protection, control and automation (WAAPCA), wide area situational awareness (WASA)
Renewable and distributed generation integration	Power conditioning equipment for bulk power and grid support, communication and control hardware for generation and enabling storage technology	Energy management system (EMS), distribution management system (DMS), SCADA, geographic Information system (GIS)
Information and communication technology integration	Communication equipment (Power line carrier, WIMAX, LTE, RF mesh network, cellular), routers, relays, switches, gateway, computers (servers)	Enterprise resource planning software (ERP), customer information system (CIS)
Transmission enhancement	Superconductors, FACTS, HVDC	Network stability analysis, automatic recovery systems
Advanced metering infrastructure	Smart meter, in-home displays, servers, relays	Meter data management system (MDMS)
Distribution grid management	Automated re-closers, switches and capacitors, remote controlled distributed generation and storage, transformer sensors, wire and cable sensors	Geographic information system (GIS), distribution management system (DMS), outage management system (OMS), workforce management system (WMS)
Electric vehicle charging infrastructure	Charging infrastructure, batteries, inverters	Energy billing, smart grid-to-vehicle charging (G2V) and discharging vehicle-to-grid (V2G) methodologies
Customer-side systems	Smart appliances, routers, in-home display, building automation systems, thermal accumulators, smart thermostat	Energy dashboards, energy management systems, energy applications for smart phones and tablets

both demand and supply. A lot of research work is going on the area of developing a automation system having facility to consumers and utilities to program their operation, on their end, as per their requirement. With implementation of automation system the utilities would be able to get the feedback of usage of power automatically with more accuracy. The obtained feedback can be effectively utilized for developing a perfect model for supply of energy on smart grid. "Do not promise what you are not going to deliver!"

Topology of the Grid

It is essential for the researchers to arrive to a near perfect topology for the smart grid robust functioning. For any system to function flawless the design parameters have be analyzed with minute detailing. Framework for the system must comply with the norms of the regulatory authorities, must adhere with the policies and should be in considerations of both the utilities and the consumers. While designing the topology the compatibility of the system devices have to be considered. Most important of all have always been communication in the smart grid. Rural regions versus dense urban areas have also to be considered while designing of the network. Consumers will definitely differ in terms of demand of energy, in terms of duration of particular amount of power, in terms of power quality and most importantly the tariff of the power.

Monitoring System

To have an effectiveness of the system the entire network has to be kept under surveillance. Keeping a strong monitoring on the network will provide the appropriate real time feedback of the system. The monitoring of entire network can also avert any possible threats during operation. Monitoring of network will include following various components.

- Load forecasting
- Network planning
- Network Operations
- Network asset maintenance, diagnostics, analytics
- Renewal integration
- Determine best practice pricing policies

Dynamic Pricing

In order to incentivize customers to participate in changing consumption dynamic pricing is a great tool. Consumers will be forced to regularize the usage of power as per the tariff. Consumers will become a part of system by indirectly having control over the consumption of energy. Pricing of energy will depend on the technology and the methodology being implemented. Every factor associated with the implementation involves the cost of the system, which has to be considered while designing the policy for tariff. Following points will also play an important role.

- Alternative clean energy sources and new materials
- Advanced power electronics and devices.
- Sensing and measurement.
- Communications.
- Advanced computing, control methodologies and intelligent technologies.
- Mature power market regulation and policies.

CONCLUSION

The literature covers major issues associated with smart power distribution and smart energy storing system. The chapter includes the challenges with which we would be confronting in practical implementation of the huge framework. Making smart grid a reality need investigation of the areas discussed above. As smart grid will become a revolution in the power distribution network. To harness the diverse capabilities of the smart grid all the aspect associated have to monitored, information or data has to interpreted and used to enhance the aspects of the system. The authors have tried to introspect the ground reality attached with the smart grid, which takes an account of both the consumers and the utilities. The methodology, technologies, discussed will throw light on very minor issues also which generally is being left out. The problems have been reviewed and a possible solution has also been discussed. The solutions mentioned here can be further researched and given a shape for its implementation. The features and advantages are great but the hurdle mainly lies in the interconnection and effective communication. Major technical and non technical issues relevant to smart grid has been uncovered and has been dealt with effectiveness. Prior existing technologies have also been compared and suggestions for shortfalls have been discussed. With all discussion and research one of the important points is awareness of the entire process. The problem in implementation mainly lies in misconception prevalent among the both consumers and the utilities. Associated members have to be involved in the policy making for their clear understanding of the benefits and features of the smart grid. Smart grid will lead to power sector revolution and more efficient system. Energy is the back bone of every nation's economy and so the priority for smart grid has be the on top. Collaboration will see a more clear vision (both government and private investors) to the smart grid implementation. Efficiency of the network will drastically improve and saving of energy will indirectly help economy. Till date the work done on the issue is commendable but a lot more has to be done. We believe that with the content covered; this chapter will be helpful for future research and innovation in smart grid as the content has precisely addressed the challenges.

REFERENCES

Anderson, R.N., Boulanger, A., Powell, W.B., & Warren, S. (2010). Adaptive Stochastic Control for the Smart Grid. *Proceedings of the IEEE*, 99(6), 1098-1115.

Atwa, Y. M., El-Saadany, E. F., Salama, M. M. A., & Seethapathy, R. (2010). Optimal renewable resources mix for distribution system energy loss minimization. *IEEE Transactions on Power Systems*, *25*(1), 360–370. doi:10.1109/TPWRS.2009.2030276

Bollen, M.H.J., Zhong, J., Zavoda, F., Meyer, J., McEachern, A., & López, F.C. (2010). Power quality aspects of Smart Grids. *Proceedings of the International Conference on Renewable Energies and Power Quality ICREPQ '10*, Granada, Spain.

Chang C. S., Wang Zhaoxia, Yang Fan, & Tan W. W. (2012, March). Hierarchical Fuzzy Logic System for Implementing Maintenance Schedules of Offshore Power Systems. *IEEE Transactions Smart Grid*, 3(1).

Chen, H., Cong, T. N., Yang, W., Tan, C., Li, Y., & Ding, Y. (2009). Progress in electrical energy storage system: A critical review. *Progress in Natural Science*, *19*(3), 291–312. doi:10.1016/j.pnsc.2008.07.014

de Boer, P., & Jillis, R. (2007). *Flow Battery (Briefing Paper). Leonardo energy*. KEMA.

Divya, K. C., & Østergaard, J. (2009). Battery energy storage technology for power systems—An overview. *Electric Power Systems Research*, *79*(4), 511–520. doi:10.1016/j.epsr.2008.09.017

Fang, X., Misra, S., Xue, G., & Yang, D. (2012). Smart grid the new and improved power grid: A survey. *IEEE Communications Surveys and Tutorials*, *14*(4), 944–980. doi:10.1109/SURV.2011.101911.00087

Fanxing, L. (2010, September). Smart Transmission Grid: Vision and Framework. *IEEE Transactions on Smart Grid*, *1*(2), 168–177. doi:10.1109/TSG.2010.2053726

Guerrero, J. M., Chandorkar, M., Lee, T., & Loh, P. C. (2013). Advanced control architectures for intelligent microgrids, part I: Decentralized and hierarchical control. *IEEE Transactions on Industrial Electronics*, *60*(4), 1254–1262. doi:10.1109/TIE.2012.2194969

J., Joy,, Jasmin, E.A., John V.R. (2013, March). Challenges of Smart Grid. *International Journal of Advanced Research in Electrical, Electronics and Instrumentation Engineering*, *2*(3).

Jiang, Z., & Dougal, R. A. (2008, July). Hierarchical microgrid paradigm for integration of distributed energy resources. *Proceedings of the IEEE Power and Energy Society 2008 General Meeting: Conversion and Delivery of Electrical Energy in the 21st Century* (pp. 1–8). doi:10.1109/PES.2008.4596185

McBee, K.D., & Simões, M.G. (2012, June). Utilizing a Smart Grid Monitoring System to Improve Voltage Quality of Customers. *IEEE Transactions on smart grid*, 3(2), 738-743.

Kahrobaeian, A; Ibrahim Mohamed, Y.A.-R. (2015). Networked-Based Hybrid Distributed Power Sharing and Control for Islanded Microgrid Systems. *IEEE Transactions on Power Electronics*, 30(2), 603 – 617.

Laskar, S. H. Mohibullah, (2012, September). Power Quality Issues and Need of Intelligent PQ Monitoring in the Smart Grid Environment. *International Journal of Emerging Technology and Advanced Engineering*, 2(9).

Llaria, A., Curea, O., Jiménez, J., & Camblong, H. (2011). Survey on microgrids: Unplanned islanding and related inverter control techniques. *Renewable Energy*, *36*(8), 2052–2061. doi:10.1016/j.renene.2011.01.010

Mahmood, H., & Jiang, J. (2012). Modeling and control system design of a grid connected VSC considering the effect of the interface transformer type. *IEEE Transactions on Smart Grid*, *3*(1), 122–134. doi:10.1109/TSG.2011.2166412

Malakhov, A., Kopyriulin, P., Petrovski, S., & Petrovski, A. (2012) Adaptation of smart grid technologies. Proceedings of the 2012 IEEE International Conference on Fuzzy Systems (FUZZ-IEEE) (pp. 1 – 6).

Massoud, A. S. (2011). Smart Grid: Overview, Issues and Opportunities. Advances and Challenges in Sensing, Modeling, Simulation, Optimization and Control. *European Journal of Control*, *17*(5-6), 547–567. doi:10.3166/ejc.17.547-567

Math, H. J. Bollen, (2000). Understanding Power Quality Problems. IEEE Press Series on Power Engineering (pp. 6-124).

Booth, A., Demirdoven, N., & Tal, H. (2010). The smart grid opportunity for solution providers. *McKinsey & Company*.

Weixiao Meng, Ruofei Ma, & Hsiao-Hwa Chen (2014). Smart grid neighbourhood area networks: a survey. Network, 28(1), 24 – 32.

Menniti, D., Sorrentino, N., Pinnarelli, A., Burgio, A., Brusco, G., & Power, B.G. (2014). In the future Smart Cities: Coordination of micro Smart Grids in a Virtual Energy District. Proceedings of the International Symposium on Electronics, Electrical Drives, Automation and Motion SPEEDAM '14 (pp. 676 – 682).

Metke, A. R., & Ekl, R. L. (2010). Security technology for smart grid networks. *IEEE Transactions on Smart Grid*, *1*(1), 99–107. doi:10.1109/TSG.2010.2046347

Pedrasa, M. A. A., Spooner, T. D., & MacGill, I. F. (2010). Coordinated scheduling of residential distributed energy resources to optimize smart home energy services. *IEEE Transactions on Smart Grid*, *1*(2), 134–143. doi:10.1109/TSG.2010.2053053

Priolkar Jayeshkumar, G., & Shet Vinayak, N. (2013, April). A Review on Protection Issues in Microgrid. *International Journal of Emerging Trends in Electrical and Electronics*, *2*(1).

Razibul, I.M., & Gabbar Hossam, A. (2012, February). Study of Micro Grid Safety & Protection Strategies with Control System Infrastructures. *Smart Grid and Renewable Energy*, 3, 1-9. Retrieved from http://www.SciRP.org/journal/sgre

Singh, T., & Vara, P.K. (2009, July). Smart metering the clouds. *Proceedings of the 2009 18th IEEE International Workshops on Enabling Technologies* (pp. 66–71).

Sirisukprasert, S. (2014). Power electronics-based energy storages: A key component for Smart Grid technology. *Proceedings of the International 2014 Electrical Engineering Congress iEECON* (pp. 1 – 7).

Zhao Tan, Peng Yang, Nehorai, A. (2014). An Optimal and Distributed Demand Response Strategy With Electric Vehicles in the Smart Grid. *IEEE Transactions on Smart Grid*, 5(2), 861 – 869.

Teleke, M. E., Baran, M. E., Huang, A. Q., Bhattacharya, S., & Anderson, L. (2009). Control strategies for battery energy storage for wind farm dispatching. *IEEE Transactions on Energy Conversion*, *24*(3), 725–732. doi:10.1109/TEC.2009.2016000

Tushar W., Saad W., Poor, H.V., & Smith D. (2012). Economics of electric vehicle charging: a game theoretic approach. *IEEE Transactions on Smart Grid*, 3(4), 1767–1778

Basics of fuel cell. (2003, August). U.S. Department of Energy, Office of Energy Efficiency and Renewable Energy.

Vaimann T., Niitsoo J., Kivipold T., & Lehtla T. (2012). Power Quality Issues in Dispersed Generation and Smart Grids. *Elektronika Ir Elektrotechnika*, 18(8).

Chapter 9
STATCOM Based Solid State Voltage Regulation for Isolated Self-Excited Induction Generator

Pallavi Thakkur
Sagar Institute of Research, Technology & Science, India

Smita Shandilya
Sagar Institute of Research, Technology & Science, India

ABSTRACT

Self-Excited Induction Generator (SEIG) offers many advantages such as low cost, simplicity, robust construction, self-protection against faults and maintenance free in today's renewable energy industry. However, the SEIG demands an external supply of reactive power to maintain the constant terminal voltage under the varying loading conditions, which limits the application of SEIG as a standalone power generator. The regulation of speed and voltage does not result in a satisfactory improvement although several studies have been emphasized on this topic in the past. To improve the performance of the SEIG system in isolated areas and to regulate the terminal voltage STATic COMpensator (STATCOM) has been modelled and developed in this dissertation. The STATCOM consists of AC inductors, a DC bus capacitor and solid-state self-commutating devices. The ratings of these components are quite important for designing and controlling of STATCOM to maintain the constant terminal voltage. The proposed generating system is modelled and simulated in MATLAB along with Simulink and sim power system block set toolboxes. The simulated results are presented to demonstrate the capability of an isolated power generating system for feeding three-phase resistive loads.

1. INTRODUCTION

The power system today is complicated network with many generating stations and load centers being interconnected through power transmission lines. An electric power system can be divided into three stages: i) generation, ii) transmission and iii) distribution. The basic structure of a power system is as shown in Figure 1. It is composed of generating plants, a transmission system and distribution system. These subsystems are interconnected through transformers T1, T2 and T3.

DOI: 10.4018/978-1-4666-9911-3.ch009

Figure 1. Typical power systems

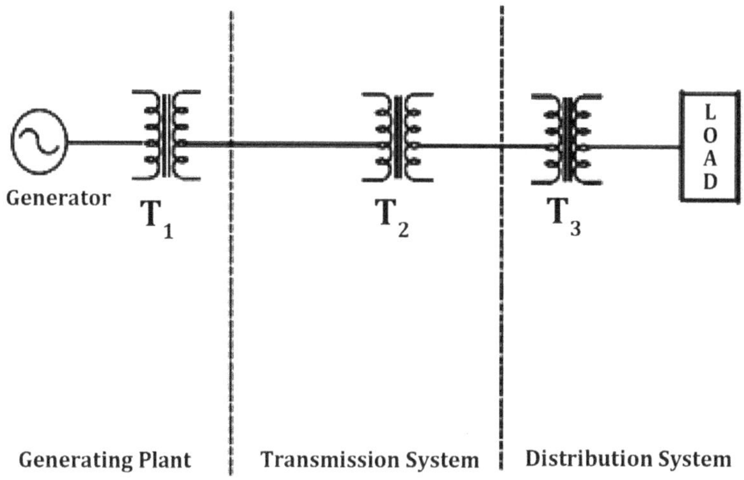

The power system is a highly nonlinear system that operates in changing conditions such as loads, generator outputs etc. When subjected to a disturbance, the stability of the system depends on the nature of the disturbance as well as the initial operating condition. The system must be capable to operate satisfactorily under these disturbing conditions and successfully meet the load demand. Also with increasing demand it becomes very important to utilize the existing transmission system assets. Flexible AC Transmission System (FACTS) controllers, based on the hasty development of power electronics technology, have been projected in recent years for better utilization of existing transmission facilities.

Now-a-days it is becoming very difficult to use it fully due to various reasons, such as environmental legislation, capital investment, rights of ways issues, construction cost of new lines, deregulation policies, etc. Electric utilities are now forced to operate their system in such a way that makes better utilization of existing transmission facilities. During the last decade, a number of control devices under the term FACTS technology have been anticipated and implemented. Application of FACTS devices in power systems, leads to better performance of system in many aspects. Voltage stability, voltage regulation and power system stability, damping can be improved by using these devices and their proper control (Haque, 2006). There are various forms of FACTS devices, some of which are connected in series with a line and the others are connected in shunt or a combination of series and shunt. The FACTS technology is not a single high power controller but rather a collection of controllers which can be applied individually or in coordination with other to control one or more of the inter related system parameters like voltage, current, impedance, phase angle and damping of oscillations at various frequencies below the rated frequency. Among all FACTS devices, static synchronous compensators (STATCOM) plays much more important role in reactive power compensation and voltage support because of its attractive steady state performance and operating characteristics. The fundamental principle of a STATCOM installed in a power system is the generation of ac voltage source by a voltage source inverter (VSI) connected to a dc capacitor. The active and reactive power transfer between the power system and the STATCOM is caused by the voltage difference across the reactance. The STATCOM can also increase transmission capacity, damping low frequency oscillation, and improving transient stability. The STATCOM is represented by a voltage source, which is connected to the system through a coupling transformer. The voltage of the source is in phase with the ac system voltage at the point of connection, and the magnitude of the voltage is controllable. The current from the source is limited to a maximum value by adjusting the voltage. Mathematical modeling and analysis of static compensator (STATCOM) is presented in (Padiyar & Devi, 1994) & (Rao, Crow, & Yang, 2000). It explains the use of STATCOM for improvement of transient stability and power transfer.

2. FACTS CONTROLLERS

The IEEE Power Engineering Society (PES) Task Force of the FACTS Working Group has defined FACTS and FACTS Controller as given below (Hingorani & Gyugyi, 1999).

Flexible AC Transmission System (FACTS): Alternating current transmission systems incorporating power electronic-based and other static controllers to enhance controllability and increase power transfer capability.

FACTS Controller: A power electronic-based system and other static equipment that provide control of one or more AC transmission system parameters.

2.1 Static Shunt Compensators: SVC and STATCOM

2.1.1 Static Var Compensator (SVC) Model

The Static Var Compensator (SVC) is a shunt device of the Flexible AC Transmission Systems (FACTS) family using power electronics to control power flow and improve transient stability on power grids The application of SVC was initially for load compensation of fast changing loads such as steel mills and arc furnaces. The application for transmission line compensators commenced in the late seventies.

Here the objectives are:

1. Increase power transfer in long lines
2. Improve stability with fast acting voltage regulation
3. Damp low frequency oscillations due to swing (rotor) modes
4. Damp sub-synchronous frequency oscillations due to torsional modes
5. Control dynamic over voltages

The SVC regulates voltage at its terminals by controlling the amount of reactive power injected into or absorbed from the power system. When system voltage is low, the SVC generates reactive power (SVC capacitive). When system voltage is high, it absorbs reactive power (SVC inductive). The variation of reactive power is performed by switching three-phase capacitor banks and inductor banks connected on the secondary side of a coupling transformer. Each capacitor bank is switched on and off by three thyristor switches (Thyristor Switched Capacitor or TSC). Reactors are either switched on-off (Thyristor Switched Reactor or TSR) or phase-controlled (Thyristor Controlled Reactor or TCR).

Figure 2 shows a single-line diagram of a static var compensator and a simplified block diagram of its control system.

The control system consists of

- A measurement system measuring the positive-sequence voltage to be controlled. A Fourier-based measurement system using a one-cycle running average is used.
- A voltage regulator that uses the voltage error (difference between the measured voltage Vm and the reference voltage Vref) to determine the SVC susceptance B needed to keep the system voltage constant

Figure 2. Single-line diagram of an SVC and its control system block diagram

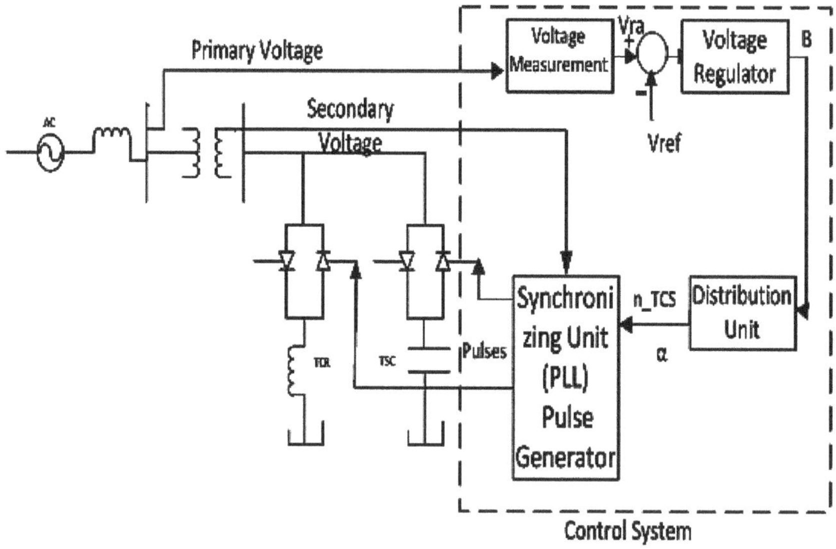

- A distribution unit that determines the TSCs (and eventually TSRs) that must be switched in and out, and computes the firing angle α of TCRs
- A synchronizing system using a phase-locked loop (PLL) synchronized on the secondary voltages and a pulse generator that send appropriate pulses to the thyristors.

The steady state control characteristics of SVC are shown in Figure 3 where ADB is the control range. OA represents the characteristic where the

Figure 3. V-I characteristics of SVC

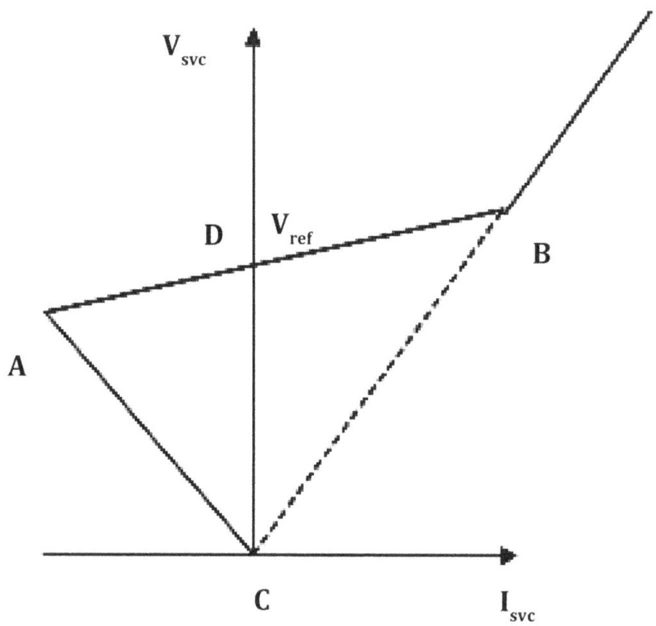

SVC hits the capacitor limit; BC represents the SVC at its inductor limit. Note that SVC current is considered positive when SVC susceptance is inductive.

Thus

$I_{SVC} = -B_{SVC}V_{SVC}$

The slope of OA is BC (susceptance of the capacitor) and the slope of OBC is BL (susceptance of the reactor). A positive slope (in the range of 1-5%) is given in the control range to (Figure 3)

- Enable parallel operation of more than one SVC connected at the same or neighboring buses and
- Prevent SVC hitting the limits frequently.

The steady state value of the SVC bus voltage is determined from the intersection of the system characteristic and the control characteristic. The system characteristic is a straight line with negative slope and is defined by

$V_{SVC} = V_{Th} - X_{Th}I_{SVC}$

where V_{Th} and X_{Th} are the Thevenin voltage and reactance viewed from the SVC bus.

Expression for Voltage and Power are given as below:

- Control Range: The SVC control range is described by (Figure 4)

$V_{SVC} = V_{ref} + X_sI_{SVC}$

where Xs is the slope of the control characteristic. V_{ref} is the SVC voltage (corresponding to point D) when $I_{SVC} = 0$.

Combining equations we get,

$$V_{SVC} = V_m = \frac{V_{Th}X_s}{X_s + X_{Th}} + \frac{V_{ref}X_{Th}}{X_s + X_{Th}} \quad (1)$$

The expression for power flow in the line is given by

$$P = \frac{V_m V \sin(\delta/2)}{Z_n \sin(\theta/2)} \quad (2)$$

Figure 4. Determination of operating point of SVC

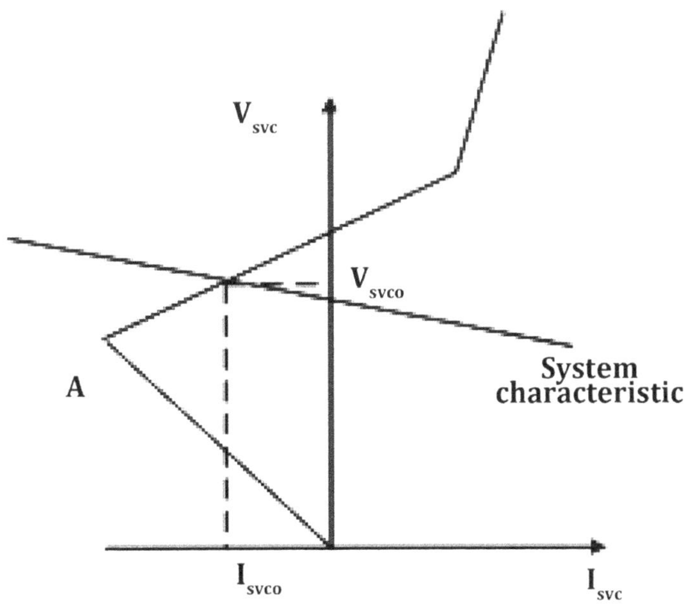

with Vref = V, it can be shown that P is given by,

$$P = kP_0 + (1-k)P_1 \tag{3}$$

where

$$P_0 = \frac{V^2 \sin\delta}{Z_n \sin\theta}, \quad P_1 = \frac{V^2 \sin(\delta/2)}{Z_n \sin(\theta/2)} \tag{4}$$

and

$$k = \frac{X_s}{X_s + X_{Th}} \tag{5}$$

2.1.2 Static Synchronous Compensator (STATCOM) Model

STATCOM, as shown in Figure 5, is a dynamic compensator consist of a set of voltage source converters (VSC) and a coupling shunt-connected transformer. It is mainly used for power systems dynamic compensation. In fact, STATCOM is a static equivalent of a synchronous compensator; however, the STATCOM is faster in absorbing or providing reactive power as there is no mechanical moving parts involved. In addition, the STATCOM offers more control flexibility in comparison with the synchronous machine. The voltage difference across the coupling transformer results in active

Figure 5. STATCOM schematic diagram

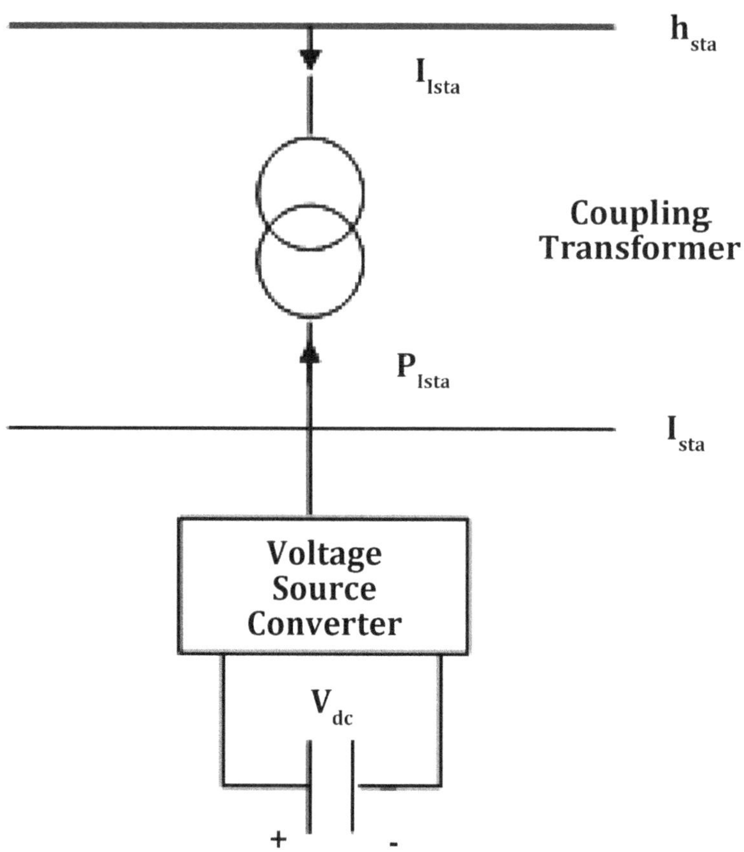

and reactive power exchanges between the network and STATCOM. The reactive power exchange is achieved by changing the voltage magnitude of voltage source. The active power exchange, used to control the DC voltage of the capacitor, in steady-state operation is zero, neglecting the VSC losses.

Static Compensator (STATCOM) apply advance power devices such as GTOs(Gate Turn Off Thyristors) or GCT (Gate Commutated Thyristors) and are able to exchange reactive current (inject or absorb) with the power system at a range of voltage levels, similar to a synchronous condenser. Thus STATCOM are able to provide voltage support to the power system in the vicinity of the bus to which it is connected. The Static Synchronous Compensator (STATCOM) is a shunt device of the Flexible AC Transmission Systems (FACTS) family using power electronics to control power flow and improve transient stability on power grids. The STATCOM regulates voltage at its terminal by controlling the amount of reactive power injected into or absorbed from the power system. When system voltage is low, the STATCOM generates reactive power (STATCOM capacitive). When system voltage is high, it absorbs reactive power (STATCOM inductive).

The variation of reactive power is performed by means of a Voltage-Sourced Converter (VSC) connected on the secondary side of a coupling transformer. The VSC uses forced-commutated power electronic devices (GTOs, IGBTs or IGCTs) to synthesize a voltage V_2 from a DC voltage source. The principle of operation of the STATCOM is explained on the figure below showing the active and reactive power transfer between a source V_1 and a source V_2. In Figure 6, V_1 represents the system voltage to be controlled and V_2 is the voltage generated by the VSC.

The control system consists of:

- A phase-locked loop (PLL) which synchronizes on the positive-sequence component of the three-phase primary voltage V_1. The output of the PLL (angle Θ=ωt) is used to compute the direct-axis and quadrature-ax-

Figure 6. Single-line diagram of a STATCOM and its control system block diagram

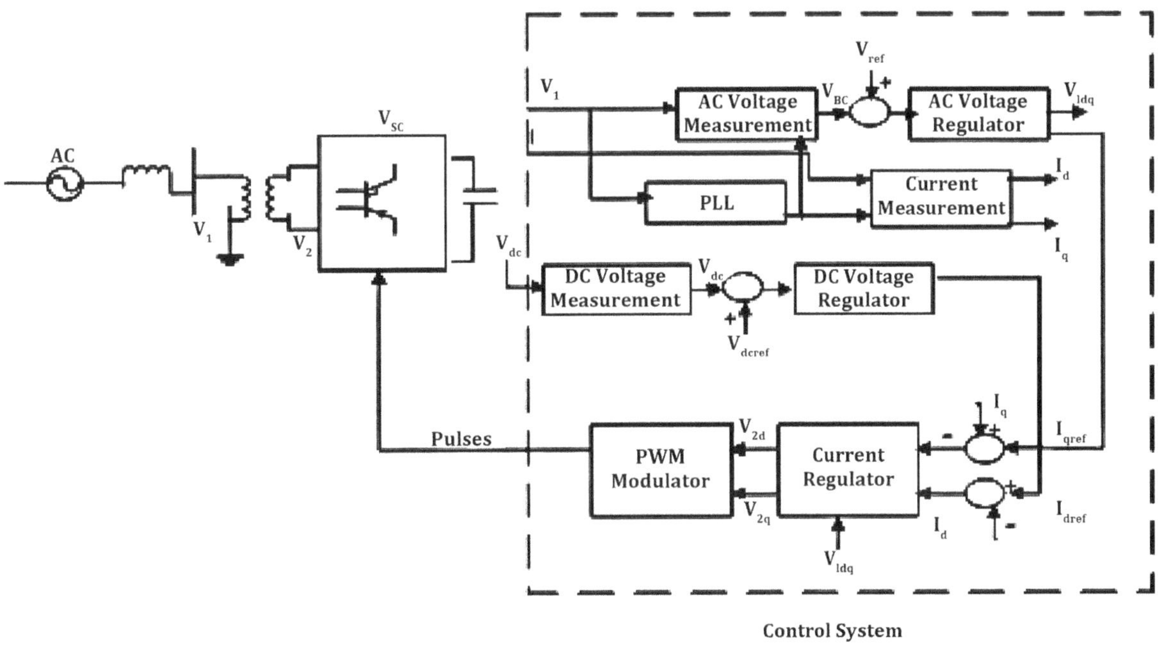

is components of the AC three-phase voltage and currents (labeled as V_d, V_q or I_d, I_q on the diagram).

- Measurement systems measuring the d and q components of AC positive-sequence voltage and currents to be controlled as well as the DC voltage Vdc.
- An outer regulation loop consisting of an AC voltage regulator and a DC voltage regulator. The output of the AC voltage regulator is the reference current I_{qref} for the current regulator (I_q = current in quadrature with voltage which controls reactive power flow). The output of the DC voltage regulator is the reference current I_{dref} for the current regulator (I_d = current in phase with voltage which controls active power flow).

An inner current regulation loop consisting of a current regulator. The current regulator controls the magnitude and phase of the voltage generated by the PWM converter (V_{2d} V_{2q}) from the Idref and Iqref reference currents produced respectively by the DC voltage regulator and the AC voltage regulator (in voltage control mode). The current regulator is assisted by a feed forward type regulator which predicts the V_2 voltage output (V_{2d} V_{2q}) from the V_1 measurement (V_{1d} V_{1q}) and the transformer leakage reactance. The reactive current injection capability of STATCOMs is illustrated in Figure 7.

3. STATIC SERIES COMPENSATORS: SSSC AND TCSC

3.1 Static Synchronous Series Compensator (SSSC-Phasor Type)

The Static Synchronous Series Compensator (SSSC) is a series device of the Flexible AC Transmission Systems (FACTS) family using power electronics to control power flow and improve power oscillation damping on power grids (Figure 8). The SSSC injects a voltage Vs in series with the transmission line where it is connected (Haque, 2006).

SSSC is one of the most important FACTS controllers. It is like a STATCOM except that the output ac voltage is in series with the line. Usually the injected voltage in series would be quite small

Figure 7. Reactive current capability of STATCOM

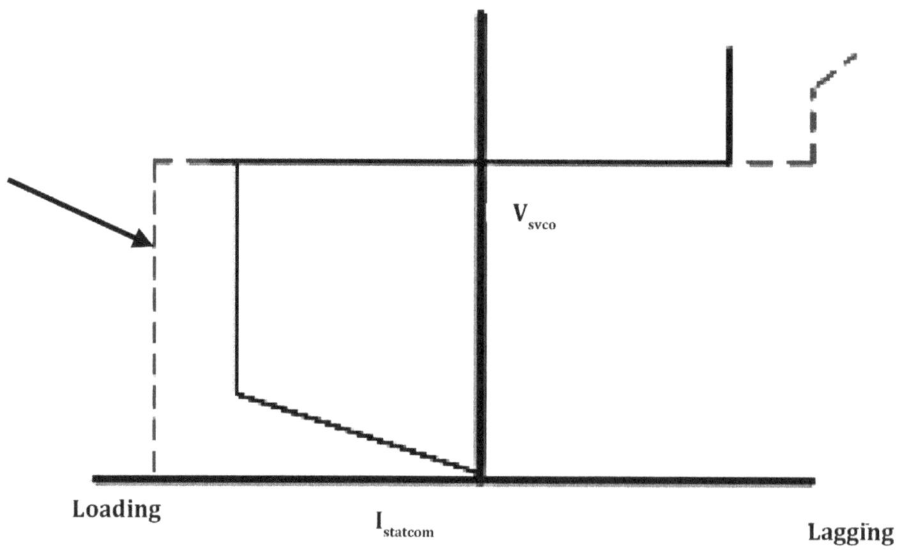

Figure 8. Single line diagram of static synchronous series compensator

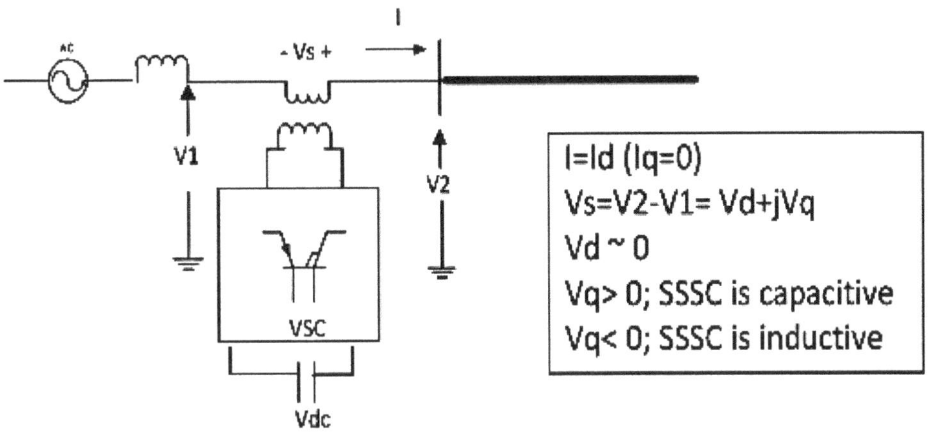

compared to the line voltage, and the insulation to ground would be quite high. The transformer ratio is tailored to the most economical converter design. Without an extra energy source, SSSC can only inject a variable voltage, which is 90 degree leading or lagging the current. The primary of the transformer and hence the secondary as well as converter has to carry full line current unless the converter is temporarily bypassed during severe line faults.

As the SSSC does not use any active power source, the injected voltage must stay in quadrature with line current. By varying the magnitude Vq of the injected voltage in quadrature with current, the SSSC performs the function of a variable reactance compensator, either capacitive or inductive (Figure 9).

The variation of injected voltage is performed by means of a Voltage-Sourced Converter (VSC) connected on the secondary side of a coupling transformer. The VSC uses forced-commutated power electronic devices (GTOs, IGBTs or IGCTs) to synthesize a voltage V_{conv} from a DC voltage source.

A capacitor connected on the DC side of the VSC acts as a DC voltage source. A small active power is drawn from the line to keep the capacitor charged and to provide transformer and VSC losses, so that the injected voltage Vs is practically 90 degrees out of phase with current I. In the control system block diagram V_d_conv and V_{q_conv} designate the components of converter voltage V_conv which are respectively in phase and in quadrature with current.

Two VSC technologies can be used:

- VSC using GTO-based square-wave inverters and special interconnection transformers. Typically four three-level inverters are used to build a 48-step voltage waveform. Special interconnection transformers are used to neutralize harmonics contained in the square waves generated by individual inverters. In this type of VSC, the fundamental component of voltage V is proportional to the voltage V_{dc}. Therefore V_{dc} has to varied for controlling the injected voltage.
- VSC using IGBT-based PWM inverters. This type of inverter uses Pulse-Width Modulation (PWM) technique to synthesize a sinusoidal waveform from a DC voltage with a typical chopping frequency of a few kilohertz. Harmonics are cancelled by connecting filters at the AC side of the VSC. This type of VSC uses a fixed DC

Figure 9. Control system block diagram of SSSC

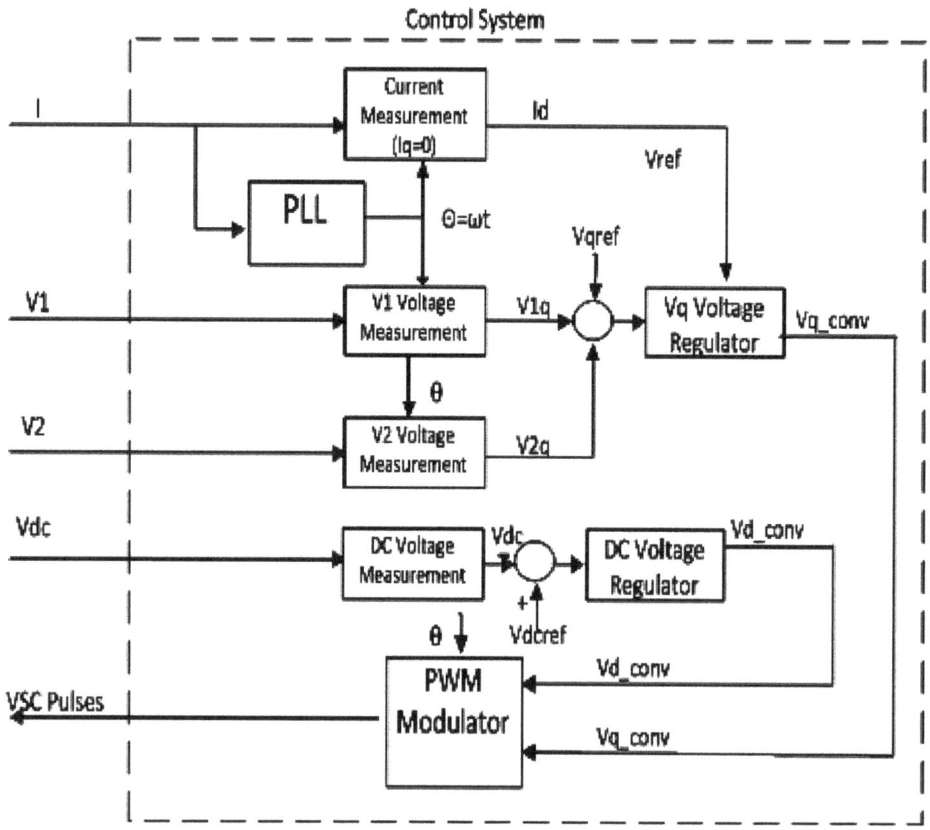

voltage V_{dc}. Voltage V_conv is varied by changing the modulation index of the PWM modulator.

The SSSC (Phasor Type) block models an IGBT-based SSSC (fixed DC voltage). However, as details of the inverter and harmonics are not represented, it can be also used to model a GTO-based SSSC in transient stability studies.

The control system consists of:

- A phase-locked loop (PLL) which synchronizes on the positive-sequence component of the current I. The output of the PLL (angle Θ=ωt) is used to compute the direct-axis and quadrature-axis components of the AC three-phase voltages and currents (labeled as V_d, Vq or I_d, Iq on the diagram).
- Measurement systems measuring the q components of AC positive-sequence of voltages V_1 and V_2 (V_{1q} and V_{2q}) as well as the DC voltage V_{dc}.
- AC and DC voltage regulators which compute the two components of the converter voltage (V_d_conv and V_q_conv) required to obtain the desired DC voltage (V_{dcref}) and the injected voltage (V_{qref}). The Vq voltage regulator is assisted by a feed forward type regulator which predicts the V_conv voltage from the Id current measurement.

As the SSSC does not use any active power source, the injected voltage must stay in quadrature with line current. By varying the magnitude V_q of the injected voltage in quadrature with current, the SSSC performs the function of a variable reactance compensator, either capacitive or inductive.

The basic scheme of SSSC is shown in figure. The compensator is equipped with a source of energy, which helps in supplying or absorbing active to or from the transmission line along with the control of reactive power flow.

SSSC, as shown in Figure 10, is similar to the variable reactance because the injected voltage and current to the circuit by this device are changing depend upon to the system conditions and the loads entering/getting out. For responding to the dynamic and transient changes created in system, SSSC utilizes the series converter.

One side of the converter is connected to the AC system and the other side is connected to a capacitor and battery which in the system we assume DC source as battery. If a dynamic change in system will be occurred, SSSC circuit works such that according to the control circuit in Figure10 the energy of battery will be converted to the ac form by converter and then injecting this voltage to the circuit the changes will be damped appropriately. To control the active and reactive powers of bus-2, the control circuit as shown in Figure 10 is utilized. For controlling the powers, first, sampling from the voltage and current is done and transformed to the dq_0 values. Active and reactive powers of bus-2 are calculated using their voltage and current in dq_0 references and compared with the determined reference and the produced error signal is given to the PI controllers. Adjusting parameters of the PI controllers, we are trying to achieve the zero signal error, such that powers can follow the reference powers precisely. Then, the output of the controllers are transformed to the abc reference and given to the PWM.

The controlled method of maintaining a quadrature relationship between the instantaneous convertor voltage and line current vectors, to provide reactive series compensation and handle SSR, can be implemented with an indirectly controlled convertor. The method of maintaining a single frequency synchronous output independent of dc terminal voltage variation requires a directly controlled convertor. Although high power directly controlled convertors are more difficult and costly to implement then indirectly controlled convertors (because their greater control flexibility is usually associated with some penalty in terms of increased losses, greater circuit complexity, and increased harmonic content in the output),

Figure 10. Configuration static synchronous series capacitor

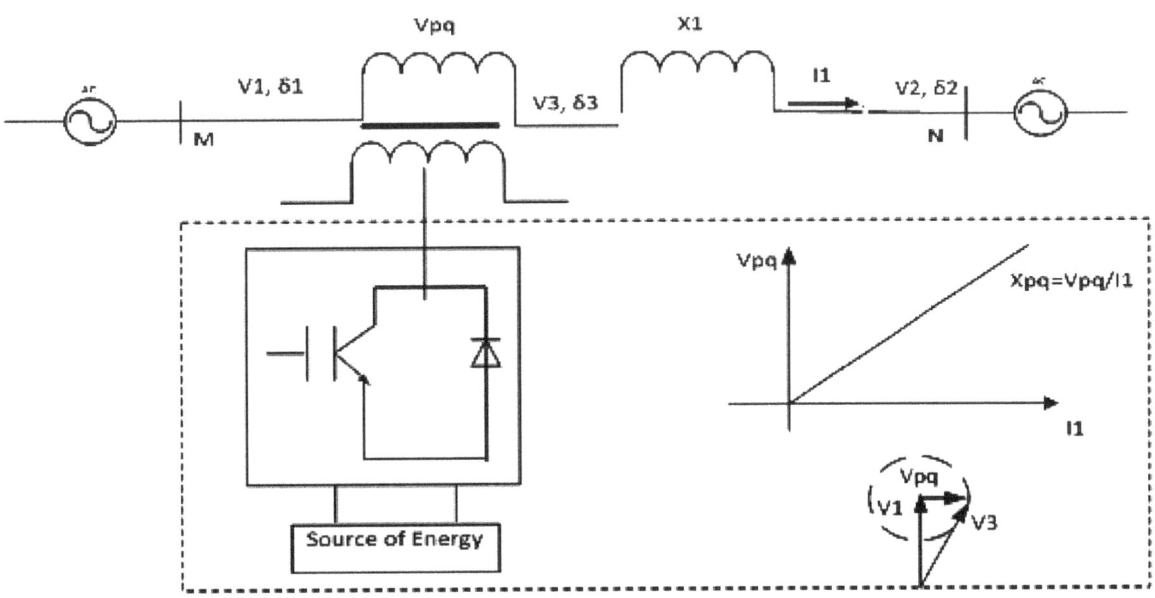

nevertheless the can be realized to meet practical utility requirements and indeed they are used along with indirectly controlled convertors, in existing installations for transmission compensation.

The input to the internal control are: the line current i, the injected compensating voltage v_q, and the reference V_q. The control is synchronized to the line current by a phase log loop which, after a $+\pi/2$ or $-\pi/2$ phase shift, provides the basic synchronizing signal θ. The phase shifter is operated from the output of a polarity detector which determines whether reference V_q is positive (capacitive) or negative (inductive). The compensating voltage v_q is controlled by a simple closed loop, the absolute value of reference V_q is compared to the measured magnitude of the injected voltage v_q and the amplified difference (error) is added as correction in the angle to the synchronizing signal θ. Depending on the polarity of angle θ, and consequently the convertor gate drive signals, will be advanced or retarded and thereby the compensating voltage v_q will be shifted with respect to the prevailing line current from its original $+\pi/2$ or $-\pi/2$ phase position. This phase shift will cause the convertor to absorb the real power from the ac system for the dc capacitor or vice-versa, supply that to the ac system from the dc capacitor. As a result, the voltage of the dc capacitor will increase or decrease, causing a corresponding change in the magnitude of the compensating voltage. Once the desired magnitude of v_q is reached (normally within 1 or 2 milliseconds), the substantially quadrature relationship between the line current and the compensating voltage gets reestablished with only remaining small, steady angular difference necessary to absorb power from the ac system to replenish the operating losses of the convertor.

3.2 Thyristor Controlled Series Capacitor (TCSC) Model

The basic thyristor-controlled-series capacitor scheme, proposed in 1986 by Vithayathil with others as a method of "rapid adjustment of network impedance". It consists of the series compensating capacitor shunted by a thyristor-controlled reactor. In a practical TCSC implementation, several such basic compensators may be connected in series to obtain the desired voltage rating and operating characteristcs. The TCSC presents a tunable parallel LC circuit to the line current that is substantially a constant alternating current source.

A typical TCSC module consists of fixed series capacitor bank (*C*) and a parallel branch with a surge reactor (*L*) and thyristor valves (*SW*), as shown in Figure 11. The reactor (*L*) in series with a bidirectional thyristor valves forms the thyristor-controlled reactor (TCR) that is fired with a phase angle θ ranging between 90° and 180° with respect to the capacitor voltage.

Figure 11. Series compensation unit

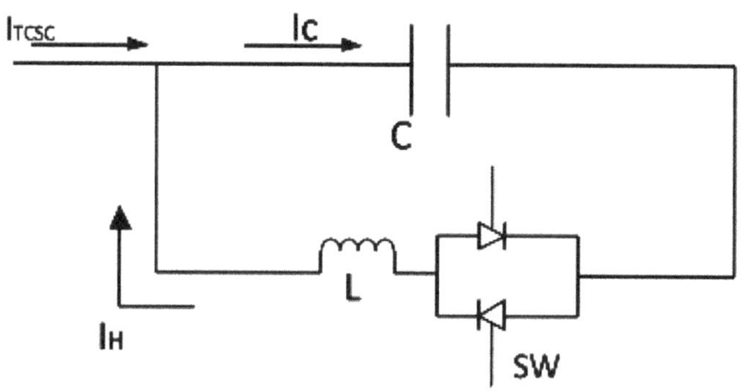

When the thyristors are blocked, the TCSC acts as a conventional series compensator with a fixed reactance X_C of interest is the TCSC vernier operation, when the thyristors are conducting, and the device can vary its net reactance as allowed by capacitor and thyristor ratings. In the vernier mode, the thyristor firing angle controls the loop current in the device. In the capacitive vernier mode, the circulating loop current adds to the line current and produces a higher capacitor current. This increases voltage drop across the capacitor, and thus increases the net series compensation as seen from the line. In the inductive vernier mode, the capacitor current flows in the direction opposite to the line current, the net TCSC reactance is inductive.

TCSC Control Scheme is explained in the following.

3.1.1. TCSC Controller

A conventional stability control scheme for TCSC is shown in Figure 12 It includes two main operational blocks as an external control and an internal control.

The function of the external control is to operate the controller to accomplish specified compensation objectives. Measured systems variables are used to define the reference for the internal control, which is usually the value of the controller reactance. The function of the internal control is to provide appropriate gate drive signals for the thyristor valves to produce the desired compensating reactance. Thus, the external control is the one that defines the functional operation of the controller. The reference X_{pf} value may also be set directly by manual control in response to an order from an energy-control centre, and it essentially represents the initial operating point of the TCSC. In Figure 12, X_m is the stability control modulation reactance value from the stability or dynamic control loop. The sum of these two values yields *Xm,* which is the final value of the reactance ordered by the external control block. The desired magnitude of the TCSC reactance, X_e, is obtained after a finite delay caused by the firing controls and the natural response of the TCSC. This delay is modelled by a lag circuit having a time constant, *T*, of typically 15-20 ms. The output of the first order lag is subject to variable limits based on the

Figure 12. TCSC model for stability studies

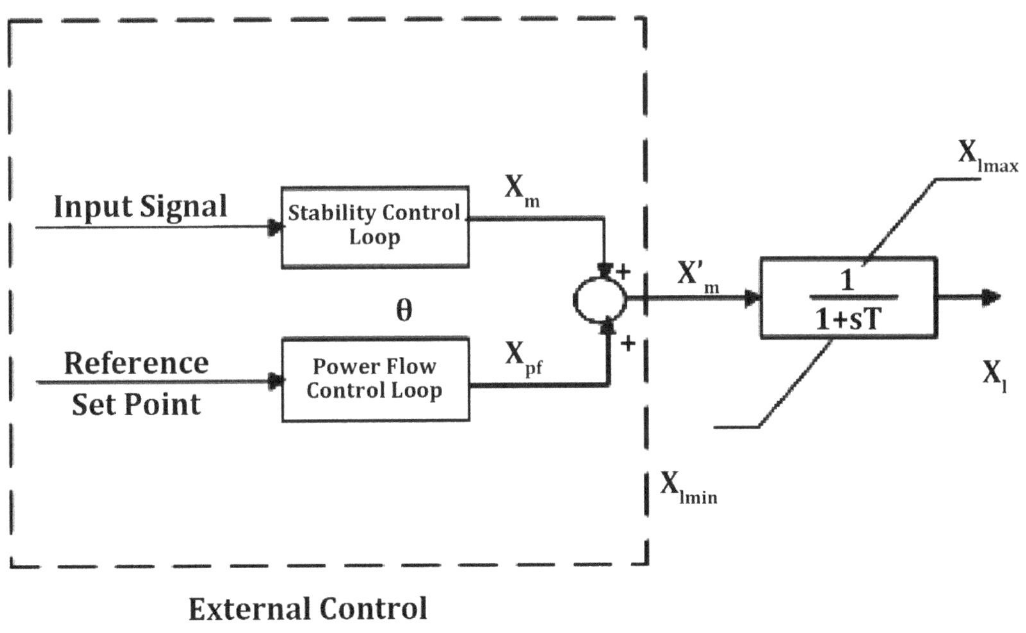

TCSC reactance-capability characteristic. The stability control loop is a pre-programmed open-loop control which is used to improve transient (or first swing) stability.

When the system is subjected to a severe transient disturbance such as a fault on transmission facilities, loss of generation, or loss of a large load, the stability control loop acts initially, leading to the maximum compensation level during a certain time, so that the synchronizing torque is increased, improving the first-swing stability response of the system. The control then provides proper modulation to damp the subsequent power oscillations. There are many studies on the damping control of TCSC based on a transfer function based linear control with a local input signal. However, this linear control is usually designed for a specific operating point and condition. They are effective at damping the oscillation under the design conditions. The general structure of the stability controller is depicted in Figure 13. It consists of a washout filter, a dynamic compensator, and a limiter. The washout filter is mainly provided to avoid the controller response to the dc offset of the input signal. The dynamic compensator consists of one or more lead-lag blocks to provide the necessary phase-lag characteristics. Finally, the limiter is used to improve controller response to large deviations in the input signal.

The input signals that can be employed for modulating TCSC reactance include the local and remote signals. The selection of the appropriate input is a fundamental issue in the design of an effective and robust controller. The local signals constitute the following: line current, active-power flow, bus voltage, and local bus frequency. In the local signals, the line current and the active power flow are found to be almost equally effective. Local bus frequency is less responsive to system power swings compared to the synthesized-voltage frequency. Alternatively, the possible remote signals constitute the following: rotor-angle/speed deviation of any remote generator, rotor angle/speed (frequency) difference across the system, and any bus voltage or phase and real-power flow on adjacent lines. Of course, every additional remote input signal used needs additional measurement devices and communication. Based on the advantages of local input signals, most controlled practical installations of TCSCs uses local measurements for controlling power flows and damping power oscillations. But TCSCs with only local input signals do not ensure optimal stability enhancement because not all oscillation modes can be correctly identified by local measurements. Especially changing grid topologies may have an adverse effect on controller performance.

Figure 13. Block diagram of conventional stability control of TCSC

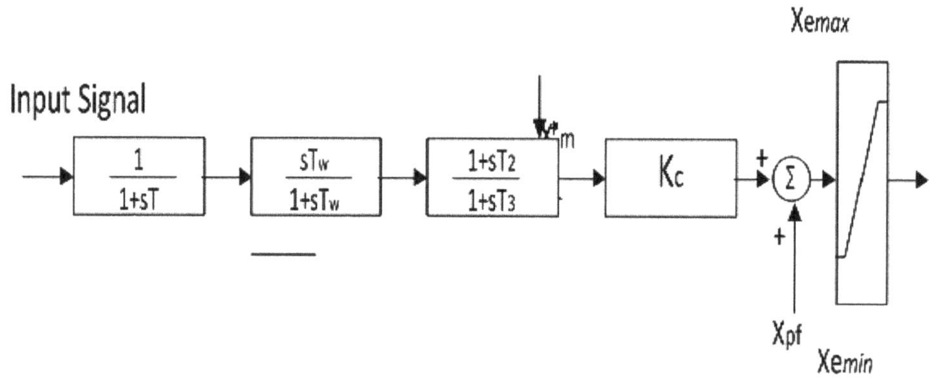

3.1.2. Damping Control Signal of TCSC

The power system representation in Figure 14 is used to model an interconnected system consisting of two areas. Each area is represented by an aggregate machine. Two interconnected systems are represented by their Thevenins equivalent generator and reactance, with the generators represented by the classical model: a constant emf behind the transient reactance and all resistances neglected. The system representation allows gaining physical insight into the effectiveness of enhancing stability through TCSC control. It is assumed that a TCSC is located on the intertie.

With stator resistance neglected, the air-gap power (*Pe*) is equal to the terminal power. In per unit quantities, the air-gap torque is equal to the air-gap power. Hence,

$$T_{e2} = P_2 = \frac{E_1^{'}E_2^{'}}{X_T + X_{TCSC}} \sin\left(\delta_1 - \delta_2\right) \quad (6)$$

$$X_T = X'_{d1} + X_1 + X'_{d2} + X_2 \quad (7)$$

where Linearizing Equations 6 and 7 about an initial operating condition represented by $\delta_1=\delta_{01}$, $\delta_2=\delta_{02}$, $X_{TCSC} = X_0$, yields

$$\Delta T_{e1} = K_s(\Delta\delta_1 - \Delta\delta_2) - K_{TCSC}\Delta X_{TCSC} \quad (8)$$

$$\Delta T_{e2} = K_s(\Delta\delta_1 - \Delta\delta_2) + K_{TCSC}\Delta X_{TCSC} \quad (9)$$

Where the synchronizing torque coefficient is:

$$K_s = \frac{E_1^{'}E_2^{'}}{X_T + X_0} \cos\left(\delta_{01} - \delta_{02}\right)$$

The power damping coefficient of the TCSC is:

$$K_{TCSC} = \frac{E_1^{'}E_2^{'}}{(X_T + X_0)^2} \sin\left(\delta_{01} - \delta_{02}\right)$$

The equations of motion of the generators in per unit are

$$\Delta\omega_1 = \frac{1}{2H_1}\left(T_{m1} - T_{e1} - K_{D1}\Delta\omega_1\right) \quad (10)$$

$$\Delta\omega_2 = \frac{1}{2H_2}\left(T_{m2} - T_{e2} - K_{D2}\Delta\omega_2\right) \quad (11)$$

where

δ_1, δ_2 = Rotor angles corresponding to machine 1 and 2;
ω_1, ω_2 = Rotor electrical speeds;
$\Delta\delta_1$, $\Delta\delta_2$ = Rotor angle deviations;
$\Delta\omega_1$, $\Delta\omega_2$ = Rotor speed deviations;
H_1, H_2 = inertia constants;

Figure 14. System for inter-area power swing study

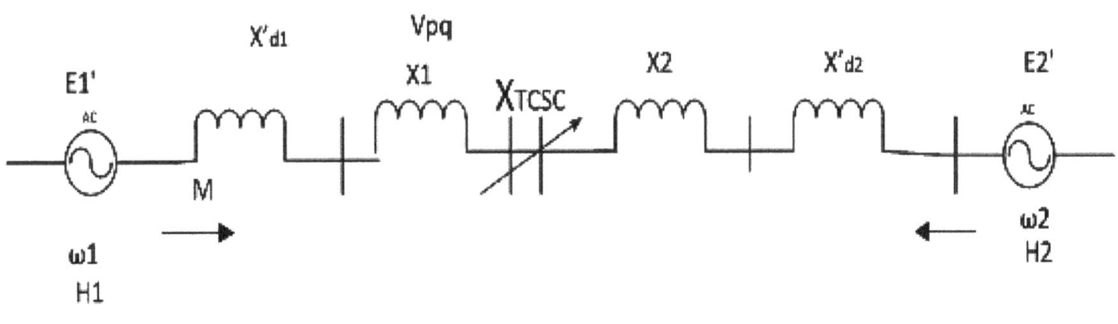

K_{D1}, K_{D2} = damping factors;
$E_1^{'}$, $E_2^{'}$ = transient q-axes voltages;
$X_{d1}^{'}$, $X_{d2}^{'}$ = d-axis transient reactances; and
X_{TCSC} = effective reactance of the TCSC.

Linearizing Equations (8), (9) and substituting for ΔT_{e1}, ΔT_{e2} by Equations (8), (9) we obtain

$$\Delta\omega_1 = \frac{1}{2H_1}\left\{\begin{array}{l}\Delta T_{m1} - K_S\left(\Delta\delta_1 - \Delta\delta_2\right)\\ + K_{TCSC}\Delta X_{TCSC} - K_{D1}\Delta\omega_1\end{array}\right\} \tag{12}$$

$$\Delta\omega_2 = \frac{1}{2H_2}\left\{\begin{array}{l}\Delta T_{m2} + K_S\left(\Delta\delta_1 - \Delta\delta_2\right)\\ - K_{TCSC}\Delta X_{TCSC} - K_{D2}\Delta\omega_2\end{array}\right\} \tag{13}$$

Equations (12) and (13) represent an additional damping torque by the TCSC in the two-machine system, i.e. $K_{TCSC}\ \Delta X_{TCSC}$. The initial line power is transferred from machine one to machine two and the initial relative angle $\Delta\delta_0=(\Delta\delta_{01}-\Delta\delta_{02})$ is positive. During a power swing, the machines oscillate at a relative speed $\Delta\omega=(\Delta\omega_1-\Delta\omega_2)$ If the receiving end machine speed is lower than the sending end-machine speed, i.e. $\Delta\omega$ is positive, the rotor angle difference $\Delta\delta=(\Delta\delta_1-\Delta\delta_2)$ is positive and machine one decelerates if the TCSC damping torque is negative. Since K_{TCSC} is positive, this means that ΔX_{TCSC} is negative. It also shows that machine two accelerates in this condition. In other words, while the sending end-machine accelerates, the TCSC control should attempt to draw more power from this machine, to create a reduction of the kinetic energy responsible for its acceleration. When $\Delta\omega$ is negative, the TCSC must decrease the power flow in the line, i.e. ΔX_{TCSC} becomes positive. The above analysis brings out an effective control strategy in damping system oscillations using the TCSC. The control signal is selected as the difference between the rotor angle deviations of the two machines $(\Delta\delta_2-\Delta\delta_1)$. For previous reasons, this signal can contribute directly to oscillation damping and enhance stability in power systems.

4. GENERAL NETWORK REPRESENTATION

The equations of a single branch in the network frame of reference are rotating at angular velocity, ω_f, with a series resistance R and inductance X/ω_0, with ω_0 being the nominal angular frequency, are given in the following:

$$\begin{bmatrix}\Delta v_D\\ \Delta v_Q\end{bmatrix} = \begin{bmatrix}R & \frac{-\omega_f}{\omega_0}X\\ \frac{\omega_f}{\omega_0}X & R\end{bmatrix}\begin{bmatrix}i_D\\ i_Q\end{bmatrix} + \frac{1}{\omega_0}\begin{bmatrix}X & \\ & X\end{bmatrix}\begin{bmatrix}pi_D\\ pi_Q\end{bmatrix} \tag{14}$$

In the time frame relevant to the voltage stability studies, electromechanical transients of generators outweigh the electromagnetic transients in the network. As a result, a widely-used basis for voltage/transient stability studies is to discount the electromagnetic transient terms in (14), which results in the network equation in the algebraic form as follows:

$$\begin{bmatrix}\Delta v_D\\ \Delta v_Q\end{bmatrix} = \begin{bmatrix}R & \frac{-\omega_f}{\omega_0}X\\ \frac{\omega_f}{\omega_0}X & R\end{bmatrix}\begin{bmatrix}i_D\\ i_Q\end{bmatrix} \tag{15}$$

Network comprising transmission lines, transformers, FACTS devices and static load demands can be represented by the nodal admittance matrix. The admittance form of (15) is:

$$\begin{bmatrix} i_D \\ i_Q \end{bmatrix} = \begin{bmatrix} G & B \\ -B & G \end{bmatrix} \begin{bmatrix} \Delta v_D \\ \Delta v_Q \end{bmatrix} \tag{16}$$

where

$$G = \frac{R}{R^2 + \left(\frac{\omega_f X}{\omega_0}\right)^2} \tag{17}$$

and

$$B = \frac{\frac{\omega_f X}{\omega_0}}{R^2 + \left(\frac{\omega_f X}{\omega_0}\right)^2} \tag{18}$$

If the angular frequency of the rotating frame of reference in which the networks equations are formed is equal to the nominal angular frequency of the system, i.e. the network frequency changes are discounted, then the ratio ω_f/ω_0 is unity and the reactance in the branch equations takes the nominal supply-frequency value. The nodal admittance equations are formed in the ordinary way. For the network systems as a whole, the network equations are assembled in the following vector/matrix form:

$$i_N = Y_{bus} v_N \tag{19}$$

where i_N, v_N are vectors of the components of nodal currents and voltages respectively and Y_{bus} is the network nodal admittance matrix expressed in real number form. If n is the number of network nodes, vector i_N and v_N are each of dimension $2n$, and Y_{bus} is a $2n\times2n$ matrix the element of which are formed from individual admittances of system elements.

As static loads are represented in the admittance matrix, nodal currents are only non- zero for the nodes which have a generator or induction motor. In power system transient studies, the idea is to model all of the dynamic elements as shunt current injectors to the network, as shown in Figure 15. However, when there is a series dynamic element in the network, including FACTS, some modifications are needed to the admittance matrix in every time-step of integration in the time-domain simulation.

In order to take advantage of scarcity of the admittance matrix, all of the network nodes are retained and generators' stator currents are transformed to the network frame of reference. Alternatively, Y_{bus} can be reduced to include only the generator nodes as follows:

$$\begin{bmatrix} i_{SN} \\ 0 \end{bmatrix} = \begin{bmatrix} Y_{SS} & Y_{SL} \\ Y_{LS} & Y_{LS} \end{bmatrix} \begin{bmatrix} v_{SN} \\ v_{LN} \end{bmatrix} \tag{20}$$

where S and L denote a set of generators or induction motors and load nodes respectively. Based on the Kron reduction, equation can be rearranged in the following way:

$$i_{SN} = Y_N v_{SN} \tag{21}$$

$$Y_N = Y_{SS} - [Y_{SL}(Y_{LL})^{-1} Y_{LS}] \tag{22}$$

5. VOLTAGE AND REACTIVE POWER CONTROL CONCEPT

Voltage changes continuously according to the varying electrical demand, transmission lines utilization, system control by the control centers, and emergency situations occurred in the system.

Figure 15. Multi-machine network representation

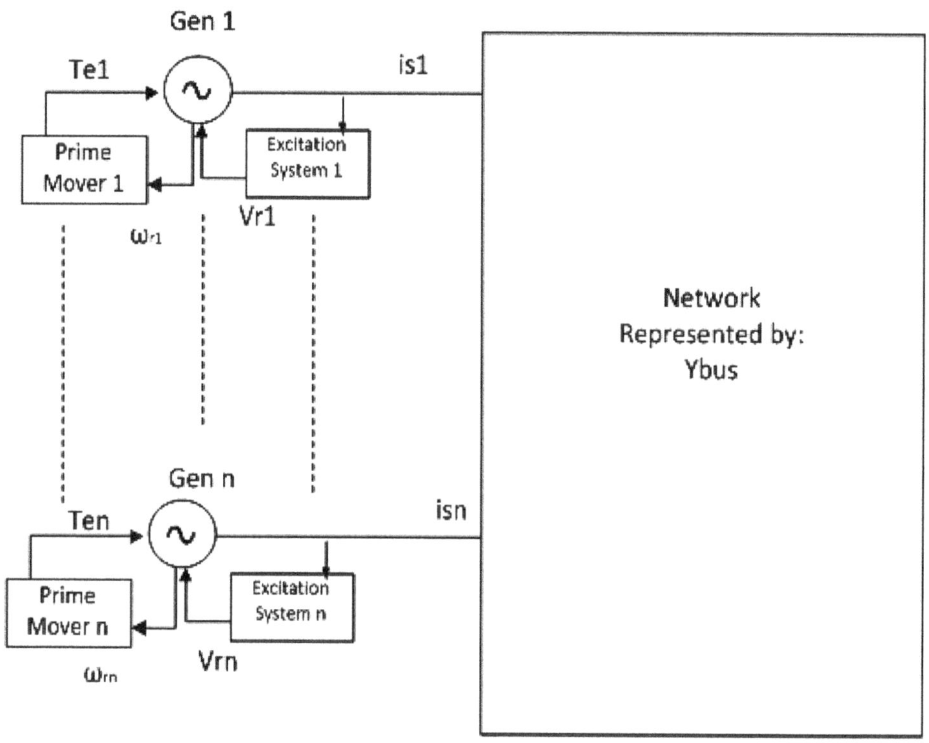

Since customers require voltage quality, at delivery points, to meet the agreed criteria, it is the control centers' responsibility to control the voltage so that it can satisfy the agreement.

Controlling the voltage is regional problems. In other words, the voltage controlling problems are needed to be solved separately by each control area. This can be achieved by providing sufficient reactive power sources for controlling voltage level as specified by the agreement. The voltage controlling problems can be divided into two situations, which are normal situation, and emergency situation. The reactive power sources are classified into two types, static and dynamic. The static type consists of shunt capacitor, and shunt reactor. The dynamic type is the source of reactive power produced by generators, FACTS (SVC, STATCOM etc.) and OLTC that operates automatically. The control centers need to organize and compile the constantly updated information concerning the availability of relevant equipment. Also the control centers have to provide procedures, in writing, required for properly controlling of the equipment in order to support normal situation, and emergency situation that may lead to voltage collapse and cascade tripping both of which can cause a blackout.

5.1 Purpose

- To control the voltage at delivery points according to the agreement without damaging customers' equipment.
- To control the power flow in the system to an optimal level in order to reduce losses.
- To control the reserve reactive power in order to ensure its sufficiency during normal and emergency conditions to prevent voltage collapse.
- To monitor the reactive power and to resolve problems in order to maintain system security so that the international electricity

purchasing can be completed according to agreements.

5.2 Requirements

- The control center of each country shall operate cooperatively to ensure that the policies and procedures are developed, maintained, and executed in order to monitor whether the control of real and reactive power in Thailand and in the neighbor countries are in accordance with the specification and having sufficient stability.
- The control center of each country shall provide, follow up, and examine reactive power absorption and injection sources to ensure the sufficiency of reactive power absorption and injection sources in their areas in order to maintain voltage levels, for all points, as specified in the standard during normal and emergency conditions.
- The control centers shall specify voltage levels or reactive power at interconnection points between the power system and the transmission system. The control centers shall notify the specification to generator operators. Normally, generators are operated in AVR mode.
- The control centers must know the conditions of all reactive power sources, both absorption and infection, in the system including the condition of AVR and PSS (if available) of all generators.
    a. When the control centers are notified that AVR cannot be operated in automatic mode, the control centers shall give orders to the generator operators in order to vary or maintain voltage or reactive power.
    b. In case where sources of reactive power supply are not available, the control centers shall have plans to remedy such the case.
- Generators must follow above requirements. However, if there are exceptions to be made for any generators, the control centers shall specify the conditions for the exceptions.
    a. The control centers shall have the list of generators, which are exempt from the reactive power or voltage requirement, within the specified responsible area.
    b. The generators, which are exempt from such requirements, are required to supply reactive power according to the following requirements:
        - The generators shall maintain voltages within 0.97-1.03 per unit at the delivery points.
        - The generators shall maintain power factor within ±0.85 per unit during maintaining voltage levels as specified above.
- The control centers shall have database containing information on reactive power sources in the responsible control area. The information consists of:
    - Reactive capability curves of generators;
    - Synchronous condensers;
    - Static VAR compensators;
    - Shunt reactor and shunt capacitors;
    - On load tap change transformers and their settings; and
    - Other reactive power sources such as transmission lines.
- The control centers shall know current MVAR values, maximum and minimum MVAR absorption/injection capabilities, and operating restriction (if applicable) of each equipment so that the levels of reserve reactive power are accurate for all time.
- Information concerning equipment, which can significantly affect neighbor power systems, shall be available online for each other, which depends on an agreement.

5.3 Procedures for Controlling Voltage and Reactive Power

- **Primary Voltage Control:** The control centers shall control primary voltage by providing specific voltage levels to generators according to the Requirement 4. The generators shall adjust the AVR which will vary the excitation of generating units in order to achieve the specified voltage levels. For other voltage control equipment such as SVCs or automatic tap changing transformers, they are considered to be a part of primary voltage control. All equipment that is used for the primary voltage control is considered to be dynamic.
- **Secondary Voltage Control:** The control centers shall employ voltage control mechanism by managing static reactive power sources, both absorption and injection, for examples, shunt reactors, capacitor bank, etc., all of which are within the considered zones in cooperate with the first step for voltage control in order to maintain specified voltage levels at essential delivery points that represent the zones. The voltage levels for different situations are given as:
    - During normal situation: 0.95 – 1.05 per unit
    - During emergency situation: 0.90 – 1.10 per unit in order to prevent damages for equipment in the system.

Power flows, both actual and potential, must be carefully controlled for a power system to operate within acceptable voltage limits. Reactive power flows can give rise to substantial voltage changes across the system, which means that it is necessary to maintain reactive power balances between sources of generation and points of demand on a 'zonal basis'. Unlike system frequency, which is consistent throughout an interconnected system, voltages experienced at points across the system form a "voltage profile" which is uniquely related to local generation and demand at that instant, and is also affected by the prevailing system network arrangements. National Grid is obliged to secure the transmission network to closely defined voltage and stability criteria. This is predominantly achieved through circuit arrangements, transformers and shunt or static compensation.

Voltage control in an electric power system is important for many reasons:

- All end-use equipment need near nominal voltage for their proper operation,
- Near nominal voltage results in near minimum transmission losses, and
- Near nominal voltages increase the ability of the system to withstand disturbances (security).

A reasonable voltage profile throughout an electric power system is associated with the ability of the system to transfer power from one location to another. When the voltage sags to low value this ability of the system is compromised. The onset of power transfer inability can be detected with sensitivity analysis of reactive power requirements versus real power load increases. This sensitivity is dependent on the characteristics of the electric load. Such sensitivity analyses have been performed using various electric load models, i.e. constant power load, constant impedance load, or combination of the two (voltage dependent load). The majority of electric loads are induction motors.

Power system operators ensure the quality and reliability of supply to the customers by maintaining the load bus voltages in their permissible limits. Any changes to the system configuration or in power demands can result in higher or lower voltages in the system. This situation can be improved by the operator by reallocating reactive power generations in the system, i.e., by adjusting transformer taps, changing generator voltages, and by switching VAR sources. Also, it is possible to minimize the system losses by reactive power redistributions in the system.

Thus, it is a twofold objective function:

1. To improve the voltage profiles.
2. To minimize the system losses.

Reactive power and voltage control is one of the ancillary services used to maintain the voltage profile through injecting or absorbing reactive power. Reactive power services play an important role as follows:

- Satisfy the requirement of reactive power load.
- Control bus voltage in a system wide.
- Decrease the network loss.
- Relieve the transmission block.
- Provide sufficient reserve to ensure the security of system in emergency.

Reactive power support is regarded as the obligation of generators or grid owners. The system operators use a reactive power optimization program in which real power loss is considered. In this situation, the reactive power suppliers and consumers cannot make their own decision on how to use reactive power efficiently.

Reactive power support may be provided by a variety of devices, including generators, synchronous condensers, shunt capacitors/reactors and static VAr compensators (SVCs). In the deregulated environment, these devices belong to different entities that have their own costs and benefits. There should be means to evaluate the reactive power resources and pricing it appropriately. The costs of the suppliers should be considered. A framework needs to be established to procure the service and remunerate the suppliers. The following part of this chapter will focus on economic analysis of reactive power service.

Voltage control and reactive power support are essential services for enabling the delivery of power in the transmission grid and for providing adequate reserves margins and appropriate dynamic response capabilities. While various equipment is able to produce and/or absorb reactive power, we focus our attention on the reactive support provided by synchronous generators. Such support depends principally on the set points assigned to the primary and, in some cases, to the secondary and tertiary voltage control systems (Haque, 2006). These settings are under the discretion of the Independent Grid Operator (IGO)1 and are typically specified by criteria formulated in an optimal power flow (*OPF*) framework. In fact, the definition of ancillary services in the U.S. regulatory structure requires reactive support and voltage control to be provided by generators, or possibly, loads (Padiyar & Devi, 1994).

Availability of a sufficient amount of reactive power support depends on the operational limits specified by the capability curves of the generation units. The ability of a generation unit to provide reactive support at a specified bus of the system depends on the "static" reactive power margin at the given operating point, which is a function of the distance to the boundary of the unit capability curve and on the capability of transporting this reactive power margin via the transmission network to the specified bus.

Power system reactive power planning is a powerful measure to maintain reactive power balance and reduce real power loss. There is tight relationship between voltage stability and reactive power planning. Therefore the reactive power planning in consideration of voltage stability is presented which can improve both of power system voltage stability and distribution of reactive power flow. However, the calculation of voltage stability limit is too complicated in those methods so that the application for bulk power system is difficult.

6. SQUIRREL CAGE INDUCTION GENERATORS

The ever increasing demand for energy, the depletion of conventional energy resources and the degradation of environmental conditions throughout the world have led governments/ scientists/researchers to explore renewable or nonconventional energy sources (e.g. wind, hydro, solar, bio-energy and micro-generation) in recent two to three decades. Renewable energy is the key to our low carbon energy future. In early 2008, European Union (EU) Commission aimed to achieve a 2020 target of deriving 20% of the EU's energy consumption from renewable sources, while the UK government proposed to achieving 15% renewable target by 2020.

The use of an induction machine as a generator is becoming more and more popular for renewable energy applications. Squirrel cage induction generators with excitation capacitors (known as SEIGs) are popular in isolated nonconventional energy systems.

As discussed in the previous chapter, the main limitation of the SEIG system is the poor voltage and frequency regulation when supplying variable loads connected to the stator terminals. However, the development of static power converters has facilitated the control of the output voltage and frequency of the induction generator.

6.1 SEIG System Configuration

The SEIG system is composed of four main items: the prime mover, the induction machine, the load and the self-excitation capacitor bank. The general layout of the SEIG system is shown in Figure 16. The hydro/wind turbine is assumed to operate with constant input power transferred to the induction generator. The real power required by the load is supplied by the induction generator by extracting power from the prime mover (turbine). When the speed of the turbine is not regulated, both the speed and shaft torque vary with variations in the power demanded by the loads. The self-excitation capacitors connected at the stator terminals of the induction machine must produce sufficient reactive power to supply the needs of the load and the induction generator.

A squirrel cage induction generator (SEIG) is more attractive than a conventional synchronous generator in this type of application because of its low unit cost, absence of DC excitation source, brushless cage rotor construction and lower maintenance requirement. A suitably sized three-phase capacitor bank connected at the generator terminals is used as variable lagging VAr source to meet the excitation demand of the cage machine and the load. The machine operated in this mode is known as a *Self-excited Induction Generator* (SEIG). However, the main drawback of the standalone SEIG is its poor voltage and frequency regulations under variable loads. A change in the load impedance directly affects

Figure 16. Schematic diagram of a standalone self-excited induction generator with constant prime mover

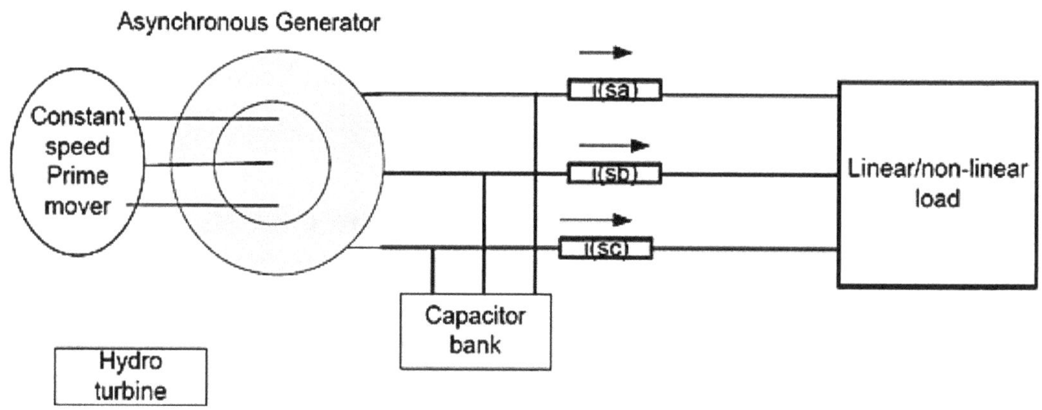

the excitation of the machine because the reactive power of the excitation capacitors is shared by both the machine and the load. Therefore, the generating voltage drops when the impedance of the load is increased resulting in poor voltage regulation. Poor frequency regulation occurs (an increase in the slip of the induction machine) when the load is increased.

Owing to increased emphasis on renewable resources (such as small hydro, wind, solar, tidal and wave) and their variable speed nature, the performance of synchronous generators can be greatly affected. Hence, SEIGs have emerged as suitable candidates for isolated power sources to convert power at the shaft of wind/hydro turbines into electrical energy.

6.2 The Self-Excitation Phenomenon

The self-excitation phenomenon of an induction machine is still under considerable attention although it is known for more than a half century. When a standalone induction machine is driven by a mechanical prime mover, the residual magnetism in the rotor of the machine induces an EMF in the stator windings at a frequency proportional to the rotor speed. This EMF is applied to the capacitors connected to the stator terminals and causes reactive current to flow in the stator windings. Hence a magnetizing flux in the machine is established. The final value of the stator voltage is limited by the magnetic saturation within the machine. The induction machine is then capable of operating as a generator in isolated locations without a grid supply.

Once the machine is self-excited and loaded, the magnitude of the steady-state voltage generated by the SEIG is determined by the nonlinearity of the magnetizing curves, the value of the self-excitation capacitance, speed, machine parameters and terminal loads. As the load and speed of the SEIG changes, the demand for lagging VArs to maintain a constant AC voltage across the machine terminals also changes. Arrillaga and Watson predicted the effect of a resistive load on the frequency of the generator where a new operating point was obtained by shifting the saturation curve and capacitive load in the terminal voltage-current characteristics. The effect of an inductive load was also studied and experimentally verified in.

6.3 SEIG System Performance

The performance characteristics of the SEIG system depend mainly on the Following:

- **The parameters of the induction machine:** The machine operating voltage, rated power, power factor, rotor speed and operating temperature and the induction machine parameters directly affect the performance of the SEIG system.
- **The Self-excitation process:** The connection of a capacitor bank across the induction machine stator terminals is necessary in the case of standalone operation of the system. The capacitor connection scheme (delta or star) and the use of fixed or controlled self-excitation capacitors have a direct impact on the performance of a SEIG system.
- **Load parameters:** The power factor, starting/maximum torque and current, generated harmonics and load type also affect the performance of the SEIG system directly.
- **Type of prime mover:** Whether the primary source is hydro, wind biomass or combinations and will affect the performance of the SEIG system is affected.

6.4 Operational Problems of the SEIG System

The main operational problem of the SEIG system is its poor voltage and frequency regulation under varying load conditions. A change in the load impedance directly affects the machine excitation. This is because the reactive power of the excitation capacitors is shared by both the induction machine and the load impedance. Therefore, the generator's voltage drops when the load impedance is increased resulting in poor voltage regulation. On the other hand, the slip of the induction generator increases with increasing load, resulting in a load dependent frequency, even if the speed of the prime mover remains constant.

Many studies have been conducted in the past to regulate the voltage and frequency of a SEIG system operating with variable loads. A high cost speed governor is generally used as a conventional SEIG controller.

A strategy of controlling voltage and frequency of a SEIG system was presented by Suarez et al. in 1999. Sliding mode controller was proposed showing controlled dynamic response and behavior of the system upon changes in generator parameters and load. However, an accurate transient model of representing instantaneous phase angle between stator voltage and external inductor current at switching instant is highly recommended in order to achieve optimal control.

Regulating voltage and frequency of a SEIG under varying load conditions by an electronicload controller (ELC) was examined by Singh et al. in 2006. The ELC consists of a chop circuit and a rectifier which produces harmonics during operation. AC current harmonics generated by ELC during operation and their impact on the performance of the SEIG were studied in.

In most of these studies, the control of the terminal voltage is generally performed by controlling a variable reactive power source and frequency regulation is achieved by regulating the speed of the prime mover by utilizing a mechanical speed governor. However, the regulation of speed and voltage does not result in a satisfactory level of performance due to the highly dynamic changes in slip of the machine and the difficulties in building a smooth variable reactive power source at low costs.

In order to contribute to a solution of the above operational problems, this thesis presents an investigation of the steady-state nonlinear behavior of the SEIG system when feeding variable inductive loads. The analysis of this highly dynamic nonlinear system reveals a variety of unique behaviors which have not been previously reported.

6.5 Types of Prime Mover

6.5.1 Hydro Energy Systems

The purpose of a hydroelectric plant is to harness power from water flowing under pressure .As such it incorporates a number of water driven prime mover known as Water Turbines Water flow under pressure having two form of energy –kinetic and potential. The kinetic energy depends on the mass of the water flowing and its velocity while the potential energy exist as a results of the difference of the water level between two points which is known as Head. The water turbine, as it is sometimes named, converts the kinetic and potential energies possessed by water in to mechanical power .The hydraulic turbine is thus a prime mover which when coupled to a generator produces electric power .The projects designed to produce electric power from water are known as 'Hydro Electric Projects'. Hydroelectric project may not be used exclusively for power generation. Sometimes, they are the off shoot of flood control and irrigation projects in which case they are known as "Multipurpose Projects", for example, bhakra Nangal and Damodar Valley Corporation projects.

Hydro power plant can be developed wherever; water continuously flowing under pressure is available. Dams constructed across flowing rivers divert the river in bounty through the turbines giving rise to such useful power. But this is not all. Water that collects in natural or artificial lakes in to high and huge mountains, due to the melting of snow and due to heavy monsoon can be led down to the turbines turbine through large pipes known as penstocks.

6.5.1.1 Essential Component of Hydro Electric Power Plant

A. Storage Reservoir

The water available from catchment area is stored in a reservoir so that it can be utilized to run the turbine for Natural reservoir or lake may be found in high and huge mountains. Water is taken off from one end of the reservoir through a tunnel built by cutting the mountain.

B. Dam with Its Control Works

Dam is a structure erected on suited site to provide for the storage of water. Dam may be built to make an artificial reservoir from a valley. Dam can be different type it can earth dam, masonry dam, wooden dam.

C. Water Ways with Their Control Works

Waterway is passage through which water is carried from the storage reservoir to the power house. It may consists of tunnel, canals, flumes, forebays, Pipes. Its function is to store temporarily, the water rejected by the plants when its loads is reduced and to meet the instantaneous increased demand of water due to sudden increase in load.

D. Power House

It is building to house the turbines, generators and other accessories for operations of machines.

E. Tail Race

It's the waterway to conduct the water discharge d from the turbines to a suitable point where it can be safely disposed of or stored to be pumped back into original reservoir.

F. Generation and Transmission of Electric Power

It consists of electrical generating machines and transformers, switching equipment and transmission lines.

6.5.1.2 Types of Hydro Turbine

A. Impulse Turbine

It's operated under a high head of water and therefore, requires comparatively less quantity of water. It is also called as free jet turbines. The water is conveyed from the reservoir in the mountains to the turbines in the power house through penstocks. The penstocks are joined to a branch pipe lower bend fitted with nozzle at the end. Water come out of nozzle in the form of a free and compact jet. The number of nozzles required for a turbine is depends on the specific speed. All the pressure water is converted in to velocity head. The impact of the water on the surface of the bucket produces a force which causes wheel to rotate thus supplying the mechanical torque on the shaft. This mechanical power is used to produce electrical power using generator.

B. Reaction Turbine

This turbine is an inward mixed flow reaction turbine i.e. the water .under pressure, enters the runner from the guide vanes towards the center in radial direction and discharges out of the runner axially. The reaction turbine operates at medium head and required medium quantity of water.

A part of the head acting on the turbine is transformed in to kinetic energy and the rest remains as pressure head. There is a difference of pressure between the guide vanes and the runner which is called the *reaction pressure* and is responsible for the motion of the runner. That is why this turbine is called as *reaction turbine.*

7. LITERATURE REVIEW

Basset and Potter first discovered the possibility of using an induction machine as a SEIG during the 1935s. An induction machine operates as a generator if an appropriate supply of inductive VArs is available to provide the machine's excitation. Self-excitation can be achieved by the connection of capacitors bank at the machine's stator terminals. The lagging VArs supplied by the capacitors is consumed by the machine's excitation, leakage reactance and the reactance of the inductive load.

Although the SEIG scheme was discovered more than eighty years ago, a large number of research papers are only increasingly focused on the investigation and applications of SEIGs in the recent two/three decades. This is due to the improving voltage and frequency control techniques and the worldwide vast attention on the development of renewable energy sources over the past thirty years (Bassett & Potter, 1935) & (Basset & Potter, 1935).

In 1982, Murthy et al. developed a mathematical model for obtaining the steady-state performance of a SEIG by using the equivalent circuit (EC) of the induction machine. The Newton-Raphson method was used to obtain the operating frequency and the magnetising reactance from the nonlinear equations of the system. A year later, Quazene et al. developed a nodal equation to solve for the operating frequency and then the magnetising reactance by using a nodal admittance technique (Murthy & Pinto, 1990) & (Murthy, Jha, Rao, 1990)

Tandon et al. presented an alternative approach to the steady-state performance analysis of standalone SEIG using balanced terminal capacitors in 1984. The predicted steady-state performance of the SEIG under different load conditions by operational ECs matched well with corresponding experimental results (Tandon, Murthy, Berg, 1984)

In 2001, Wang and S.C .Kuo analyzed results of both voltage regulation and current-harmonic suppression of a self excited induction generator (SEIG), under unbalanced and/or nonlinear loading conditions using a current-controlled voltage source inverter (CC-VSI) are presented (Kuo & Wang, 2001). An analysis of the steady-state performance of SEIGs driven by both regulated and unregulated turbines was presented in. The steady-state performance of a standalone SEIG with a transformer connected to its terminal was analyzed in. The function of the transformer was to step up the stator terminal voltage or to supply a different voltage to the load. At relatively high rotational speeds, the saturated transformer was able to absorb reactive power and regulate the terminal voltage. However, such a transformer introduced an additional nonlinearity into the system, which complicates the analysis process.

In the past few decades, other papers have been focused on the analysis of the transient voltage built up of SEIGs. By taken into account the magnetic saturation nonlinear effect, the influence of the terminal capacitors on the transient behavior of an induction machine was studied by Smith et al. in 1968. The tests showed that it was not advisable to reconnect a capacitor excited induction generator to the bus bar, unless the interruption is of short period. The results also indicated that the saturation of the main flux path does not affect the transient behavior of the machine without self-excited capacitors.

The self-excitation voltage built up process in a standalone SEIG was analyzed by Wang et al. in 1999. They examined the SEIG under unbalanced excitation capacitor conditions (by the sudden switching of one excitation capacitor or two excitation capacitors). The results showed that the SEIG could maintain the self-excitation process and generate the other two phase voltage when one phase capacitor was switched off. However, the SEIG voltage collapses when two phase capacitors were cut off.

Wang et al. in year 2000 presented a comparison study of different shunt capacitor configurations of an isolated SEIG feeding an induction motor load. Their analysis showed that the short-shunt connection provided better voltage regulation, whereas the long-shunt connection might cause unwanted oscillations. The eigen value technique was employed to examine the unstable operating conditions of the SEIG system. The dynamic behavior of the SEIG was experimentally investigated by Levy in 1997. However, under variable speed conditions with a fixed capacitor bank connected to the stator terminals, testing leaded to voltage collapsing and the de magnetization of the machine.

A general analysis of a three-phase SEIG with asymmetrical connected loads and excitation capacitors by adopting the symmetrical components method was presented.

In 2004, Bhim Singh, *Senior Member, IEEE*, S. S. Murthy, *Senior Member, IEEE*, and Sushma Gupta described the Analysis and Design of STATCOM-Based Voltage Regulator for Self-Excited Induction Generators. This paper represented the design of SEIG has poor voltage regulation and it requires adjustable reactive power source with varying load to maintain constant terminal voltage. The required reactive power can be provided by a STATCOM consisting of ac inductors.

In 2006, Bhim singh *Senior Member IEEE* and Gaurav Kumar Kasal presented the solid state voltage regulator also known as static compensator (STATCOM) to control the voltage of constant speed prime mover driven isolated Asynchronous generators for feeding 3-phase 4-wire loads.

In 2008, Bhim singh *Senior Member IEEE* and Gaurav Kumar Kasal presented solid state voltage and frequency controller for a standalone wind power generating system, paper deals with the voltage and frequency controller of a wind turbine driven isolated asynchronous generator. The proposed voltage and frequency controller consists of an insulated gate bipolar junction transistor based voltage source converter alongwith battery energy storage system at its dc link.

Many of these articles have been focused on analyzing the steady-state and transient performance of SEIGs from the design and operational point of view. This thesis, on the other hand, is focused on studying and analyzing the comparative analysis of renewable energy source for self-excited induction generator, the presence of the capacitor bank together with the resistive and inductive load at the terminal of the induction generator produce the voltage drop at the load. Voltage regulation scheme is applied in feedback path in the case of hydro power as an input, voltage and frequency regulation scheme is required in the case of wind power as mechanical input (Singh, & Kasal, 2008) & (Singh, Murthy, & Gupta, 2004)

8. MODELING OF STATCOM AND SIMULATION RESULTS

The SEIG systems are the isolated systems and successfully generating electrical power at remote/isolated/hilly areas. However, poor voltage and frequency are the major bottleneck of the SEIG system. Due to recent advancement in the solid-state-commutating devices, voltage and frequency regulation of SEIG system is possible using static noiseless voltage/frequency regulator. Such regulator can continuously provide variable reactive power to the SEIG condition to maintain the constant terminal voltage under the variable loading. This system is known as STATic COMpensator (STATCOM). As the SEIG systems are mostly small in size, therefore injected harmonic may pollute the generated voltage of the SEIG. The modeled STATCOM has the capacity to eliminate the injected harmonics, provide load balancing and supply the reactive power to the load and generator.

In this chapter, a simple mathematical modeling has been developed for the transient and steady state analysis of the SEIG-STATCOM system under balanced/unbalanced resistive, RL load and the obtained simulated results demonstrated that the developed SEIG-STATCOM behaves like an ideal power supply system under different types of load.

8.1 Initial Excitation of SEIG

The parameters of a 7.5 kW, 50 Hz, 4-pole, 415 V, 20 A squirrel cage induction machine with 3-5 kVAR balanced capacitors are used to test the

SEIG based hydro system and given in the Table 1. The equivalent circuit parameters referred to the stator are given in the Table 2. The prime mover of the generator is characterized by the Equation 23

$$K_1 - K_2 \bullet \omega_r \tag{23}$$

where ω_r: generators rotor speed and the constant coefficients (i.e. K_1 and K_2) are specified as K1 = 1520 and K2 =10. Figure 17 depicts the simulated results for the voltage build up of 7.5 kW rated SEIG. A three phase balanced Δ-connected capacitor bank of 94 μF per phase is connected across the stator terminals of the SEIG to generate 230 V with the rotor speed 1500 rpm. The SEIG reaches its final build up point at 0.3 s and SEIG based hydro system is able to feed load after 0.3 s. Figure 17 shows the waveforms of the generated stator voltage (Vs), stator current at full load (Is) and active (P) and reactive power (Q) flow. The reactive power (Q) represents the variation in kVAr supplied by the capacitor bank. Figure 18 shows the rotor speed, frequency of generated quantity and electromechanical torque of the SEIG with 94 μF capacitor bank.

8.2 SEIG System Performance without STATCOM

To understand the importance of the STATCOM, for successful operation of the SEIG system in remote/hilly areas SEIG system is tested without the STATCOM. Figure 19 illustrates the results when variable linear resistive load are fed by the SEIG system. Figure 19 shows the generated voltage (V_G), generated current (I_G), current supplied by the capacitor bank (I_C), load current (I_L), rotor speed of SEIG (ω) and electromechanical torque (T_e) respectively. The SEIG system takes 0.7 s to build-up the final voltage (415 V), hence the load is applied after the 0.7 s. At 1.0 s, three-phase linear resistive load of about 2.5 kW is applied to the SEIG system. At the sudden application of the load, the generated voltage is reduced. At 1.3 s, step change in resistive load from 2.5 kW to 5.0 kW has been done and it is noticed that the terminal voltage of the generator is further reduced. At 1.6 s, the resistive load is again increased to 7.5 kW (Full load), which results diminishing of the generated voltage up-to zero. Hence, it can be concluded that SEIG system performance is very much affected by the variation in the load and cannot supply the reliable power in remote areas.

To ensure the power quality of the SEIG system, harmonic spectrum analysis has been performed. Figure 20 depicts the harmonic spectrum analysis of the generated voltage of the SEIG at the application of the 2.5 kW resistive load, which is 4.78%. Figure21 depicts the harmonic spectrum analysis of the load current, which shows the load current THD is 4.60%.

Table 1. Rating of induction machine operating as a SEIG

Sr. No.	Quantity	Symbol	Value
1.	Power	P_h	7.5 kW
2.	Connection	Δ	
3.	Line Voltage	V_h	415 V
4.	Line Current	..	11 A
5.	Frequency	f_h	50 Hz
6.	No. of Pole	P	4

Table 2. Per-phase equivalent circuit parameters of induction machine operating as a SEIG

Sr. No.	Quantity	Symbol	Value
1.	Stator resistance	R_s	2.48 Ω
2.	Rotor resistance	R_r	3.96 Ω
3.	Stator reactance	X_{ls}	5.27 Ω
4.	Rotor reactance	X_{lr}	5.27 Ω
5.	Base Impedance	X_b	5.27 Ω
6.	Magnetizing resistance	R_m	1592.07 Ω
7.	Magnetizing reactance	X_m	140.4 Ω

Figure 17. Generated voltage, stator current and power at 94 μF

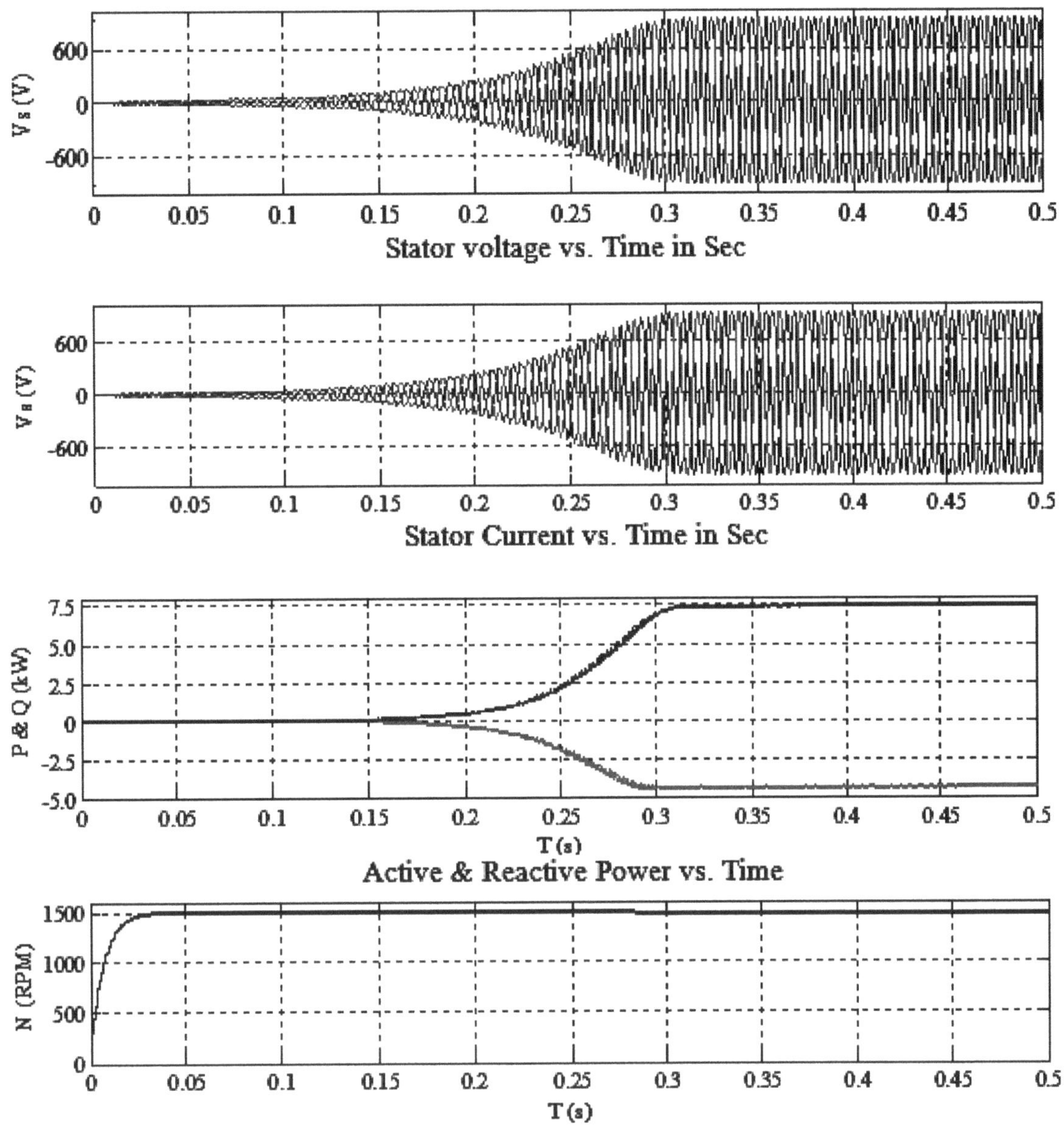

8.3 SEIG System Performance with STATCOM

As discussed in the previous section, the SEIG system requires the additional source of reactive power to maintain the terminal voltage constant under the variation in the load. Hence, to regulate the terminal voltage and frequency of the SEIG system, STATCOM has been developed.

8.3.1 Performance of SEIG-STATCOM under Balanced/Unbalanced Resistive Load

Figure 22 shows the transient waveforms of the terminal SEIG voltages (V_G), SEIG line currents (I_G), DC bus voltage (V_{DC}), load current (I_L), current supplied by the STATCOM (I_S), current supplied by the capacitor bank (I_C), generator

Figure 18. Rotor speed, frequency and torque at 94 μF

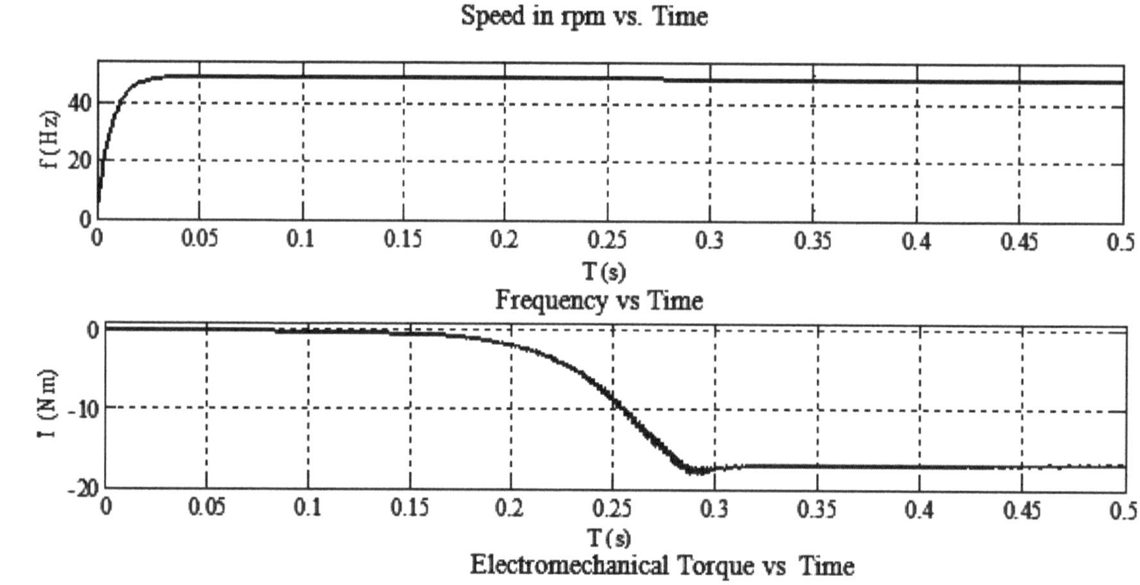

speed (ω) and electro-mechanical torque (Te), demonstrating the response of the STATCOM in regulating the SEIG terminal voltage supplying the linear pure variable resistive load. At 1.0 s, a three phase resistive load of about 2.5 kW is applied on the SEIG-STATCOM system. At the application of three phase resistive load, the generated voltage remains constant but the generated current and STATCOM current increase to supply the active and reactive power to load respectively. A small dip in the DC voltage of the capacitor connected at STATCOM is observed at the application of the load. At 1.3 s, the resistive load is changed to 5.0 kW. As the load is increased, the generated current and STATCOM current further increases to maintain the constant terminal voltage and supplies the required active and reactive power to the load. At 1.6 s, the load is changed to 7.5 kW (Full load) and variations are observed.

To analyze the power quality of the developed system, harmonic spectrum analysis has been performed. Figure 23 demonstrates the harmonic spectrum analysis of the terminal voltage of the SEIG-STATCOM system feeding the resistive load. Figure 23 shows the harmonic spectrum content of the regulated terminal voltage using STATCOM is 1.88%. Without STATCOM, the THD of terminal voltage was 4.78%. Figure 24 shows the harmonic spectrum analysis of the load current supplied by the SEIG-STATCOM system and the THD is 2.75%. However, without STATCOM the harmonic spectrum content was 4.60%. Hence, it can be concluded that the STATCOM reduces the injected harmonics of the system.

From the simulated results, it has been found that the SEIG terminal voltage remains constant, with the sinusoidal feeding of the three-phase resistive loads. The generator voltage drops with the increase in the load and at the full load the terminal voltage totally diminished. But with the STATCOM, the SEIG system is not affected with the variation in the load. Hence, the developed system is capable to maintain the constant terminal voltage and also acts as a load balancer. The

Figure 19. SEIG system performance without Statcom under variation in linear resistive load

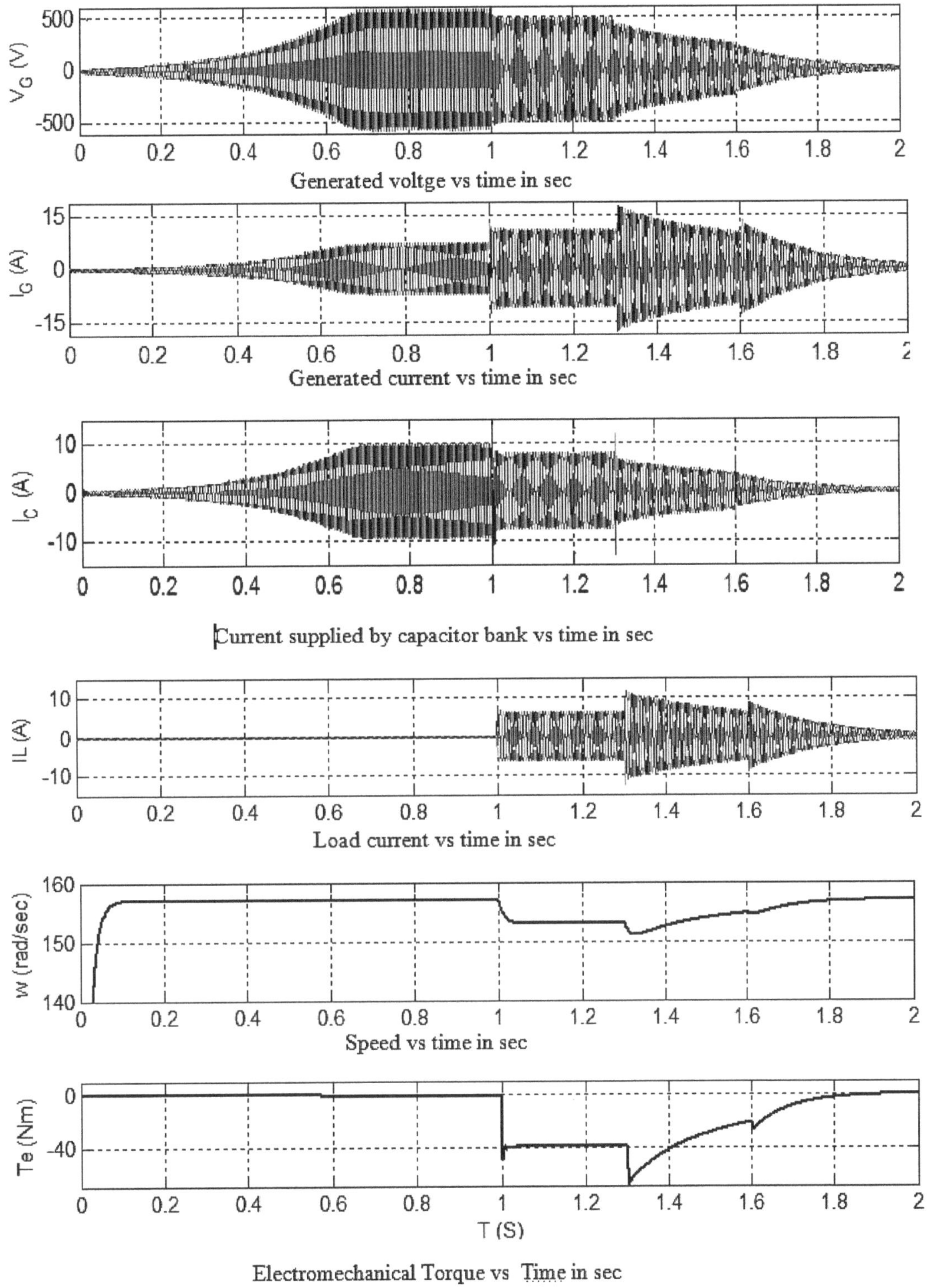

Figure 20. Harmonic spectrum analysis at the application of linear resistive load: (a) Analyzed generated voltage waveform (b) Harmonic spectrum content

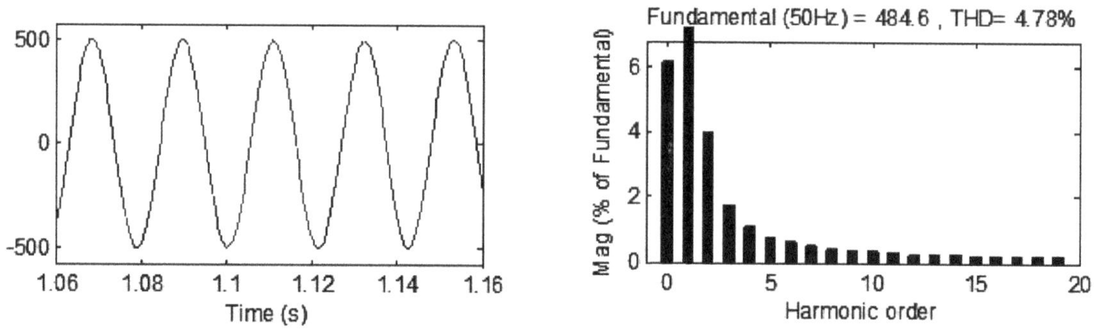

ab

Figure 21. Harmonic spectrum analysis of load current: (a) Analyzed load current waveform (b) Harmonic spectrum content

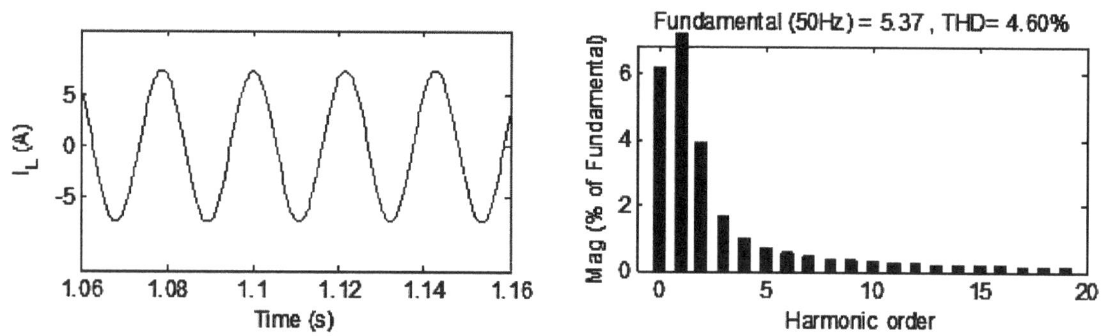

ab

STATCOM is also able to reduce the generated harmonic content of system and improves the power quality of the system. Thus, the developed system is efficient and able to put into the service in the remote/hilly areas.

9. CONCLUSION

The Self-Excited Induction generator (SEIG) is now a matured technology for electrification in remote/hilly areas. The drawbacks of the SEIG system are poor voltage and frequency regulation under the variation in the load. Hence, the SEIG system requires an adjustable reactive power source with varying loads to maintain a constant terminal voltage. In this dissertation, a three-phase insulated-gate-bipolar-transistor based current-controlled voltage source inverter based STATCOM has been modeled and developed in MATLAB software to compensate the reactive power for SEIG as well as harmonic content elimination with the varying load to maintain a constant voltage at the terminal.

The SEIG-STATCOM system has been mathematically modeled and simulated. The obtained simulated results have shown quiet satisfactory operation of the developed system for isolated areas. The results also demonstrated that the proposed STATCOM with the controller is suitable

Figure 22. Waveforms of the SEIG-STATCOM system supplying variable linear pure resistive load

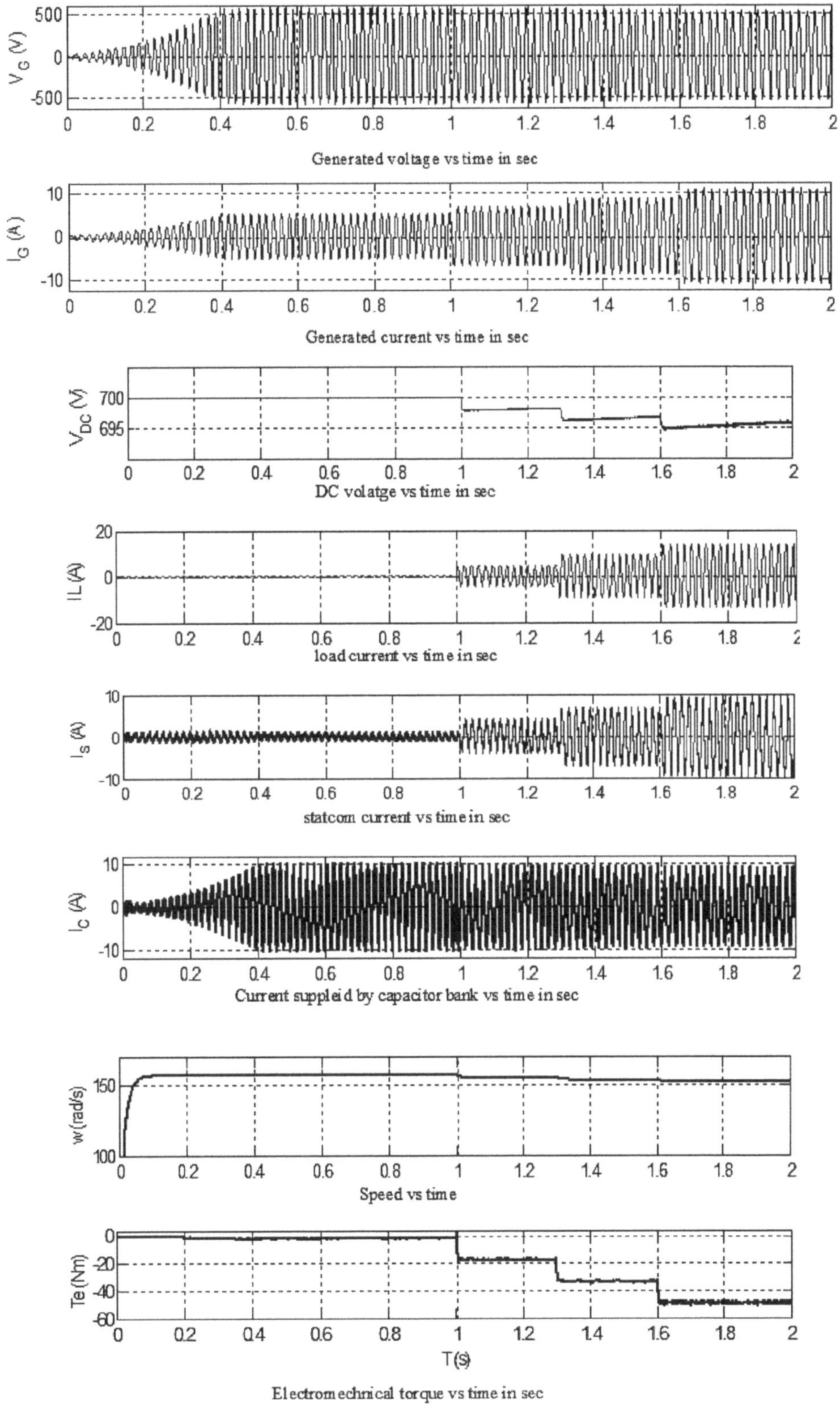

Figure 23. Harmonic spectrum analysis at the application of resistive load to SEIG-STATCOM system: (a) Analyzed generated voltage waveform (b) Harmonic spectrum content

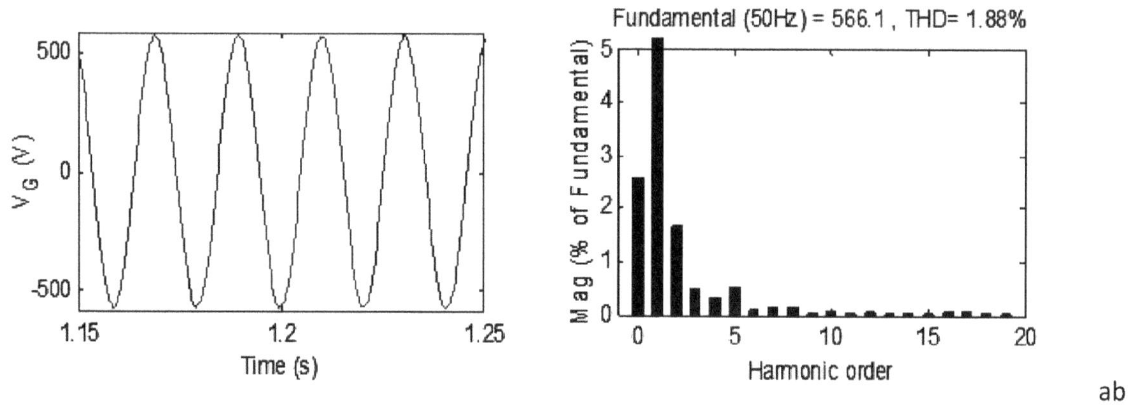

Figure 24. Harmonic spectrum analysis of load current of SEIG-STATCOM system: (a) Analyzed load current waveform (b) Harmonic spectrum content

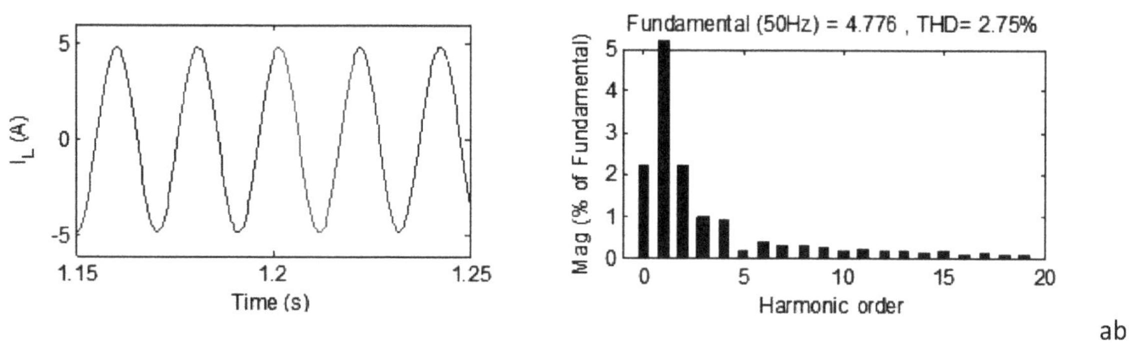

for voltage regulation, harmonic elimination and load balancing. The advantages of the proposed SEIG-STATCOM system are listed here as

- The proposed system has simpler circuit and less complexity.
- The system has higher efficiency, as the numbers of switches are less.
- The improvement in the reliability of the system is evident because it has full control over the active and reactive power.
- Fast transient response within fraction of seconds (15 ms).
- Reduced mechanical and electrical stress in power generators.

The proposed system is investigated under all possible electrical conditions and it can be concluded that the proposed system has better dynamic performance. The harmonic spectrum analysis of the system has also been performed to prove the better power quality. With the STATCOM, the generated voltage and load current THD are 1.88% and 2.75% respectively which are less than 5% (limit imposed by the IEEE-519 standard). While without the STATCOM, generated voltage and load current THD was 4.78% and 4.60% respectively. Hence, it can be concluded that the power quality of the SEIG has been improved by introducing the STATCOM.

REFERENCES

Abdel-Halim, M. A., Almarshoud, A. F., & Alolah, A. I. (2001). Control of grid connected induction generator using naturally commutated AC voltage controller. *Proceedings of the Canadian Conference on Electrical and Computer Engineering* (Vol. 2, pp. 839-843). IEEE.

Abu-Elhaija, W.S., & Muetze, A. (2013). Self-Excitation and Stability at Speed Transients of Self-Excited Single-Phase Reluctance Generators. *IEEE Transactions on Sustainable Energy*, 4(1), 136-144.

Ahuja, R. K., Murthy, S. S., & Chaudhary, J. K. (2012). Steady state analysis of three phase self excited induction generator using MATLAB GUI to feed single phase loads with three capacitor configuration. *Proceedings of the 2012 IEEE 5th India International Conference on Power Electronics IICPE* (pp. 1–6). doi:10.1109/IICPE.2012.6450481

Al-Saffar, M., Nho, E. C., & Lip, T. (1998). Controlled Shunt Capacitor Self Excited Induction Generator. *Proceedings of the 33rd IAS Annual meeting on Industry applications conference* (Vol.2, pp. 1486 – 1490).

Anderson, M. D., & Carr, D. S. (1993). Battery energy storage technology. *Proceedings of the IEEE*, *81*(3), 475–479. doi:10.1109/5.241482

Bassett, E.D. & Potter, F.M. (1935). Capacitive excitation for induction generators. *Electrical Engineering*, 54(5), 540-545. doi:10.1109/EE.1935.6540048

Ceraolo, M. (2000). New dynamic models of lead-acid Batteries. *IEEE Transactions on Power Systems*, *15*(4), 1184–1190. doi:10.1109/59.898088

Godoy Simoes, M., & Farret, F. A. (2004). *Renewable Energy Systems: Design and Analysis with Induction generators*. Boca Raton, FL: CRC Press.

Haque, M. H. (2006). Damping improvement using facts devices. *Electric Power Systems Research*, *76*(9-10). doi:10.1016/j.epsr.2005.11.001

Hingorani, N. G., & Gyugyi, L. (1999). *Understanding FACTS: concepts and technology of flexible ac transmission systems*. NY: IEEE Press. doi:10.1109/9780470546802

Joshi, D., Sandhu, K. S., & Soni, M. K. (2006). Constant voltage constant frequency operation for self excited induction generators. *IEEE Transactions on Energy Conversion*, *21*(1), 228–234. doi:10.1109/TEC.2005.858074

Kuo, S.-C., & Wang, L. (2001). Analysis of voltage control for a self-excited induction generator using a current-controlled voltage source inverter (CC-VSI). *Generation, Transmission and Distribution, IEE Proceedings*, 148(5), 431-438. doi:10.1049/ip-gtd:20010477

Li Wang & Chang-Min Cheng. (2000). Excitation capacitance required for an isolated three-phase induction generator supplying a single-phase load. *Proceedings of the Power Engineering Society Winter Meeting 2000* (Vol. 1, pp. 299-303).

Luiz, A. C. Lopes & Almeida, R.G. (2006). Wind –Driven Self –Excited Induction Generator With Voltage and Frequency Regulated by a Reduced-Rating Voltage Source Inverter. IEEE Trans. on Energy conversion, 21(2).

Rashid, M.H. (2003). *Power electronics Circuits, Devices and Applications* (3rd ed.).

Murthy, S.S., Jha, C.S., & Rao, P.S.N. (1990). Analysis of grid connected induction generators driven by hydro/wind turbines under realistic system constraints. *IEEE Transactions on Energy Conversion*, 5(1), 1-7. doi:10.1109/60.50804

Murthy, S.S., & Pinto, A.J.P. (2005, September). A Generalized Dynamic and Steady State Analysis of Self Excited Induction Generator (SEIG) Based on MATLAB. *Proceedings of the Eighth International Conference on Electrical Machines and Systems ICEMS '05* (Vol. 3, pp. 1933-1938).

Padiyar, K. R., & Devi, A. L. (1994). Control and simulation of static condenser. *Proc. of the 9th Annual Applied Power Electronics Conf. Expo.*

Ramirez, J.M., & Torres, M.E. (2007). An electronic load controller for self excited induction generators. *IEEE Trans. Energy Conversion,* 21(2), 546-548.

Rao, P., Crow, M. L., & Yang, Z. (2000). STATCOM Control for Power System Voltage Control Applications. *IEEE Transactions on Power Delivery*, *16*, 18–23.

Singh, B., & Kasal, G.K. (2006). Solid state voltage regulator for isolated asynchronous Generators. *Proc. of India International Conference on Power Electronics.*

Singh, B., Al-Haddad, K., & Chandra, A. (1998). Harmonic elimination, reactive power compensation and load balancing in three phase four wire electric distribution system supplying non linear loads. *Electric Power Systems Research*, *44*(2), 93–100. doi:10.1016/S0378-7796(97)01187-5

Singh, B., Murthy, S.S., Gupta, S. (2004). Analysis and Design of STAT-COM Based Voltage Regulator for Self Excited Induction Generator. *IEEE Trans. on Energy conversion, 19(4).*

Singh, B., & Murthy, S. S., Sushma Gupta (2006). STATCOM-Based Voltage Regulator for Self-Excited Induction Generator Feeding Nonlinear Loads. IEEE Transactions on Industrial Electronics, 53(5).

Singh, B., Murthy, S. S., & Gupta, S. (2005). Transient analysis of self excited induction generator with electronic load controller supplying static and dynamic loads. *IEEE Transactions on Industry Applications*, *41*(5), 1194–1204. doi:10.1109/TIA.2005.855047

Singh, B., Murthy, S. S., & Gupta, S. (2006). Analysis and design of electronic load controller for self excited induction generator. *IEEE Transactions on Energy Conversion*, *21*(1), 285–293. doi:10.1109/TEC.2005.847950

Singh, B., & Shilpakar, L. B. (1998). Analysis of a novel solid state voltage regulator for a self excited induction generator. *IEE Proceedings on Generation, Transmission and Distribution*, *145*(6), 647–655. doi:10.1049/ip-gtd:19982357

Singh, G. K. (2004). Self excited induction generator research-a survey. *Electric Power Systems Research*, *69*(2-3), 107–114. doi:10.1016/j.epsr.2003.08.004

Singh, B., & Kasal, G.K. (2008). Solid State voltage and frequency Controller for a Stand Alone Wind Power Generating System. *IEEE Trans. on Power Electronics*, 23(3).

Tandon, A. K., Murthy, S. S., & Berg, C. J. (1984). Steady State Analysis of capacitor self-excited induction generators. *IEEE Transactions on Power Apparatus and Systems*, *103*(3), 612–617. doi:10.1109/TPAS.1984.318748

Youssef, K. H., & Wahba, M. A., Yousef, Hasan A., Sebakhy, O.A. (2008). A new method for voltage and frequency control of stand-alone self-excited induction generator using PWM converter with variable DC link voltage. *Proceedings of the American Control Conference 2008* (pp. 2486-2491). doi:10.1109/ACC.2008.4586864

Zhuang, Y., & Menzies, R. W. (1995). Effect of DC Capacitance of a STATCOM on the dynamic performance at a weak HVDC terminal. Proc of IEEE WESCANEX '95 (pp. 289-293).

Chapter 10
Artificial Neural Network (ANN) in Network Reconfiguration for Improvement of Voltage Stability

Dipu Sarkar
National Institute of Technology, Nagaland, India

Joyanta Kumar Roy
MCKV Institute of Engineering, West Bengal, India

ABSTRACT

Issues related to power system voltage levels have become increasingly important issue during last two and half decades. In power networks, low voltage situations may result in the loss of stability, voltage collapse and eventually to cascading power outages. Large number of incidents of voltage collapse has been reported in different countries across the globe. A simple indicator that has the potential in real time, i.e. L indicator has been used to find voltage profile at different switching condition and simulated using ANN in network reconfiguration for the improvement of voltage stability. A method for improving voltage stability in a power network comprising of multiple lines and switches has been suggested in this chapter based on system reconfiguration approach. ANN based fast and efficient methodology has been developed to obtain the optimum switching combination to achieve best voltage stability. The proposed scheme has been tested on an IEEE 14-bus system.

1. INTRODUCTION

Intelligent algorithms are algorithms with an element of artificial intelligence. They can decide their next course of action based on the current state of the system. The ability to take self-guided decisions is called *heuristic*, which comes from experience. Some algorithms follow human defined heuristics and some follow nature's heuristics. Those following human defined heuristics are called Artificially Intelligent Algorithms, and those following nature's, are called Nature Inspired Algorithms. Hill-Climbing (HC), Depth First Search (DFS), Minimal Spanning Tree (MST), etc.

DOI: 10.4018/978-1-4666-9911-3.ch010

are some of the artificially intelligent algorithms, while Genetic Algorithms (GA), Artificial Neural Networks (ANN), Particle Swarm Optimization (PSO) etc. are some naturally intelligent algorithms. The advantage of such algorithms is that they make use of human or natural intelligence to reach a solution in a short time preventing the investigator from taking exhaustive approaches.

The breakthrough of artificial intelligence in the electric power system which started in the mid-eighties has resulted in the development of numerous expert systems and knowledge-based system tools for solving various power system problems (Huneault, Rosu, Manoliu, & Galiana, 1994). Lot of works in the field of power system protection, distribution planning, design and management have been reported which utilized artificial intelligence. Over the last decade, a number of papers have been reported in the field of distributed automation using Hardware based systems and as well as various soft computational techniques (Warwick, Ekwue, Arthur & Aggarwal, 1997). The present discussion proposes the applications of ANN for optimal operation of a power system at highest voltage stability and minimum active and reactive power losses.

2. OVERVIEW OF VOLTAGE STABILITY ISSUES IN GENERAL

Issues related to power system voltage levels have become increasingly important during last two and half decades. In power networks, low voltage situations may result in the loss of stability, voltage collapse and eventually to cascading power outages. Large number of incidents of voltage collapse has been reported in different countries across the globe (Taylor, 1994) & (Aboytes & Arroys, 1987).

It is a well-established fact that there is a strong relationship between real power transmission (MW) and rotor angle (δ), and reactive power transmission (MVAR) and the voltage (Chakrabarti & Mukhopadhyay, 1989) & (Aboytes & Arroys, 1987). In other words, the availability of MW is dictated by machine angle, which in turn is decided by the input to the prime mover. On the other hand, voltage magnitude is related to MVAR availability at that point of time. Voltage instability can be ascribed to lack of VAR support necessary to maintain the voltage profile within specified levels (Davidson, 1975) at certain load condition.

Voltage stability is concerned with maintaining 'acceptable voltage levels' all system buses under normal conditions as well as when the system is being subjected to a disturbance. The main causes of voltage instability are lack of reactive power resources, heavy loading and severe contingencies. During instability, voltage magnitude of some of the system buses decrease gradually and then rapidly to reach the collapse point (Aboytes & Arroys, 1987). The excess load on the transmission lines, large distance of the generating sources, and lack in local reactive compensation (Chakrabarti & Mukhopadhyay, 1989) are the also caused the voltage instability.

Some of the researchers have been developed a number of voltage stability indicators in the last two decades. Due to this fact, accurate preventive control actions can be taken easily with the help of voltage stability index. Voltage stability assessment can be done with the help of different indicators (L-index indicator, V-Q sensitivity indicator, Voltage Collapse Proximity Indicator (Warwick, Ekwue, Arthur, & Aggarwal, 1997), which involves numerous load flow solutions under both normal and abnormal operating conditions like outage of transmission lines, outage of some generating source etc.

Maintaining voltage stability and voltage levels at all important nodes in a large and heavily loaded power network are critical and challenging tasks for power engineers. A number of different approaches have been traditionally used for improvement of voltage stability. The improvement of voltage stability is achieved by incorporating additional hardware installation. These include i) shunt capacitor ii) tap-changing transformers iii) FACTS devices etc. at proper position at the transmission line and load side of a power network. Voltage Stability enhancement can be done using Network Reconfiguration Technique. The following sections give a brief idea of Network Reconfigure Technique.

3. NETWORK RECONFIGURATION AND VOLTAGE STABILITY IMPROVEMENT

System reconfiguration means restructuring the power lines which connect various buses in a power system. Restructuring of specific lines leads to alternative system configurations. System reconfiguration can be accomplished by placing line interconnection switches into network. Opening and closing a switch connects or disconnects a line to the existing network. If there be N switches in a network, there will be 2N possible switching combinations. Reconfiguring of power network can also provide voltage level improvement and voltage stability improvement. Improving voltage stability by network reconfiguration involves study of switching options which enhances voltage stability under a given load and generation condition. The challenge in the proposed method however lies with the task to find the optimum switching pattern that would maximize the overall voltage stability of the system.

Manual Reconfiguration at an every operational period is a difficult task as the decision has to be based on operating load, optimized power loss, consideration of load ability and voltage stability. It is however imperative that for a large networks, manual reconfiguration is not feasible. The present work encompasses applications of Artificial Neural Network (ANN) to accomplish the task.

4. CONCEPT OF MANUAL NETWORK RECONFIGURATION TECHNIQUE THROUGH A MODIFIED IEEE 8- BUS SYSTEM

In a power system, reconfiguration process is used to reroute the power by changing the status of sectionalizing switches and altering the topological structure under both normal and abnormal conditions. The reconfiguration of a power system is performed to achieve objectives, such as power loss minimization, voltage stability improvement in some event.

In the present section, a 8 bus system is considered. The standard system has been modified by adding two lines. In the network 15 numbers of lines are present. For network reconfigurations, sectionalizing switches are placed at each of the line in series. Opening and closing switch connect and disconnect a line from the existing network. In the present network there are 16 switches. The total possible configuration will be 2^{16}. Each possible configuration is not feasible at practical point of view, while some configuration may have better performance in some event such as fault, over loading condition etc.

In the next section, the effect of reconfiguration for a standard system has been simulated. The performances of some arbitrary feasible configuration have been studied guided by two important parameters voltage stability and over all active and reactive power losses of the system.

5. EFFECT OF RECONFIGURATION ON VOLTAGE STABILITY, ACTIVE AND REACTIVE POWER LOSS OF THE OVERALL SYSTEM

The primary objectives of power suppliers are continuous, economical and quality power supply to customer. In other words, quick restoration of supply is desirable after the occurrence of fault maintaining voltage stability and voltage levels at all important nodes in a large and heavily loaded Power network. A number of different approaches have been reported in papers for improvement of voltage stability.

In the proposed reconfiguration technique, there may be large number of feasible alternative combinations of lines and switch in a model network for fixed loading condition. Attempt has been made to find out the best possible combination of switches & lines which results in highest voltage stability occurring with least amount of active and reactive power loss of the overall system. The present work reveals that careful and optimum selection of power network lines and inter connection switches can provide the best voltage stability solution for a given loading condition.

In this section illustration of one case study has been presented by IEEE-modified 8 bus system. The IEEE –modified 8 bus system overall voltage stability index has been calculated for each of the load bus for different configuration. Over all aggregate gives the performance of voltage stability index of the overall system.

6. A TYPICAL CASE STUDY

6.1 Voltage Stability Indicator (L indicator)

The objective of the present topic is to make familiar with a simple indicator that has the potential to be implemented in real-time. The L indicator (Kessel, & Glavitsch, 1986) has been used to find voltage profile at different switching condition.

6.2 Computation of Voltage Stability using two bus model & L-indicator

Let,

V_G : Generator voltage
S_G : Power delivered by generator
I_G : Current delivered by the generator.
Y_Q : Shunt admittance of the transmission line.
Y_L : Series admittance
I_D : Load Bus Current
V_D : Load Bus voltage
S_D: Power at Load Bus
δ_0: Generator Bus voltage angle.
δ_D: Load Bus voltage angle.
δ: Difference of angle between Generator Bus voltage angle and Load Bus voltage angle.

Using simple circuit laws, from the Figure 1:

$$I_D = V_D Y_Q + \left(V_D - V_G\right) Y_L = \frac{S_D}{V_D} \quad (1)$$

$$S_D = V_D^2 Y_Q + V_D^2 Y_L - V_D V_G Y_L = V_D^2 Y_{11} + V_0 V_D Y_{11} \quad (2)$$

where,

Figure 1. A 2 bus system with a single Generator and Single Load

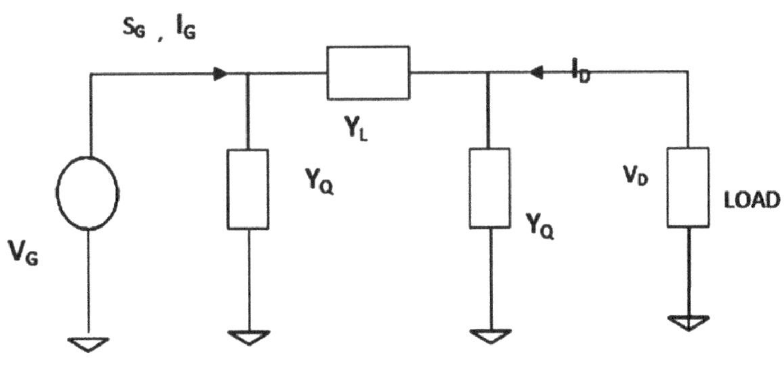

$$Y_{11} = Y_Q + Y_L$$

and

$$Y_{11} = -\frac{Y_L}{Y_L + Y_Q} V_G \tag{3}$$

Here, $\bar{V}_D$ is the load bus voltage. Let us consider,

$$\frac{S_D}{\bar{Y}_{11}} = a + jb$$

Then, equation (2) becomes:

$$\frac{S_D}{Y_{11}} = a + jb = V_D^2 + V_0 V_D = V_D^2 + V_0 V_D \cos\left(\delta_0 - \delta_D\right) + jV_0 V_D \sin\left(\delta_0 - \delta_D\right) \tag{4}$$

$$\cos' = \cos\left(\delta_0 - \delta_D\right) = \frac{a - V_D^2}{V_0 V_D} \tag{5}$$

$$\sin' = \sin\left(\delta_0 - \delta_D\right) = \frac{b}{V_0 V_D} \tag{6}$$

Squaring (5) and (6) and then adding together,

$$V_0^2 V_D^2 = \left(a - V_D^2\right)^2 + b^2 = a^2 - 2aV_D^2 + V_D^4 + b^2 \tag{7}$$

Solving this equation we get,

$$V_D = \sqrt{\left(\frac{V_0^2}{2} + a \pm \sqrt{\left(\frac{V_0^2}{4} + aV_0^2 - b^2\right)}\right)} = \sqrt{\left(\frac{S_D}{Y_{11}}\left(r \pm \sqrt{r^2 - 1}\right)\right)} \tag{8}$$

From the above equation this can be concluded that,

$$\sqrt{\frac{V_0^4}{4} + aV_0^2 - b^2} = 0$$

When: the voltage at load bus will collapse. Since $V_0 = -\frac{\bar{Y}_L}{\bar{Y}_L + \bar{Y}_Q}\bar{V}_0$ it can specified and supported by generator. Therefore, when

$$\sqrt{\frac{V_0^4}{4} + aV_0^2 - b^2} = 0$$

The voltage at load bus will be sustained.

When, $\frac{V_0^4}{4} + aV_0^2 - b^2 < 0$, the voltage cannot be sustained. Therefore, the threshold of voltage collapse is expressed by

$$a = \frac{b^2}{V_0^2} - \frac{V_0^2}{4}. \tag{9}$$

Considering: $\frac{S_D}{\bar{Y}_{11}} = a + jb$, we can deduce the active power P_D and reactive power Q_D of

$$\bar{S}_D\left(\bar{S}_D = P_D + JQ_D\right)$$

to be expressed as follows:

$$P_D = \mathrm{Re}\left(\bar{Y}_{11}\left(a + jb\right)\right)$$
$$Q_D = \mathrm{Im}\left(\bar{Y}_{11}\left(a + jb\right)\right)$$

From eqn. (2) we get:

$$\frac{S_1}{{V_D}^2 Y_{11}} = 1 + \frac{V_0}{V_D}$$

So, we define an indicator L for voltage collapse as:

$$L = \left|1 + \frac{V_0}{V_D}\right| = \left|\frac{S_1}{{V_D}^2 Y_{11}}\right| \tag{10}$$

When the load is zero ($S_1 \equiv 0$) then L=0. If the voltage of bus1 collapses then L=4. The indicator of the voltage stability of the load bus j will be easily obtained as:

$L_j = |\ 1 + V_{0j} / V_j | = S_j / {V_j}^2 Y_{jj}$,

where V_{0j} is generator bus voltage V_j bus voltage at jth bus

Thus the voltage stability indicator L varies between 0 (no collapse) to 1 (voltage collapse).

Value of L closer to' 0' indicates more stable operation for the system where as L values closer to 1 indicates greater risk of voltage instability.

6.3 Computational Aspects

A fast and efficient method of calculating L index value of each bus and also the L index average value for the whole system for various switching patterns has been developed and implemented. Here, for each of the configuration, Newton Raphson Method (Kessel, & Glavitsch, 1986) Load Flow has been run with corresponding admittance matrix.

The basic algorithm can be described as follows:

1. Run the load flow analysis program to obtain the values of bus voltages and complex powers.

Estimate:

$L_j = |\ 1 + V_{0j} / V_j | = S_j / V_j 2 Y_{jj}$,

Where V_{0j} is generator bus voltage V_j bus voltage at jth bus S_j is jth bus power.

3. Estimate L_{avg} value for the system for each switching pattern.
4. Analyze the results and find out which switching combination gives the lowest value of L_{avg}., i.e. the best voltage stability.

The main advantages of this algorithm are the small amount of computation time needed, and its limited requirement of information about the system.

6.4 Discussion of the Results and Inferences

The case study has been conducted on a modified IEEE 8 bus system (Figure 2). The standard IEEE 8 bus system has been modified with the addition of 2 power lines with line connection/disconnection switches. Also connection/disconnection switches have been placed in series with the existing lines. The system data are given in Table 1 and Table 2. It was assumed that loads are constant during each switching operation. The system under study is illustrated in figure 3. By closing or opening a switch, the line can be added or removed from the system respectively. '0' indicates switch is open, '1' indicates switch is closed. In normal condition or base configuration of the given system S_3=0, S_4=0, S_8=0, S_{15}=0 and all the other switches are closed. Total 201 pos-

Figure 2. A schematic diagram showing modified IEEE 8-bus system with line inter connecting switches (Switch open /close status are sown corresponding to normal configuration)

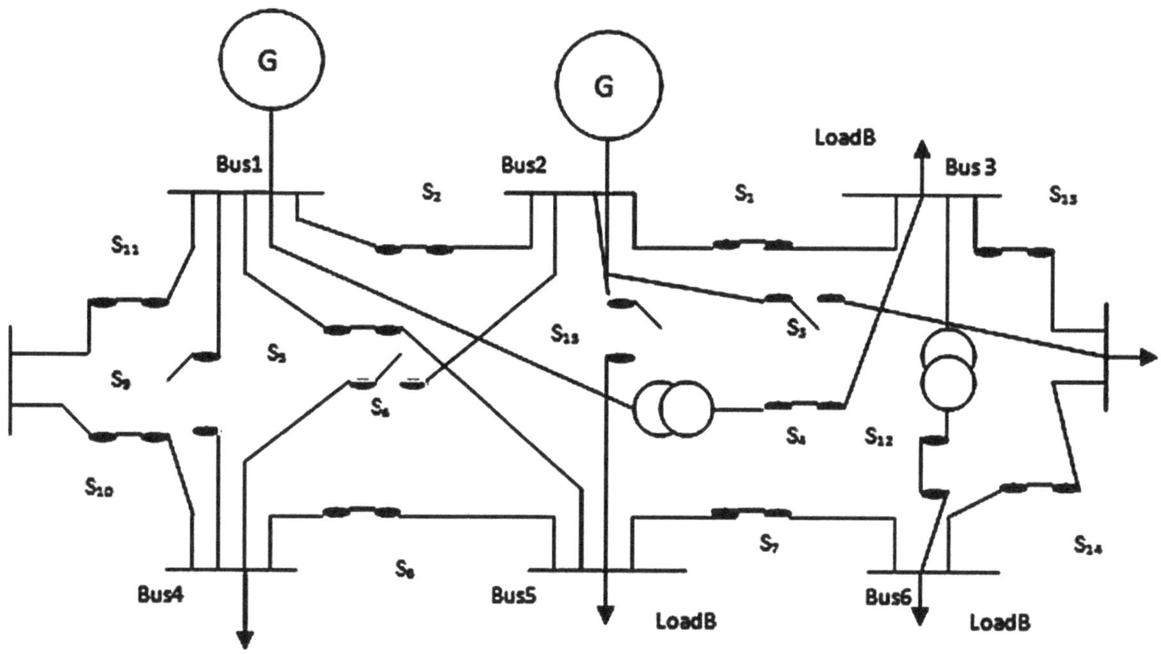

Table 1. Bus-data for the modified IEEE 8 bus system

Bus No.	Bus Voltages		Loads		Generator Data				Static Mvar +Qc/-Qc
	Magnitude (p.u.)	Angle (Degree)	MW	MVAR	MW	MVAR	Q_{min}	Q_{max}	
1	4.04	0.0	0.0	0.0	0.0	0.0	0.0	0.0	0.0
2	4.00	0.0	0.0	0.0	150.0	75.0	0.0	0.0	0.0
3	4.00	0.0	100.0	10.0	0.0	0.0	0.0	0.0	0.0
4	4.00	0.0	45.0	25.0	0.0	0.0	0.0	0.0	0.0
5	4.00	0.0	40.0	25.0	0.0	0.0	0.0	0.0	0.0
6	4.00	0.0	35.0	10.0	0.0	0.0	0.0	0.0	0.0
7	4.00	0.0	60.0	35.0	0.0	0.0	0.0	0.0	0.0
8	4.00	0.0	45.0	25.0	0.0	0.0	0.0	0.0	0.0

sible switching combinations have been studied and each combination has been designated by a unique configuration number. For example the configuration code 00 indicates normal condition. The proposed system under study has been analyzed to evaluate voltage stability index for each bus of the system as well as overall stability from average L index of the given network. Few sample cases of lines and switching combinations, their corresponding system configuration code and computed L index values have been presented in the Table 3.The computed average

Table 2. Line-data for the modified IEEE 8 bus system

Lines	Starting Bus	End Bus	R (p.u)	X (p.u)	B/2 (p.u)	Tap Setting Value
L1	1	2	0.04	0.30	0.01	4.00
L2	3	2	0.03	0.20	0.01	4.00
L3	2	7	0.03	0.15	0.01	4.00
L4	1	3	0.0	0.40	0.01	4.05
L5	1	5	0.04	0.20	0.01	4.00
L6	4	5	0.03	0.15	0.01	4.00
L7	5	6	0.06	0.30	0.01	4.00
L8	2	4	0.04	0.20	0.01	4.00
L9	1	4	0.04	0.02	0.01	4.00
L10	8	4	0.03	0.20	0.01	4.00
L11	8	1	0.04	0.30	0.01	4.00
L12	3	6	0.0	0.20	0.01	4.02
L13	3	7	0.06	0.15	0.01	4.00
L14	7	6	0.02	0.20	0.00	4.00
L15	2	5	0.04	0.20	0.01	4.00

Figure 3. Variation of Average L-index value for different system configurations under fixed loading condition

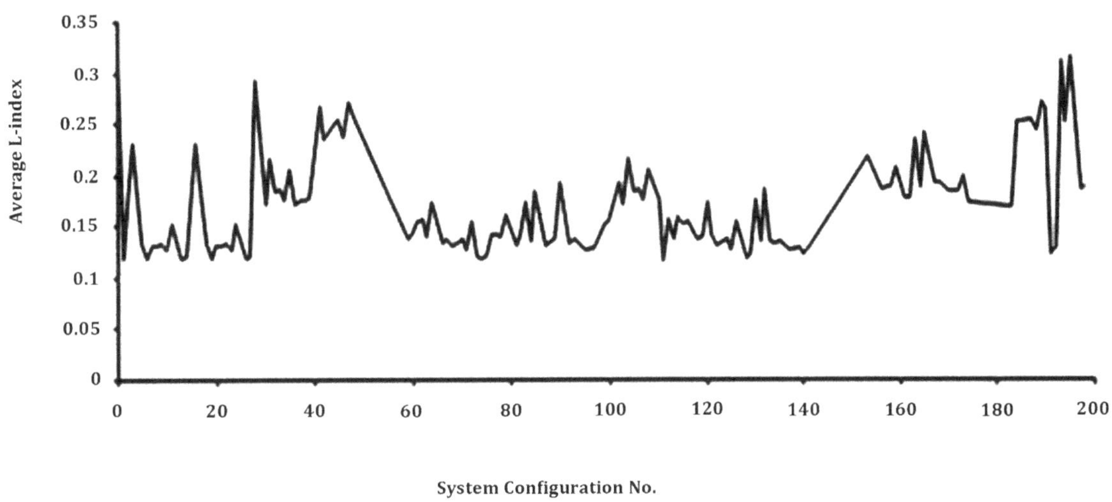

L-index value for normal system configuration (00) is 0.3037 in Table 3. That many alternative system configurations result lower value of L_{avg} as compared to normal configuration and hence can improve overall voltage stability. The best possible switching option for the given loading condition is obtained for the configuration code '111' corresponding to switch status of S_1, S_2 and S_{14} open

Table 3. Few samples of the computed L-index values for normal and modified system configurations

Sys Config No	Switch Status: '0' indicates switch open, 1 indicate switch closed															Computed Values of L-index for different buses (Bus 1&2 have been omitted as these are slack bus & voltage controlled buses respectively)						L_{avg}
	S_{15}	S_{14}	S_{13}	S_{12}	S_{11}	S_{10}	S_9	S_8	S_7	S_6	S_5	S_4	S_3	S_2	S_1	B_3	B_4	B_5	B_6	B_7	B_8	
000*	0	1	1	1	1	1	0	0	1	1	1	0	0	1	1	.313	.139	.227	.427	.562	.155	.303
102	1	1	1	1	1	1	1	1	1	1	0	1	1	0	0	.308	.097	.138	.270	.214	.128	.192
103	1	1	1	1	1	1	1	1	1	0	1	1	1	0	0	.293	.074	.096	.248	.204	.112	.171
104	1	1	1	1	1	1	1	1	0	1	1	1	1	0	0	.345	.067	.057	.397	.277	.108	.216
105	1	1	1	1	1	1	1	0	1	1	1	1	1	0	0	.294	.118	.104	.252	.206	.140	.185
106	1	1	1	1	1	1	0	1	1	1	1	1	1	0	0	.295	.121	.106	.252	.205	.144	.187
107	1	1	1	1	1	0	1	1	1	1	1	1	1	0	0	.288	.067	.085	.242	.201	.179	.177
108	1	1	1	1	0	1	1	1	1	1	1	1	1	0	0	.295	.115	.103	.251	.205	.263	.205
110	1	1	0	1	1	1	1	1	1	1	1	1	1	0	0	.335	.080	.092	.256	.184	.116	.177
111**	1	0	1	1	1	1	1	1	1	1	1	1	1	0	0	.304	.082	.097	.283	.184	.118	.117
112	0	1	1	1	1	1	1	1	1	1	1	0	1	1	0	.221	.094	.125	.213	.174	.128	.157
113	1	1	1	1	1	1	1	1	1	0	1	0	1	0	1	.200	.074	.085	.194	.165	.113	.138
114	1	1	1	1	1	1	1	1	0	1	1	0	1	0	1	.244	.068	.057	.275	.203	.109	.159
115	1	1	1	1	1	1	1	0	1	1	1	0	1	0	1	.202	.113	.095	.198	.167	.138	.152
116	1	1	1	1	1	1	0	1	1	1	1	0	1	0	1	.204	.118	.098	.201	.168	.143	.155
117	1	1	1	1	1	0	1	1	1	1	1	0	1	0	1	.198	.063	.077	.190	.163	.179	.145
118	1	1	1	1	0	1	1	1	1	1	1	0	1	0	1	.206	.076	.08	.185	.162	.114	.137
119	1	1	1	0	1	1	1	1	1	1	1	0	1	0	1	.212	.077	.082	.194	.156	.115	.139
120	1	1	0	1	1	1	1	1	1	1	1	0	1	0	1	.203	.113	.095	.199	.167	.261	.173
121	1	0	1	1	1	1	1	1	1	1	1	0	1	0	0	.206	.078	.086	.216	.151	.115	.142
122	1	1	1	1	1	1	1	1	1	1	1	1	1	0	0	.173	.074	.049	.217	.166	.112	.132
124	1	1	1	1	1	1	0	1	1	1	1	1	1	0	0	.150	.139	.077	.162	.142	.156	.137
125	1	1	1	1	1	0	1	1	1	1	1	1	1	0	0	.150	.054	.077	.162	.142	.179	.127
126	1	1	1	1	0	1	1	1	1	1	1	1	1	0	0	.150	.125	.077	.062	.142	.277	.155

* Configuration '00' stands for normal system configuration

** Configuration '111' provides least value of L_{avg} and best system voltage stability

and all remaining switches closed. However, there are few switching patterns and corresponding system configurations which decrease the value of L_{avg}, and should therefore be avoided by all means. Variation of L_{avg} value with the change in system configuration is illustrated in Figure 3.

7. APPLICATION OF ARTIFICIAL NEURAL NETWORK (ANN) IN NETWORK RECONFIGURATION FOR IMPROVEMENT OF VOLTAGE STABILITY

It has been described in the previous section that voltage stability of a system can substantially improve by reconfiguration of the network topology. It has also been stated that in a large power network comprising of multiple lines, objective of network reconfiguration is to obtain the optimal combination of these lines which result improvement in voltage stability. Usually this is being achieved by opening/closing of sectionalizing switches placed in series with a power lines. Thus the primary task is to derive the switching state of these power lines (On/Off) which will improve system voltage stability in a given operating condition. However, in a practical large system switching combinations can be so large that it is almost impossible for an operator to determine which of this huge number of switches combination is optimal one. However, this optimal selection can be efficiently done using artificial neural network (ANN).

In this section an ANN based fast and efficient methodology has been proposed to select the optimal switching combinations to achieve the highest voltage stability in a large multi bus system. The proposed method has been successfully implemented and verified on IEEE 14 bus system and results indicate that ANN based optimal selection of switching states can improve voltage stability.

8. APPLICATION OF ANN IN VOLTAGE STABILITY ENHANCEMENT USING NETWORK RECONFIGURATION

Neural networks are often perceived as powerful function approximation. They have excellent ability to approximate continuous functions based upon a few available sample values, i.e. ANN can evaluate and approximate a smooth curve, even if the data supplied to the network is discrete. Hence while approximating any mathematical functions, neural networks do not require the entire information, but requires only a few discrete data points lying on the function's graph and sampled at appropriate intervals over the whole range of the function. The network reconfiguration problem has extremely nonlinear relationship between the input status of dynamic switches and corresponding voltage stability indicator of the overall system. A neural network based function approximation can be used to model this complex non-linear relationship. A multi-layer Perception model with error back propagation algorithm has been chosen for estimation and prediction of voltage stability for different system configuration. The block diagram of the proposed ANN based system voltage stability estimation technique has been shown in Figure 4.

In the present case study on the IEEE 14-bus system, each of the power line is assumed to be connected through series sectionalizing switches.

Figure 4. ANN block with switch status as input and Voltage-Stability Indicator as output

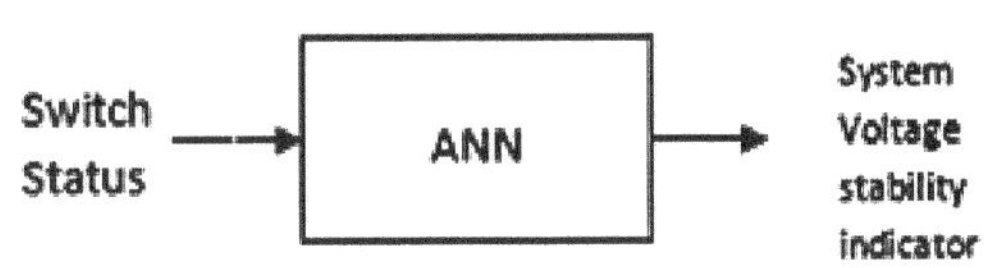

During the course of reconfiguration the following two important criteria must be maintained:

1. No buses can be left out of service.
2. During each operation, active power flow through a line should be within maximum power transfer capability of the line.

There are 20 lines in the test system and hence total number of possible network configuration would be 2^{20}. The best voltage stability condition will result from a unique switching configuration out of these 2^{20} combinations. However it would be impractical to study each of this huge number of probable combinations as repetitive Load Flow solution would take unacceptably long time. Thus for online operation repetitive Load Flow solution is not a practical solution to the problem. In the proposed method optimum switching state for a given operating condition is obtained by using ANN technique. Only three hundred sets of Load Flow Solution data are collected offline by Newton Raphson Load Flow analysis of the IEEE 14 bus system for different switching configurations. These data are used as training set for the ANN. A 20 bit binary code is generated for each network configuration, which serves as the input vector to the ANN. Corresponding voltage stability indicator is used as the Target output, which has to be predicted by the ANN. This process begins with the input vector code '000000000000000' corresponding to first probable configuration and ends with the code '11111111111111111111' corresponding to last probable configuration. The three hundred training samples are selected judiciously for a broad based representation of the probable system configurations. Few samples of training patterns are presented in Table 4. The trained ANN is finally exploited to predict voltage stability indicator values for unknown system configurations. From the predicted voltage stability indicator values the minimum value of voltage stability indicator is marked which corresponds to best voltage stability of the overall system. The

Table 4. Bus-data for the IEEE 14 bus system

Bus No.	Bus Voltages		Loads		Generator Data				Static Mvar +Qc/-Qc
	Magnitude (p.u.)	Angle (Degree)	MW	MVAR	MW	MVAR	Q_{min}	Q_{max}	
1	1.060	0.0	0	0.0	0.0	0.0	0.0	0.0	0.0
2	1.045	0.0	21.7	12.7	40.0	0.0	- 40.0	50.0	0.0
3	1.010	0.0	94.2	19.0	0.0	0.0	0.0	40.0	0.0
4	1.000	0.0	47.8	-3.9	0.0	0.0	0.0	0.0	0.0
5	1.000	0.0	7.6	1.6	0.0	0.0	0.0	0.0	0.0
6	1.070	0.0	11.2	7.5	0.0	0.0	- 6.0	24.0	0.0
7	1.000	0.0	0.0	0.0	0.0	0.0	0.0	0.0	0.0
8	1.090	0.0	0.0	0.0	0.0	0.0	- 6.0	24.0	0.0
9	1.000	0.0	29.5	16.6	0.0	0.0	0.0	0.0	19
10	1.000	0.0	9.0	5.8	0.0	0.0	0.0	0.0	0.0
11	1.000	0.0	3.5	1.8	0.0	0.0	0.0	0.0	0.0
12	1.000	0.0	6.1	1.6	0.0	0.0	0.0	0.0	0.0
13	1.000	0.0	13.5	5.8	0.0	0.0	0.0	0.0	0.0
14	1.000	0.0	14.9	5.0	0.0	0.0	0.0	0.0	0.0

corresponding switching status code will represent the optimum system configuration for the given operating condition. A flow chart showing the proposed method has been shown in Figure 5.

9. TOPOLOGY OF THE ANN

Input layer, hidden layer, and an output layer.

9.1 The Input Layer

In the present problem, input to the ANN is the various possible switching states. Since, there are 20 lines, each having a sectionalizing switch, there would be 20 switches altogether. Thus the input layer comprises of 20 neurons.

9.2 The Hidden Layer

Computational power of the ANN is a result of the addition of hidden layers. But there are no general guidelines to determine the number of hidden layers and neurons. Many applications have proved that ANN's with one single-hidden layer possess sufficient capability of capturing complicated relations between input and output variables. Our studies demonstrated that 1 hidden layer with 40 hidden neurons are quite sufficient for estimation of voltage stability index for the purpose of enhancement of voltage stability. The weighted input value is limited between two limits using an activation function f (x), where 'x' is the summation of weighted inputs. In general tan sigmoidal function of the type: $(e^{x} + e^{-x})/(e^{x} - e^{-x})$ or log sigmoidal functions of the type: $1/(1-e^{-x})$ are used as activation functions. In the present work a tan sigmoidal activation functions have been used for all neurons in the hidden layer. Sigmoidal activation functions are most suitable for any nonlinear mapping of the input-output combinations, as they have noise immunity for low inputs, normal outputs for range inputs, and saturation for large inputs, ensuring stability of the systems.

Figure 5. Flow chart showing ANN based determination of optimal switching configuration

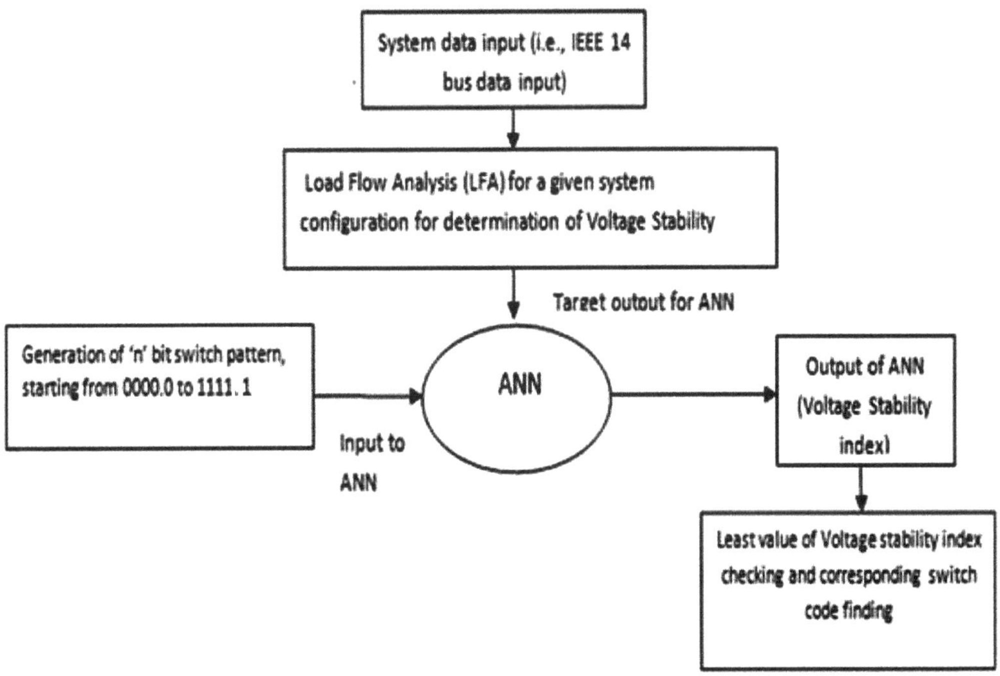

9.3 The Output Layer

Determination of the output layer is quite straight forward. The number of output layer neurons is one because, only voltage stability index has been determined as output of ANN. The linear activation function is used for the output layer neurons.

10. CASE STUDY RESULTS AND INFERENCE

As mentioned, a case study has been conducted on the standard IEEE 14 bus system as shown in Figure 6. The relevant data for the system are presented in Table 4 and Table 5. The lines are connected between the buses through sectionalizing switches. The loads on the buses are assumed to be constant during the process of restructuring. Under normal condition of the system, all the sectionalizing switches are assumed to be in closed position (switch status '1'). Opening of a switch will now disconnect a Power line from the system and this will lead to a new network configuration. As said earlier, three hundred such alternative system configurations have been studied in the present work and voltage stability indicator has been computed for each different network configuration. The training data have been collected from the Load Flow simulations of the

Figure 6. IEEE 14 bus system

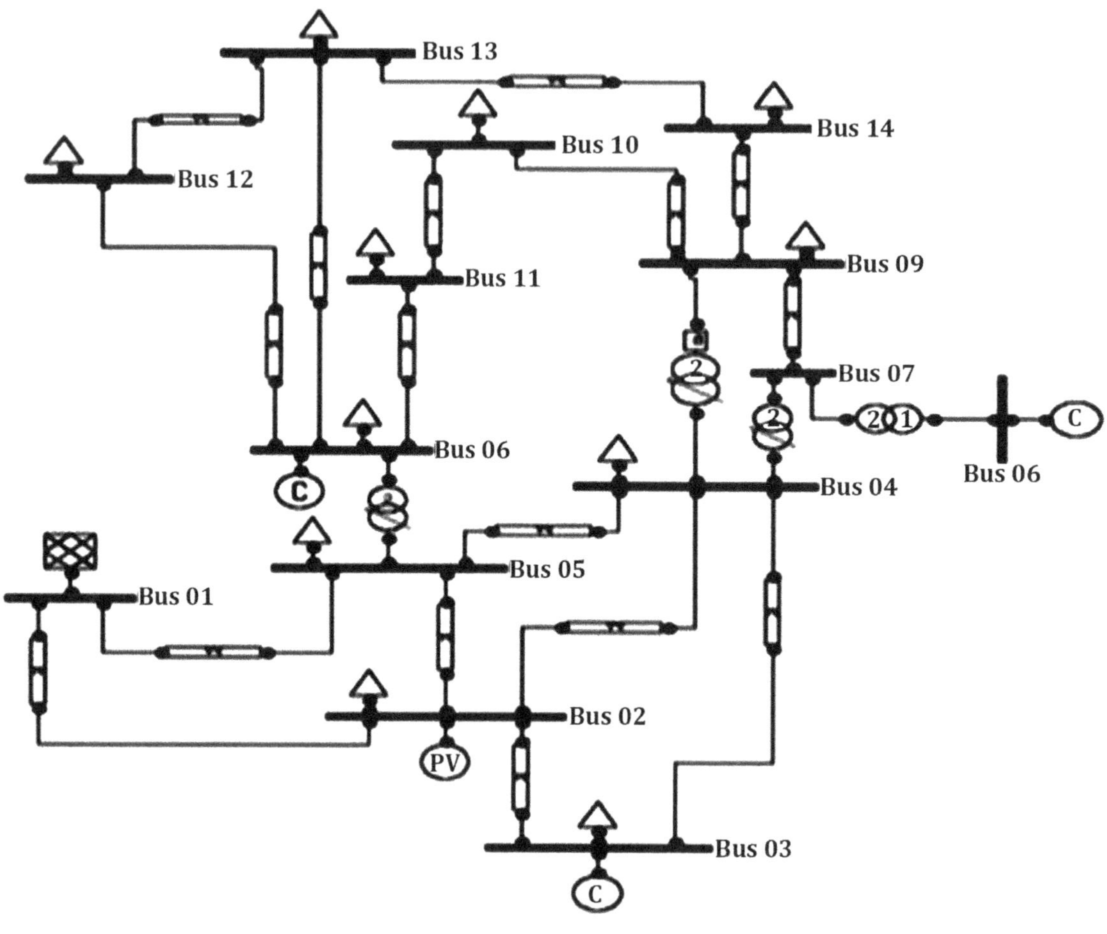

Table 5. Few samples of the computed L-index values for system configurations found from load flow solution for ANN Training

Switch Status: '0' indicates switch open, '1' indicates switch closed																				**L-index (computed)**
S_{20}	S_{19}	S_{18}	S_{17}	S_{16}	S_{15}	S_{14}	S_{13}	S_{12}	S_{11}	S_{10}	S_9	S_8	S_7	S_6	S_5	S_4	S_3	S_2	S_1	
1	1	1	1	1	1	1	1	0	0	1	1	1	0	0	0	1	1	1	1	0.4197
1	1	1	1	1	1	1	0	1	0	1	1	1	0	0	0	1	1	1	1	0.4207
0	1	1	1	1	1	1	0	1	1	1	1	1	0	0	0	1	1	1	1	0.4166
1	1	1	1	0	0	1	1	1	1	1	1	1	0	0	0	1	1	1	1	0.4122
1	1	1	0	1	0	1	1	1	1	1	1	1	0	0	0	1	1	1	1	0.4130
1	1	1	0	1	1	1	0	0	1	1	1	1	1	0	0	1	1	1	1	0.4924
1	1	0	1	1	1	1	0	0	1	1	1	1	1	0	0	1	1	1	1	0.4464
1	0	1	1	1	1	1	0	0	1	1	1	1	1	0	0	1	1	1	1	0.4799
0	1	1	1	1	1	1	0	0	1	1	1	1	1	0	0	1	1	1	1	0.4123
1	1	1	0	0	1	1	0	1	1	1	1	1	1	0	0	1	1	1	1	0.4866
1	0	1	1	0	1	1	0	1	1	1	1	1	1	0	0	1	1	1	1	0.3933

IEEE-14 bus system. A few samples of voltage stability indicator data corresponding to different switch combinations and corresponding system configuration have been presented in the Table 6 to train the ANN model.

The ANN comprise of 20 input neurons to match the dimension of the input data (switching status of the 20 sectionalizing switch), 10 hidden layer and a single output layer to predict the overall voltage stability of the system. The target output (computed voltage stability indicator) vs. the ANN output (predicted voltage stability indicator) graph for training data sample is presented in Figures 7 and 8. The graph illustrates excellent learning of the training data samples. After the training is complete, the generalization ability of the network and its prediction accuracy was tested and 11 samples of unknown switching combination and corresponding voltage stability indicator, as shown in Table 7. The prediction accuracy is found to be very good.

Finally, after training of the ANN by 300 known sets of configuration has been completed and prediction efficiency of the trained ANN has been verified with unseen data, the ANN is exploited to predict VSI (Voltage Stability indicator-L index) for all the probable 2^{20} system configurations. The least value of VSI (best voltage stability) has been searched for and the corresponding system configuration is marked as the optimum configuration which will lead to best possible voltage stability under the given operating condition. The switching status and system configuration for best voltage stability as predicted by the ANN has been presented in Table 7.

11. OTHER APPLICATIONS OF NETWORK RECONFIGURATION

Apart from voltage stability solution, Network Reconfiguration technique can be utilized for loss reduction, congestion management, Power Quality and Reliability Improvement and distributed generation system. In the following section some of the techniques have been described to meet the above mention objective using Network Reconfiguration which was reported by the previous researchers:

Table 6. Line-data for the IEEE 14 bus system

Lines	Starting Bus	End Bus	R (p.u.)	X (p.u.)	B/2 (p.u.)	Tap Setting Value
L1	1	2	0.01938	0.05917	0.02640	1.000
L2	1	5	0.05403	0.22304	0.02460	1.000
L3	2	3	0.04699	0.19797	0.02190	1.000
L4	2	4	0.05811	0.17632	0.01870	1.000
L5	2	5	0.05695	0.17388	0.01700	1.000
L6	3	4	0.06701	0.17103	0.01730	1.000
L7	4	5	0.01335	0.04211	0.00640	1.000
L8	4	7	0.0000	0.20912	0.0000	0.978
L9	4	9	0.0000	0.55618	0.0000	0.969
L10	5	6	0.0000	0.25202	0.0000	0.932
L11	6	11	0.09498	0.19890	0.0000	1.000
L12	6	12	0.12291	0.25581	0.0000	1.000
L13	6	13	0.06615	0.13027	0.0000	1.000
L14	7	8	0.0000	0.17615	0.0000	1.000
L15	7	9	0.0000	0.11001	0.0000	1.000
L16	9	10	0.03181	0.08450	0.0000	1.000
L17	9	14	0.12711	0.27038	0.0000	1.000
L18	10	11	0.08205	0.19207	0.0000	1.000
L19	12	13	0.22092	0.19988	0.0000	1.000
L20	13	14	0.17093	0.34802	0.0000	1.000

Figure 7. Target output-vs- ANN output for the training data Solution for ANN Training

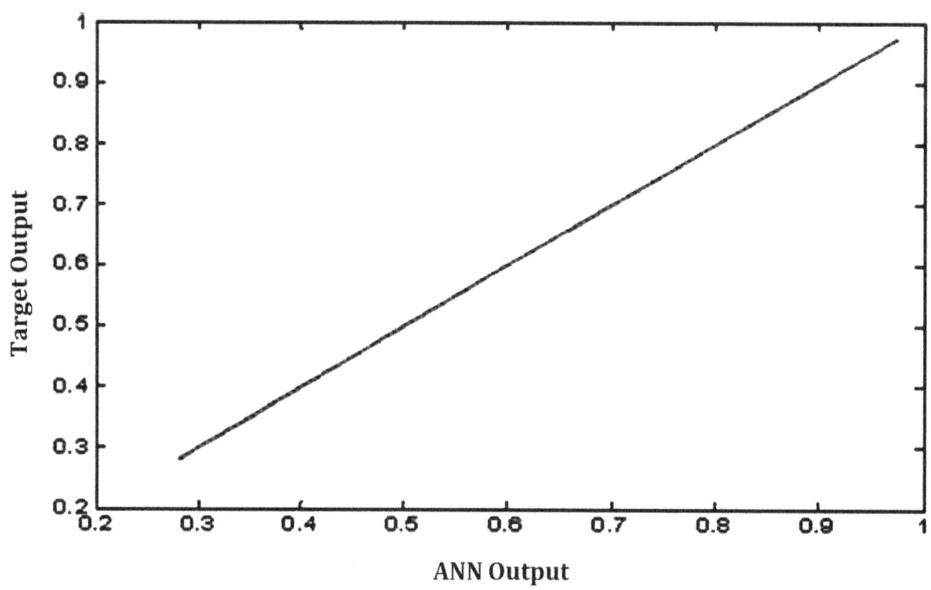

Figure 8. Training error curve for the ANN

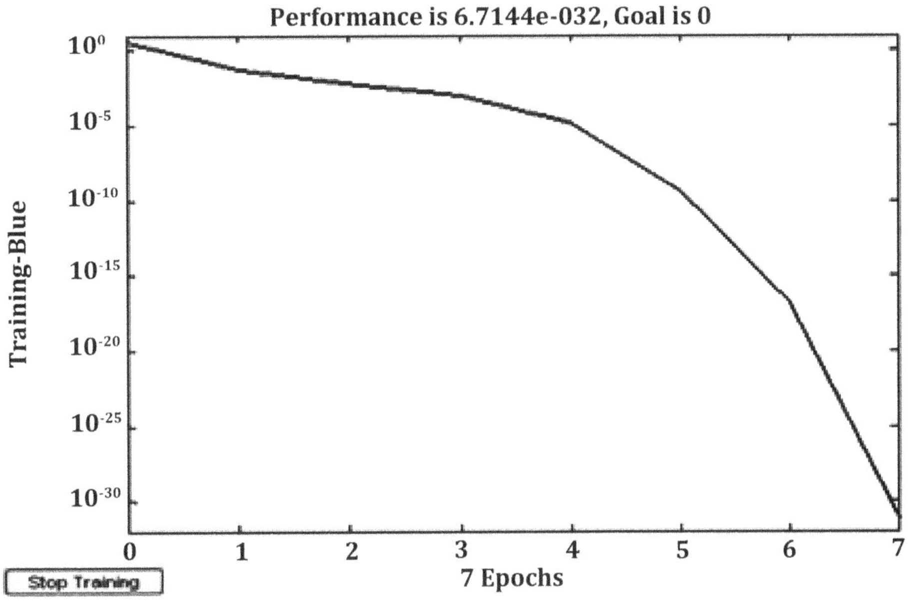

Table 7. Few samples of the L-index values predicted by ANN for different system configurations

Switch Status: '0' indicates switch open, '1' indicates switch closed																				L-index
S_{20}	S_{19}	S_{18}	S_{17}	S_{16}	S_{15}	S_{14}	S_{13}	S_{12}	S_{11}	S_{10}	S_9	S_8	S_7	S_6	S_5	S_4	S_3	S_2	S_1	
1	1	1	0	0	0	1	1	1	1	1	1	1	1	1	1	1	1	1	1	0.3053
1	1	0	0	1	0	1	1	1	1	1	1	1	1	1	1	1	1	1	1	0.3122
1	0	0	1	1	0	1	1	1	1	1	1	1	1	1	1	1	1	1	1	0.3166
0	0	1	1	1	0	1	1	1	1	1	1	1	1	1	1	1	1	1	1	0.3162
1	1	1	0	0	0	1	1	1	1	1	1	1	1	1	1	1	1	1	1	0.3053
*0	0	1	1	1	1	1	1	1	0	1	1	1	1	1	1	1	1	1	1	*0.2858
1	1	1	1	1	0	1	0	0	1	1	1	1	1	1	1	1	1	1	1	0.4992
1	1	1	1	0	1	1	0	0	1	1	1	1	1	1	1	1	1	1	1	0.3087
1	1	1	1	0	0	1	0	1	1	1	1	1	1	1	1	1	1	1	1	0.3402
1	1	1	0	0	1	1	0	1	1	1	1	1	1	1	1	1	1	1	1	0.3983
1	0	0	1	1	1	1	0	1	1	1	1	1	1	1	1	1	1	1	1	0.3298

*00111111101111111111 is the best switching configuration found by ANN

11.1 Loss Reduction

This is one of the objective for which more attention has been paid in the domain of Network Reconfiguration (TzongSua, Changb, Morrison, Gao, & Kundur, 2005). Many soft Computational and optimization techniques such as ANN, GA, Ant Colony, and Particle Swarm have been applied. In optimization approaches (Ant Colony, Particle Swarm) to obtain the configuration for minimal system power loss, considering the operating constraints as bus voltage violation and line

under a certain load pattern is taken into care. In case of ANN (HoyongKim, YunseokKo, & Jung, 1993) and GA (Vulasala, Sirigiri, & Thiruveedula, 2009) approaches some of the pre requisite system data are required for training and initial population respectively.

11.2 Congestion Management

Line load ability limit can be changed by Network Reconfiguration Method (Granelli, Montagna, Zanellini, Bresestib, Vailati, & Innorta, 2006).

11.3 Real-Time Reconfiguration of Distribution Network with Distributed Generation

This is a new methodology (Hossein, Shariatkhah, Haghifam, JavadSalehi, & Moser, 2012) where distributed generation is incorporated to the grid by automatic reconfiguration. The distributed generation wind turbines, solar photovoltaic panels and small hydropower are incorporated to the grid by means of reconfiguration process. Each distributed generating unit is connected to the grid through series switch which is remotely controlled by SCADA. The real-time reconfiguration process can be done by heuristic algorithm to obtain the best possible settings at significant improvement on network performance, reducing losses and increasing reliability.

11.4 Voltage Sag and Reliability Enhancement

Voltage Sag is one of the important issues of Power Quality in Power System. The bus which is far away from the fault point will have low severity due to the effect of Voltage sag propagation. It effect on the system component and reliability. Frequent Line faults are the main reason of voltage sags. During such kind of disturbances, system restoration and reliability improvement (Shareef, Ibrahim, Salman, Mohamed, & Ling Ai, 2014) and voltage sag minimization can be done by Network Reconfiguration.

12. PARALLEL NEURAL NETWORK TECHNIQUE IN ELECTRICAL ENGINEERING: NEURAL ENSEMBLES OR COMMITTEE MACHINES

Decisions can be taken based on switching outcomes from independently different neural networks working in parallel. This arrangement is called Neural Ensemble or Committee Machine, where each neural network unit is an expert independently trained (Haykin, 1999). They can be unique ones following different learning approaches and transfer functions.

Each neural network unit is fed with the same data pattern parallel, and switching combination output is awaited. Output of each such unit is then combined to produce the ultimate output of the Ensemble or Committee. This approach ensures more accomplished predictions (Figure 9).

13. CONCLUSION

A method for improving voltage stability in a power network comprising of multiple lines and switches has been suggested in this chapter based on system reconfiguration approach. ANN based fast and efficient methodology has been developed to obtain the optimum switching combination to achieve best voltage stability.

The results demonstrate the effectiveness of the proposed algorithm. It is observed that Voltage stability has been varying in wide range for different topological structure of test network which can be made by manually or with the help of ANN. In the present study a method for improving voltage stability in a power network

Figure 9. Parallel network implementation

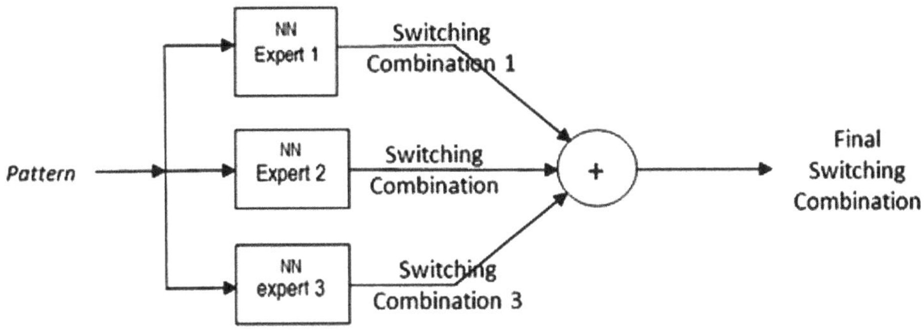

comprising of multiple lines and switches has been suggested based on system reconfiguration approach. ANN based fast and efficient methodology has been developed to obtain the optimum switching combination to achieve best voltage stability. The proposed scheme has been tested on an IEEE 14-bus system comprising of 20 power lines connected through interconnecting and sectionalizing switches, making the total number of possible network configuration to be 220. The ANN is trained with just 300 samples of Load Flow solution data corresponding to 300 judiciously chosen system configurations. After training, the ANN successfully predicts the optimum system configuration and the corresponding switching states, which would lead to best voltage stability. This scheme is easy to implement in any power distribution network using modern SCADA and PLC based automation. The network performance in regards voltage and other electrical parameters can be optimized or visualized for manual or intelligent process through development of SCADA package for power system conditioning and monitoring using the developed ANN based approach as a future scope of work.

ACKNOWLEDGMENT

The authors like to express deepest gratitude to Prof. (Dr.) Chandan Kumar Chanda and Prof. (Dr.) Abhinandan De for valuable guidance in this study. The authors are also highly grateful to Prof. T. N. Saha, Ex-Professor of IIT, Kharagpur. The authors are also highly grateful and deeply indebted to Mr. Sanjay Goswami, Asst. Professor, Computer Application Department, Narula Institute of Technology, Agarpara for his tireless help throughout the is research and writing this manuscript.

REFERENCES

Aboytes, F., & Arroys, G. (1987). Security Assessment in the Operation of Longitudinal Power Systems. *IEEE Transactions on Power Systems*, 1.

Aoki, K., Nara, K., Itoh, M., Satoh, T., & Kuwabara, H. (1989). A New Algorithm for Restoration in Distribution Systems. *IEEE Transactions on Power Delivery*, *4*(3), 1832–1839. doi:10.1109/61.32679

M. Arun & P. Aravindha Babu (2010). Fuzzy based reconfiguration algorithm for voltage stability enhancement of distribution systems. *Expert Systems with Applications: An International Journal archive*, 37(10), 6974-6978.

Baran, M.E., & Wu, F.F. (1989). Network Reconfiguration in Distribution Systems for Loss Reduction and Load Balancing. *IEEE Transactions on Power Delivery*, *4*(2), 1401–1407. doi:10.1109/61.25627

Baran, M. E., & Wu, F. F. (1989). Network Reconfiguration in Distribution Systems for Loss Reduction and Load Balancing. *IEEE Transactions on Power Delivery*, *4*(2), 1401–1407. doi:10.1109/61.25627

Borozan, V., Rajicic, D., & Ackovski, R. (1997). Minimum Loss Reconfiguration of Unbalanced Distribution Networks. *IEEE Transactions on Power Delivery*, *12*(1), 435–442. doi:10.1109/61.568268

Broadwater, R. P., Khan, A. H., Shaalan, H. P., & Lee, R. E. (1993). Time Varying Load Analysis to Reduce Distribution losses through Reconfiguration. *IEEE Transactions on Power Delivery*, *8*(1), 294–300. doi:10.1109/61.180349

Canizares, C.A., De Souza, A.Z., & Quintana, V.H. (1996). Comparison of performance indices for detection of proximity of voltage collapse. IEEE Transactions on Power Systems, 11(3), 1441-1450.

Chakrabarti, A., & Mukhopadhyay, A.K. (1989). Operating Problem in Longitudinal Power System. *Proc. IEEE International TENCON Conference*, India (pp. 932-935). doi:10.1109/TENCON.1989.177084

Chakrabarti Abhinandan De, A., Kothari, D.P., & Mukhopadhyay, A.K. (2010). *An Introduction to Reactive Power Control and Voltage Stability in Power Transmission System*. PHI publishers.

Chen, Y.L. (1996). Weak Bus-Oriented optimal Multi-objective VAR Planning. *IEEE Transactions on Power Systems*, *11*(4), 1885–1890. doi:10.1109/59.544659

Chiang, D., & Rene, J. J. (1992). Optical Network Reconfiguration in Distribution System: Part2: Solution Algorithms and Numerical Results. *IEEE Transactions on Power Delivery*, *5*(3), 1568–1574. doi:10.1109/61.58002

Chiang, H.D., & Jumeau, R.J. (1990). Optimal network reconfigurations in distribution systems: part 1: a new formulation and solution methodology. *IEEE Transactions on Power Delivery*, *5*(4), 1902–1909. doi:10.1109/61.103687

Chiang, H.D., & Jumeau, R.J. (1990). Optimal network reconfiguration in distribution systems: Part2: Solution algorithms and numerical results. *IEEE Transactions on Power Delivery*, *5*(3), 1568–1574. doi:10.1109/61.58002

Chiang, H. D., & Rene, J. J. (1990). Optical Network Reconfiguration in Distribution System: Part1: A New Formulation and Solution Methodology. *IEEE Transactions on Power Delivery*, *5*(4), 1902–1908. doi:10.1109/61.103687

Chiang, H. D., Wang, C. S., & Flueck, A. S. (1997). Look-ahead Voltage and Load Margin Contingency Selection Functions for Large Scale Power Systems. *IEEE Transactions on Power Systems*, *12*(1), 173–180. doi:10.1109/59.574937

Ching-Tzong Su, Chung-Fu Chang, Morrison, G.K., Gao, B., & Kundur, P. (2005). Voltage stability analysis using static and dynamic approaches. *IEEE Transactions on Power Systems*.

Civanlar, S., Grainger, J. J., Yin, H., & Lee, S. S. H. (1998). Distribution Feeder Reconfiguration for Loss Reduction. *IEEE Transactions on Power Delivery*, *3*(3), 1217–1223. doi:10.1109/61.193906

Davidson, D.R. (1975). Long Term Dynamic Response of Power Systems: An Analysis of Major Disturbances. *IEEE Trans. on Power Systems*, 94(3).

A.A. Edris, (1987). Controllable VAR Compensation: A Potential Solution to Loadability Problems of Low Capacity Power Systems. *IEEE Trans. on Power Systems*, 2(3).

Ganesh, V., Sivanagaraju, S., & Thiruveedula, R. (2009). Feeder Reconfiguration for Loss Reduction in Unbalanced Distribution System Using Genetic Algorithm. *International Journal of Electrical and Electronics Engineering*, 3(12), 744-762.

Gao, B., Morison, G., & Kundur, P. (1992). Voltage Stability Evaluation Using Modal Analysis. *IEEE Transactions on Power Systems*, *7*(4), 1529–1542. doi:10.1109/59.207377

Gomes, F.V., Carneiro, Jr, S., Pereira, J.L.R., & Vinagre, M.P. (2006). A new distribution system reconfiguration approach using optimum power flow and sensitivity analysis for loss reduction. IEEE Transaction on Power Systems, 21(4), 1616-1623.

Goswami, S. K., & Basu, S. K. (1992). A New Algorithm for the Reconfiguration of Distribution Feeders for Loss Minimization. *IEEE Transactions on Power Delivery*, *7*(3), 1484–1491. doi:10.1109/61.141868

Granelli, G., Montagna, M., Zanellini, F., Bresestib, P., Vailati, R., & Innorta, M. (2006). Optimal network reconfiguration for congestion management by deterministic and genetic algorithms. *Electric Power Systems Research*, 76, 549–556.

Greene, S., Dobson, I., & Alvarado, F. L. (1997). Sensitivity of the loading margin to voltage collapse with respect to arbitrary parameters. *IEEE Transactions on Power Systems*, *12*(1), 262–272. doi:10.1109/59.574947

Gyugyi, L. (1994). Dynamic Compensation of AC Transmission Lines by Solid-state Synchronous Voltage Sources. *IEEE Transactions on Power Delivery*, *9*(2), 904–911. doi:10.1109/61.296273

HadiSaadat. (1998). *Power Systems Analysis*. Mg-H College.

Haykin, S. (1999). *Neural Networks – A Comprehensive Foundation* (2nd ed.). New Delhi: Pearson.

Hingorani, N. G., & Gyugyi, L. (2000). *Understanding FACTS*. New York: IEEE Press.

Hoyong Kim, Yunseok Ko, & Kyung-Hee Jung (1993). Artificial neural-network based feeder reconfiguration for loss reduction in distribution systems. *IEEE Transactions on Power Delivery*, 8(3), 1356 – 1366.

Hsiao, Y. T., & Chiem, C. Y. (2000). Enhancement of restoration service in distribution systems using a Combination Fuzzy-GA Method. *IEEE Transactions on Power Systems*, *15*(4), 1394–1400. doi:10.1109/59.898118

Hsiao, Y. T., & Chien, C. Y. (2000). Enhancement of Restoration Service in Distribution Systems using a Combination Fuzzy-GA Method. *IEEE Transactions on Power Systems*, *15*(4), 1394–1400. doi:10.1109/59.898118

Y. Y. Hsu, C.C. Liu & C. T. Huang, (1988). Application of Power System Stabilizers and Static VAR Compensators on a Longitudinal Power System. *IEEE Trans. on Power Systems*, 3(4).

Huneault, M., Rosu, C., Manoliu, R., & Galiana, F. D. (1994). A study of knowledge engineering tools in power engineering applications. *IEEE Transactions on Power Systems*, *9*(4), 1825–1832. doi:10.1109/59.331437

Grainger, J.J. & Stevenson, W.D. (1993). Power System Analysis (T.M.H. Edition).

Jasmon, G. B., & Lee, L. H. C. C. (1993). New contingency ranking technique incorporating a voltage stability criterion. *IEE Proceedings. Generation, Transmission and Distribution*, *140*(2), 87–90. doi:10.1049/ip-c.1993.0012

Johnson, W. A. (1980). EHV Operating Problems Associated with Reactive Control (Paper no 80 SM 513-2). IEEE PES Summer Meeting. doi:10.1109/TPAS.1981.316611

Jung, K.-H., Kim, H., & Ko, Y. (1993). Network Reconfiguration algorithm for Automated Distribution Systems based on Artificial Intelligence Approach. *IEEE Transactions on Power Delivery*, *8*(4), 1933–1941. doi:10.1109/61.248305

M. A. Kashem et al. (2000). Network Reconfiguration for Enhancement of Voltage stability in Distribution Network. *IEEE*, 147(3), 171 – 175.

Kessel, P., & Glavitsch, H. (1986). Estimating the voltage stability of a power system. *IEEE Transactions on Power Delivery*, *PWRD-1*(3), 346–354219. doi:10.1109/TPWRD.1986.4308013

Kim, H., Ko, Y., & Jung, K.-H. (1993). Artificial neural network based feeder reconfiguration for loss reduction in distribution systems. *IEEE Transactions on Power Delivery*, *8*(3), 1356–1366. doi:10.1109/61.252662

Larsen, E., Miller, N., Nilsson, S., & Lindgren, S. R. (1992). Benefits of GTO-Based Compensation Systems For Electric Utility Applications. *IEEE Transactions on Power Delivery*, *7*(4), 2056–2062. doi:10.1109/61.157009

Lee, S.-J., Lim, S.-I., & Ahn, B.-S. (1998). Service Restoration of Primary Distribution Systems Based on Fuzzy Evaluation of Multi-Criteria. *IEEE Transactions on Power Systems*, *13*(3), 1156–1163. doi:10.1109/59.709114

Lin, W. M., & Chin, H. C. (1997). Preventive and Corrective Switching for Feeder Contingencies in Distribution Systems with Fuzzy Set Algorithm. *IEEE Transactions on Power Delivery*, *12*(4), 1711–1716. doi:10.1109/61.634195

Mansour, Y., Xu, W., Alverado, F., & Rinzin, C. (1994). SVC Placement Using Critical Modes of Voltage Instability. *IEEE Transactions on Power Systems*, *9*(2), 757–763. doi:10.1109/59.317664

Mendosa, J., Lopez, R., Morales, D., Lopez, E., Dessante, P., & Moraga, R. (2006). Minimal loss reconfiguration using genetic algorithms with restricted population and addressed operators: Real application. IEEE transactions on power systems, 21(2), 948-954.

Mendoza, J., Lopez, R., Morales, D., Lopez, E., Dessante, P., & Moraga, R. (2006). Minimal Loss Reconfiguration using Genetic Algorithms with Restricted Population and Addressed Operators: Real Application. *IEEE Transactions on Power Systems*, *21*(2), 948–954. doi:10.1109/TPWRS.2006.873124

Miller, T. J. E. (Ed.). (1982). *Reactive Power Control in Electric Power System*. John Wiley& Sons.

Miranda, V., Ranito, J. V., & Proenca, L. M. (1994). Genetic Algorithms in Optimal Multistage Distribution Network Planning. *IEEE Transactions on Power Systems*, *9*(4), 1927–1933. doi:10.1109/59.331452

Miranda, V., Ranito, Z. V., & Proenca, L. M. (1994). Genetic algorithm in optimal multistage distribution network planning. *IEEE Transactions on Power Systems*, *9*(4), 1927–1933. doi:10.1109/59.331452

Mohammad-Hossein, Shariatkhah, Mahmoud-Reza Haghifam, JavadSalehi, & Moser, A. (2012). Duration based reconfiguration of electric distribution networks using dynamic programming and harmony search algorithm. *Electrical Power and Energy Systems*, 41, 1–10.

Morrison, G.K., Gao, B., & Kundur, P. (1993). Voltage stability analysis using static and dynamic approaches. IEEE Transactions on Power Systems, 8(3), 1159-1171.

Morton, A. B., & Mareels, I. M. Y. (2000). Efficient Brute-Force solution to Network Reconfiguration problem. *IEEE Transactions on Power Delivery*, *15*(3), 996–1000. doi:10.1109/61.871365

Nara, K., Shiose, A., Kitagaowaand, M., & Ishihara, T. (1992). Implementation of Genetic Algorithm for Distribution System Loss Minimum Reconfiguration. *IEEE Transactions on Power Systems*, *7*(3), 1044–1051. doi:10.1109/59.207317

Peponis, G. J., Papadopoulos, M. P., & Hatziargyriou, N. D. (1995). Distribution Network Reconfiguration to Minimize Resistive Line Losses. *IEEE Transactions on Power Delivery*, *10*(3), 1338–1342. doi:10.1109/61.400914

PrabhaKundur. (1951). *Power System Stability and Control*. EPRI Power System Engineering Series, McGraw Hill.

N.C. Sahoo & K. Prasad, (2006). A fuzzy genetic approach for network reconfiguration to enhance voltage stability in radial distribution systems. *Energy Conversion Management*, 47(18-19), 3288-3306.

Schmidt, H.P., Ida, N., Kagan, N., & Guaraldo, J.C. (2005). Fast Reconfiguration of Distribution Systems Considering Loss Minimization. *IEEE Transactions on Power Systems*, *20*(3), 1311–1319. doi:10.1109/TPWRS.2005.846180

Shareef, H., Ibrahim, A.A., Salman, N., Mohamed, A., & Ling Ai, W. (2014). Power quality and reliability enhancement in distribution systems via optimum network reconfiguration by using quantum firefly algorithm. *Electrical Power and Energy Systems*, 58, 160–169.

Shirmohammadi, D., & Hong, H. W. (1989). Reconfiguration of Electric Distribution Networks for Resistive Line Loss Reduction. *IEEE Transactions on Power Delivery*, *4*(2), 1492–1498. doi:10.1109/61.25637

Shirmohammadi, D., & Hong, H.W. (1989). Reconfiguration of electric distribution network for resistive line losses reduction. *IEEE Transactions on Power Delivery*, *4*(2), 1492–1498. doi:10.1109/61.25637

Taylor, C.W. (1994). Power System Voltage Stability. New York, USA: McGraw Hill.

Taylor, T., & Lubkeman, D. (1990). Implementation on Heuristic Search Strategy for Distribution Feeder Reconfiguration. *IEEE Transactions on Power Delivery*, *5*(1), 239–246. doi:10.1109/61.107279

Van Custem, T., & Mailhot, R. (1997). Validation of a Fast Voltage Stability Analysis Method on The Hydro-quebec System. *IEEE Trans. on Power Systems*, 12(1), 282-288.

Van Cutsem, T., & Vournas, C. (1998). *Voltage Stability of Electric Power systems*. Boston, USA: Kluwer Academic Publishers. doi:10.1007/978-0-387-75536-6

VAR Management Task Force of IEEE. (1984). VAR Management –Problem Recognition and Control. *IEEE Trans. on Power Systems*, 103(8).

Wu, J.S., Tomsovic, K.L., & Chen, C.S. (1991). A Heuristic Search Approach to Feeders Switching Operation for Overload, faults, unbalanced flow and maintenance. IEEE Transactions on Power Delivery, 6(4), 1579-1586.

Wagner, T. P., Chikhani, A. Y., & Hackam, R. (1991). Feeder Reconfiguration for Loss Reduction: An Application of Distribution Automation. *IEEE Transactions on Power Delivery*, *6*(4), 1922–1931. doi:10.1109/61.97741

Wagner, T. P., Chikhani, A. Y., & Hackam, R. (1991). Feeder Reconfiguration for Loss Reduction: An Application of Distribution Automation. *IEEE Transactions on Power Delivery*, *6*(4), 1922–1933. doi:10.1109/61.97741

Warwick, K., & Ekwue, A. (Eds.). (1997). *Artificial Intelligence Techniques in Power Systems*. K. Warwick Arthur Ekwue Raj Aggarwal, Institution of Electrical Engineers. doi:10.1049/PBPO022E

Weedy, B.M., & Cox, B.R. (1968). Voltage Stability of Radial Power Links. *Proc. of IEEE (Vol. 115)*.

Zhao, B., Guo, C.X., & Cao, Y.J. (2005). A Multiagent-Based Particle Swarm Optimization Approach for Optimal Reactive Power Dispatch. *IEEE Transactions on Power Systems*, *20*(2), 1070–1078. doi:10.1109/TPWRS.2005.846064

Zhou, Q., Shirmohammadi, D., & Liu, W.-H. E. (1997). Distribution Feeder Reconfiguration for Service Restoration and Load Balancing. *IEEE Transactions on Power Systems*, *12*(2), 724–729. doi:10.1109/59.589664

Chapter 11

Global and Local Clustering-Based Regression Models to Forecast Power Consumption in Buildings

Gonzalo Vergara
University of Valencia, Spain

Juan J. Carrasco
University of Valencia, Spain

Jesus Martínez-Gómez
University of Castilla-La Mancha, Spain

Manuel Domínguez
University of León, Spain

José A. Gámez
University of Castilla-La Mancha, Spain

Emilio Soria-Olivas
University of Valencia, Spain

ABSTRACT

The study of energy efficiency in buildings is an active field of research. Modeling and predicting energy related magnitudes leads to analyze electric power consumption and can achieve economical benefits. In this study, classical time series analysis and machine learning techniques, introducing clustering in some models, are applied to predict active power in buildings. The real data acquired corresponds to time, environmental and electrical data of 30 buildings belonging to the University of León (Spain). Firstly, we segmented buildings in terms of their energy consumption using principal component analysis. Afterwards, we applied state of the art machine learning methods and compare between them. Finally, we predicted daily electric power consumption profiles and compare them with actual data for different buildings. Our analysis shows that multilayer perceptrons have the lowest error followed by support vector regression and clustered extreme learning machines. We also analyze daily load profiles on weekdays and weekends for different buildings.

DOI: 10.4018/978-1-4666-9911-3.ch011

INTRODUCTION

Nowadays, building consumption is estimated to be about 40% from total energy consumed in developed countries (U.S. Department of Energy, 2010). The International Energy Agency (IEA) claims that building sector is one of the most interesting sectors to invest in energy efficiency. From the economical point of view, it is possible to reduce the energy consumption in 1509 Mtoe (1 toe = 1 tonne oil equivalent = 11.63 MWh) in year 2050. This fact also implies important environmental benefits: if the energy demand is reduced from buildings, the carbon dioxide (CO_2) emissions could be dramatically reduced too. It can be estimated that 12.6 Gt of CO_2 could not be emitted to the atmosphere in 2050 (I.E.A. International Energy Agency, 2013).

An important part of this consumption is due to the Heating, Ventilating and Air Conditioning (*HVAC*) systems. For instance, in the USA, *HVAC* systems are 50% in building consumption and 20% in total electric energy consumption (Pérez Lombard, Ortiz, & Pout, 2008).

European Commission has initiated actions to reach a 20% reduction in energy consumption until 2020. It would imply 30 billion Euros savings per year (European Commission, 2010). The prediction of energy demand in buildings is an important tool to obtain economical and environmental benefits, allowing to:

- Improve the knowledge of electric power demand.
- Optimize *HVAC* systems.
- Contract the optimal electric power requirements that fit buildings needs.

In the literature, there is a great variety of techniques that can be used in prediction of energy demanding tasks: Linear regression (*LR*) is applied to predict monthly electrical consumption in large public buildings (Ma, Yu, Yang, & Wang, 2010), autoregressive integrated moving average (*ARIMA*) models let failure detection in electronic equipment (Bian, Xu, Li, & Xu, 2007), also, the building occupancy is considered to improve predictions (Newsham & Birt, 2010). Artificial neural networks (*ANN*) have been widely applied to study energy consumption in buildings: minimization of the energy to air condition an office-type facility (Kusiak, Li, & Tang, 2010), energy of office buildings with day lighting (Wong, Wan, & Lam, 2010) and air condition heating dependence on electrical consumption (Ekici & Aksoy, 2009). Also, *ANN* are applied to analyze power consumption in subway station design in (Leung & Lee, 2013) and support vector regression (*SVR*) is utilized in (Zhao & Magoulès, 2010) to predict electrical consumption in large buildings using simulated dataset.

For this study we used real datasets from 30 different buildings of the University of León (Spain). Linear time series and machine learning techniques, as Multilayer Perceptrons (*MLP*), *SVR* and *ELM* were utilized to model and predict active power of each building. We observe that multivariate non-linear methods as *MLP*, *SVR* and *ELM* give better results than classical autoregressive (*AR*) or *LR* models.

ANALYZED DATASETS

From the technical code of Spanish buildings (Ministerio de Fomento, 2010), the city of León (Spain) is placed in E1 climate zone at 346 m above sea level. This zone has severe climate conditions in winter as well as in summer, hence the electrical consumption due to *HVAC* equipment is high. Table 1 shows some statistics of climate variables, the temperature goes from -3.69ºC in winter to 38.95ºC in summer.

The data was recorded simultaneously for 30 buildings placed at campus of the University of León from March 1, 2010 to March 31, 2011. The buildings purposes are of different nature: teaching, research and services as administration

Table 1. Basic statistics of ambience temperature, relative humidity and solar radiation at campus of University of León

Statistic	Temperature (°C)	Relative Humidity (%)	Solar Radiation (W/m²)
Mean	13.84	66.30	163.86
Maximum	38.95	99.73	996.13
Minimum	-3.69	8.09	0
Standard deviation	7.33	19.49	244.85

and dining halls. These buildings have different features, e.g., size, occupation level and electric equipment.

The available datasets contain information about environmental, electric and time information variables. The environmental variables were acquired from a single meteorological station, meanwhile, the rest of variables were separately acquired from each building. The variables involved in this study are: hour, day, month, year, ambient temperature, relative humidity, solar radiation, working day, active power and standard deviation of active power.

Data Processing

The datasets corresponding to 30 buildings have been registered each two minutes for 13 months, reaching to 285120 instances for each building. In order to obtain an hourly measure for each variable, the mean was calculated reducing the number of instances to 9504 per building.

Buildings have different energy characteristics; therefore we start by grouping them according to their behavior. For this task, a principal component analysis (*PCA*) is done for power related variables: minimum, maximum, mean, range, standard deviation and coefficient of variation for all buildings. After this, over the obtained *PCA* space a *k-medoids* clustering were performed. The best choice for the number of groups, in terms of mean silhouette index, was three groups (Figure 1). These three groups have 4, 10 and 16 buildings

Figure 1. Second principal component versus first principal component after PCA for 30 buildings. Clustering with k-medoids is shown where ellipses correspond to different buildings group: Group 1 (13=Biology), group 2 (6=Philosophy) and group 3 (25=Dormitory)

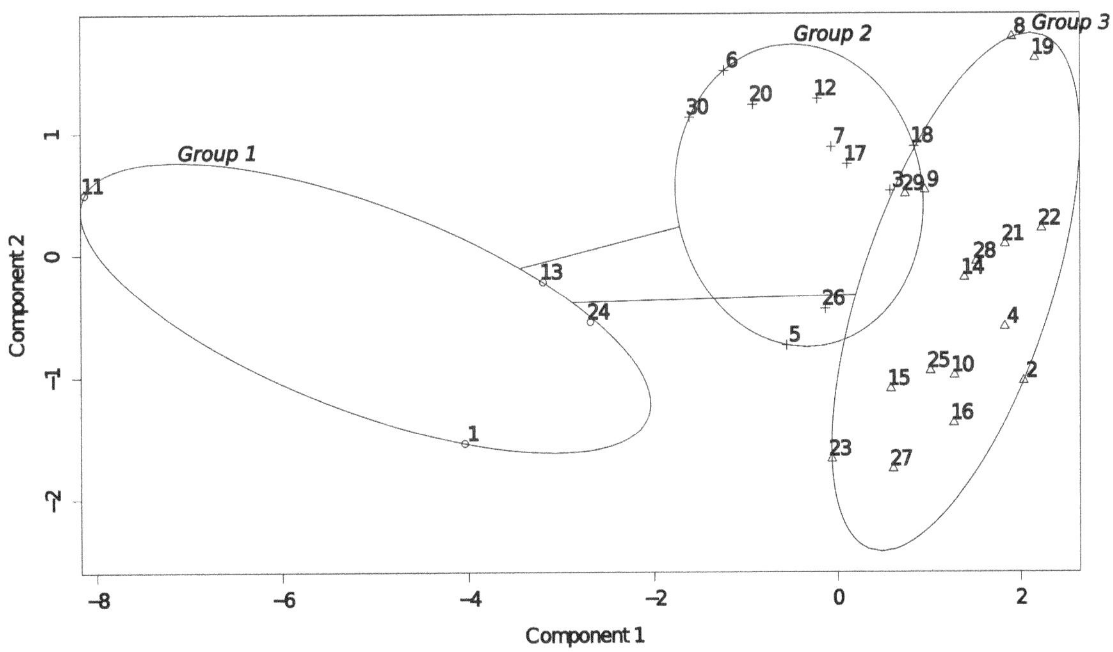

each one. As a descriptive analysis, we found that first group of buildings has high hourly power (100kW) and low power variation (30%), second group has medium power (30kW) and high power variation (90%) and third group has low power (10kW) and high power variation (100%). The three selected prototypes representing each group are the faculties of *Biology* and *Philosophy* and the *Dormitory*. Their floor space, mean active power and standard deviation of active power are shown in Table 2. These buildings have different consumption profiles, so they are useful to test different models. For instance, *Biology* building has a high electrical consumption whereas *Dormitory* building has low consumption but high deviation, having half floor space than the former. The active power histograms of these buildings are shown in Figure 2, appearing a principal peak close to 50kW and a secondary one at 100kW for *Biology* building. A similar profile at 10kW and 50kW is found in the *Philosophy* building but a different histogram appears in Dormitory where a principal peak is located in 10kW and a smoother curve dies in 45kW.

Table 2. Description of medoids buildings for different groups with floor space and some statistics of active power

Building	Usable Space (m^2)	Statistic	Active Power (kW)
Biology	12480	Mean	72.54
		Minimum	26.51
		Maximum	218.80
		Standard deviation	31.70
Philosophy	14700	Mean	27.73
		Minimum	5.78
		Maximum	167.25
		Standard deviation	26.78
Dormitory	5500	Mean	14.12
		Minimum	2.44
		Maximum	102.71
		Standard deviation	9.25

In the case of the *AR* univariate technique, the only analyzed variable is the active power. Missing values imply to choose datasets separated in time with more than 400 instances leading to seven, five and nine time series for the faculties of *Biology*, *Philosophy* and *Dormitory* respectively. This fact will be an impediment for modeling and prediction in some days as it will be discussed in results section.

For multivariate techniques, variables considered were: day, month, year, working day, exterior temperature, relative humidity and exterior solar radiation. The variable hour has to be preprocessed because of the discontinuity between 23:00 and 0:00. In order to avoid this it is common to unfold the hour in two new variables:

$$\left(h_x, h_y\right) = \left(\sin\left(\frac{2\pi h}{24}\right), \cos\left(\frac{2\pi h}{24}\right)\right) \tag{1}$$

where h means hour.

In addition, datasets have been randomly rearranged 10 times and these arrangements are used for each multivariate technique. Inputs for all variables were standardized.

UTILIZED TECHNIQUES

Physical Models

Physical models are utilized to study thermal behavior in buildings from heat transfer equations, geometrical parameters and construction information (Liao & Dexter, 2004). Nevertheless, machine-learning techniques can fit a dataset achieving a high accuracy in prediction without any additional hypothesis. The main advantage of these methods is the low level of problem domain

Figure 2. Active power histograms for group representative buildings: Biology, Philosophy and Dormitory

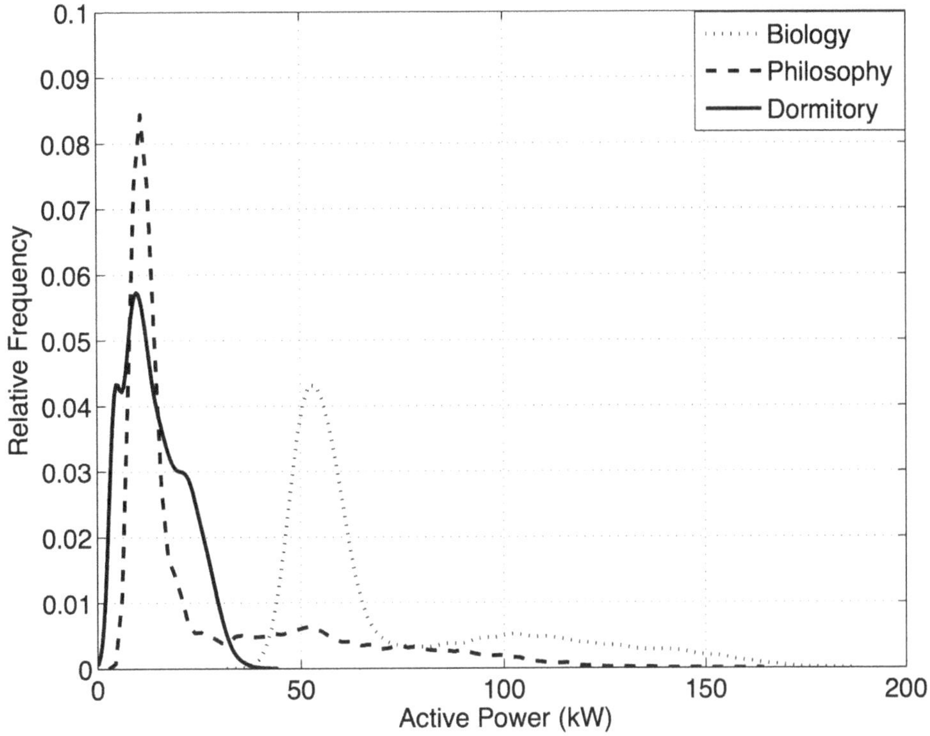

knowledge needed, e.g. physical parameters of the system, to construct the predictive models (Foucquier, Robert, Suard, Stéphan, & Jay, 2013).

In certain cases, linear techniques, as *AR* or *LR*, could be satisfactory (Soliman & Al-Kandari, 2010). These methods are easy to program and their computational cost is low. When linear techniques are not enough, it is recommended to use non-linear techniques as *MLP* (Bishop, 2006; Haykin, 2009). *MLP* have proven robustness and good performance in a broad variety of applications (Paliwal & Kumar, 2009; Vellido, Lisboa, & Vaughan, 1999). Another kind of non-linear successful methods used in regression problems are *SVR*. In the following, mentioned techniques are briefly described:

Clustering

Clustering algorithms (Webb & Copsey, 2011) are commonly utilized in data mining with the objective of grouping patterns to discover some kind of subjacent structure in data. In general, a cluster is a region in the space of characteristics where there is a high density of patterns. We can utilize different criteria to do clustering:

- **Physical Variables:** As a first approximation, the physical variable of interest is selected, in our case: active power, day type, hour, etc. After this, the training set is divided for different values of the chosen variable. As an example, data could be divided in day or night by means of variable solar radiation. Also, data may be divided in two or more sets using the variable active power. For this purpose histogram of power for each building give useful information (Figure 2).
- **Clustering Algorithms:** The algorithms of automatic clustering have more than fifty years and his origin belongs to differ-

ent subjects. There exist two types of clustering algorithms. Hierarchical algorithms start with each point in its own cluster. Clusters are then combined based on their closeness and combination stops when further combination leads to clusters that are undesirable. Partitional algorithms involve point assignment. Points are considered in some order, and each one is assigned into which it best fits. This process is normally preceded by a short phase in which initial clusters are estimated. Variations allow occasional combining or splitting of clusters, or may allow points to be unassigned if they are too far from any of the current clusters (outliers). Algorithms for clustering can assume a Euclidean space where it is possible to summarize a collection of points by their *centroid* (the average of the points).

$$d = \sqrt{\sum_{i=1}^{m} \left(x_i - y_i\right)^2} \quad (2)$$

where $(x_1, \ldots, x_m)$ and $(y_1, \ldots, y_m)$ are two random m-dimensional points.

- **K-Means Algorithm:** The *k-means* is a partitional algorithm born in different disciplines and due to (Lloyd, 1982), (Ball & Hall, 1965) and (Macqueen, 1967). The objective is to divide dataset into k clusters which minimizes the sum of distances to the center of *j-th* cluster or *centroid* (Webb & Copsey, 2011):

$$S_j = \sum_{i=1}^{N} z_{ji} \left\| x_i - m_j \right\|^2 \quad (3)$$

where: $z_{ji} = 1$ if x_i belongs to j-th cluster and $z_{ji} = 0$ otherwise.

$$m_j = \frac{1}{n_j} \sum_{i=1}^{N} z_{ji} x_i$$

is the mean position of j-th cluster

$$n_j = \sum_{i=1}^{N} z_{ji}$$

The mean sum of distances for k clusters is

$$\frac{1}{N} \sum_{j=1}^{k} S_j$$

which is the function to minimize.

The algorithm consists of:

  - **Step 1:** Take k random points as initial *centroids*.
  - **Step 2:** Associate each point to the cluster it belongs or nearest cluster.
  - **Step 3:** Calculate the coordinates of *centroids*.
  - **Step 3:** Repeat steps two and three until the coordinates of *centroids* remain unchanged.

- **K-Medoids Algorithm:** The *k-medoids* algorithm is a partitional algorithm related to the *k-means* algorithm but the centers are in this case data points or *medoids*. This method is more robust to noise and outliers than the *k-means* because it minimizes a sum of pair wise dissimilarities instead of a sum of squared Euclidean distances.
- **Silhouette Index:** Usually the number of clusters is unknown or it is difficult to select it adequately for the dataset. The silhouette (Rousseeuw, 1987) give an idea of the adjustment of a point to a cluster. Some definitions for the *i-th* point are:

- *a(i)* is the mean distance of *i-th* point to each point of the cluster whom *i-th* point belongs.
- *d(i,C)* is mean distance from *i-th* point to each point of cluster *C*.
- *b(i)* is the minimum of *d(i,C)* for all clusters.

With these definitions, silhouette for *i-th* point is defined as follows:

$$s(i) = \frac{b(i) - a(i)}{\max\{a(i), b(i)\}} \tag{4}$$

this equation shows that $-1 \leq s(i) \leq 1$. If the nearest cluster is known as the neighbor of *i-th* point, *s(i)* = 1 indicates that *i-th* point is well grouped and *s(i)* = -1 that *i-th* point is better grouped in the neighbor cluster. If *s(i)* = 0 then *i-th* point is in the frontier between two clusters. Evaluation of silhouettes let us to calculate the optimal number of clusters for the dataset.

Univariate Techniques

- **Autoregressive Models**: *AR* models are random processes that can be applied to build models aimed at predicting the current output of the system based on a linear combination of its past outputs (Harris & Sollis, 2003). The formula that defines an *AR* process of order *p* is:

$$x(t) = c + \sum_{i=1}^{p} \varphi_i \, x(t-i) + \varepsilon(t) \tag{5}$$

where *c* is a constant, φ_i are the parameters of the model and *ε(t)* is white noise. The *AR* parameters can be calculated using different methods as least-squares, Yule-Walker or Burg's methods (Kay, 1988). In this study, we have utilized the forward-backward least squares method.

Multivariate Techniques

- **Linear Regression**

A regression expression establishes a relationship between a dependent variable **y** and some independent variables **X** (Weisberg, 2005). We can express this in matrix notation:

$$y = X\beta + e \tag{6}$$

where **y** is a column vector of *n* predictions defined as $\mathbf{y} = [y_1 \ldots y_n]^t$, **X** is a matrix of order *n* x *(p +* 1 *)* defined as $\mathbf{X} = [\mathbf{x}_1 \ldots \mathbf{x}_n]^t$, where t indicates transpose, $\mathbf{x}_i = [1\, x_i^1 \ldots x_i^p]$, β is a column vector of *p* + 1 regression coefficients defined as $\beta = [\beta_0 \ldots \beta_p]^t$ and **e** is a column vector of *n* errors.

The estimator $\hat{\beta}$ of β is obtained with (Weisberg, 2005):

$$\hat{\beta} = (X^t X)^{-1} X^t y \tag{7}$$

- **Artificial Neural Networks**

In this study we focus on the mathematical approach of artificial neural models related to statistical models either linear or non-linear (Haykin, 2009; Bishop, 2006). In this approach, an artificial neural system can be described as it is shown in Figure 3.

Three main parts can be observed:

- Neural model. It is the structure used to solve a certain problem. This model can be either linear or non-linear, it can have one or more than one outputs, it can consist of a combination (linear, non-linear, hierarchical, etc.) of simple neural models, etc. The performance of this model depends on several parameters that determine the complexity of the neural model.
- Cost function. It provides an evaluation about the quality of the solution obtained

Figure 3. Schematic of a neural model

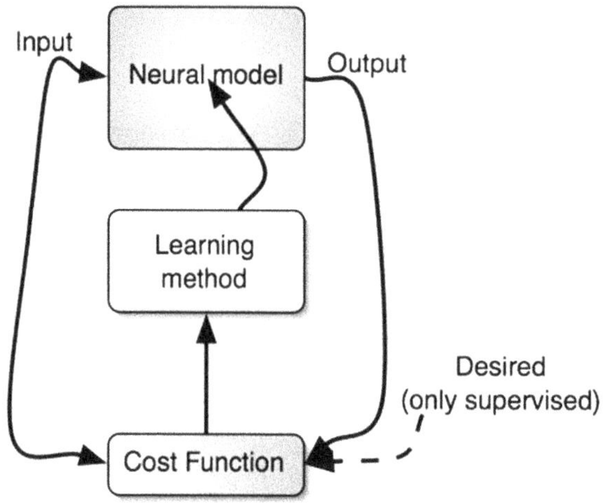

by the neural model. If an external signal to carry out this evaluation is available, then the neural model is called supervised model. If an external signal is not available, then the neural model is called unsupervised model (Haykin, 2009).

- Learning algorithm. The task of the learning algorithm is to obtain those parameters of the neural models that provide a solution to the tackled problem. If this solution does not exist, then the task of the learning algorithm is to find an optimal solution according to the criterion set by the cost function. Therefore, the learning algorithm depends on the chosen cost function.

Summarizing, there are three basic elements in an artificial neural model. Since all these elements can change independently to the others, there is a huge amount of models/cost functions/ learning algorithms (Arbib, 2003). In this chapter, we focus on the most widely used neural model, the Multilayer Perceptron (*MLP*) trained by supervised algorithms (in the case of unsupervised training, the resulting model is the so-called Nonlinear Principal Component Analysis, *NPCA*). We also analyze the most common cost functions associated with *MLPs*. For a further analysis of the most important neural models, supervised and unsupervised, the reader is encouraged to consult the excellent text (Haykin, 2009).

- **Multilayer Perceptron**

An *MLP* is a neural network made up by multiple, similar, non-linear processing units, termed neurons by analogy to the integrate-and-fire action of neural cells in biological nervous systems. Each neuron carries out a many-to-one mapping its inputs to a single scalar output. A general schematic of a neuron model is shown in Figure 4.

- **Neurons:** This neuron model is formed by the following components:
    - **Parameters:** These parameters are usually known as synaptic weights. They are used in combination with the inputs according to a certain function. Considering input vectors and coefficient vectors, the most used functions are the scalar product and the Euclidean distance.

Figure 4. General schematic of a neuron

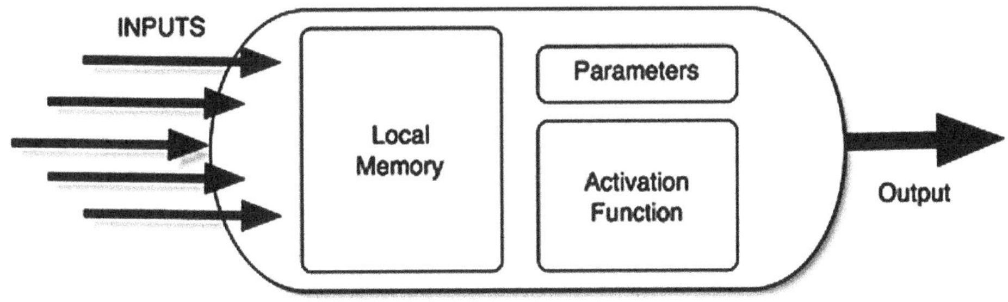

- **Activation function:** This function is very important since the neural network capabilities to solve complex problems stem from it. It is a non-linear function whose argument is the aforementioned combination between synaptic weights and input vector. The most used functions are the sign function, the hyperbolic tangent, the sigmoidal function (hyperbolic tangent modified by restricting the range of values between 0 and 1), and the Gaussian function.
- **Local memory:** This component is used when neural networks are designed to time series modeling. Local memories can store either previous inputs or previous outputs, so that the neuron "remembers" previous behaviors. There are many different possibilities to implement this kind of memories (Weigend & Gershenfeld, 1993).

In the remainder of the chapter, we will focus on the neuron model shown in Figure 5:

This model is formed by the following components:

- **Sum function:** This component computes the sum of the product between input vectors and synaptic weights. Let be $\mathbf{w}=[w_0, w_1, \ldots, w_k]$ the synaptic weight vector and $\mathbf{x}=[1, x_1, \ldots, x_k]$ the input vector, this sum can be given by the scalar product of both vectors. From a biological point of view, this function represents the action of the inputs produced in the axons of the biolog-

Figure 5. Schematic of the most used neuron model

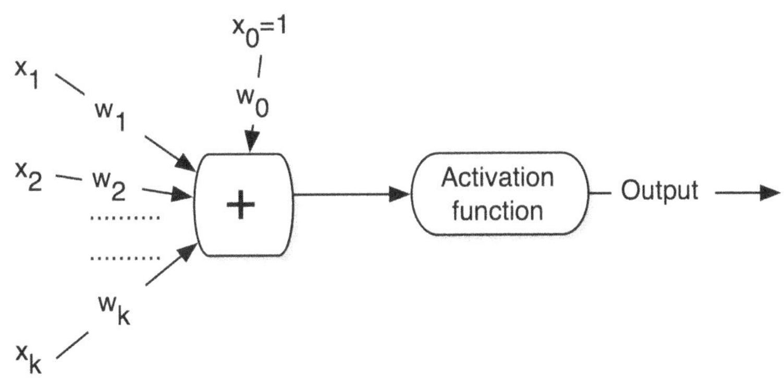

ical neurons (Arbib, 2003). The coefficient w_0 plays a relevant role in the processing of the neuron, and it is called threshold or bias.

- **Activation function:** Among all the possible choices for this activation function, the following ones should be emphasized:
  - **Sign function:** It was the first used activation function in an artificial neural model (Arbib, 2003). This function proposes a crisp separation of the input data, so that they are classified as ± 1. This function is defined as follows:

$$f\left(x\right)=\begin{cases}-1, x<0\\ 1, x\geq 0\end{cases} \qquad (8)$$

  - **Sigmoidal function:** Since many learning algorithms need a differentiable activation function, and the sign function cannot be differentiated in the origin, the sigmoidal function was proposed. This function is given by the following expression:

$$f(x)=\frac{a}{1+e^{-b\cdot x}} \qquad (9)$$

where *a* stands for the amplitude of the function and *b* is the slope in the origin; the higher the value of *b*, the closer the sigmoidal function to the sign function (but with the outputs ranging between 0 and *a*). The bipolar version of the sigmoidal function (outputs ranging between –*a* and *a*) is given by the hyperbolic tangent:

$$f(x)=a\ \frac{1-e^{-b\cdot x}}{1+e^{-b\cdot x}} \qquad (10)$$

  - **Gaussian function:** An alternative activation function is used in a different type of neural networks, known as Radial Basis Function (*RBF*). Its expression is given by:

$$f(x)=K_1\cdot e^{-K_2\cdot(x-c)^2} \qquad (11)$$

being K_1, K_2 and *c*, values which determine the amplitude of the Gaussian function, its width and its centre, respectively. The main characteristic of this activation function, which makes different from others, is its locality. The function tends to zero from a certain value x on; this way, only in a reduced range of input values, the output of the activation function is considerably different from zero. Gaussian functions have their centre and width values typically heuristically determined from the data.

- **Architecture**

The architecture of a neural model gives information about how neurons are arranged in the neural model. In the case of *MLP*, as shown in Figure 6, neurons are arranged in layers: one input layer, one output layer, and one or more than one hidden layers (Arbib, 2003). This arrangement in layers is due to the fact that the outputs of the neurons of a certain layer are used as inputs to the neurons of next layer (network without feedback) and/or to neurons of previous layers (networks with feedback or recurrent networks).

The number of neurons in both input and output layers is given by the problem to solve, whereas the number of hidden layers, and the neurons in each layer should be chosen by the designer of the neural network. It is shown that an *MLP* with one hidden layer is a universal approximator of continuous functions; if the function to model is not continuous, then two hidden layers become necessary (Cybenko, 1988). Although there are several rules to assess the approximately optimal number of hidden neurons, there is no precise methodology to determine this number exactly.

Figure 6. Multilayer perceptron. Dotted arrows stand for recurrent neural systems

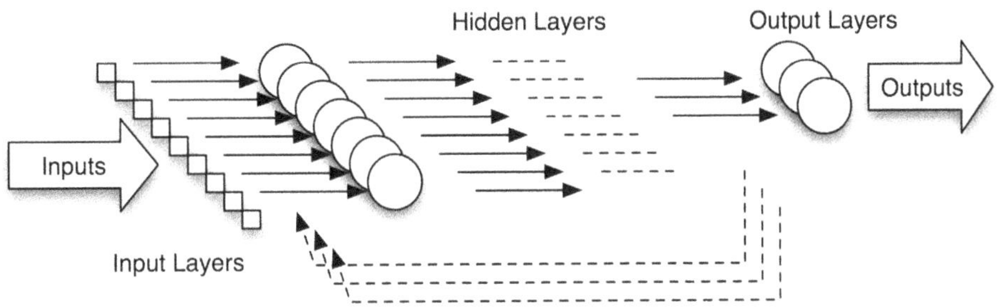

Therefore, trial-and-error procedures are usually carried out to estimate the number of hidden neurons, often relying on cross-validation or other evaluation procedures.

- **Classical learning algorithms**

A learning algorithm is a procedure that adapts the parameters and/or architecture of the neural model in order to solve a certain problem. The algorithms to adapt the architecture of the *MLP* are known as pruning methods or growing methods, depending on the strategy to carry out this adaptation (Haykin, 2009). Pruning methods have been more widely used than growing methods.

Learning algorithms to adapt the parameters of the neural model tend to be based on the minimization of the cost function chosen to solve the posed problem. The most important characteristics that a learning algorithm must show are: efficiency, robustness, independence on the initial conditions, generalization capabilities, scalability with the size and complexity of the data.

- **Backpropagation algorithm (*BP*)**

The *BP* was the first algorithm used for *MLP* adaptation, and it is still the most known and used nowadays. The goal pursued by the algorithm is to obtain the minimum of the cost function, which is denoted as *J*. The actual solution of *J* is computationally unfeasible in many applications. Moreover, the solution of *J* in a certain instant may be uninteresting in many cases since the properties of the input signals to the neural model may be time-dependent and the neural models must adapt to these changes. An iterative solution is proposed:

$$w_{n+1} = w_n + \Delta w_n \tag{12}$$

where **w** are the parameters of the neural model and the subscript stands for the time instant. Many algorithms are designed to accomplish Equation 12; within these algorithms, there are two possibilities for learning (Haykin, 2009):

- **On-Line:** The *MLP* is fed during all the training process with the input of every pattern and the corresponding desired output. Then, the error is measured and the synaptic weights are adapted depending on this error by using the chosen algorithm.
- **Batch:** In this case, the error committed by the neural model for all the patterns is computed. Then, the model parameters are adapted depending on the average error for all the patterns. The computation of all the outputs of the neural model for all the available input patterns is known as epoch.

One of the keys of Equation 12 is to find out the optimal increase/decrease in the parameters that enables to find the minimum of the function. A low computational burden approach is based on a geometrical analysis of the problem. This analysis is based on finding the direction of the minimum from a certain point w_n in the parameter space of the neural model. Since the function gradient points to the direction of the function maximum, the approach will be based on finding the function minimum by moving the synaptic weights in the opposite gradient direction (Bishop, 2006; Luenberger, 1984):

$$w_{n+1} = w_n - \alpha \cdot \nabla_{w_n} J \tag{13}$$

where α is the so-called learning rate or adaptation constant.

The *BP* algorithm, based on Equation 13, is a gradient-descent algorithm which back propagate the error signals from the output layer to the input layer, thus optimizing the values of the synaptic weights through an iterative process. Therefore, two stages can be considered:

- **Feed-forward propagation:** The output of the neural network is obtained, and then, the error is computed by comparing this output with the desired signal.
- **Backpropagation:** Depending on the errors committed in the output layer, the algorithm optimizes the values of the synaptic weights by means of error backpropagation from the output layer to the input layer, and through the hidden layers.

The gradient of the cost function must be computed with respect to the parameters of the neural network, i.e., the synaptic weights. The following notation will be used: $y_s^{L-1}; z_s^{L-1}$ stands for the output of the sum function and the activation function, respectively (neuron s and layer L-1). Therefore, taking all these facts into account:

$$\begin{gathered} \frac{\partial J}{\partial w_{ij}^L} = \frac{\partial J}{\partial y_i^L} \cdot \frac{\partial y_i^L}{\partial w_{ij}^L} = \\ \delta_i^L \cdot \frac{\partial}{\partial w_{ij}^L} \sum_s w_{is}^L \cdot z_s^{L-1} = \delta_i^L \cdot z_j^{L-1} \end{gathered} \tag{14}$$

where $\delta_i^L = \frac{\partial J}{\partial y_i^L}$ is known as local gradient and can be obtained as follows:

$$\begin{gathered} \delta_i^L = \frac{\partial J}{\partial y_i^L} = \sum_m \frac{\partial J}{\partial y_m^{L+1}} \cdot \frac{\partial y_m^{L+1}}{\partial y_i^L} = \\ \sum_m \delta_m^{L+1} \cdot \frac{\partial y_m^{L+1}}{\partial y_i^L} \Rightarrow \\ \delta_i^L = \sum_m \delta_m^{L+1} \cdot \frac{\partial}{\partial y_i^L} \cdot \sum_j w_{mj}^{L+1} \cdot z_j^L \Rightarrow \delta_i^L \\ = \frac{dz_i^L}{dy_i^L} \cdot \sum_m \delta_m^{L+1} \cdot w_{mi}^{L+1} \end{gathered} \tag{15}$$

Therefore, an iterative procedure is used to compute local gradients of first layers from local gradients of last layers. The output layer gradient is given by:

$$\delta_k = \frac{\partial J}{\partial e_k} \cdot \frac{de_k}{dz_k} \cdot \frac{dz_k}{dy_k} = -\frac{\partial J}{\partial e_k} \cdot \frac{dz_k}{dy_k} \tag{16}$$

Output gradient depends on the error and activation functions. A very attractive advantage of using a sigmoidal/hyperbolic tangent activation function stems from the fact that their derivatives can be expressed by using the own activation functions (Haykin, 2009).

In spite of the advantages offered by the BP algorithm, it also shows a number of drawbacks that should be known (Haykin, 2009; Bishop 2006; Arbib, 2003):

- **Neuron saturation:** Since the derivative of the activation function appears in the weight update, and this derivative equals zero for the most used activation functions (hyperbolic tangent and sigmoidal) in the function extremes, weights are not updated in these zones although the model is committing an error.

- **Weight initialization:** Weight initialization is basic in order to achieve a good modeling. Since the learning algorithm is based on finding the minimum of the error function that is closest to a certain initial point, this minimum may be a local minimum, not a global one, and therefore this initial point is fundamental for assuring that the network finally achieves a global minimum. Weight initialization affects algorithm behavior in three main factors: a) Convergence speed; the convergence speed depends not only on the learning rate, but also on the initial distance to the minimum. b) Minimum achieved; the algorithm finds the closest minimum to the initial point, which may be a local minimum. c) Neuron saturation; large values of the weights can involve a neuron saturation.
- **Plateaux:** Weight update is proportional to the derivative of the error function. This derivative equals zero or a very low value on the flat parts of the error function. Therefore, weights are hardly updated.
- **Choice of the learning rate:** Too high values can involve instabilities while too low values can make the converge speed very slow. There are many algorithms that propose strategies to find an optimal learning rate. They tend to be based on the following claim: "the value of the learning rate should be high far away from the minimum and should be low near the minimum"
- **Early stopping:** There are different criteria to carry out the learning stopping, such as, to fix a number of epochs in advance, an error threshold, to find plateaus in the cost function, etcetera. Data are usually split into two sets: a training set and a generalization set. The former is used to train the network, whereas the latter is used to check the behavior of the network with patterns different from those used in the training process. The goal is to achieve a model with good generalization properties. The generalization error tends to decrease as the learning process is progressing until a certain epoch, in which the generalization error starts to increase because the network is over fitting the patterns of the training data set. When this change in the tendency of the generalization error is observed, the learning must be stopped. There are also different criteria to decide the rate of patterns that should be assigned to each data set: 66% of the patterns to the training set, a small percentage if there is a large amount of data available, and there are also dynamic processes to carry out the selection of these sets (Haykin, 2009; Bishop, 2006).
- **Architecture choice:** The number of hidden neurons and layers is a difficult choice, being determined in many cases by using trial-and-error procedures. Nonetheless, there also pruning and growing methods that are used to find optimal structures (Reed, 1993).

In spite of *BP* algorithm and its variants are usually chosen to train *MLPs*, more complex algorithms have also been proposed. These algorithms tend to show a higher computational burden and faster convergence speed than *BP*.

- **Levenberg-Marquardt (*LM*) algorithm**

The Levenberg-Marquardt (*LM*) algorithm was proposed with the quadratic cost function (Bishop, 2006):

$$J = \frac{1}{2}\sum_k e_k^2 \tag{17}$$

being e_k the committed error for pattern k and appearing the factor ½ for the sake of simplicity (after differentiation, all the constant terms are cancelled). If the vector of components e_k is denoted by **e**, and if small perturbations of the synaptic weights are considered (Bishop, 2006):

$$e\Big|_{new} = e\Big|_{old} + L\left[w\Big|_{old} - w\Big|_{new}\right] \quad (18)$$

where **L** is a the following matrix

$$L_{st} = \frac{\partial e_s}{\partial w_t} \quad (19)$$

The cost function shown in Equation 17 can be written as:

$$J = \frac{1}{2}\left\| e\Big|_{old} + L\left[w\Big|_{old} - w\Big|_{new}\right]\right\|^2 \quad (20)$$

The minimization of this function with respect to the new weights leads to the following weight update:

$$\mathbf{w}\Big|_{new} = \mathbf{w}\Big|_{old} - \left(\mathbf{L}^t\mathbf{L}\right)^{-1}\mathbf{L}^t\,\mathbf{e}\Big|_{old} \quad (21)$$

The matrix **L** is easy to be obtained since it only needs the first derivatives of the cost function. This procedure depends on the requirement about small changes of the synaptic weights in Equation 18, so that if this condition is not true, then the algorithm can become instable. *LM* algorithm solves this problem forcing the change of the weights to be small by means of a cost function:

$$J = \frac{1}{2}\left\| e\Big|_{old} + L\left[w\Big|_{new} - w\Big|_{old}\right]\right\|^2 + \lambda\left\| w\Big|_{new} - w\Big|_{old}\right\|^2 \quad (22)$$

The minimization of Equation 22 with respect to the new weights leads to:

$$w\Big|_{new} = w\Big|_{old} - \left(L^t L + \lambda I\right)^{-1} L^t\, e\Big|_{old} \quad (23)$$

being **I** the identity matrix. If λ takes a very high value, then the *BP* algorithm is obtained. There are iterative procedures to obtain this factor (Bishop, 2006).

The main drawback of this algorithm lies in the need of saving the inverse matrix of **L**. The size of this matrix is the square of the number of synaptic weights of the network. Therefore, this method is not a good choice when the number of weights is large (Nelles, 2001).

- **Support Vector Machines**

The objective in this study is to find a function that relates the ensemble of input variables, defined by vectors $\mathbf{x}_k$ with desired output. The basic idea inside *SVR (Support Vector Regression)* is to define an interval in such a way that the function adjusts data less than a desired value ε. After this, a "tube" is defined around desired signal and our predictions must lie inside this "tube". In this methodology it is take into account if the error exceeds the value ε.

This approximation let us to treat a regression problem as a classification one where output can be inside or outside the tube (Bishop, 2006). The definition of slack variables ξ_i indicate when *SVM* model fail in classification, in our case when regression function gives an output that is outside the "tube" (Figure 7).

As the same way as in *SVM* it can be defined a cost function that regularizes the coefficients of regression function and is given by:

$$\frac{1}{2}\left\|\mathbf{w}\right\|^2 + C\sum_i\left(\xi_i + \hat{\xi}_i\right) \quad (24)$$

where **w** is the weights vector and C is a penalty factor. The conditions that must be accomplished are:

Figure 7. Support vector regression. Points are around the "tube" coloured in grey

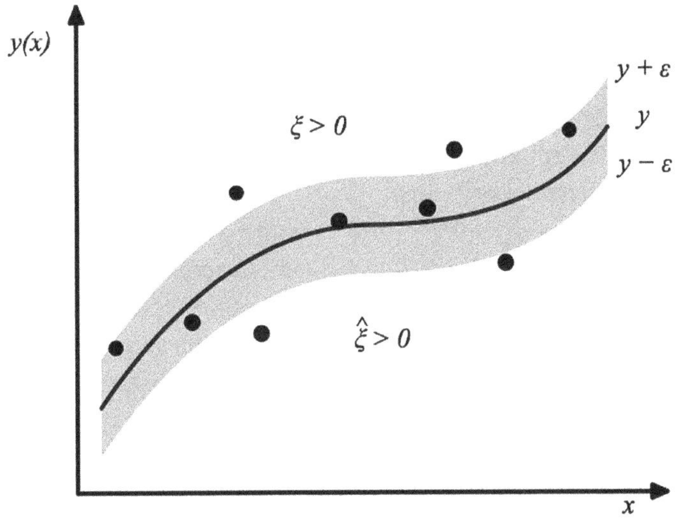

$$\begin{aligned} &y_i - w \cdot x - b \le \varepsilon + \xi_i \\ &w \cdot x + b - y_i \le \varepsilon + \hat{\xi}_i \\ &\xi_i, \hat{\xi}_i \ge 0 \end{aligned} \tag{25}$$

Equation 24 has two parts, in first part the squared norm of the weights vector is minimized to regularize the regression function and in the second part the values outside the tube are considered. The factor *C* determine the influence of errors compared with the objective of found the regularized function. The problem of optimization with restrictions (Equation 25) can be transformed, using Lagrange multipliers α_i, in the so-called *dual problem* defined by the equation (Bishop, 2006):

$$\begin{aligned} &-\frac{1}{2}\sum_{i,j}\left(\alpha_i - \hat{\alpha}_i\right)\left(\alpha_j - \hat{\alpha}_j\right)x_i \cdot x_j - \\ &\varepsilon\sum_i\left(\alpha_i + \hat{\alpha}_i\right) + \sum_i y_i\left(\alpha_i - \hat{\alpha}_i\right) \end{aligned} \tag{26}$$

where α_i and $\hat{\alpha}_i$ are the Lagrange multipliers that verify $\sum_i (\alpha_i - \hat{\alpha}_i) = 0$ and $\alpha_i,\ \hat{\alpha}_i \in [0, C]$.

To have more flexibility in the model it is usual to use *kernels* (Bishop, 2006), (Haykin, 2009). This means a mapping between input variables and another space implementing non-linear relationships. With *kernels*, the error function to minimize is defined by (Bishop, 2006):

$$\begin{aligned} &-\frac{1}{2}\sum_{i,j}\left(\alpha_i - \hat{\alpha}_i\right)\left(\alpha_j - \hat{\alpha}_j\right)k\left(x_i, x_j\right) \\ &-\varepsilon\sum_i\left(\alpha_i + \hat{\alpha}_i\right) + \sum_i y_i\left(\alpha_i - \hat{\alpha}_i\right) \end{aligned}, \tag{27}$$

with the same conditions for equation 26. Comparing equations 26 and 27 the difference is in the term corresponding to the *kernel* $k(x_i, x_j)$. The most commonly used *kernel* is the Gaussian one defined by (Bishop, 2006):

$$k(x_i, x_j) = e^{-\frac{\|x_i - x_j\|^2}{2\sigma^2}} \tag{28}$$

in this function there is another parameter σ to adjust.

- **Extreme learning machines:** The algorithm called extreme learning machine (*ELM*) is due to (Huang, Zhu, & Siew,

2006) and it is based on a *MLP* with *H* hidden neurons with input weights randomly initialized. The *MLP* can be considered as a linear system and the output weights can be obtained analytically with the pseudoinverse matrix of *H* hidden outputs neurons for a training set. So, given a set of *N* input vectors, there exist β_i, $\mathbf{w}_i$ and b_i with

$$\sum_{j=1}^{H} \beta_j \rho(w_j x_i + b_j) = t_i \quad (29)$$

with i = 1, . . ., *N* and $\mathbf{t}_i$ is the *i*-th target vector. This equation can be expressed in matrix form as

$$H\beta = T \quad (30)$$

where **H** is the $N \times H$ matrix of outputs hidden layers, β is a $H \times m$ matrix of output weights and T is a $N \times m$ matrix of *N* targets. The *MLP* training consists in a minimum squares problem with optimal weights $\hat{\beta} = \mathbf{H}^{\dagger}\mathbf{T}$, where † indicates pseudoinverse of Moore-Penrose (Huang, Zhu, & Siew, 2006). This training method is fast and efficient but it is necessary to introduce the number of hidden neurons.

As an alternative to the minimum square method there exist some regularization methods. (Tibshirani, 1996) proposed the *Least Absolute Shrinkage and Selection Operator* (*LASSO*) method. For $\lambda \geq 0$, *LASSO* finds the parameters α_0 and α doing minimum the expression

$$\frac{1}{N}\sum_{i=1}^{N}(y_i - \alpha_0 - x_i\alpha^t)^2 + \lambda\sum_{j=1}^{p}\left|\alpha_j\right| \quad (31)$$

where α_0 is an scalar and α is a vector of dimension p. If λ increases, the number of α components decreases. This algorithm let us to reduce the number of variables utilized in the training process.

EXPERIMENTAL EVALUATION

In this section, we explain the methodology for univariate and multivariate techniques. The process of training, validation and test is defined as well as the error measures used. We have calculated errors and show them for three representative buildings. Furthermore, we represent graphics with 24 hour predicted consumed power for some days.

Experimental Setting

In this study, each arrangement is split in three sets: training (55%), validation (15%) and test (30%). The training set is utilized to determine the parameters of the model, the validation set is used to control the model flexibility, preventing overfitting, and the test set shows the model ability to generalize and predict.

The measures of error utilized are: *Mean Absolute Error (MAE)* (Equation 32) which is not as sensitive to outliers as *Root Mean Squared Error (RMSE)* (Equation 33):

$$MAE = \frac{\sum_{i=1}^{N}\left|y_i - \hat{y}_i\right|}{N} \quad (32)$$

$$RMSE = \sqrt{\frac{\sum_{i=1}^{N}(y_i - \hat{y}_i)^2}{N}} \quad (33)$$

where y_i are real data, $\hat{y}_i$ are predictions and *N* is the number of instances.

Univariate Methods

Autoregressive methods have the advantage of fast calculations but the disadvantage of missing values. In this study, when there is a missing value the time series is cut so there are some series for each building at different times. In aim to have enough data in each time interval, the criterion to save a time series is to have more than 400 records. This procedure gives a set of time series between seven and nine series for each one of the 30 buildings.

For model selection, in the case of *AR* methodology we used some information criteria as *Akaike's Information Criterion (AIC)* (Akaike, 1973):

$$AIC = -2\, ln\, L + 2k \quad (34)$$

where *L* is the likelihood and *k* the number of model parameters. Also it is used the corrected *AICc* (Hurvich & Tsai, 1989):

$$AIC_c = AIC + \frac{2k\left(k+1\right)}{N-k-1} \quad (35)$$

Nevertheless, the models with lowest test errors were found using the *Bayesian Information Criterion (BIC)* up to order 50:

$$BIC = -2\, ln\, L + k \cdot ln\, N \quad (36)$$

Multivariate Methods

Multilayer Perceptrons have been calculated trying up to 100 runs for each configuration to find minima in error function. This procedure is done to avoid local minima (Haykin, 2009). The activation functions utilized were: hyperbolic tangent for hidden layers and linear function for output layers. One (*1LP*) and Two Layer Perceptrons (*2LP*) have been analyzed for 2, 4, . . ., 20 neurons in the hidden layer.

Final model for one building is constructed selecting the arrangement with minimum *MAE* of corresponding validation set. After this, the *MAE* for each test set is calculated using models to predict consumed power and comparing it with real data.

RESULTS AND ANALYSIS

No Clustering

- **Errors:** *MAE* and *RMSE* for test sets are shown in Table 3 for different models and three representative buildings. In the case of *AR* models, where different series must

Table 3. Mean active power, MAE and RMSE for different models and buildings. The error is calculated for test set and all units are kW. Models are: AR = autoregressive (minimum / mean / maximum), LR = linear regression, 1LP = one layer perceptron, 2LP = two layer perceptron, SVR = support vector regression

MAE	Power	*AR*	*LR*	*1LP*	*2LP*	*SVR*
Biology	73.06 ± 30.75	3.36 / 5.16 / 7.85	6.97	2.88	**2.81**	3.81
Philosophy	27.84 ± 26.74	1.48 / 3.35 / 4.51	5.97	1.96	**1.80**	2.27
Dormitory	14.22 ± 7.44	0.62 / 2.19 / 3.40	2.46	1.55	**1.48**	1.82
RMSE						
Biology	73.06 ± 30.75	4.94 / 7.89 / 12.69	9.76	4.04	**3.92**	5.78
Philosophy	27.84 ± 26.74	2.85 / 5.52 / 7.10	8.95	3.03	**2.81**	3.85
Dormitory	14.22 ± 7.44	1.28 / 3.20 / 4.44	3.44	2.28	**2.26**	2.72

be taken due to lost values, minimum and maximum *MAE* for different series is presented as well as mean *MAE*. Meanwhile *LR* models have the highest errors, they can be considered as a base line, *AR* results depend on evaluated date. The best results are given by *MLP*, for two and one hidden layers. With similar errors and in second place are *SVR* models. The main inconvenience of *MLP* methodology is their computational cost, which is the highest. In the case of *Dormitory* building, relative differences in errors between methods are lower than in bigger buildings.

- **Power Profiles:** Daily power profiles were obtained calculating firstly active power at 1:00 for each building. After introducing this value as a new input, prediction is calculated for 2:00. This process is done iteratively up to 24:00. In the case of multivariate models, actual values of the rest of variables are utilized as input to predict hourly active power.

AR prediction (Figure 8 left) is accurate for three representative buildings. These models tend to slightly underestimate and overestimate successively real active power for each hour in a day. It must be taken into account the low information required to construct this kind of models as well as the low computational cost. Nevertheless, the main disadvantage of this technique is that missing values impede the prediction of active power in times following the missing value. For these models predictions are restricted to some days if there are untreated missing values.

Results of linear regression are shown (Figure 8 right) for three representative buildings. They are the only models that cannot predict typical two peaks profile, giving a smoother consumption profile than the real one. Predictions are far from real values and hours with higher consume are strongly underestimated.

In the case of *2LP* (Figure 9 left) results are very similar to one *1LP*. For *Biology* building, prediction is very accurate and slightly better than *SVR* (Figure 9 right) as we can see from 12 to 18h. For *Philosophy* building, *SVR* profile underestimate more than *SVR* mainly for the first peak hence the best prediction corresponds to *2LP* (Figure 9). For *Dormitory* building, where power profiles show more than two peaks, there is overestimation around midday as well as underestimation at the end of the day and *SVR* wins to *2LP* from 12 to 24 h just for this building.

Finally, in Figure 11, there is a comparison of four best models during a weekend in the *Dormitory* building. This building is different, from the consumption point of view, to *Biology* and *Philosophy*. It is occupied seven days a week and the daily power profiles have three or more peaks. In this case best results correspond to *SVR* models. For Saturday, 01/15/11, *SVR* draw accurately the power base line but peaks are not predicted by models, but for Sunday, 01/16/2011 prediction of *SVR* is better than *2LP*. and MT models is more accurate than MLP or SVR profiles.

Clustering

In this section we describe the results obtained after applying clustering with linear regression, *SVR* and *ELM*.

- Clusterezd linear regression
    - **Global Ensemble:** Initially *k-means* algorithm is applied for each arrangement from two to ten clusters. *Centroids* are initialized 50 times, it is calculated highest silhouette for

Figure 8. Prediction of daily power profiles by AR (left column) and LR (right column) models for Biology, Philosophy and Dormitory buildings on Wednesday, 10/27/10

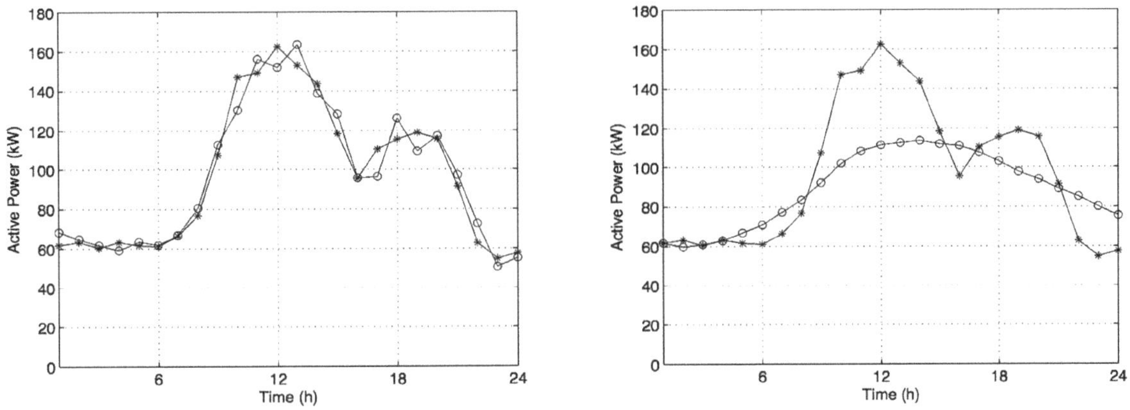

(a) *Biology*

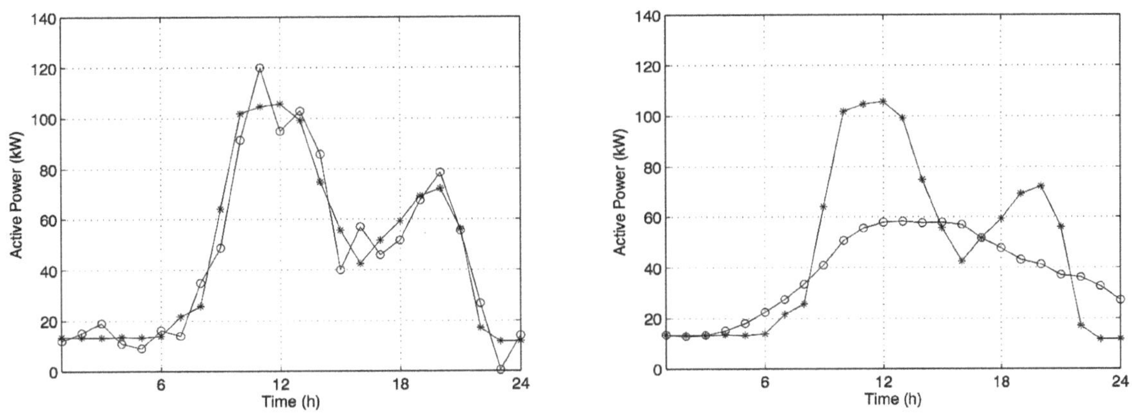

(b) *Philosophy*

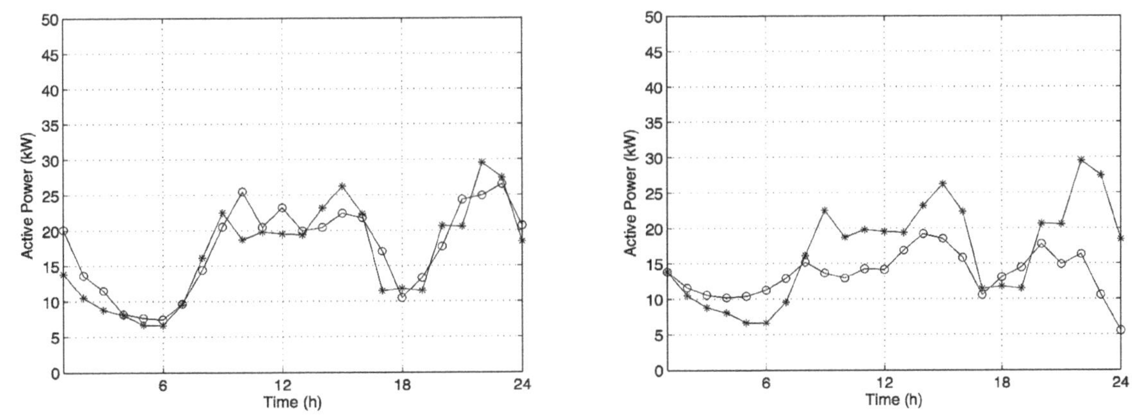

(c) *Dormitory*

Figure 9. Prediction of daily power profiles by 2LP (left column) and SVR (right column) models for Biology, Philosophy and Dormitory buildings on Wednesday, 10/27/10

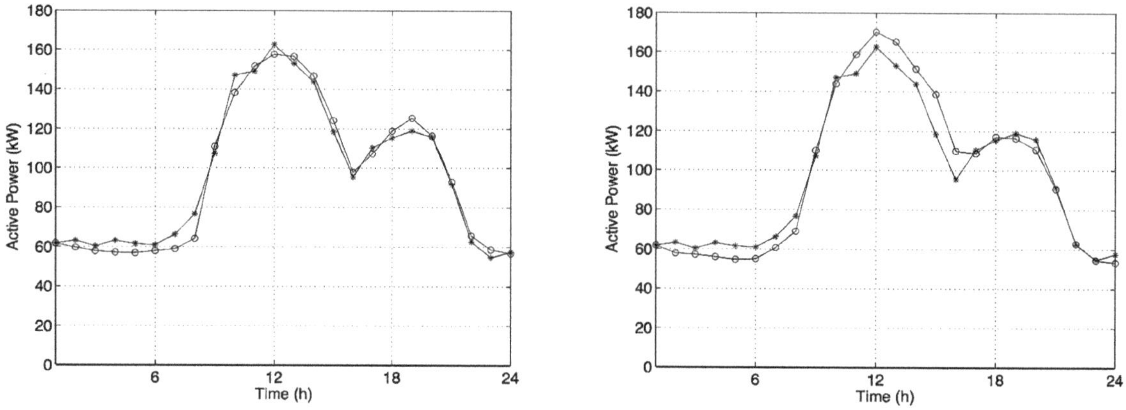

(a) *Biology*

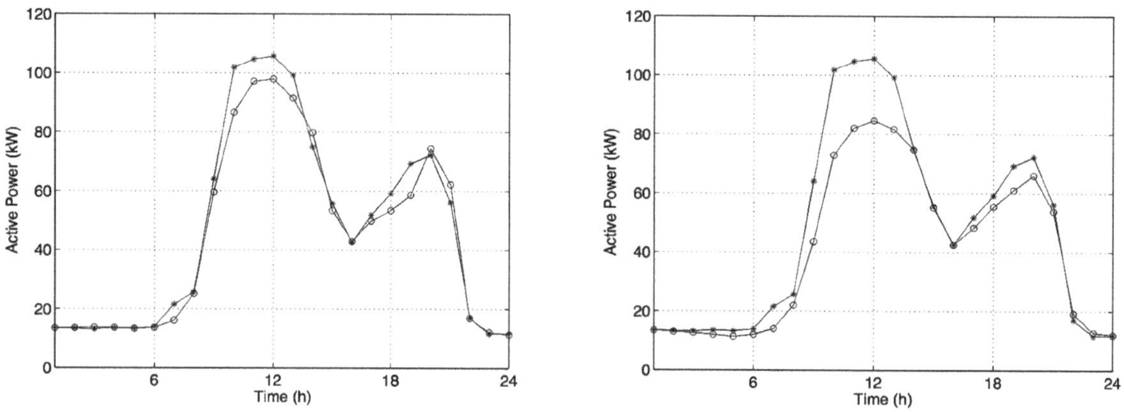

(b) *Philosophy*

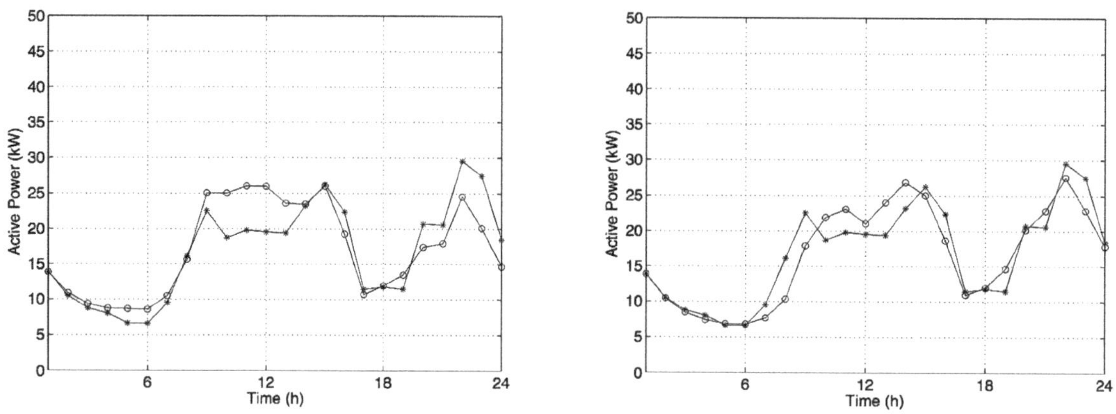

(c) *Dormitory*

Figure 10. Prediction of daily power profiles by CLR (left column) and CELM LASSO (right column) models for Biology, Philosophy and Dormitory buildings on Wednesday, 10/27/10.

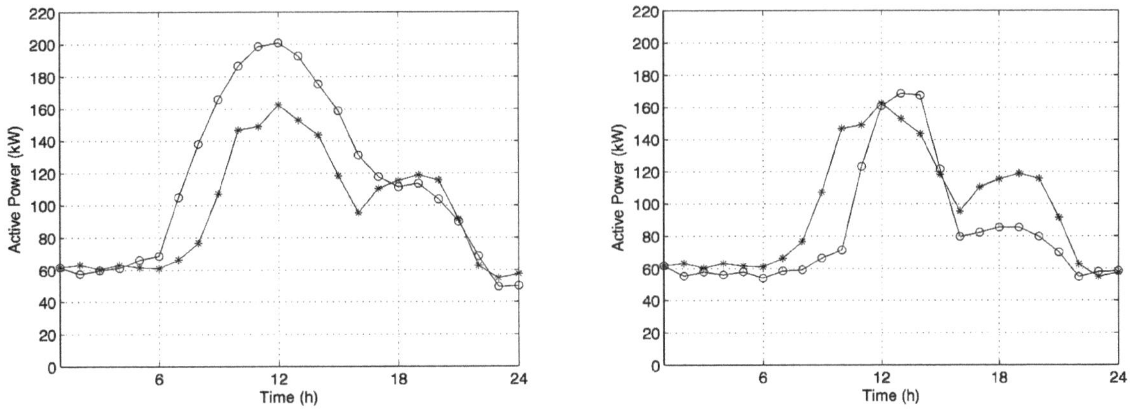

(a) *Biology*

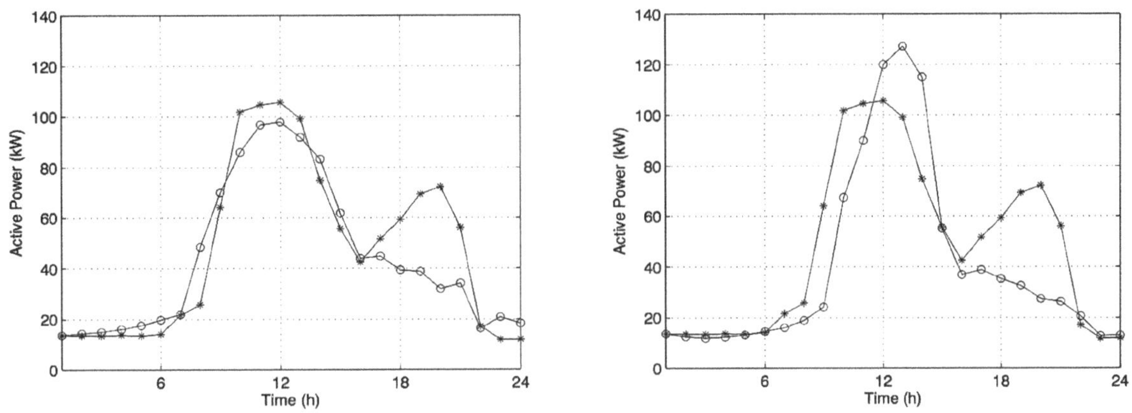

(b) *Philosophy*

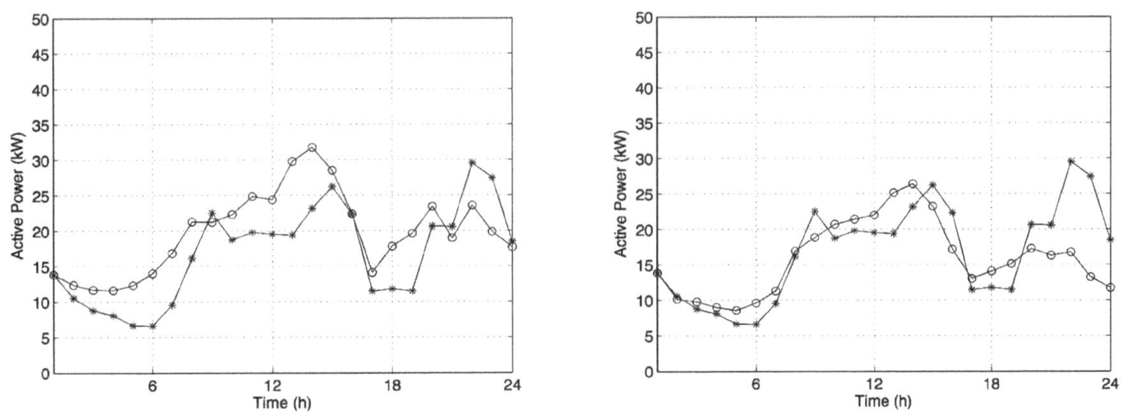

(c) *Dormitory*

Figure 11. Prediction of daily power profiles by different models on Saturday, 01/15/11 (left column) and Sunday, 01/16/11 (right column) models for Dormitory building

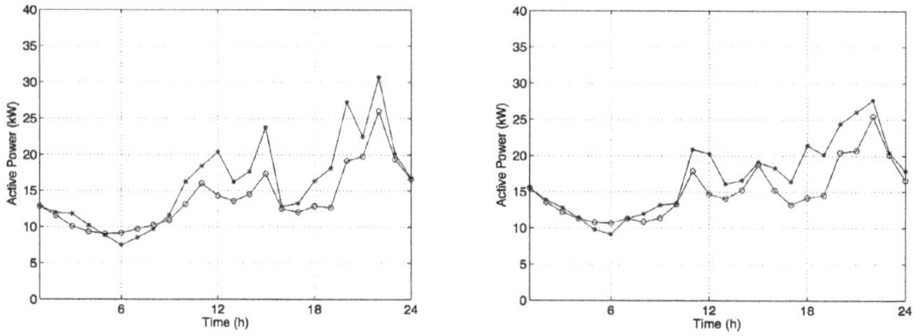

(a) *Two Layer Perceptron.*

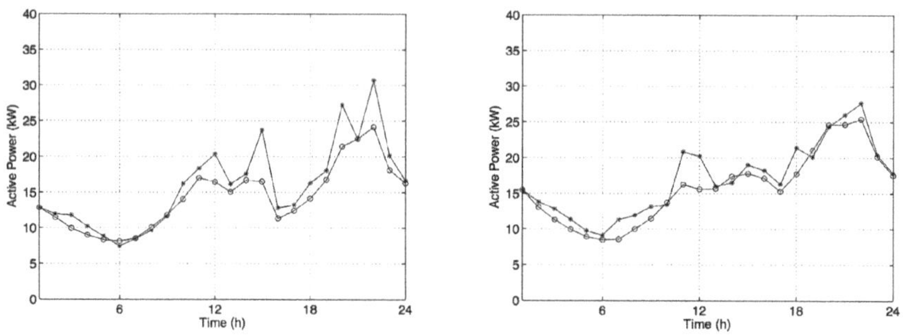

(b) *Support Vector Regression.*

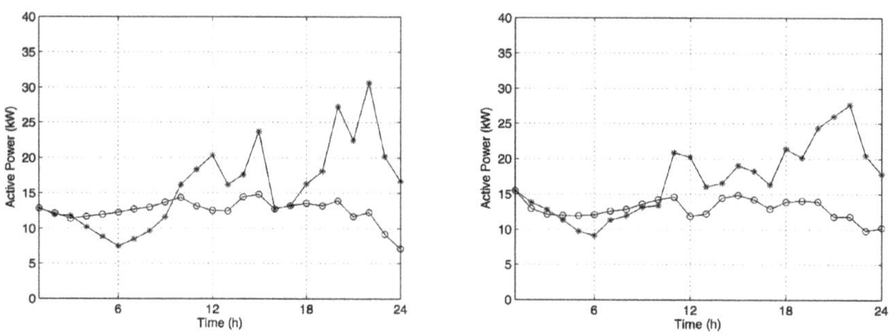

(c) *Clustered Linear Regression TP.*

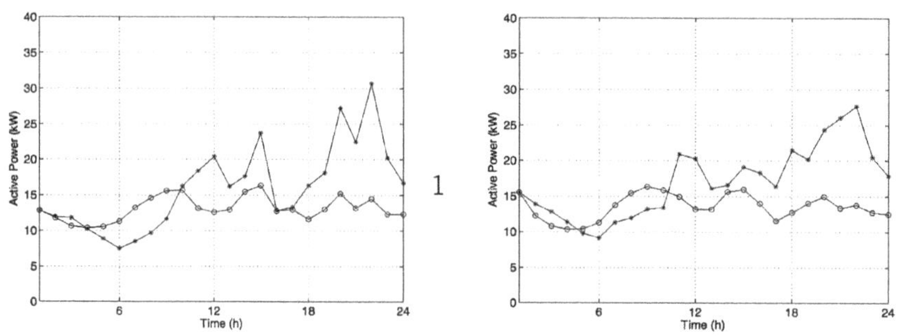

(d) *Clustered Extreme Learning Machines + LASSO.*

each arrangement and computed the mean to choose the best configuration building-number of clusters. Best *silhouette* value is near 0.6 for the three buildings which corresponds to two clusters, where one cluster is data with solar radiation zero or near to zero.

When comparing clustered linear regression with linear regression models, first models show lower values of error as seen in Table 4. In all buildings *MAE* with clustering is lower than results without clustering. The highest difference occurs for *Philosophy* building and the lowest for *Dormitory*. Results show that clustering improves linear regression but is no sufficient if compared to nonlinear methods as *MLP* or *SVR*.

- **Segmentation:** After the above results, the next step is to divide the training, validation and test sets in different ways or criteria. Different tested segmentations are:

T: Divide dataset in power terciles.
TP: Divide dataset in power terciles and third tercile in two power ensembles.
TH: Divide dataset in power terciles and third tercile in two ensembles by hour, before and after 16h.
TK: Divide dataset in power terciles and third tercile by *k-means* algorithm.
Q: Divide dataset in power quartiles.

Clustered linear regression results with different segmentations are shown in Table 5. For example, T model for *Dormitory* has similar errors than more sophisticated *SVR* (Table 4). Also *TH* has better results than *SVR* for *Biology* and *Dormitory*. In the case of *TK* models, results are similar but time to train prototypes is higher (1h average) because of the *k-means* calculations involved.

Table 4. MAE of clustered and non clustered linear regression and support vector regression for Biology, Philosophy and Dormitory buildings. Two layer perceptron results are presented to compare

MAE	Power	*2LP*	*LR*	*CLR*	*SVR*	*CSVR*
Biology	73.06 ± 30.75	2.81	6.97	6.53	3.63	**3.60**
Philosophy	27.84 ± 26.74	1.80	5.97	5.05	2.27	**2.20**
Dormitory	14.22 ± 7.44	1.48	2.46	2.46	**1.70**	1.73

Table 5. MAE of linear regression with different segmentation models for Biology, Philosophy and Dormitory buildings. Non clustered linear regression results are presented to compare

MAE	Power	*LR*		*T*	*TP*	*TH*	*TK*	*Q*
Biology	73.06 ± 30.75	6.97	*Min*	1.72	1.72	1.75	2.15	1.39
			Mean	4.49	3.89	3.32	**3.23**	3.84
			Max	9.65	8.54	6.67	8.16	8.08
Philosophy	27.84 ± 26.74	5.97	*Min*	0.62	0.63	0.65	1.21	0.52
			Mean	3.75	3.03	2.81	**2.57**	3.30
			Max	8.84	7.25	6.37	10.11	8.26
Dormitory	14.22 ± 7.44	2.46	*Min*	0.58	0.58	0.58	1.15	0.42
			Mean	1.71	**1.29**	1.63	1.50	1.38
			Max	3.00	2.26	2.93	3.46	2.68

To validate the predictive power of linear models with clustering it is interesting to represent real vs. predicted values of active power for representative buildings. Here it can be seen that profiles are not so good for Wednesday, 10/27/10. Figure 10 shows how power prediction is worst than *2LP* or *SVR* (Figure 9), in the case of *Biology* building prediction is, from 7 to 16h, much higher than real data. For *Philosophy* building, the first peak in power (12h) is well predicted but not the second one (19h).

It is studied the significance of variables for linear models. These variables are for all the buildings, hour and power at previous hour; in addition, for *Philosophy* and *Dormitory*, type of day and solar radiation.

- Clustered support vector regression

As in the previous section, firstly global ensemble is analyzed using *k-means* from two to ten clusters initializing *centroids* 50 times. *SVM* models are constructed and errors calculated and compared to *SVM* models without clustering (Table 4). In this case, clustering technique gives very similar results in terms of *MAE*.

- Clustered extreme learning machines
    - **Global Ensemble:** As done in linear regression, *ELM* technique is applied to global ensemble for each building. Different numbers of hidden nodes are tested and errors of validation sets vary near 10% for different arrangements. This variation is shown for all the buildings and the criterion to select number of nodes is to choose mean *MAE* of validation set for each one of ten arrangements.

In Table 6 models are presented in terms of number of nodes and *MAE* of test set for each building. Results are as poorer as non clustered linear regression (Table 4).

- **Terciles:** Because of the poor results of *ELM* for global ensembles of each building next step is to divide set in terciles by active power and apply *k-means* to the third tercile.

The optimal number of nodes is 180 for *Biology*, 240 for *Philosophy* and 60 for *Dormitory*. In Table 6, *MAE* for test set are presented and results are similar as given by *MLP* with two hidden layers, *Dormitory* has low *MAE* but is higher for the other two buildings. When observing power profiles for selected days (Figure 10) second peak is not well predicted for *Biology* and *Philosophy* buildings and predictions are over real consumed power at 12h. Nevertheless, *Dormitory* behavior, with some peaks in a day, is as well predicted as does *MLP* except around 22h.

Although analysis have been repeated using *LASSO* regularization, this process needs more time to compute. The optimal number of nodes is lower than analysis without regularization but *MAE* are lower and the higher reduction corre-

Table 6. MAE of test sets for ELM, CELM and CELM regularized with LASSO models for Biology, Philosophy and Dormitory buildings. Numbers in parentheses indicate number of nodes in the hidden layer. Two layer perceptron results are presented to compare.

MAE	Power	*2LP*	*ELM*	*CELM*	*CELM LASSO*
Biology	73.06 ± 30.75	2.81	4.68 (800)	3.03 (180)	**2.97** (90)
Philosophy	27.84 ± 26.74	1.80	3.57 (1300)	2.17 (240)	**2.17** (180)
Dormitory	14.22 ± 7.44	1.48	2.16 (600)	1.45 (60)	**1.28** (220)

sponds to *Dormitory* building where value decrease form 1.45 to 1.28 improving 15% the best two hidden layers perceptron result. In prediction results are similar as seen without regularization and even better in the time interval between 0 and 12h.

CONCLUSION

The interest of global and local techniques to forecast power consumption in buildings is explained in this chapter. The benefit of this methodology is not only economical and energetic but also environmental (Pérez Lombard, Ortiz, & Pout, 2008). With this objective, linear and non-linear models were applied to electrical data from buildings belonging to the University of León (Spain). Firstly we utilize a principal component analysis on variables related to active power to arrange buildings in three different groups. After this, we applied different machine learning techniques and we calculated and constructed daily power profiles. We have divided the different methods in two categories: global and local methods.

- **Global Methods:** Autoregressive models (*AR*) obtain good results in terms of *MAE* despite they are the simplest and the only univariate approximations analyzed. These models do not utilize the information contained in time nor environmental variables. With linear regression (*LR*) or *ELM* models (Huang, Zhu, & Siew, 2006) we take into account these variables but results show the highest errors and *LR* models are unable to predict typical daily peaks in load profiles. When linear methods are not accurate enough then non-linear methods are utilized (Foucquier, Robert, Suard, Stéphan, & Jay, 2013). Support vector regression (*SVR*) give good prediction results giving the most accurate model in the case of *Dormitory* building. Nevertheless multilayer perceptrons (*MLP*) give the lowest *MAE* and *RMSE* errors and in this case two hidden layers configurations are better than one being their main counterpart the long computing times.
- **Local Methods:** After applying different clustering techniques (Bishop, 2006) in terms of load magnitude we generally find a reduction in errors. In linear regression models the results for the third cluster, that corresponds to higher loads, are poor and prediction profiles have load peaks higher than actual values. A detailed analysis shows that lowest *MAE*s in clustered linear methods correspond to *TK* model for groups 1 (*Biology*) and 2 (*Philosophy*) and *TP* for group 3 (*Dormitory*). The case of *SVR* models is an exception where errors remain unchanged after applying clustering methods. In the case of clustered *ELM* models the prediction of the principal peak at 12h is better for *Biology* building than clustered linear models. Nevertheless the opposite occurs in the case of *Philosophy* building. For *Dormitory* building *TP* clustering errors are similar to *CELM* ones but the use of *LASSO* regularization gives the best results in terms of error for *Dormitory* building and profiles are well predicted.

If we compare local versus global methods we find that in the case of clustered linear models the errors decrease 15% for *Philosophy* and 7% for *Biology* building but results are not improved for *Dormitory* building. For support vector machines the improvement is very low for groups one and two and clustering gives the same errors for *Dormitory*. It is for extreme learning machines (*ELM*) models where clustering reduces more the errors and the number of hidden nodes needed to learn models. The reduction in error is higher than 21% for the three groups. In the case of the *Dormitory* building the error equals perceptrons with two hidden layers and if we apply regularization via de *LASSO* (Tibshirani, 1996) the error reduces again for the *CELM* model.

Although local models have some improvements in terms of *MAE* error load profiles do not reach the precision of global models. The best prediction in load profiles of *Dormitory* building stands for global *SVM* models and perceptrons with two hidden layers. In a second place we have clustered extreme learning machines. In the case of *Biology* and *Philosophy* buildings the best prediction of load profiles corresponds to perceptrons with two hidden layers followed by *SVM* models but we have not found a local model with the performance at the level of global models.

REFERENCES

Akaike, H. (1973). Maximum likelihood identification of Gaussian auto-regressive moving-average models. *Biometrika*, *60*, 255–266.

Arbib, M. (Ed.). (2003). *The Handbook of Brain Theory and Neural Networks*. MIT Press.

Ball, G. & Hall, D. (1965). ISODATA: A novel method of data analysis and pattern classification. Stanford Research Institute. Menlo Park.

Bian, X., Xu, Q., Li, B., & Xu, L. (2007). Equipment Fault Forecasting Based on a Two-level Hierarchical Model. *Proceedings of the IEEE International Conference on Automation and Logistics* (pp. 2095-2099). doi:10.1109/ICAL.2007.4338921

Bishop, C. M. (2006). *Pattern Recognition and Machine Learning (Information Science and Statistics)*. New York: Springer-Verlag.

Buildings Energy Data Book. (2010). U.S. Department of Energy. Retrieved from http://buildingsdatabook.eren.doe.gov/DataBooks.aspx

Código Técnico de la Edificación. (2010). Ministerio de Fomento, Gobierno de España. Retrieved from: http://www.codigotecnico.org

Cybenko, G. (1989). Approximations by superpositions of sigmoidal functions. *Mathematics of Control, Signals, and Systems*, *2*(4), 303–314. doi:10.1007/BF02551274

Ekici, B. B., & Aksoy, U. T. (2009). Prediction of building energy consumption by using artificial neural networks. *Advances in Engineering Software*, *40*(5), 356–362. doi:10.1016/j.advengsoft.2008.05.003

Energy Performance Certification of Buildings. (2013). International Energy Agency. Retrieved from http://www.iea.org/publications/freepublications/publication/buildings_certification.pdf

European Commission. (2010). Energy 2020: A strategy for competitive, sustainable and secure energy. Brussels. Retrieved from http://ec.europa.eu/energy/strategies/2010/2020_en.htm

Foucquier, A., Robert, S., Suard, F., Stéphan, L., & Jay, A. (2013). State of the art in building modeling and energy performances prediction: A review. *Renewable & Sustainable Energy Reviews*, *23*, 272–288. doi:10.1016/j.rser.2013.03.004

Harris, R., & Sollis, R. (2003). *Applied Time Series Modeling and Forecasting*. Hoboken, NJ: John Wiley & Sons Inc.

Haykin, S. (2009). *Neural Networks and Learning Machines*. Prentice Hall.

Huang, G., Zhu, Q. & Siew, C. (2006). Extreme learning machine: Theory and applications. *Neurocomputing*, *70*, 489–501.

Hurvich, C. M., & Tsai, C. L. (1989). Regression and time series model selection in small samples. *Biometrika*, *76*(2), 297–307. doi:10.1093/biomet/76.2.297

Kay, S. M. (1988). *Modern Spectral Estimation: Theory and Application*. Englewood Cliffs, NJ: Prentice Hall.

Kusiak, A., Li, M., & Tang, F. (2010). Modeling and optimization of *HVAC* energy consumption. *Applied Energy*, *87*(10), 3092–3102. doi:10.1016/j.apenergy.2010.04.008

Leung, P. C. M. & Lee, E.W.M. (2013). Estimation of electrical power consumption in subway station design by intelligent approach. *Applied Energy* 101 C 634-643.

Liao, Z., & Dexter, A. L. (2004). A simplified physical model for estimating the average air temperature in multi-zone heating systems. *Building and Environment*, *39*(9), 1013–1022. doi:10.1016/j.buildenv.2004.01.034

Lloyd, S. P. (1982). Least squares quantization in PCM. *IEEE Transactions on Information Theory*, *28*(2), 129–137. doi:10.1109/TIT.1982.1056489

Luenberger, D. G. (1984). *Linear and Nonlinear Programming*. Addison-Wesley.

Ma, Y., Yu, J. Q., Yang, C. Y., & Wang, L. (2010). Study on power energy consumption model for large-scale public building. *Proceedings of the 2nd international workshop on intelligent systems and applications* (pp. 1-4). doi:10.1109/IWISA.2010.5473608

Macqueen, J. (1967). Some methods for classification and analysis of multivariate observations. *Proceedings of the 5th Berkeley Symposium on Mathematical Statistics and Probability* (pp. 281-297).

Nelles, O. (2001). *Nonlinear System Identification. From Classical Approaches to Neural Networks and Fuzzy Models*. Springer.

Newsham, G. R., & Birt, B. J. (2010). Building-level occupancy data to improve ARIMA-based electricity use forecasts. *Proceedings of the 2nd ACM workshop on embedded sensing systems for energy-efficiency in building* (pp. 13-18). New York, USA. doi:10.1145/1878431.1878435

Paliwal, M., & Kumar, U. A. (2009). Neural networks and statistical techniques: A review of applications. *Expert Systems with Applications*, *36*(1), 2–17. doi:10.1016/j.eswa.2007.10.005

Pérez Lombard, L., Ortiz, J., & Pout, J. (2008). A review on buildings energy consumption information. *Energy and Building*, *40*(3), 394–398. doi:10.1016/j.enbuild.2007.03.007

Rousseeuw, P. J. (1987). Silhouettes: A graphical aid to the interpretation and validation of cluster analysis. *Journal of Computational and Applied Mathematics*, *20*(0), 53–65. doi:10.1016/0377-0427(87)90125-7

Soliman, S.A. & Al-Kandari A.M. (2010). Electric Load Modeling for Long-Term Forecasting. *Electrical Load Forecasting* (pp. 353-406).

Tibshirani, R. (1996). Regression Shrinkage and Selection Via the Lasso. *Journal of the Royal Statistical Society. Series B. Methodological*, *58*, 267–288.

Vellido, A., Lisboa, P. J. G., & Vaughan, J. (1999). Neural networks in business: A survey of applications (1992–1998). *Expert Systems with Applications*, *17*(1), 51–70. doi:10.1016/S0957-4174(99)00016-0

Webb, A. R., & Copsey, K. D. (2011). *Statistical Pattern Recognition*. Wiley. doi:10.1002/9781119952954

Weigend, A. S., & Gershenfeld, N. A. (Eds.). (1993). *Time Series Prediction: Forecasting the Future and Understanding the Past*. Addison-Wesley.

Weisberg, S. (2005). *Applied Linear Regression*. Hoboken, NJ: Wiley. doi:10.1002/0471704091

Wong, S. L., Wan, K. K. W., & Lam, T. N. T. (2010). Artificial neural networks for energy analysis of office buildings with daylighting. *Applied Energy*, *87*(2), 551–557. doi:10.1016/j.apenergy.2009.06.028

Zhao, H. X., & Magoulès, F. (2010). Parallel support vector machines applied to the prediction of multiple buildings energy consumption. *Journal of Algorithms & Computational Technology*, *4*(2), 231–249. doi:10.1260/1748-3018.4.2.231

Chapter 12
Building Automation Systems: Recent Trends, Design and Development

Jignesh G. Bhatt
Dharmsinh Desai University, India

ABSTRACT

Building Automation Systems (BAS) have started attracting focus of global research community and demonstrating rapid growth potential due to their transformation from the legacy stand-alone security systems into intelligent computerized wired or wireless communication technology based solutions. Amid rising concerns over energy and security globally, BAS demonstrates commendable outperformance with promising research and commercial opportunities. Rising energy costs, stringent energy requirements and vibrant, but abnormal-unpredictable swings in environment, etc. all have contributed immensely as driving forces for BAS developments. This chapter presents the learning outcomes from author's research work and experiences in terms of instrumentation engineer's perspective towards the focused area nicely supported by literature review as well as practical hands-on exposure. The chapter judiciously puts forth historical and modern perspectives in overall development of BAS for reader's digest.

1. INTRODUCTION

Building Automation System (BAS) is a data acquisition and control system that incorporates various functionalities provided by the control system of a building. As mentioned by (Kastner, Neugschwandtner & Newman, 2005), in general, BAS are also known by other names such as: Energy Management Control Systems (EMCS), Building Management Systems (BMS), Building Energy Management Systems (BEMS), Facility Management Systems (FMS) and so on. As presented by (Granzer, Kastner, Neugschwandtner & Praus, 2006), Modern BAS is a computerized, intelligent network of electronic devices, designed to monitor and control lighting, internal climate and other systems in a building resulting in optimized energy usage, safety, security, information, communication and entertainment facilities. (Kastner, Neugschwandtner & Newman, 2005) has also mentioned that Home Automation System (HAS) or domotics is possibly the system believed to have the highest level of functional similarity to building automation. But the former has significantly different characteristics. While BAS focuses on financial benefits, the domotics

DOI: 10.4018/978-1-4666-9911-3.ch012

focuses on comfort and mental peace. Domotics are of comparatively quite smaller scale, with low costs of equipment as well as commissioning and maintenance.

As per (Bhatt, 2009), BAS have been deployed to improve the indoor climate as well as to reduce the operational costs. As mentioned by (Kastner, Neugschwandtner & Newman, 2005) and (Bhatt & Verma, 2010), originally, BAS mostly configured to operate with HVAC systems. To further enhance overall building management, reduce costs incurred in lighting, safety, security, entertainment, etc. of the building and for supervision of asset movements, BAS are being employed. (Bhatt & Verma, 2015) have mentioned that BAS, in general, maintains the internal climate of building within a specified range by regulating temperature and humidity, regulates lighting based on parameters like occupancy, ambient light and timing schedule, monitors system performance and device failures; and generates audio visual email and/or text notifications to building O&M staff. The BAS reduces building energy consumption and thereby, reduces operational and maintenance costs as compared to an uncontrolled building. A building controlled by a BAS is often referred to as an Intelligent Building. Typically, the functionalities like entertainment, communication and information exchange need high data rates, while the features like automation, safety and security need low data rates but low latency, high network reliability and data security. Moreover, the BAS designs for residential buildings must be cost-effective and affordable to common citizens and easy to operate without specialized training.

The chapter begins with introduction section covering preliminary information, and next, background section including recent trends, historical data in brief regarding evolution as well as literature review covering review of recent works. The next section, Technological study includes technical details regarding BAS such as generalized and functional architectures, device classes, as well as major wired and wireless communication technologies. Design and development section covers design objectives, functionalities being implemented, architectural modularity and data communication in a building, Solutions and Recommendations section starts with clarifying differences between security system and building automation system. Later, the same section presents in-depth coverage of wired solutions and wireless solutions. In this section, wired as well as wireless versions of home security system and building automation system developed by author have been covered. Future research directions section throws light over possible extensions and future scope of the work presented along with upcoming opportunities. Finally, the chapter ends with conclusions, acknowledgments and list of references cited. Major attractions of this chapter includes, simple but efficient drafting, justification to the focused topic and extensive discussions supported by interesting diagrams and more importantly, the work contribution by the author of this chapter.

2. BACKGROUND

2.1 Recent Trends

Recently, almost all countries have been observed revamping their infrastructure facilities, especially the electric power grids. In many developing countries, in synchronization with developed countries, conventional electric grids are being replaced by smart grids. As presented by (Bhatt, 2013) and (Bhatt, Shah & Jani, 2014), Smart Grid (SG) has been referred as integrated network to converge an existing power grid with an information and communication infrastructure. It has caused the dynamic change and new business contexts in the utility infrastructure, especially in demand side. The conventional energy customers have extended their role as the energy prosumer (producer+consumer), who can generate distributed electricity with renewable energy resources and make profits from selling back the energy to power exchange market. The success of the Smart Grid depends on the interoperability to make flexible power infrastructure, efficient controlling system, network integration, and business models to motivate energy prosumers.

As more and more home appliances and consumer electronics have been deployed, power consumption in home area tends to grow. Although advanced Integrated Circuit (IC) chipset and hardware technology have been found to enhance the energy efficiencies of home appliances and consumer electronics, the current energy crisis and green house effect require more efficient energy management in all areas. Similar advances in electronics and device management technologies have ability to enhance the efficiency of energy consumption devices in the commercial market. However, the current energy crisis has exacerbated the need for more optimized energy management, users' endeavor to use less power and save money, especially with regard to peak demand rates could result into drastic improvements. To address this situation, a home network, context management and Internet technology could be combined to enhance total energy management.

As presented by (Bhatt & Verma, 2015), HAN is a control system with a complete set of home communications, home automation equipment, home securities and other functions based on advanced equipment and scientific management mode, using a variety of communication networks, combined automation and communication technology. It contains a network that connects home elements like sensors, appliances and thermostat. The purpose of HAN is building comfortable, safe, energy efficiency, economical and environment-friendly living environment, especially during present situations of energy shortage and global warming elicited by high-energy consumption model. In the efforts to reduce carbon emission and energy wastage, smart home energy management is the key, which is also an integral part of the Smart Grid.

The key driver of the building automation system market has been the promise of increased user comfort at reduced O&M costs. To achieve such objectives, BAS designs carry out optimization of control schemes of Heating, Ventilation and Air Conditioning (HVAC) systems, lighting, and equipment automation. Thus achieved improvements in energy efficiency could also contribute to environmental protection. For such reasons, relevant regulatory authorities may sometimes make it compulsory to use BAS in upcoming realty projects.

Further cost reduction could be achieved by providing centralized access to building service systems via a centralized monitoring and control center. This allows abnormal or faulty conditions to be identified, isolated and rectified quite quickly as soon as they occur and with minimum human intervention, especially when the site is remotely accessible. Intelligently designed graphical user interfaces and data management schemes could further enhance operator comfort and efficiency as well as results in optimized budgetary allocation and planning for building management.

In addition to these immediate cost benefits, other indirect benefits could be realized due to higher than expected staff productivity or by the enhanced apparent value of the automated building that could be the "prestige factor," for the contractor, proprietor and occupant of the building. Although investment in building automation systems could result in slightly higher overall initial construction cost of a building, their use could be quite justified by the features, services and energy optimization they offer, as soon as the entire building becomes operational.

As per (Kastner, Neugschwandtner & Newman, 2005) and (Granzer, Kastner, Neugschwandtner & Praus, 2006), Home Automation Network or Home Area Network (HAN) has been the network within the premises of a house or building enabling devices and loads to communicate with each other and dynamically respond to externally sent signals (i.e. price). This type of network, characterized by a low data rate requirement, provides the communication infrastructure behind the meter. As a result, smart grid applications could be extended into the home premises and Energy Management Systems (EMS) capabilities could be enabled. Further benefits derived from the realization of such HANs have been anticipated for customers, utilities and society as a whole. Overall, Energy savings and user satisfaction have been two major design considerations for modern HAN systems.

2.2 Brief History of Evolution

Historically as suggested by (Yuwono, 2009), in order to reduce the crimes like home robbery and theft, an online home control system has been developed as presented by (M. Kim; S. Kim; J. Kim & Y. Yoo, 2011) in order to monitor empty residences using electrical network as communication media and PC based man-machine interface. While, telephone, internet, leased line, intercom, radio frequency, short message service from mobile phone were the other interesting features; as presented by (Dickmann, 2011), utilization of X-10 technology to make electrical appliances easily controllable by Short Messaging System (SMS), where LabVIEW has been utilized to configure operational interface with ease by (Lien, Bai, Chen & Hung, 2009), provided another possible option with transmission media for home control system.

The home control systems were developed for monitoring and controlling household appliances using the sensors installed in a house or a building by (Son, Pulkkinen, Moon & Kim, 2010). The doors and windows have been monitored by occupancy and contacts operation to ascertain access privilege and to prevent theft or burglary, smoke/gas has been detected to report fire/gas leak alerts. In such cases, the system shall send a signal to the computer, so that the owner can remotely control corresponding actions in the home, such as turning on/off a light, an alarm, a fan, or any other devices based on necessity, from various installed transducers that communicate through powerline based electrical network.

Advanced technology has shown many improvements of earlier technology such as X10 home automation by (Y. Lai, Rodrigues, Huang, Wang & C. Lai, 2012). The combination of X10 and personal computer makes possible to remotely control home appliances by using a computer as a man-machine interface by (Wang, Sun, Zhou & Dai, 2012) with the help of LabVIEW software for the purposes of energy conservation, security and user-friendly operation.

The development of automatic control systems has resulted in intelligent buildings with the automation of HVAC and lighting systems in order to provide a healthy, comfortable and productive indoor environment with a minimum of energy use, and to control them in an efficient and rational way by (Langhammer & Kays, 2012).

2.3 Literature Review

Design and implementation of MAC protocol for smart grid HAN environment has been put forth by (M. Kim; S. Kim; J. Kim & Yoo, 2011). The smart grid communication network has been divided into three parts; HAN (Home Area Network), NAN (Neighborhood Area Network), and WAN (Wide Area Network). IEEE 802.15.4 and IEEE 802.11 MAC protocols, which are based on the CSMA/CA mechanism, have been assumed to be the candidates for the MAC layer of HAN and corresponding latency and energy efficiency problems have been discussed. A tree-based TDMA MAC protocol has been re-designed which could be applied in SmartGrid HAN. From the experimental validation, the performance of the proposed protocol has been verified with respect to the network latency, high packet transmission rate and energy efficiency.

The advantages of the digitalSTROM® topology have been presented by (Dickmann, 2011). The paper has been elaborating on the modulation schemes that have been optimized to lend themselves to miniaturized implementation and quantified the level of interference caused by typical home appliances.

An embedded remote monitoring and controlling power socket (RMCPS) has been developed by (Lien, Bai, Chen & Hung, 2009) with high suitability for automatic and power management of home electric appliances. Without needs for new layout, the design has advantages such as low cost, low electricity consumption, small volume and convenient installation to replace the PC with a Web server construction. The RMCPS has been consisting of three modules: the Essential Module, the Power Line Communication (PLC) Module and the Detection Module. A remote user connection via internet has been allowed to an embedded power socket for monitoring and control of home electric appliances.

(Son, Pulkkinen, Moon & Kim, 2010) presented a concept of a home energy management system (HEMS) based on power line communication. Smart metering and power line communication have been leveraged to provide detailed information of energy consumption patterns and intelligent controlling to appliances at home. Proposed HEMS could provide easy-to-access information on home energy consumption in real time, intelligent planning for controlling appliances, and optimization of power consumption at home. The system has been consisting of three modules: an advanced power control planning engine, a device control module, and a power resource management server. The implemented prototype system claimed reduction in the cost of power consumption by about 10%.

(Y. Lai, Rodrigues, Huang, Wang & C. Lai, 2012) proposed a transmission protocol with two-way recognition and control service for a home grid. Based on this transmission protocol, a user could plug in a household appliance into a wall socket and then smart meter could read the electronic energy information. When this information has been captured, the features of electrical appliances have been transferred to the home energy management platform through the power line or wireless communications interface. The home energy management platform automatically could classify the electronic energy features of different appliances to identify them, which has then been used to build a control list of electric appliances on the management platform, thus, could achieve automation services with appliance intercommunication. This transmission mechanism could be built in the real-time electronic appliance interconnecting recognition and control service of the home grid and the results have been reported.

An optimal dispatching model has been developed by (Wang, Sun, Zhou & J. Dai, 2012) as Smart Home Energy Management System (SHEMS) with distributed energy resources (DERs) and intelligent domestic appliances. Multi-objective optimization has been achieved between the saving money and living comfortable and the math models of various components have been investigated to come up with the new concept "load value" for quantitative measure of users' comfort. Later, the control strategies have been set up with demand response and the parameters of optimal dispatching model have been adjusted by the load characteristic in this system. The simulation of application of the model to one house with Photovoltaic (PV), wind turbine (WT), storage battery and time-of-use prices, has proved that the energy management system and optimal dispatching model could make users of smart home live in a comfortable and economical way.

A methodology for the evaluation of wireless smart homes and home automation networks (HANs) in indoor scenarios has been presented by (Langhammer & Kays, 2012). The methodology has been applied to investigate the performance of actual wireless technologies that could be used to interconnect devices in a smart home environment. The technologies have been compared with each other by investigating the performance in realistic European indoor scenarios. Additionally, they have been evaluated with respect to their reliable indoor coverage range. To provide a decision guideline, which wireless standard is most suitable for a certain smart home scenario, has been the main intention of this paper.

An embedded system without any new additional wiring has been developed by (Lien, Bai, Chen & Hung, 2008) for home power management. By using Power Line Communication (PLC) technology, electric home appliances could be controlled and monitored through domestic power lines. A PPCOM (PLC Power-Controlled Outlet Module) has been described which could integrate the multiple AC power sockets, the power measuring module, the PLC module and a microcontroller into a power outlet to switch the power of the sockets on/off and to measure the power consumption of plugged-in electric home appliances. An embedded home server has also been designed which could support the Web page user interface, thus allowing the user to easily control and monitor the electric home appliances by means of the Internet. Additionally, the field experiments have been reported to demonstrate the practical implementation with adequate results.

New Building Energy Management Systems (BEMS) strategies have been presented by (Missaoui, Warkozek, Bacha & Ploix, 2012) to support Demand Side Management. A real-time platform: A Power-Hardware-In the-Loop (PHIL) test bench has been utilized for validation. Physical devices have been introduced into the simulated system to check whether their support for the control strategies.

Design of smart home device descriptions and standard practices for demand response and load management "Smart Energy" applications needed in a smart energy based residential or light commercial environment have been presented by (Han & Lim, 2010). The control application domains included in this initial version have been sensing device control, pricing and demand response and load control applications. Smart home interfaces and device definitions have been introduced to allow interoperability among ZigBee devices produced by various manufacturers of electrical equipment, meters, and smart energy enabling products. The proposed home energy control systems design has been introduced that provides intelligent services for users and we demonstrate its implementation using a real testbed.

A smart home computing platform architecture has been presented by (Al-Ali, El-Hag, Dhaouadi & Zainaldain, 2011) extends smart home network to be compatible for smart grid integration. The proposed hardware architecture along with a tailored software algorithm has empowered utility companies and homeowners to communicate bidirectionally with home appliances via a public mobile network to monitor and manage power consumption of home appliances. The platform has also enabled homeowners to operate, monitor and control their home appliances locally via a control panel and remotely via their mobile phones.

(Kawade & Nekovee, 2009) focused to assess the feasibility of using the TV White Space (TVWS) spectrum for home networking services and to compare the performance with that of other license-exempt spectral bands, namely 2.4 GHz and 5 GHz band. Limitations of TVWS in presence of interference have been studied and the operating conditions when the system performance would start to become unacceptable have been identified. Using analytical and simulation techniques, the outperformance of TVWS against typical 5 GHz and 2.4 GHz has been shown for low to medium traffic loadings (around 2 Mb/s per home) at significantly lower energy requirements with necessary tradeoffs and limitations. For heavy traffic loadings (6 Mb/s and above per home) in dense deployment densities, it has been shown that TVWS band could be used as complementary interface for congestion relief instead.

Overview of the application of WSNs for electric power systems has been included by (Erol-Kantarci & Mouftah, 2011), along with their opportunities and challenges and opens up future work in many unexploited research areas in diverse smart-grid applications. Performance of an in-home energy management (iHEM) application has been then evaluated and later compared with an optimization-based residential energy management (OREM) scheme which has been based on an objective is to minimize the energy expenses of the consumers. It could be observed that iHEM has been decreasing energy expenses, reducing the contribution of the consumers to the peak load, reducing the carbon emissions of the household and its savings have been close to OREM. On the other hand, iHEM application has been more flexible as communication between the controller and the consumer has been allowed for utilizing the wireless sensor home area network (WSHAN). Performance of iHEM has been evaluated under the presence of local energy generation capability, prioritized appliances, and for real-time pricing. It has also been shown that iHEM has been reducing the expenses of the consumers for each case. Furthermore, it has also been shown that packet delivery ratio, delay, and jitter of the WSHAN has been improving as the packet size of the monitoring applications with that of WSHAN has been decreasing.

3. TECHNOLOGICAL STUDY

For *security* concerns, securing connection points for remote access is of specifically importance. Since they often allow access to management level functions, attacks on them will have higher probabilities of overall system wide effects. BAS field networks are exposed to inside attacks as well, especially when run through publicly accessible spaces. Open media such as wireless and powerline signaling further increase vulnerability, since access to the medium could be gained in an unobtrusive manner. Further, the shift to open systems reduces the knowledge barrier for intruders.

3.1 Generalized Architecture, Functional Architecture and Device Classes

3.1.1 Generalized Architecture (Kastner, Neugschwandtner, Soucek, & Newman, 2005)

The overall generalized architecture for a typical BAS could be split into three major levels depending upon functional hierarchy implemented as shown in Figure1.

3.1.1.1 Field Level

At the field level, interaction with the physical world has been taking place. Field related data have been collected (including measurement, counting, metering, etc.) and suitably converted for transmission to remote location and further processing. Likewise, parameters of the field have also been physically controlled (including switching setting, positioning, etc.) in response to commands received from the remotely located centralized control system.

3.1.1.2 Automation Level

Automation level has been assigned to automatic control, including all kinds of autonomously executed sequences. This level has been configured to operate on data furnished by the field level, establishing logical connections and control loops. In this level, the processing entities may also communicate values of more global interest to each other. This type of process data exchange has been referred to as *horizontal* communication. In addition, required aggregated values for *vertical* access by the management level have also been prepared in this level. The accumulation of historical data at specified rates (*trending*) have also been included here.

Figure 1. Overall generalized architecture of a typical BAS (Kastner, Neugschwandtner, Soucek, & Newman, 2005)

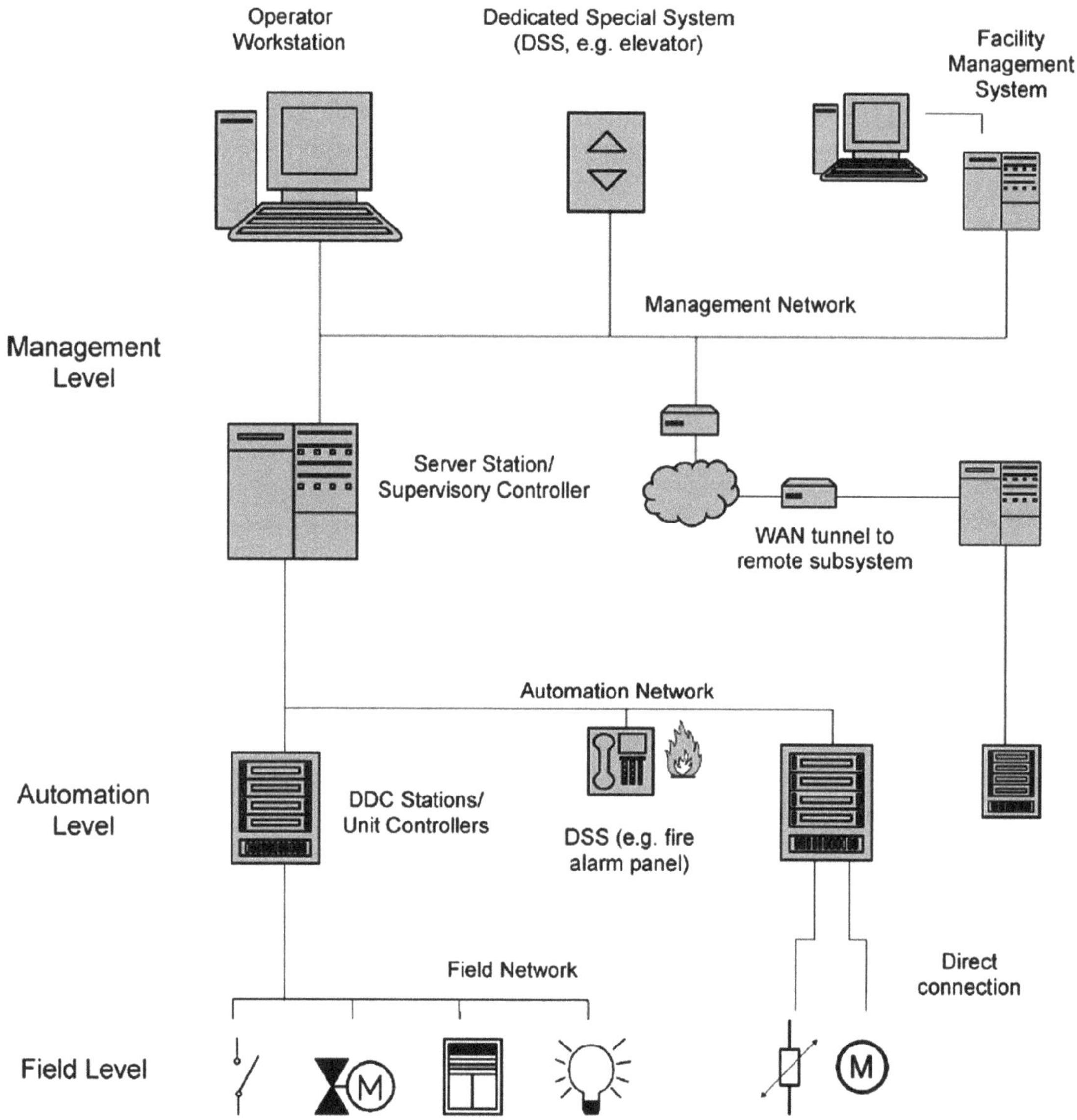

3.1.1.3 Management Level

At the management level, information from throughout the entire system has been made accessible. A unified interface has been presented to the operator for manual intervention. Vertical access to automation-level values has been provided, including the modification of parameters such as schedules. Alerts have been generated for exceptional situations like technical faults or critical conditions. Long-term historical data storage with the possibility to generate reports and statistics has also been considered as part of this level.

3.1.2 Functional Architecture (Granzer, Kastner, Neugschwandtner & Praus, 2006)

Each BAS could be categorized in various subsystems according to the basic functionalities it incorporates as described below:

3.1.2.1 Automation Subsystem

This subsystem has also been popularly known as automatic control subsystem or energy saving subsystem.

1. **Internal Climate Control:** Heating, Ventilation and Air Conditioning (HVAC) solutions including temperature and humidity control have been implemented using this subsystem and has been generally one of the most important aspects for a homeowner. Remote as well as local control of systems for ambient temperature, humidity, heating and air conditioning has been allowed and availed using this subsystem.
2. **Lighting Control:** Lighting of building has been one of the major contributors to electricity bill that could be reduced significantly by controlling switching frequency and ON/OFF timings of electric lights in a prefixed scheduled or on the basis of ambient light intensity. Lighting control systems could involve aspects related to controlling electric lights. Natural lighting control involves controlling window shades, LCD shades, draperies and awnings. Recent advances include use of RF technology in order to avoid wiring for switches and integration with third party home automation systems for centralized control. The under mentioned strategies could be useful:
    a. To keep all the lights of the house 'Extinguished' in general.
    b. Automation of switched off / ignition in every point of light.
    c. Regulation of the illumination according to the level of ambient luminosity.
    d. Day timing based lighting control is also employed for energy efficiency.
3. **Water Level Control:** Control of water level of underground and/or overhead tanks has been quite crucial– for routine activities and to tackle abnormal occurrences like fire. Hence, availability of sufficient water should have been ensured by automatic water-level control (under automation subsystem) to switch ON/OFF the filling pump and discharge pump as necessary.

3.1.2.2 Security Subsystem

Under this subsystem, the consumer could select and watch cameras live using internet to their home or business. Security cameras could be controlled, allowing the user to observe activity around a house or business right from monitor or touch panel. Security systems could also include motion sensors for detection of unauthorized movement and annunciation of the user through security system or via cell phone.

Particular security camera channels could be selected for viewing, controlling camera movement and for monitoring activity. Motion sensing feature could also be configured that will detect unauthorized movement and generate alerts via audio-visual annunciation. Audio-video recording of areas under monitoring could also be carried out. Past recordings could also be stored on the server and replayed on demand. Under mentioned functionalities could be incorporated:

1. Energy conservation by switching off lights when not required.
2. Security of residents from intruders.
3. Safety against fire and gas leakage.
4. Remote operation (control) of any electric appliance from a central location within the building as well as from outside the building.

5. Connectivity to the internet to meet information, e-services and communication needs of the residents.
6. Access to a central digital entertainment library.
7. Simple and user friendly graphical user interface (GUI).
8. Low cost of the solution.
9. Simplicity, future expandability and interoperability.

This category also includes control and distribution of security cameras for surveillance, etc.

3.1.2.3 Entertainment Subsystem

1. **Audio Entertainment:** This category includes audio switching and distribution. Audio switching determines the selection of an audio source. Audio distribution allows an audio source to be heard in one or more rooms. This feature is often referred to as 'multi-zone' audio. Major companies associated with Audio Distribution include three components that allow the consumers to listen to audio throughout their homes, or businesses:
   a. CAT 5e/ CAT6 cable from Audio central unit.
   b. 2 sets of speaker cabling (4ply from amplifier, and 2 ply from key pad to ceiling or wall speakers).
   c. A keypad to control your volume and sources.
2. **Video Entertainment:** Similarly, the video subsystem could also include video switching and distribution upon user demand, and hence, the corresponding feature has been often referred to as 'multi-zone' video. Video door entry system could also be integrated with the TV screens, allowing the user to view the entrant as seen by a door camera. This includes video switching and distribution, allowing a video source to be viewed on multiple TVs. This feature is often referred to as 'multi-zone' video. Integration of the intercom to the telephone or of the video door entry system to the television set, allowing the residents to view the door camera automatically.

3.1.2.4 Communication Subsystem

As an integral part of this subsystem, an intercom system could be installed to allow communication via a microphone and speaker between multiple rooms. Following features could be added:

1. Remote Control: using Intranet / Internet / PDA with wireless connectivity.
2. Alarm annunciation.
3. Inter-person communications.
4. Ubiquity in the external control as much internal, remote control from the Internet, PC, wireless controls (e.g. PDA with Wi-Fi), electrical equipment.
5. Transmission of alarms.
6. Intercommunications.
7. Control of home robots, using if necessary demotic electric beacon.
8. Home robot communication (i.e. using Wi-Fi) with the demotic network and other home robots.

3.1.3 Device Classes (Lewis, 2004)

Based on views presented in and above cited materials, BAS devices and their functionality have been classified in three categories as–

3.1.3.1 Sensor, Actuators and Controllers (SAC)

These devices have been installed generally at field level. Such devices interact directly with the physical environment of the building and are responsible for data acquisition and control of the behavior of the internal climate of the building. Additionally, they may include controller functionality.

3.1.3.2 Inter Connecting Devices (ICD)

These devices have been utilized to link different types of communication networks and network segments meaningfully together to fulfill specific objectives. Representatives of this device class could enable BAS devices to interact either using the same network protocol or via different network protocols.

3.1.3.3 Configuration and Management Devices (CMD)

These devices have been employed to configure and maintain the BAS. Typical CMDs could implement tasks including the change of control parameters, setting up of new automation tasks as well as monitoring, logging and archiving of process data values. In most cases, representatives of the CMD class could be located at the backbone level. Still, some representatives could also be found at the field/automation level (e.g., for on-site device configuration). Since CMDs have often being controlled by humans, they should have provision for simple and user-friendly graphical user interface (GUI).

In BAS, an interface (e.g. a gateway or more generally an ICD) is required to establish the communication between controllers and building systems due to their different communication rates and data formats. The continued use at the corporate level to ensure the communication between controllers and plants and building components determines proper and effective operation of this ICD. As the BAS leverages information between the central computerized control centre and various access stations or terminals through different network protocols that utilize different data formats, it is necessary to deploy such an ICD. BAS use this data to exchange information between management level and field level through an open protocol.

3.2 Communication Technologies for BAS (Lewis, 2004)

The overall building environment could be split into different networks as depicted in a typical BAS operated building as shown in Figure 2. For ascer-

Figure 2. Various networks in a typical BAS operated Smart Home. (Lewis, 2004)

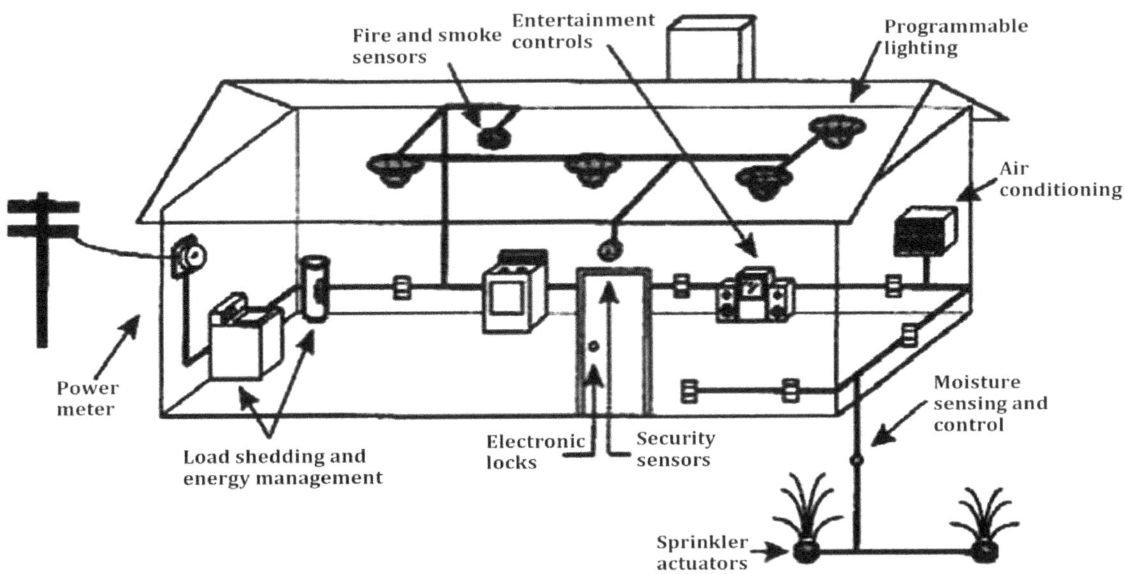

taining the operational integrity between various BAS subsystems deployed with different network technologies, the overall role of data communication in BAS becomes significant and critical. In real-time building operations, it is necessary for control laws to communicate with real building subsystems (like HVAC systems). Recently several network protocols (BACnet, LonWorks, MODBUS, etc.) have been developed for a better exploitation of the BAS communication network.

3.2.1 Major Wired Technologies for BAS

3.2.1.1 RS-485

Recommended Standard 485 (RS-485) is a standard for serial binary data signals connecting between a DTE (Data Terminal Equipment) and a DCE (Data Communication Equipment). It is commonly used for communication using serial ports. Generally, RS-485 operates on differential balanced line over twisted pair flexible signal cable and follows topologies such as point-to-point, multi-drop, multi-point, etc. RS-485 can handle maximum 32 devices (32 drivers and 32 receivers) spread over maximum distance of 1200 meters or 4000 feet approximately. It operates within -7 to +12 V range (±6V typical) and provides maximum baud rates in the range of 100 kbps to 10 Mbps. RS-485 specify only the electrical characteristics of the driver and the receiver. It does not specify or recommend any data protocol. Ideally, the two ends of the cable shall have a termination resistor connected across the two wires to prevent data corruption due to multiple data edges. Termination resistors have also been used to reduce electrical noise sensitivity due to the lower impedance. The value of each termination resistor should be equal to the cable impedance (typically, 120 ohms for twisted pairs). EIA-485 is employed in Master-Slave arrangement in full duplex operation.

RS-485 is used as the physical layer underlying many standard and proprietary automation protocols including MODBUS, PROFIBUS, etc. It is used with PLCs, on factory floors, etc. Since it is differential, it resists electromagnetic interference from motors and welding equipment. RS-485 is also used in building automation as the simple bus wiring and long cable length is ideal for joining remote devices. It may be used to control video surveillance systems or to interconnect security control panels and devices such as access control card readers. Although many applications use EIA-485 signal levels, the speed, format, and protocol of the data transmission is not specified by EIA-485 which is its main limitation (RS-485, 2015).

3.2.1.2 MODBUS

In the late 1970s, Modicon Inc. developed MODBUS protocol. It is positioned in application layer, which is 7th layer of the ISO/OSI reference model and supports client-server communications among PLCs and other networked devices. It follows the basic OSI model, although not all layers are used. The MODBUS protocol defines the methods for a PLC to obtain access to another PLC, for a PLC to respond to other devices, and means for detecting and reporting errors. The protocol supports other protocols such as asynchronous master-slave transmission. Variants of MODBUS, MODBUS Plus and MODBUS/TCP were also developed to take advantage of the internet based supporting tools, hardware, and software. Modbus is often used to connect a supervisory computer with a remote terminal unit (RTU) in supervisory control and data acquisition (SCADA) systems (MODBUS, 2015).

Since Modbus protocol is just a messaging structure, it is independent of the underlying physical layer. It is traditionally implemented using RS232, RS422, or RS485 over a variety of media (e.g. fibre, radio, cellular, etc.).A total of 247 different slaves can be addressed. The typical data rate is 19.2 Kb/s. A mode of transmission over TCP/IP is also defined, in which every node can be both client and server.

Main motivations behind its extensive use have been:

1. It is openly published and royalty-free.
2. Relatively easy industrial network to deploy.
3. It moves raw bits or words without placing many restrictions on vendors.

Major limitations have been:

1. Modbus was designed in the late 1970s to communicate to PLCs; the number of data types is limited to those understood by PLCs at the time. Large binary objects are not supported.
2. No standard way exists for a node to find the description of a data object, for example, to determine if a register value represents a temperature between 30 and 175 degrees.
3. Since Modbus is a master/slave protocol, there is no way for a field device to "report by exception" (except over Ethernet TCP/IP, called open-mbus) - the master node must routinely poll each field device, and look for changes in the data. This consumes bandwidth and network time in applications where bandwidth may be expensive, such as over a low-bit-rate radio link.
4. Modbus is restricted to addressing 247 devices on one data link, which limits the number of field devices that may be connected to a master station (once again Ethernet TCP/IP proving the exception).
5. Modbus transmissions must be contiguous which limits the types of remote communications devices to those that can buffer data to avoid gaps in the transmission.

Although never formally standardized, Modbus can definitely be regarded as an industry proven and open protocol. This protocol was designed in the late 1970s and is currently supported by most programmable logic controllers (PLCs) in some form. Implementation of the Modbus protocol is license-free, which makes it especially interesting for integration and interfacing between BAS and other systems. It still is supported to some extent by numerous BA controllers, especially for the purpose of HVAC controller-to-controller-communication (e.g., with chillers). Moreover, Modbus is also present in devices belonging to other building disciplines, like electricity meters or fire alarm systems as well as many industrial instruments and automation systems.

3.2.1.3 PLC

Power Line Communication (PLC) carries data on a conductor that is also used simultaneously for AC electric power transmission or electric power distribution to consumers. It is also popularly known as other names such as Power Line Carrier Communication (PLCC), Power-line Digital Subscriber Line (PDSL), Power Line Networking (PLN), mains communication, power-line telecommunications, etc. (Power-line communication, 2015).

Typically home-control power-line communication devices operate by modulating in a carrier wave of between 20 and 200 kHz into the household wiring at the transmitter. The carrier is modulated by digital signals. Each receiver in the system has an address and can be individually commanded by the signals transmitted over the household wiring and decoded at the receiver. These devices may be either plugged into regular power outlets, or permanently wired in place. Since the carrier signal may propagate to nearby homes (or apartments) on the same distribution system, these control schemes have a "house address" that designates the owner.

The "universal powerline bus", introduced in 1999, uses pulse-position modulation (PPM). The physical layer method is a very different scheme from the X10. LonTalk, part of the LonWorks home automation product line, was accepted as part of some automation standards.

Since the mid-1980s, there has been a surge of interest in using the potential of digital communications techniques and digital signal processing. The drive is to produce a reliable system which is cheap enough to be widely installed and able to complete cost effectively with wireless solutions.

Applications of mains communications vary enormously, as would be expected of such a widely available medium. One natural application of narrow band power-line communication is the control and telemetry of electrical equipment such as meters, switches, heaters and domestic appliances. A number of active developments are considering such applications from a systems point of view, such as demand side management. In this, domestic appliances would intelligently co-ordinate their use of resources, for example limiting peak loads.

A wide range of power-line communication technologies have been needed for different applications, ranging from home automation to Internet access which is often called broadband over Power Lines (BPL). Most PLC technologies limit themselves to one type of wires (such as premises wiring within a single building), but some can cross between two levels (for example, both the distribution network and premises wiring). Typically transformers prevent propagating the signal, which requires multiple technologies to form very large networks. Various data rates and frequencies are used in different situations.

A number of difficult technical problems are common between wireless and power-line communication, notably those of spread spectrum radio signals operating in a crowded environment. Radio interference, for example, has long been a concern of amateur radio groups.

Power-line communications systems operate by adding a modulated carrier signal to the wiring system. Different types of power-line communications use different frequency bands. Since the power distribution system was originally intended for transmission of AC power at typical frequencies of 50 or 60 Hz, power wire circuits have only a limited ability to carry higher frequencies. The propagation problem is a limiting factor for each type of power-line communications.

The main issue determining the frequencies of power-line communication is laws to limit interference with radio services. Many nations regulate unshielded wired emissions as if they were radio transmitters. These jurisdictions usually require unlicensed uses to be below 500 kHz or in unlicensed radio bands. Some jurisdictions (such as the EU), regulate wire-line transmissions further. The U.S. is a notable exception, permitting limited-power wide-band signals to be injected into unshielded wiring, as long as the wiring is not designed to propagate radio waves in free space.

Data rates and distance limits vary widely over many power-line communication standards. Low-frequency (about 100–200 kHz) carriers impressed on high-voltage transmission lines may carry one or two analog voice circuits, or telemetry and control circuits with an equivalent data rate of a few hundred bits per second; however, these circuits may be many miles long. Higher data rates generally imply shorter ranges; a local area network operating at millions of bits per second may only cover one floor of an office building, but eliminates the need for installation of dedicated network cabling.

The modulation generally used in the system is amplitude modulation. The carrier frequency range is used for audio signals, protection and a pilot frequency. The pilot frequency is a signal in the audio range that is transmitted continuously for failure detection.

A *PLC carrier repeating station* is a facility, at which a power-line communication (PLC) signal on a powerline is refreshed. Therefore the signal is filtered out from the powerline, demodulated and modulated on a new carrier frequency, and then re-injected onto the powerline again. As PLC signals can carry long distances (several 100 kilometers), such facilities only exist on very long power lines using PLC equipment.

PLC has been one of the technologies used for Automatic Meter Reading (AMR) and Automated Metering Infrastructure (AMI), which have been mandatory and critical subsystems of smart grid.

Power-line communications technology can use the electrical power wiring within a home for home automation: for example, remote control of lighting and appliances without installation of additional control wiring.

The Distribution Line Carrier (DLC) System technology used a frequency range of 9 to 500 kHz with data rate up to 576 Kbit/s. In 2009, a group of vendors formed the Powerline Intelligent Metering Evolution (PRIME) alliance. As delivered, the physical layer is OFDM, sampled at 250 kHz, with 512 differential phase shift keying channels from 42–89 kHz. Its fastest transmission rate is 128.6 kilobits/second, while its most robust is 21.4 Kbit/s. It uses a convolutional code for error detection and correction. The upper layer is usually IPv4.

In 2011, several companies including distribution network operators (ERDF, Enexis), meter vendors (Sagemcom, Landis & Gyr) and chip vendors (Maxim Integrated, Texas Instruments, STMicroelectronics) founded the G3-PLC Alliance to promote G3-PLC technology. G3-PLC is the low layer protocol to enable large scale infrastructure on the electrical grid. G3-PLC may operate on CENELEC A band (35 kHz to 91 kHz) or CENELEC B band (98 kHz to 122 kHz) in Europe, on ARIB band (155 kHz to 403 kHz) in Japan and on FCC (155 kHz to 487 kHz) for the US and the rest of the world. The technology used is OFDM sampled at 400 kHz with adaptive modulation and tone mapping. Error detection and correction is made by both a convolutional code and Reed-Solomon error correction. The required media access control is taken from IEEE 802.15.4, a radio standard.

Power line communications can also be used in a home to interconnect home computers and peripherals, and home entertainment devices that have an Ethernet port. Adapters allowing for such connectivity are often marketed as "Ethernet Over Power" (EOP). Powerline adapter sets plug into power outlets and establish an Ethernet connection using the existing electrical wiring in the home. This allows devices to share data without the inconvenience of running dedicated network cables.

The most widely deployed powerline networking standard is from the HomePlug Powerline Alliance. HomePlug AV is the most current of the HomePlug specifications and was adopted by the IEEE 1901 group as a baseline technology for their standard, published 30 December 2010. HomePlug estimates that over 45 million HomePlug devices have been deployed worldwide.

3.2.1.4 BACnet

The Building Automation and Control Networking (BACnet) Protocol was developed by American Society of Heating, Refrigerating and Air-Conditioning Engineers (ASHRAE), to specifically address the needs of building automation and control systems of all sizes and types. While the data exchanged in BACnet is encapsulated in objects accessed by services calls, BACnet can ensure the interoperability between devices of different suppliers. The use of BACnet is free of any licenses or fees (BACnet, 2015).

While BACnet messages can, in principle, be conveyed over any network, a small number of network types were standardized for BACnet's use and such types are chosen to cover a range of speed and throughput. They are Ethernet, ARCNET, Master–Slave/Token-Passing (MS/TP), LonTalk, and Point-to-Point (PTP). MS/TP addresses connectivity over twisted pairs using RS-485 signaling while PTP supports dial-up communications and other point-to-point applications using RS-232 and, possibly, modems or other data communication equipment. The MS/TP RS-485 medium provides a low-cost, well-established means for communication up to 78.4 kb/s and is useful with traffic loads such as would typically be experienced with unitary or application specific controllers. BACnet/IP and BACnet over Ethernet are more suited to communications involving higher data volumes. BACnet over wireless Ethernet has been around for years, proven largely at trade show exhibitions, but not yet formally in practice.

3.2.1.5 LonWorks

LonWorks is not specifically designed for automated building networks, but is also used in other industries. Now it is gaining a high importance use in BAS architecture. While supporting communications on a variety of LANs or allowing different components to coexist in the same network, LON (or LonWorks) communicates via Standard Network Variable Types (SNVT) to facilitate interoperability by providing a well-defined interface. Basically a proprietary communication protocol is called LonTalk, created by ECHELON Corporation. A chip is required for any device that uses LON.

The LonWorks system has been originally designed by Echelon Corp. as an event-triggered control network system. The term EIA-709 is used to refer to the standardized model of this communication protocol (LonWorks, 2015).

It supports a variety of different communication media and different wiring topologies. Since it was designed as a generic control network, many protocol parameters are free to choose for the designer. To achieve interoperability, a number of communication *channel* profiles were defined. These still include a variety of Twisted-Pair (TP), Powerline (PL), Fiber Optic (FO) and RF (Radio Frequency) solutions; albeit no standard interoperability profile exists for these. The most popular channel for building automation purposes is the 78.1 kb/s free topology TP profile (FT-10), which allows physical segments of up to 500 m using low-cost TP cable. A variant providing link power (LP-10) is also available. Often the 1.25 Mb/s bus topology TP (TP-1250) or sometimes FO is used as a backbone to connect the lower speed FT-10 buses.

3.2.1.6 EIB/KNX

The European Installation Bus (EIB) is a Fieldbus designed to enhance electrical installations in homes and buildings of all sizes by separating the transmission of control information from the traditional mains wiring. EIB is based on an open specification maintained until recently by EIB Association (EIBA). In 2002, EIB was merged with Batibus and EHS (European Home System). The new KNX standard seeks to combine their best aspects. The target of this merger was to create a single European *home and building electronic system* standard. Likewise, EIBA joined forces with the European Home Systems Association and Batibus Club International to form Konnex Association. Still, the EIB system technology continues to exist unchanged as a set of profiles within KNX, frequently referred to as EIB/KNX (EIB/KNX, 2015).

Regarding physical media, EIB already provided the choice of dedicated twisted-pair cabling and powerline transmission as well as a simple form of IP tunneling. RF communication and advanced IP tunneling were added under the KNX umbrella (albeit are not yet published within the context of EIB). The KNX specification also includes additional TP and PL variants which could be used for future devices.

The main EIB/KNX medium is the twisted-pair cabling variant now known as KNX TP1. The single twisted pair carries the signal as well as 29 V DC link power. Data is transmitted using a balanced base band signal with 9600 b/s. TP1 allows free topology wiring with up to 1000 m cable length per physical segment. Up to four segments can be concatenated using bridges (called *line repeaters*), forming a *line*. Medium access on TP1 is controlled using CSMA with bit-wise arbitration on message priority and station address with provision of four priority levels.

KNX RF uses a sub-band in the 868 MHz frequency band reserved for short-range devices (tele-command, control, telemetry and alarms) by European regulatory bodies which is limited by a duty cycle requirement of less than one percent.

Particular attention was given to minimizing hardware requirements. To this end, KNXRF does not only support bidirectional communication, but transmit-only devices as well. This reduces cost for simple sensors and switches without status indicators. KNXRF devices communicate peer-to-peer. On the management level, IT standards prevail for connectivity, while on the automation level EIA-485 is very popular.

Field buses which are well-established in factory and process automation (like Interbus, CAN-based protocols as DeviceNet or CANOpen, and Profibus DP16) are largely irrelevant in BAS; except for occasional use in "plant room network" controller-to-controller communication (specifically including variable frequency drives for fans and pumps).

3.2.2 Major Wireless Technologies for BAS

3.2.2.1 Bluetooth

Bluetooth is based on a wireless radio system designed for short-range and cheap devices to replace cables especially for computer peripherals. Bluetooth was initiated in 1998 and standardized by the IEEE as Wireless Personal Area Network (WPAN) specification wide IEEE 802.15.1 standard. Bluetooth is a short range RF technology aimed at facilitating communication of electronic devices between each other and with the Internet, allowing for data synchronization that is transparent to the user. Supported devices include PCs, laptops, printers, joysticks, keyboards, mice, cell phones, PDAs, and consumer products. Mobile devices are also supported. Discovery protocols allow new devices to be hooked up easily to the network. Bluetooth uses the unlicensed 2.4 GHz ISM band, can transmit data up to 1Mbit/s, can penetrate solid non-metal barriers, and has a nominal range of 10m that can be extended to 100m. A master station can service up to 7 simultaneous slave links. Forming a network of these networks, e.g. a piconet, can allow one master to service up to 200 slaves (Bluetooth, 2015).

Two connectivity topologies are defined in Bluetooth: piconet and scatternet. A piconet is a WPAN formed by a Bluetooth device serving as a master in the piconet and one or more Bluetooth devices serving as slaves. A frequency-hopping channel based on the address of the master defines each piconet. All devices participating in communications in a given piconet are synchronized using the clock of the master. Slaves communicate only with their master in a point-to-point fashion under the control of the master. The master's transmissions may be either point-to-point or point-to-multipoint. Also, besides in an active mode, a slave device can be in the parked or standby modes so as to reduce power consumptions. A scatternet is a collection of operational Bluetooth piconets overlapping in time and space. Two piconets can be connected to form a scatternet. A Bluetooth device may participate in several piconets at the same time, thus allowing for the possibility that information could flow beyond the coverage area of the single piconet. A device in a scatternet could be a slave in several piconets, but master in only one of them.

Currently, Bluetooth development kits can be purchased from a variety of suppliers, but the systems generally require a great deal of time, effort, and knowledge for programming and debugging. Forming piconets has not yet been streamlined and is unduly difficult.

3.2.2.2 Wi-Fi

IEEE ratified the IEEE 802.11 specification in 1997 as a standard for Wireless Local Area Network (WLAN). Current versions of 802.11 (i.e. 802.11b) support transmission up to 11Mbit/s. Wireless Fidelity (Wi-Fi), as it is popularly known, is useful for fast and easy networking of PCs, printers, and other devices in a local environment, e.g. office or home or public places like airports, hotels, cafes. It allows users to surf the Internet at broadband speeds when connected to an access point (AP) or in ad hoc mode. Current PCs and laptops as purchased have the hardware to support Wi-Fi. Purchasing and installing a Wi-Fi router and receivers is within the budget and capability of home PC enthusiasts.

Wi-Fi includes IEEE 802.11a/b/g standards for wireless local area networks (WLAN). The IEEE 802.11 architecture consists of several components that interact to provide a wireless LAN that supports station mobility transparently to upper layers. Wi-Fi enabled stations are able to communicate directly without any AP. Because this type of network is often formed without pre-planning, for only as long as the network is required, this type of operation is often referred to as an ad hoc network (Wi-Fi, 2015).

3.2.2.3 ZigBee

ZigBee is a cost-effective, IEEE 802.15.4 standard-based wireless networking technology defined by ZigBee Alliance, specifically designed to replace the proliferation of individual remote controls, and with characteristics like low data rates, low power consumption, self-healing and multi-hop communications results in providing features like low cost, robustness with security and reliability (ZigBee, 2015).

Although ZigBee presents similarities with Bluetooth, it is simpler, has a lower data rate, spends most of its time snoozing, and targets building automation as the main application. By taking advantages of those characteristics, this technology is suitable for remote control of equipment that require low level of complexity such as home lighting systems, HVAC equipment, other household electrical appliances, etc. With those features, ZigBee represents a promising technology for the development of commercial wireless BAS.

The focus of IEEE 802.15.4 and ZigBee has been to provide general purpose, easy-to-use and self-organizing wireless communication for low cost and low power embedded devices. Based on the OSI (Open System Interconnection) reference layered model, the IEEE 802.15.4, a PHY and MAC layer standard, defined a 250 kb/s direct sequence spread spectrum (DSSS) radio, operating in the 2.4 GHz unlicensed industrial, scientific and medical (ISM) band. While IEEE 802.15.4 defines the physical and the MAC layer, ZigBee defines the layers above. The IEEE 802.15.4 physical layer specifies 3 different frequency bands: 868-868.6 MHz (1 channel, 20 kb/s), 902-928 MHz (10 channels, 40 kb/s) and 2.40-2.48 GHz (16 channels, 250 kb/s) all using PSK (phase shift keying) modulation. On top of this PHY and MAC layer standard, several proprietary and standards-based sensor network systems emerged.

Wireless sensor networks (WSN) are mesh type networks of small size, battery powered sensor nodes (motes) that sense the environmental parameters, perform signal processing computations, and communicate with other nodes using the onboard radio module. Typically, sensor nodes transmit the measured data to a base station using multi-hop communication over an RF link.

Wireless sensor networks have been an active research topic in recent time. The release of IEEE 802.15.4 standard and ZigBee technology helped to bring the technology out of research labs and accelerated the availability of various commercial reliable products that are designed based on standards-based networking and RF solutions are used in applications such as home automation, building automation, industrial automation, habitat monitoring, utility metering, etc.

WSN based BAS can offer many benefits. The main benefit is that the installation costs of BAS with WSN are much lower than that of traditional wired BAS versions, due to wires being replaced by RF link. New buildings can therefore be equipped with the WSN based BAS. It is also possible to expand an existing wired BAS to increase the sensor coverage. The reduced installation investment promotes to include more number of sensors and hence enhance the spatial resolution. This enhanced spatial resolution results in more fine-grained measurements and better control. WSNs enable setting up of systems for temporary measurements. Hence, an adhoc network can be set up to take measurements during a limited time in order to measure, optimize and evaluate the effect of the optimization strategy planned. Additionally, WSN motes can also be easily installed inaccessible or inconvenient locations and still they are able to perform continued and reliable operation for long time periods.

3.3 Design and Development

3.3.1 Design Objectives (Lewis, 2004)

1. Energy conservation by switching off lights when not required based on occupancy, daylight and schedule.
2. Security of residents from intruders.
3. Safety against fire and gas leakage.
4. Remote operation (control) of any electric appliance from local control panel using LCD keypad or remotely within vicinity of control panel using wireless keychain touchpad (limited functionality).
5. Connectivity to the telephone line to serve alarms and updates to the users via voice alerts.
6. Simple and user-friendly operation.
7. Low cost of the solution.
8. Simplicity and future expandability.

3.3.2 Functionalities (Lewis, 2004)

1. Reporting alarms and status of different areas under its coverage.
2. Alarmreportingbyaudio-visualannunciation.
3. Provides control of various devices via LCD keypad and/or via GUI and using wireless keychain touchpad (limited functionality).
4. Remote monitoring and control facility from any PC in intranet/internet and within vicinity of control panel using wireless keychain touchpad (limited functionality)
5. Live video of view (with audio) captured by IP camera and its live streaming-recording facility
6. Entertainment panel provides the user to listen continuous broadcast of live radio via GUI
7. Entertainment panel provides unique video-on-demand facility via GUI
8. Recording of alarms in MS Excel format and maintaining history of events/alarms
9. Remote monitoring and control facility from any PC in intranet/internet.

3.3.3 Architectural Modularity

Modern BAS architecture has been modular in nature and could be classified in three major modules as described below:

3.3.3.1 Field Level Modules

These modules have also been known as Device Level Modules and consist of sensors, actuators and controllers. Such field devices are ordinary ones, which have been inexpensive, easy to interface and readily available from a large number of vendors. They could function directly in the physical environment and have been deployed for data acquisition from and control of the behavior of that environment.

Smoke sensors, glass break detectors, PIR sensors, door-window sensors, LPG detectors, etc. have been included in this class. Digital I/O modules have also been included here that could perform the function of data acquisition from sensors.

3.3.3.2 Interconnecting Modules

These modules have been used to link various networks and/or network segments together to facilitate meaningful application development. These modules could enable various BAS modules to interact with same protocol or by conversion of protocol. These modules have also been used to provide network range extension, isolation as well as interfacing at various levels of ISO/OSI Model.

3.3.3.3 Management and Configuration Modules

The main purpose of these modules has to configure and manage various BAS functionalities. These modules could be accessed locally or remotely. Such modules have been deployed for monitoring, controlling, logging as well as archiving the process data values. These devices provide functionalities at the backbone level and their another major function has to provide user-friendly GUI for building monitoring and control using data collected from above mentioned both types of modules. These modules could also generate audio-visual alarm annunciation with some extra useful features for user comfort.

3.3.4 Data Communication in a Building

Building Automation Systems (BAS) are used to both improve the indoor climate in buildings and to reduce the operational costs. Originally, BAS mostly consisted of heating, ventilation and air-conditioning (HVAC) systems. To further increase management and reduce costs, the lighting, safety, security, entertainment and supervision of asset movement have been integrated into BAS.

For ascertaining the operational integrity between various BAS subsystems deployed with different network technologies, the overall role of data communication in BAS becomes significant and critical. In real-time building operations, it is necessary for control laws to communicate with real building subsystems (like HVAC systems). Recently several network protocols (BACnet, LonWorks, MODBUS, etc.) have been developed for a better exploitation of the BAS communication network.

4. SOLUTIONS AND RECOMMENDATIONS

4.1 Security System vs. Building Automation System (Lewis, 2004)

Security has always remained a core and focused subsystem of entire BAS. Therefore, in such an understanding, security system could be considered as an important and mandatory section of overall BAS.

Comparatively smaller size, home security systems mainly facilitate important and much needed functionalities including automation, safety and security as well as communication subsystems with limited features as compared to full-fledged BAS. It is important to note here that such home security systems work with relatively lesser functionalities and does not provide any functionalities with high bandwidth requirements including high speed, high throughput communication, entertainment, etc. unless otherwise designed or manufactured in a specially customized form.

4.1.1 Development of Wired Solutions

4.1.1.1 Wired Home Security System

1. **Introduction:** Wired home security systems are cost-effective miniature and affordable versions of large scale BAS for domestic

usage. Wired home security system are generally designed and implemented based on popular wired network technologies such as RS485, PLC, BACnet, etc. Sometimes wired ethernet is also preferred, if feasible.

2. **Usefulness:** As compared to versions designed based on industrial protocols (MODBUS) or large scale BAS protocols (BACnet/LonWorks or EIB/KNX), versions based on proprietary designs are available in markets for home security solutions. Such versions being of significantly low cost and ease of availability in good quantities with local vendors result in quick deployable and easy maintainable or replaceable solutions for satisfying low priority, low budget requirements. One such a kind of home security solution has been developed and reported below:
3. **Implementation:** The photograph of implementation of wired home security system has been shown in Figure 3. Sensors are installed as per design requirements on panel as well as at necessary field locations and wiring is drawn between field mounted sensors and panel.
4. **Strengths and Limitations:**
    a. Strengths:
        i. Cost effective and cheap
        ii. Quick deployable
        iii. Easy configuration and modification within scope
        iv. Easy availability in local markets
    b. Limitations:
        i. Close ended design; hence root level modifications not possible, changes can be made within permissible scope.
        ii. Maintenance rarely possible; only replacement is made in case of failure.
        iii. As being manufactured by small vendors after sales support and spares supply channels are poor.
        iv. No standardization in design and models, alterations are found quite frequent.
5. **Results and Discussions:** The system reacts by sounding an audible alarm, by changing the display in the LCD display on keypad as well as by sending voice alerts on phone for the following events:
    a. Detection of movement.
    b. Opening or closing of a door or window.
    c. Impact or breakage of glass.
    d. Detection of fire due to smoke or leakage of gas.
    e. Request of urgent help by pressing of panic button or foot switch.
    f. Detection of empty or full water tank.

The system can be operated both by using LCD keypad locally as well using keychain touchpad remotely and all the major functions can be carried

Figure 3. Photograph of implementation of Wired Home Security System (Bhatt, 2009; Bhatt & Verma, 2015)

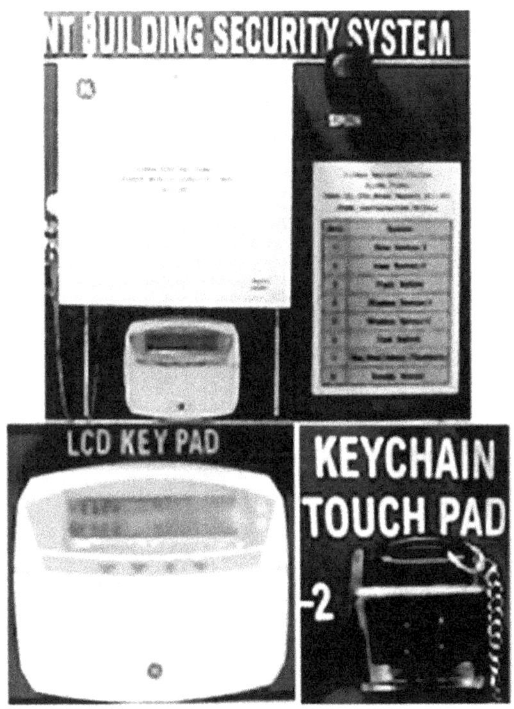

out quickly. The system is also accepting inputs of limited variety of proprietarily designed wireless sensors via add-on card and acts on them similarly as on normal wired sensors.

4.1.1.2 Wired Building Automation System

1. Usefulness

For continuous monitoring of large scale residential as well as commercial complexes, where along with safety and security provisions, communication and entertainment are essential requirements, for such buildings this implementation can work as a comprehensive solution.

Managements of such premises can conveniently afford higher, but justified cost of this solution without many issues and can afford to recruit skilled manpower to build, operate and maintain the system for long efficient, proven and reliable service operation.

2. Implementation

The photograph of implementation of wired BAS has been shown in Figure 4. The complete panel has been divided in different areas like room, corridor, lawn, entrance, water tank, etc. for a residential building model and actual sensors along with LED actuators are installed in corresponding areas as per design requirements. Some sensors are also installed in field at different locations and wires are drawn between field mounted sensor and the panel.

3. Description

Room is monitored for occupancy and open/close status of Door and Windows and controlled for electrical appliances like lights, fan, air conditioner, heater, television and geyser. Corridor is controlled for lights. Lawn and corridor lights are controlled based on daylight sensing and controls

Figure 4. Photograph of implementation of Wired Building Automation System (Bhatt, 2009; Bhatt & Verma, 2015)

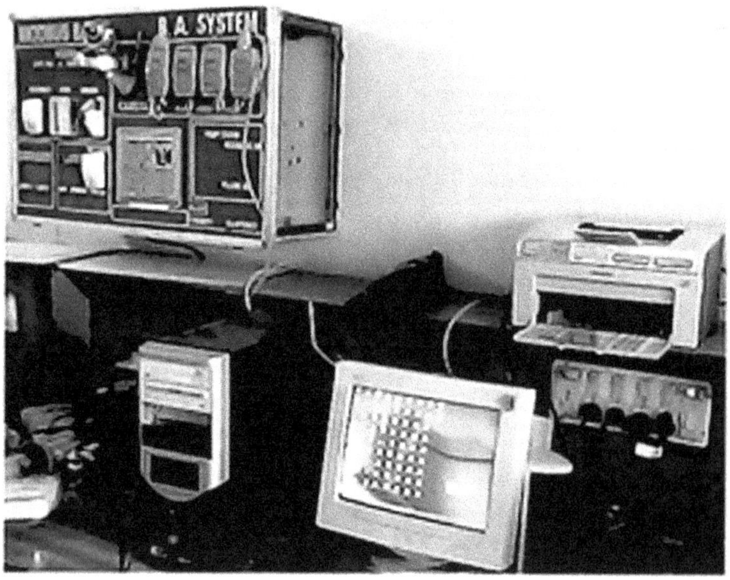

for sprinkler and fountain are also provided. Water tank level indication along with controls for filling pump and discharge pump are provided. IP camera is installed at the main entrance for live video surveillance and monitoring of entry and movement at the main entrance. MODBUS modules used in the design are installed and wired on the same panel to form EIA-485 network. EIA-485 to EIA-232 convertor module is installed and connection with PC is established. The panel is prepared with professional touch and pictorial representation of LabVIEW GUI along with network connections make it easy to understand and interesting.

The front panel of operational GUI has been shown which is prepared by integrating ICP-CON LabVIEW driver, NAPOPC server and LabVIEW Graphical Programming Language version 8.0.

4. Front Panel GUI Features:

   a. Protection against unauthorized access using password in local as well as remote mode of operation
   b. Instant audio-visual annunciation by blinking + color changing (Green to Red) combined with alarming sound.
   c. Monitoring and control both provided using common panel.
   d. Facility to view as well as maneuver the IP camera as necessary with facilities like zoom, snapshot, recording, patrolling and even configuration editing with suitable authorization from the panel itself.
   e. Design of panel is simple and easy to understand and use.

Entertainment panel is an integral part of the same virtual instrument and it serves on-demand audio-video entertainment using high speed wired ethernet protocol. As shown, live streaming FM radio channels or songs-on-demand as well as movies-on-demand can be played as per user's wish.

The remote version of front panel is accessible via intranet/internet and hence provides remote monitoring and/or control facility. The only difference between the local and remote front panel is the description of the project in header and recommended software requirements (not shown) in footer for making this panel operational at the remote end are at the bottom of the panel, else the remaining complete panel remains exactly same and operates in the same way. Although, control can be done from any one point of access only, while monitoring of the same building can be done from any no. of points in the world geographically distributed.

5. Strengths and Limitations:
   a. Strengths:
      i. Open ended design; hence root level modifications are possible, changes can be made considering the design specification requirements.
      ii. Maintenance though quite less frequent; though possible and even replacement is also easy.
      iii. As being manufactured by internationally reputed MNC vendors, hence after sales support and spares supply channels are reliable and trustworthy.
      iv. Easy availability in local and international markets
      v. Complete standardization in design and models, alterations are less frequent. Compatibility issues are quite rarely found.
      vi. Quick deployable.
      vii. Easy configuration and modification within scope.
   b. Limitations:
      i. Maintenance rarely possible; only replacement is made in case of failure.

ii. Not cheap as compared to home automation version, however investment is justified in long run of service.
iii. Installation, configuration and hook up takes significantly longer time, hence cannot be deployed quickly (especially as adhoc network), but such time delays are one time and justified by superior performance.
iv. Highly skilled manpower is required for installation and maintenance.

6. **Results and Discussions:** The system reacts by sounding an audible alarm, by changing the display in the computer GUI as well as by sending alerts on phone/email for the following events:
   a. Detection of movement.
   b. Opening or closing of a door or window.
   c. Impact or breakage of glass.
   d. Detection of fire due to smoke or leakage of gas.
   e. Detection of movement in "Areas under surveillance" via IP camera.
   f. Detection of empty or full water tank.

Turning off of unused electrical appliances remotely from any computer via internet/intranet just by clicking button saves time and money for utilities consumption bill. History of all the activities is recorded in the PC for later troubleshooting or diagnosis purposes.

The design of EIA-485/MODBUS based BAS implemented in laboratory employs high speed Ethernet at layer-3 for establishing connection between servers and clients. Data acquisition and control in layer-1 use EIA-485 and MODBUS protocols, which are industry standard. In this layer, application is implemented using LabVIEW GUIs. For BAS software, OPC is selected because of interoperability, vendor independence and other merits.

This system is designed to run online 24*7 and hence for reliable operation in case of power supply failure uninterruptible power supply for PC as well as standby dc power supply using rechargeable batteries for sensors, actuators and EIA-485 modules has been included in the implementation for reliable operation.

For automation, MODBUS protocol is used for data communication. Although not formally standardized, it is regarded as an open protocol. The physical layer of communication system is an RS-485 two-wire network, chosen for its simplicity, low cost and adequate data bandwidth. At the management level, the high-bandwidth Ethernet technology is used. Remote operation of the BAS is provided by the TCP/IP-based Intranet/Internet.

The BAS is divided into three layers as indicated in Figure 4, which shows only one residence with a room, a corridor, a water tank and a lawn with main entrance monitored and controlled from the control room. The central servers for OPC, Camera, GUI and Entertainment as well as Remote Workstation that are at the top forming the Configuration or Management Modules (Layer-3) are connected to the Ethernet backbone and to the Internet. A PC based Main Controller and IP camera, are at the Middle Layer (Layer-2). The Main Controller has features like TCP/IP support, Integrated OPC server and Multiple Communication Ports. An RS-485 to RS-232 Convertor Module is also at the same layer (Layer-2) as an Interfacing Module. This module performs dual functions of Protocol Conversion and Isolation. Various sensors and digital I/O modules forms the third and last layer (Layer-1). This layer also consists of digital input and output (I/O) modules connected on an RS-485 network in multi-drop configuration and further to the main controller via RS-485 to RS-232 Convertor Module. These digital I/O modules perform data acquisition from sensors and avail control outputs of Main Controller to the actuators.

4.1.2 Development of Wireless Solutions

4.1.2.1 Wireless Home Security System

1. **Introduction:** Wireless home security systems are cost-effective miniature and affordable versions of large scale BAS for domestic usage. Wireless home security system are generally designed and implemented based on popular wireless network technologies such as proprietary RF, ZigBee, etc. Sometimes wireless ethernet (Wi-Fi) is also preferred, if feasible.
2. **Usefulness:** As compared to versions designed based on wireless protocols for business applications (Wi-Fi) or that are used for PAN applications (Bluetooth/ZigBee), versions based on proprietary designs are available in markets for wireless home automation solutions. Such versions being of significantly low cost and ease of availability in good quantities with local vendors result in quick deployable and easy maintainable or replaceable solutions for satisfying low priority, low budget wireless home automation requirements. One such a kind of wireless home automation solution has been reported below has been validated by implementing in the laboratory.
3. **Implementation:** The photograph of implementation has been shown in Figure5. Sensors are installed as per design requirements on actual field locations and integration of sensors is carried out by detection and configuration of sensors one by one with the alarm panel.
4. **Strengths and Limitations:**
   a. Strengths:
      i. Cost effective and cheap.
      ii. Quick deployable.
      iii. Easy configuration and modification within scope.
      iv. Easy availability in local markets.
   b. Limitations:
      i. Close ended design; hence root level modifications not possible, changes can be made within permissible scope.

Figure 5. Photograph of implementation of Wireless Home Security System (Bhatt, 2009)

    ii. Maintenance rarely possible; only replacement is made in case of failure.
    iii. As being manufactured by small vendors after sales support and spares supply channels are poor.
    iv. No standardization in design and models, alterations are found quite frequent.

5. **Results and Discussions:** The system reacts by sounding an audible alarm, by changing the display in the LCD display on keypad as well as by sending voice alerts on phone for the following events:
    a. Detection of movement.
    b. Opening or closing of a door or window.
    c. Impact or breakage of glass.
    d. Detection of fire due to smoke or leakage of gas.
    e. Request of urgent help by pressing of panic button or foot switch.
    f. Detection of empty or full water tank.

The system can be operated both by using LCD keypad locally as well using keychain touchpad remotely and all the major functions can be carried out quickly.

4.1.3 Wireless Building Automation System

4.1.3.1 Usefulness

For continuous monitoring of large scale residential as well as commercial complexes, where along with safety and security provisions, communication and entertainment are essential requirements, for such buildings this implementation can work as comprehensive solution.

Managements of such premises can conveniently afford higher, but justified cost of this solution without many issues and can afford to recruit skilled manpower to build, operate and maintain the system for long efficient, proven and reliable service operation.

4.1.3.2 Implementation

The photograph of implementation of wireless BAS has been shown in Figure 6. The complete panel has been divided in different areas like room, corridor, lawn, entrance, water tank, etc. for a residential building model and actual sensors along with LED actuators are installed in corresponding areas as per design requirements. Some sensors are also installed in field at different locations and wires are drawn between field mounted sensor and the panel.

Figure 6. Photograph of implementation of Wireless Building Automation System (Bhatt, 2009)

Room is monitored for occupancy and open/close status of Door and Windows and controlled for electrical appliances like lights, fan, air conditioner, heater, television and geyser. Corridor is controlled for lights. Lawn lights are controlled based on daylight sensing and controls for sprinkler and fountain are also provided. Water tank level indication along with controls for filling pump and discharge pump are provided. IP camera is installed at the main entrance for live video surveillance and monitoring of entry and movement at the main entrance. ZigBee based Crossbow WSN motes used in the design are installed and wired on the field mounted sensors and actuators to form wireless mesh network among them. MIB520 gateway is installed and connection with PC is established via USB. The panel is prepared with professional touch and pictorial representation of LabVIEW GUI along with network connections make it easy to understand and interesting.

The front Panel of operational GUI has been prepared by integrating Crossbow MDA320 LabVIEW driver, XServe and LabVIEW Graphical Programming Language version 8.0. The GUI is prepared in symmetry with the actual field plan layout and having similarity in its architecture with our previous work on wired BAS for simplicity and better understanding.

Front Panel GUI Features:

1. Protection against unauthorized access using password in local as well as remote mode of operation
2. Instant audio-visual annunciation by blinking + color changing (Green to Red) combined with alarming sound.
3. Monitoring and control both provided using common panel.
4. Facility to view as well as maneuver the IP camera as necessary with facilities like zoom, snapshot, recording, patrolling and even configuration editing with suitable authorization from the panel itself.
5. Design of panel is simple and easy to understand and use.

Entertainment panel is an integral part of the same virtual instrument and it serves on-demand audio-video entertainment using high speed wired ethernet protocol. As shown, live streaming FM radio channels or songs-on-demand as well as movies-on-demand can be played as per user's wish.

The remote version of front panel has been accessible via intranet/internet and hence provides remote monitoring and/or control facility. The only difference between the local and remote front panel is the description of the project in footer and recommended software requirements (not shown) for making this panel operational at the remote end are at the bottom of the panel, else the remaining complete panel remains exactly same and operates in the same way. Although, control can be done from any one point of access only, while monitoring of the same building can be done from any no. of points in the world geographically distributed.

4.1.3.3 Strengths and Limitations

1. Strengths:
    a. Open ended design; hence root level modifications are possible, changes can be made considering the design specification requirements.
    b. Maintenance though quite less frequent; though possible and even replacement is also easy.
    c. As being manufactured by internationally reputed MNC vendors, hence after sales support and spares supply channels are reliable and trustworthy.
    d. Easy availability in local and international markets.
    e. Complete standardization in design and models, alterations are less frequent. Compatibility issues are quite rarely found.
    f. Quick deployable.
    g. Easy configuration and modification within scope.

2. Limitations:
   a. Maintenance rarely possible; only replacement is made in case of failure.
   b. Not cheap as compared to home automation version, however investment is justified in long run of service.
   c. Installation, configuration and hook up takes significantly longer time, hence cannot be deployed quickly (especially as adhoc network), but such time delays are one time and justified by superior performance.
   d. Highly skilled manpower is required for installation and maintenance.

5. FUTURE RESEARCH DIRECTIONS

Enhancement of each and every facet is critically important for rapid growth and easy mass acceptance of BAS. Focus should be provided on making BAS more and more user-friendly and a part of one's routine life. Therefore, serious efforts should be made to make BAS rugged, highly accurate and reliable as well as extremely user-friendly.

Topics of future research works in the area of BAS include, but are not limited to:

1. Quick deployable BAS.
2. Adhoc BAS for public functions organized in large areas.
3. Network capacity/reach scalability enhancements without imperfections.
4. improvements insecurity, robustness, reliability.
5. High speed data communication and live media streaming.
6. Enriched GUI development and user-friendliness enhancements.
7. Self-troubleshooting solutions (self-healing abilities).
8. Novel Applications, architectures and topologies evolvements.
9. Special BAS developments for Elderly and Physically challenged humans.

6. CONCLUSION

In modern contexts, BAS is a rapidly growing phenomenon. Emerging wired/wireless communication technologies are expected to transform the present partially digitally networked houses into Smart, Intelligent and Adaptive Homes. Higher and higher penetration of automation will lead to increased device intelligence and they would be performing their routine tasks in vivid situations autonomously and without much human intervention. Such Smart homes will provide high level of energy optimization, security, communication and entertainment facilities. Home security system could be suitable for small building and low budget applications, while for large building and big budget applications, full scale building automation systems could be deployed.

Upcoming buildings could be made equipped with wired solutions presented in forms of Home Security System and Building Automation System. However, as re-wiring of existing buildings may be inconvenient and costly, and hence in such cases powerline based communication systems could be utilized. In fact, a mixed-network protocol at the device level could be the best solution. Overall Building Automation System could be envisaged as a rich mix of various communication technologies– both wired and wireless type. Wireless home security system could be suitable for small building and low budget applications since beginning of their construction phase, while for extending range of applications for existing large buildings already equipped with wired solutions and big budget applications, wireless solutions could be deployed.

ACKNOWLEDGMENT

The author thankfully acknowledges kind help, guidance and timely cooperation provided by his then supervisor and guide Prof. (Dr.) H. K. Verma, Department of Electrical Engineering, IIT Roorkee. The author also expresses his grateful thanks with appreciations to all the Faculty and staff Members, Management and Office Bearers of Indian Institute of Technology Roorkee (IIT Roorkee), Roorkee– 247 667, Uttarakhand, India and Dharmsinh Desai University (DDU), Nadiad– 387 001, Gujarat, India, for their active support and timely help in development of the work presented in the chapter.

REFERENCES

Al-Ali, A. R., El-Hag, A. H., Dhaouadi, R., & Zainaldain, A. (2011, April). Smart home gateway for smart grid. *Proceedings of the 2011 International Conference on Innovations in Information Technology IIT* (pp. 90-93). IEEE. doi:10.1109/INNOVATIONS.2011.5893876

BACnet. (2015). *Wikipedia*. Retrieved 24 May, 2015 from http://en.wikipedia.org/wiki/BACnet

Bester, J., Robertson, I., & Groenewald, B. (2009). Wireless Control Network for Building Automation using IEEE 802.15.4. *Proceedings ICUE International Conference, Theme: Energy Management*, Cape Town, South Africa.

Bhatt, J., Shah, V., & Jani, O. (2014). An instrumentation engineer's review on smart grid: Critical applications and parameters. *Renewable & Sustainable Energy Reviews*, *40*, 1217–1239. doi:10.1016/j.rser.2014.07.187

Bhatt, J., & Verma, H. K. (2015). *Design and Development of Wired Building Automation Systems*. Energy and Buildings.

Bhatt, J.G. (2009, November). *Design and Implementation of Building Automation Systems*. M.Tech. Project Report, Department of Electrical Engineering, Indian Institute of Technology Roorkee, Roorkee, India.

Bhatt, J.G. (2013, December). Literature Survey Report for Doctoral (Ph.D.) Research [Doctoral Thesis]. Pandit Deendayal Petroleum University, Gandhinagar, India.

Bhatt, J. G., & Verma, H. K. (2010). RS-485/MODBUS based intelligent building automation system using LabVIEW.

Bluetooth. (2015). *Wikipedia*. Retrieved 24 May, 2015 from http://en.wikipedia.org/wiki/Bluetooth

Dickmann, G. (2011, April). DigitalSTROM: A centralized PLC topology for home automation and energy management. *Proceedings of the 2011 IEEE International Symposium on Power Line Communications and Its Applications ISPLC* (pp. 352-357). IEEE.

EIB/KNX. (2015). *Wikipedia*. Retrieved 24 May, 2015 from http://en.wikipedia.org/wiki/KNX_(standard)

Erol-Kantarci, M., & Mouftah, H. T. (2011). Wireless sensor networks for cost-efficient residential energy management in the smart grid. *IEEE Transactions on* Smart Grid, *2*(2), 314–325.

Granzer, W., Kastner, W., Neugschwandtner, G., & Praus, F. (2006). *A modular architecture for building automation systems*.

Han, D. M., & Lim, J. H. (2010). Smart home energy management system using IEEE 802.15.4 and zigbee. *IEEE Transactions on* Consumer Electronics, *56*(3), 1403–1410.

Kastner, W., Neugschwandtner, G., Soucek, S., & Newman, H. M. (2005). Communication systems for building automation and control. *Proceedings of the IEEE*, *93*(6), 1178–1203. doi:10.1109/JPROC.2005.849726

Kawade, S., & Nekovee, M. (2009, December). Wireless options for high data-rate indoor users: Cognitive access to TV white space. *Proceedings of the 2009 First UK-India International Workshop on Cognitive Wireless Systems UKIWCWS* (pp. 1-6). IEEE.

Kim, M., Kim, S., Kim, J., & Yoo, Y. (2011, August). Design and implementation of MAC protocol for SmartGrid HAN environment. *Proceedings of the 2011 IEEE 11th International Conference on Computer and Information Technology CIT* (pp. 212-217). IEEE. doi:10.1109/CIT.2011.78

Lai, Y. X., Rodrigues, J. J., Huang, Y. M., Wang, H. G., & Lai, C. F. (2012). An intercommunication home energy management system with appliance recognition in home network. *Mobile Networks and Applications*, *17*(1), 132–142. doi:10.1007/s11036-011-0345-z

Langhammer, N., & Kays, R. (2012). Performance evaluation of wireless home automation networks in indoor scenarios. *IEEE Transactions on* Smart Grid, *3*(4), 2252–2261.

Lewis, F. L. (2004). *Smart Environments: Technologies, Protocols, and Applications. In Wireless Sensor Networks* (pp. 11–46). John Wiley.

Lien, C. H., Bai, Y. W., Chen, H. C., & Hung, C. H. (2009, January). Home appliance energy monitoring and controlling based on power line communication. *Proceedings of the Digest of Technical Papers International Conference on Consumer Electronics ICCE '09* (pp. 1-2). IEEE. doi:10.1109/ICCE.2009.5012315

Lien, C. H., Chen, H. C., Bai, Y. W., & Lin, M. B. (2008, May). Power monitoring and control for electric home appliances based on power line communication. *Proceedings of the Instrumentation and Measurement Technology Conference Proceedings IMTC '08* (pp. 2179-2184). IEEE. doi:10.1109/IMTC.2008.4547409

LonWorks. (2015). *Wikipedia*. Retrieved 24 May, 2015 from http://en.wikipedia.org/wiki/LonWorks

Missaoui, R., Warkozek, G., Bacha, S., & Ploix, S. (2012, October). Real time validation of an optimization building energy management strategy based on power-hardware-in-the-loop tool. *Proceedings of the 2012 3rd IEEE PES International Conference and Exhibition on Innovative Smart Grid Technologies ISGT Europe* (pp. 1-7). IEEE. doi:10.1109/ISGTEurope.2012.6465791

MODBUS. (2015). *MODBUS Specifications and Implementation Guidelines*. Retrieved from.http://www.modbus.org

Power-line communication. (2015). *Wikipedia*. Retrieved 24 May, 2015 from http://en.wikipedia.org/wiki/Power-line_communication

RS-485. (2015). *Wikipedia*. Retrieved 24 May, 2015 from http://en.wikipedia.org/wiki/RS-485

Son, Y. S., Pulkkinen, T., Moon, K. D., & Kim, C. (2010). Home energy management system based on power line communication. *IEEE Transactions on* Consumer Electronics, *56*(3), 1380–1386.

Wang, J., Sun, Z., Zhou, Y., & Dai, J. (2012, May). Optimal dispatching model of smart home energy management system. Proceedings of the 2012 IEEE Conference on Innovative Smart Grid Technologies-Asia ISGT Asia (pp. 1-5). IEEE.

Wi-Fi. (2015). *Wikipedia*. Retrieved 24 May, 2015 from http://en.wikipedia.org/wiki/Wi-Fi

Yuwono, R. Y. (2009). X10 Protocol Man Machine Interface Implementation using Labview. *Journal of Applied Sciences*, *9*(19), 3562–3568. doi:10.3923/jas.2009.3562.3568

ZigBee. (2015). *Wikipedia*. Retrieved 24 May, 2015 from http://en.wikipedia.org/wiki/ZigBee

Chapter 13
Contemporary Energy Management Systems and Future Prospects

Amir Manzoor
Bahria University, Pakistan

ABSTRACT

The transformation of electric grid into smart grid has improved management of available resources and increased energy efficiency. Energy management systems (EMS) play an important role in enhancing user participation in control of energy management. Using such systems, consumers can obtain information about their energy consumption patterns and shape their energy consumption behaviors for efficient energy utilization. Contemporary EMS utilizes advanced analytics and ICT to provide consumers actionable feedback and control of energy management. These systems provide high availability, an easy-to-use user interface, security, and privacy. This chapter explores the contemporary EMS, their applications, classifications, standards, and frameworks. The chapter defines a set of requirements for EMS and provides feature comparison of various EMS. The chapter also discusses emerging trends and future research areas in EMS.

1. INTRODUCTION

Energy management is a term that has a number of meanings. In this chapter, we define energy management as saving energy in businesses, public-sector/government organizations, and homes. According to this definition, energy management is the process of monitoring, controlling, and conserving energy in a building or organization. Typically, this involves the following steps:

1. Metering energy consumption and collecting the data
2. Finding opportunities to save energy, and estimating how much energy each opportunity could save
3. Taking action to target opportunities of saving energy
4. Tracking progress by analyzing meter data to see how well energy-saving efforts have worked

DOI: 10.4018/978-1-4666-9911-3.ch013

Energy management, as the means to controlling and reducing energy consumption, is important because it enables consumers to:

1. Reduce costs – this is becoming increasingly important as energy costs rise.
2. Reduce carbon emissions and the environmental damage that they cause - as well as the cost-related implications of carbon taxes and the like. Organization may be keen to reduce their carbon footprint to promote a green, sustainable image. Not least because promoting such an image is often good for the bottom line.
3. Reduce risk – the more the energy consumed, the greater the risk that energy price increases or supply shortages could have serious impact on profitability, or even continuation of business. With energy management, the risk can be reduced by reducing demand for energy and by controlling it making it more predictable.

Today's global environment is characterized by soaring energy prices and increasing environmental regulations. To remain competitive in today's global and competitive economy, a quick analysis and close control of operating costs is critical. A common approach is to lowering energy costs while reducing energy consumption and minimizing the environmental impact of energy use. In USA, various policy initiatives (such as the Energy Policy Act, the American Competitiveness Initiative and the Advanced Energy Initiative) have been taken as part of an aggressive strategy for tackling the long-term energy challenges (US Energy Department, 2005). Under the Energy Policy Act, businesses can get deductions for new or renovated buildings that save 50% or more of projected annual energy costs for heating, cooling, and lighting compared with model national standards. Businesses can also get partial deductions for efficiency improvements to individual lighting, HVAC and water heating, or envelope systems (US Department of Housing, 2006). The US Energy Policy Act of 2005 requires that federal buildings be energy efficient and strongly recommends that private enterprises follow suit. Its strong emphasis is on implementing energy-efficient and savings processes and equipment. The American Recovery and Reinvestment Act provide the details and envision a modernized, more energy-efficient America through targeted investments in updating infrastructure. Sweeping privatization, deregulation, and re-regulation initiatives have radically altered the competitive landscape of the electric utility business. Deregulation in wholesale power markets has resulted in a tenfold increase in the number of interchange transactions. The combined effects of weather and forced outages have created congestion driving wholesale spot prices to all-time highs. Competition is fierce and there exist great security challenges to security of critical infrastructure and IT systems used to manage power systems. The electric utilities are facing greater challenges having a critical balance among cost reduction, maximization of existing assets, and maintenance of adequate security margins (Bose, Wu, Wierman, & Mohsenian-Rad, 2014). Energy management is a common practice for larger buildings. Energy management at homes is a recent trend. Many home users still follow a haphazard energy management. However, most of the energy management principles that apply to businesses and other organizations are also applicable to homes (Energylens.com, 2015).

Evolution of the smart grid has attracted energy professional worldwide. It is a modern power grid capable of supporting two-way communication between energy providers and consumers. Its significant features include fine-grained metering, control, and feedback, and enhanced energy efficiency, and management of available energy resources. Energy management systems (EMS) are an important tool to control home energy consumption. EMS often integrated with home automation systems and allows consumers to understand, control, and optimize energy consumption e.g.

energy consumption control during peak hours. EM systems today provide advanced features with simplified low-cost installation procedures. These systems provide integration with users' monitoring and actionable feedback through social networking sites, mobile apps, and web-based portals. It is expected that a widespread adoption of EMS will bring more efficient consumption behavior and benefit the energy sector as a whole.

Energy management systems today require less specialized instrumentation, less manual configuration, and less interaction with the customer to be useful. It is due to the advancements in many technologies such as non-intrusive load monitoring (NILM) techniques, pervasive availability of sensors, cloud computing and mobile platforms, and increased use of social networking sites. These technological advancements have made energy consumption data more meaningful for consumers, increased EMS efficiency, enhanced system analytics, and motivated behavioral changes in consumers. An NILM system (Hart, 1992) does not require a voltage or current sensor at every load, and it determines the operating schedule of the major electric loads in a building at the power service entrance. Compared with the traditional load monitoring system, NILM systems offer simple hardware installation. Therefore, an NILM system is a wide load monitoring scheme and a powerful energy management system (Hart, 1992; Leeb, 1993;Cole & Albicki, 2000;Hong & Chou, 2012).

Modern EMS can provide real-time visualization of energy data; secure custom energy dashboards and kiosks to view energy reports, consumption patterns, and resource usage etc. Site managers, building engineers or maintenance personnel can quickly and intuitively navigate and discover opportunities for improvement. These dashboards are customizable for different viewing and reporting needs. Users can enter comments to highlight success or flag problem areas. A connectivity layer, such as ICONICS GENESIS64, is used by EMS to acquire energy meter data, through any available networking, from a variety of meters such as electric, gas, fuel oil, steam, or chilled water. EMS can interface to the smart grid for rate and other supply information and provide aggregation and recording of consumption data. This recorded consumption data can be used for long-term archiving and continuous analysis and comparison. Using rate model configuration tools, users can enter any rate model that their utility contract defines to automatically derive and record costs. This cost data can be used for comparison to budgets and past performance. This data can also be used for validating the actual utility bills. Modern EMS can provide a 3-D view of a building and that allows faster and easier navigation of the building and get their desired control zone.

The objective of this chapter is to explore the contemporary energy management systems, their classifications, requirements, standards, and future prospects of energy management systems.

2. CLASSIFICATIONS OF EMS

Based on their applications, EMS can be divided into following broad classes.

2.1 Building Energy Management Systems (BEMS)

BEMS is a complex system that provides a heterogeneous network platform. This platform provides application-programming interfaces (API's), intelligent algorithms for dynamic control and system optimization, and interoperable system modules. BEMS performs upper level functions such as advance monitoring, smart control and decision-making (Lazos, Sproul, & Kay, 2015; Brewer & Johnson, 2010; Karnouskos, 2011; Zhao, Suryanarayanan, & Simoes, 2013; Sundramoorthy, Cooper, Linge, & Liu, 2011; Yang & Li, 2010). BEMS are supported by home automation systems.

The market for building energy management systems (BEMSs) continues to grow. This growth is attributed to technological advances as well as increased customer awareness of benefits of BEMS. Digital controls are becoming commonplace in buildings and major companies are prioritizing energy efficiency and sustainability in their strategies. The regulatory requirements regarding energy consumption are becoming stringent for commercial buildings. The situation presents an ideal environment for BEMS adoption.

The use of cloud-based BEMS is producing fundamental changes the way BEMS is installed and business model of BEMS operates. BEMS today are focused not only on energy visualization and energy analytics but also on demand response and property management. This has broadened the capabilities of BEMS and bringing a number of new companies in the market. It is forecasted that the BEMS market will grow from just under $1.8 billion in annual revenue in 2012 to nearly $5.6 billion in 2020 (Nastu, 2013).

2.2 Enterprise Energy Management System (EEMS)

Enterprise Energy Management System collects and organizes all data related to energy consumptions and costs. EEMS provides a complete view of an entire facility's operations, not just individual components. Use of EEMS provides several benefits such as lower energy consumption, improved space comfort, extended equipment life, better use of staff, energy purchase savings, and demand reduction (Farhangi, 2014).

Early EEMS supported strategic energy management by capturing energy performance data to form a baseline to set and track energy performance goals. Today, EEMS provides many functions that support strategic energy management. These functions include modelling and forecasting, benchmarking, energy use and cost analysis and measurement and verification. Organizations can combine these functions to develop more powerful information tool to gain comprehensive understanding of current energy performance and plan/monitor the performance measures.

EEMS identifies energy savings opportunities by applying business intelligence and predictive modeling algorithms to facilities' energy and control systems data. This approach saves money through smarter central plant management, detection of abnormal building behavior and improved operational planning capability. All of this enables better maintenance decisions, enhanced awareness of energy consumption, and provides detailed management insight that is not possible with other solutions. This insight into a facility's energy requirements can be further augmented through the optimization of energy resources for facilities that have a central utility plant. Optimizing the supply of steam, chilled water, and electricity to the lowest cost can further drive down energy costs while maintaining a reliable supply of utilities to the facility.

Some notable vendors of EEMS include Honeywell, Siemens, GridPoint, Rockwell Software, PowerLogic, and Lockheed Martin.

In general, EEMS are divided into following broad categories (Forth & Tobin, 2002).

2.2.1 Commercial EEMS

Commercial customers can use these systems to control procurement and consumption of electrical energy. These systems can provide aggregated data, profile energy usage patterns, and perform variable analysis of utility rate. These systems can also verify energy bills and allocate energy costs to various entities such as clients, departments.

2.2.2 Energy Services EEMS

These systems are used by energy service providers to perform reporting, performance contracting, and consulting services to large numbers of customers. These customers may have a large number of facilities separated by large distance.

2.2.3 Grid Enterprises EEMS

These systems are used by utility companies to enhance their demand response capability. These systems provide high-speed effective communication with customers, load control, and operations verification. These systems can also automatically acquire energy logs from sites for speedy billing. Some vendors of grid EMS include Alstom, GridPoint, and Toshiba.

2.2.4 Manufacturing EEMS

These systems provide energy data collection and processing, breakdown prediction, alarm and monitoring, and energy consumption prediction and analysis (Dong & De-jiang, 2010). Two notable vendors of manufacturing EMS are Siemens and Honeywell.

Table 1 provides feature comparison of various enterprise energy management systems.

2.3. Home Energy Management Systems (HEMS)

Home Energy Management System (HEMS) can be defined as encompassing any product or service that monitors, controls, or analyzes energy in the home. This definition includes residential utility demand response programs, home automation services, personal energy management, data analysis, and visualization, auditing, and related security services (Amer, Naaman, M'Sirdi, & El-Zonkoly, 2014).

HEMS provides management of energy use to reduce energy consumption while maintaining a satisfactorily comfortable lifestyle in the home. HEMS uses weather, sensor, and other types of information to identify unnecessary energy use and controls appliances accordingly. HEMS can also provide best use of electric power e.g. by operating a washer-dryer when excess power is available from solar panels.

The weather-dependent renewable energy sources is producing increased amount of power. To maintain a reliable supply of electric power, the installation and optimum operation of batteries will be very important. HEMS can provide visualization of electric power use and advice on more efficient use of electric power. HEMS can also coordinate with external power systems, information systems, and the local community.

The current Home Energy Management Systems (HEMS) market is valued at $1.5 billion in 2013, and includes utility residential demand response (DR), connected home subscriptions,

Table 1. Feature comparison of enterprise energy management systems

Feature	Commercial EEMS	Energy Services EEMS	Grid Enterprise EEMS	Manufacturing EEMS
Procurement Control	√			
Control of energy consumption	√	√	√	√
Data collection and aggregation	√	√	√	√
Energy usage profiling	√			
Energy costs analysis	√			
Reporting	√	√	√	√
Communication with customer		√	√	√
Monitoring	√	√	√	√

software services, and associated hardware (PRNewsire.com, 2014). While market is still in its early stages of development, the two leading companies combined already have more than 2 million customers with monthly subscription. This demonstrates huge market potential of HEMS. Many large HEMS vendors exist (such as Alarm.com, Tendril, Opower, Vivint, and ADT) and many new companies are entering into the market. The future opportunities in this market revolve around software and bundled service solutions.

Despite the promising HEMS market, the use of HEMS is still limited. One possible issue is that technological advances in electrical grid have tended to benefit the utilities more than they benefit consumers. For example, it is expected that by the end of 2015, approximately 45% of all U.S. households will be served by smart meters but as few as 10% of those meters will be enabled for two-way communications via HEMS (Szablya, 2012). Due to the lack of interactive connection with the energy grid, taking full benefits from HEMS won't be possible for users. To boost use of HENMS, it would be necessary that users be able to obtain real-time information on energy costs and usage, which will let users save money by using power when its costs are lowest.

There are a number of ways to divide the HEMS market. One is to split solutions into utility and non-utility sectors (Greentechmedia, 2013).

2.3.1. Utility HEMS

Utility HEMS are solutions that include both the utility and the homes enrolled in the HEMS program. A detailed economic analysis, custom-tailored to each utility, is carried out. The outcome of this analysis is a single-load-per-home demand response program. Demand response systems are in the low-to-mid hundreds of dollars range. A one-time hardware purchase with an annual subscription/integration service for energy management is typically done by the utilities. Utilities prefer communications-agnostic solutions that can integrate with their existing software platform. Top players in the utility area include EcoFactor, Sequentric, EnergyHub, Calico Energy Solutions, Comverge, Consert, and Aclara.

2.3.2. Non-Utility HEMS

The entire focus of these solutions is on the home customer and integration of various smart-home services. There is no detailed economic analysis for utility customers. Instead, a smart marketing campaign is launched to target homeowners' needs. A general vendor approach is to first offer home owners a security solution and then HEMS as an add-on sale. The cost of HEMS for home owners range between mid-hundreds to low-thousands. The homeowners pay a monthly fee of $15-$60 for energy management, automation, and security services(Greentechmedia, 2013).

3. REQUIREMENTS OF ENERGY MANAGEMENT SYSTEMS

EM systems must provide advanced and versatile functionality while keeping the installation simple and running cost low. The systems should integrate with users' daily activities and offer actionable feedback. In order to facilitate effortless energy monitoring and control, an EMS must meet certain requirements (Saini & Prakash, 2014). These requirements are defined below.

3.1. Monitoring

The energy management system must be able to provide energy consumption information for various time intervals, such as daily and weekly, consistently over long period. (Fischer, 2008) noted that the feedback is most successful when it is provided frequently and over a long period. Consumers can then directly relate near-real-time information with their energy use actions (Weiss, Mattern, Graml, Staake, & Fleisch, 2009).

3.2. Disaggregation

Often, consumers have a misperception about the energy consumed by individual appliances (Froehlich, Findlater, & Landay, 2010), which can be corrected by providing disaggregated data for different appliances. The system must be able to provide information about energy consumption by individual appliances. Consumers can also benefit greatly from information about the real-time impact of specific appliances being powered on or off (Fitzpatrick & Smith, 2009; D'Alessandro, Tonello, Monacchi, & Elmenreich, 2014). The disaggregated data also highlights the impact of long-term changes such as switching to an energy-efficient appliance. Many EM systems use indirect load sensing methods to provide disaggregated information based on specific current and voltage waveform "signatures" of individual appliances (Kim, Schmid, Charbiwala, & Srivastava, 2009).

3.3. Availability and Accessibility

The system must be able to provide information to the consumer all the time through an easy-to-use user interface. This user interface can be either a physical device or remote access through a web or mobile portal. The user interface can also push the information to the consumers in form of a text message to the consumer's smart phone.

3.4 Information Integration

An EMS must be able to integrate other types of information with the information provided by the EMS itself. Some example of such information includes indoor temperature, humidity, peers' consumption data etc. An EMS must also be able interpolate the information collected at different time intervals. An EMS also utilizes semantic web technologies for this purpose.

3.5. Affordability

The EMS must also provide an easy and simple installation and configuration with minimum professional help needed. The system must also provide simple maintenance, minimal energy consumption, and low running cost.

3.6. Control

The EMS should be able to provide remote, programmable, and automatic control of devices. Generally, the consumer is expected to perform necessary control operations manually. However, a digital control option or automated actions are more effective.

3.7. Cyber-Security and Privacy

With an ever-changing spectrum of plausible threats ranging from disgruntled employees to cyber-terrorists, sound IT security has become an ongoing process that requires effective utilization of the latest technologies. An EMS system must provide authentication of all transactions to ensure that consumers' data and control operations are secure, and not accessible to third parties without explicit consent. In addition to more fundamental security mechanisms, such as firewalls and user ID/password protection, EMS can also provide advanced security measures including centralized management of end users, user authentication, data/communication encryption, and file tampering detection.

3.8. Intelligence and Analytics

Many consumers lack a deep understanding of electrical systems and have limited time to make energy-related decisions (Bartram, Rodgers, & Muise, 2010). This limits consumers' capability to directly control and manipulate all appliances all the time. An EMS should be able to perform intelligent predictive actions that balance energy consumption and consumer comfort. A variety of techniques, such as machine learning, human–computer interaction, and "big data" analytics, can be used in this regard. This reduces the onus on the consumer to directly control and manipulate all appliances all the time (Paradiso, Dutta, Gellersen, & Schooler, 2011).

4. REVIEW OF ENERGY MANAGEMENT SYSTEMS

In this section, a review of several EMS will be presented along with evaluation of their features in the light of requirements of EMS presented in previous section.

4.1. PERSON

This system, implemented by (Yang & Li, 2010), is based on Pervasive Service-Oriented Networks (PERSON) framework. This system consists of three sub-systems: Application programing interfaces (APIs) that support information exchange, services that support modularity and inter-operability, and intelligent algorithms for dynamic control and system optimization. This system provides data monitoring and disaggregation. The system is also able to monitor and control the energy consumption but does not provide any web or mobile user interface. The system can aggregates others' usage information and other information such as temperature, humidity, luminance, and motion sensor data. The system is low cost and consumes low power. Devices can be controlled via manual remote control. However, the system provides no security and privacy.

4.2. WattDepot

This system, implemented by (Brewer & Johnson, 2010), is based on an open source Internet-based service-oriented framework for energy management. It offers collection, aggregation, analysis, and visualization of energy data. Its architecture is specified around three services: sensors (collection of data from various brands of energy meters), servers (storage of data), and clients (analysis and display of data). Consumers can configure the source to be monitored and the type of data displayed. Each source can be labelled as public or private. Software sensors retrieve power consumption/generation data from different classes of devices, servers collect data from sensors, store them in a relational database, and provide them through APIs, and clients, which either present data to the final user using different visualization techniques or provide input data for analysis tools. All data requests are authenticated before being stored on the server. The system provides monitoring function and it is possible to implement data disaggregation. The system provides a web-based user interface. Information integration is provided through automatic interpolation of information. The system is open source and freely available. While this system uses a simple privacy model, it lacks any intelligence and automatic or programmable controls. WattDepot has mainly been used in research and as a prototyping and simulation mechanism by local Hawaiian electric companies. It has also been used in a dorm energy-reduction competition. So far, WattDepot has not yet been used for commercial purposes (Riveiro, Johansson, & Karlsson, 2011).

4.3. ViridiScope

This system, implemented by (Kim, Schmid, Charbiwala, & Srivastava, 2009), is an easy-to-install EMS for long-term fine-grained monitoring of domestic energy consumption. It provides estimation of real-time appliance-level energy use and can integrate information from various sources. The system uses heterogeneous sensors placed near each appliance. These sensors use machine-learning algorithms to learn and estimate the energy consumption of every appliance. This method does not need any external calibration and requires little human intervention. The system provides monitoring and data disaggregation capability. The system can aggregate various types of information such as magnetic, acoustic, and light information. The system requires indirect sensors however; no in-line installation is required. The major drawbacks of the system are lack of intelligent control of devices and security. The system provided no security and privacy. In addition, you need to place a sensor near each appliance to be monitored.

4.4. Digital Home Energy Management System (DEHMS)

This system, proposed by (Sundramoorthy, Cooper, Linge, & Liu, 2011), provides real-time information regarding energy use, historical use, energy- saving tips, and comparison with average use of all system users. The systems monitor user experiences and preferences by collecting data from various appliances and sensors. The data is aggregated and sent to the server. The server than sends this data to the web services for use by higher-level services. The system provides monitoring and data dis-aggregation capability. The system provides a web-based user interface for real-time display of data. It can integrate information from sensors, electric supply, and gas supply lines. The system requires sensors to operate and provide no control of devices or security/privacy. The system also lacks any intelligence features.

4.5. NOBEL (Energy Crowdsourcing System)

This system, proposed by (Karnouskos, 2011), crowd source high-quality energy use data and context information from consumer mobile phones to help build decision support systems, and profile analytics and prediction. The data collected can be detailed consumption data as well as device-specific disaggregated information. An enterprise version of the system can provide centralized device management, data management, and authentication of all transactions. The system provides monitoring and data dis-aggregation capability. A mobile phone app is used to provide user interface. The major drawbacks of the system include lack of consumer information privacy and lack of integration of information from various sources. In addition, the system provides limited intelligence capabilities.

4.6. Adaptive Living Interface System (ALIS)

This system, developed by (Bartram, Rodgers, & Muise, 2010), is a distributed system that consists of energy dashboards, monitoring and awareness tools, device controls, and a feedback system. The system provides monitoring but no data-disaggregation capability. The system can integrate historical use and community usage data. The Adaptive Living Interface System (ALIS) facilitate solar-powered living by supporting informed energy use decisions. Using ALIS, users can control the house, and receive feedback and incentives through a set of distributed interfaces. The various interactive components help to reduce the complexity of solar-powered living. These components encourage net-zero living through ubiquitous support to the resident in energy use decisions. ALIS provide feedback on how an occupant's actions are affecting their energy footprint. Using a smart-phone or any web-enabled device, the user can easily control lighting and temperature. User can also access tools to monitor and analyze real-time, historical, and forecasting data on the energy production and consumption in the home. However, the system extensive installation effort, doesn't provide intelligence or support for security and privacy.

4.7. Multi-Level EMS

This system, developed by (Dong & De-jiang, 2010), is a multi-level system that provides energy management, monitoring, multi-level alarms, and breakdown expert system for breakdown information, analysis, and restore. The systems utilize GIS (Geographic Information System) for breakdown prediction, alarm, and processing sub-system. The multi-level alarms help in breakdown judgment.

4.8. Smart EMS

(Xu, Jia, Guan, & Shen, 2013) developed a system to manage and optimize the storage devices together with other energy systems, in order to provide the required environment with minimal cost over automotive painting shops.

4.9. Mobile Feedback System

This system, developed by Weiss, et al., (2009), provides monitoring and data dis-aggregation capability. The system provides interactive and readily available feedback to the user via smart phone. The system can integrate historical information. The system provided high availability using a mobile phone app that is used to access the system. The major drawbacks of the system are lack of control, security and privacy, and intelligence features.

4.10. EnergyWiz

This system, developed by Petkov, et al., (2011), provides monitoring but no data dis-aggregation capability. The user interface is provided through a mobile phone app. The system can integrate historical usage and user information from peers, social network friends, and EnergyWiz users. The system provides no control, security and privacy, and intelligence capabilities.

5. ENERGY MANAGEMENT SYSTEM SOFTWARE

The available EMS software can be divided into two categories: large and small/medium. The large software solutions are applicable to diverse large commercial entities and provide little room for advanced customization. They offer broad features, robust mathematical analysis, and standard integration with corporate software. Powerful interfacing tools and a wide range of data presentations are also common features. Some examples of such software solutions include Energy CAP and SAP Energy Management Software. The small and medium software solutions, frequently oriented to home and tertiary buildings, offer closer customization, module approach to address specific problem, and agile response to customer needs. Many times, companies offer large software solutions also offer a scaled down version of their product. This scaled down product has all of the features of large software but intended for companies without large IT resources. One example of such software is EnergyCAP Express software from EnergyCAP Corporation. Some companies, such as C3, also offer a residential version of their EMS.

5.1. EMS Software Special Requirements

There exist 3C's of energy namely cost, consumption, and carbon. All EMS should address these 3C's which represents common requirements with commercial buildings and facilities. Industrial manufacturing concerns may have some specialized requirements. Following are some of these requirements.

5.1.1. Collection and Monitoring of Multiple Types of Energy Usage

Besides electricity, industrial manufacturing concerns use many other types of energy such as natural gas, water, compressed air, steam, and inert gases. The communication protocols used by many industrial sensors include HART or Foundation Fieldbus. These protocols are not generally supported by commercial EMS products.

5.1.2. Making Sense of Energy Usage Aggregation

To make sense for a manufacturing concern, aggregate energy usage, in appropriate manner, must be provided by EMS. This aggregation may be in terms of departments or facilities. Some manufacturing concerns may require usage aggregation on multi-department and/or multi-location basis. Such degree of flexibility in aggregating energy usage is not a common feature of EMS.

5.1.3. Normalization of Energy Usage Data

Each industrial concern require normalization of energy usage against certain key production information. An automobile manufacturer would require energy usage aggregated against number of vehicles produced i.e. energy usage per vehicle produced. A food manufacturer would require energy usage aggregated against pallets shipped. The source of data for providing such aggregation would be organizational Enterprise Resource Planning (ERP) system such as SAP. Therefore, EMS should able to integrate with ERP system.

5.1.4. Increased Hardware Communication

In industrial concerns, the communication requirements with hardware devices are very broad. There exist many standard commercial energy management system communications protocols. Two examples of such protocols are BACnet and Modbus. In addition of supporting these protocols, EMS for manufacturing concerns also need to integrate with programmable logic controllers (PLC's) produced by many different manufacturers. That require EMS to provide additional support of communication protocols such as EthernetIP, ModbusTCP, Profibus, ProfiNet etc.

5.1.5. Collection of Weather Data

Industrial EMS must also be able to collect weather data, using communications tools such as web services. This is a feature also available in commercial EMS. A manufacturing concern may also require EMS to aggregate energy usage against degree-days.

5.1.6. Many Pieces of Equipment

Industrial concerns have a large number of equipment in the facility. Each equipment has a different best practice to operate. For example, the best practice for a chiller may be based on the operational efficiency.

5.1.6. Scalability of EMS

Industrial concerns do grow, updated and changed to meet the changing production requirements. An EMS must be able to adapt to the changed environment. That would means the EMS should have software tools that can be easily maintained and modified by the internal personnel without seeking costly help from EMS manufacturer for minor software updates. Not using custom programming code and scripting can result in significant reduction in costs.

5.1.7. Increased Data Display

An industrial concern may have a large number of sensors and data tags. An EMS would need to collect and log a large number of data samples on a daily basis. This logging of data should be performed at relatively high speeds. This requirement is generally beyond the capabilities of EMS that relay on SQL-based databases.

5.1.8. Redundant Data

Another common requirement of industrial concerns is redundant data collection and redundant data logging. This is due to the fact that some large industrial concerns are highly dependent on the information systems. As such, the EMS for an industrial concern must provide redundancy so that no data is lost in case the server fails.

5.1.9. Interface with Other Systems

Most industrial concerns deploy an equipment maintenance and/or asset management system in operation. An EMS for industrial concern must interface with these system to organize equipment in industry standard hierarchies using standards such as ISA-95.

5.2 EMS Software Features

Current EMS offer many features such as data acquisition, analytics, Software-as-a-Service, financial analysis, multi-level energy consumption monitoring, management and aggregation of information from different sources, aggregate information for analytics, evaluation and mixing of information, data mining, and integration with existing corporate software to enable communication and interaction. These systems provide a graphical interface to present information such as costs, billing, trends, etc. More advanced systems offer Artificial Intelligence (AI)-based analysis capabilities, and automated energy optimization, energy-related maintenance. These EMS are also based on a modular structure that provides many independent plug-n-play functions. These function include optimization functions data mining and analytics functions, auto-response functions etc. EMS can also provide detailed customizable reports about energy usage.

EMS can be focused on consumption management, sustainability and green issues, and financial gains and losses. When choosing an EMS, consumers need to pay attention. An EMS specifically developed for consumption management may not be an ideal tool for either sustainability and carbon control or reducing environmental footprint. The business software integration can also be overwhelming for companies that do not want or need it. For some companies, a cheap and simple EMS is easy to use. For example, Credit360 EMS is more focused on reducing a company's environmental footprint PowerLogic ION Enterprise Energy Management solution is focused on financial aspects of energy consumption. SAP Energy Management Software is focused on sustainability and streamlining processes. The JouleX EMS is focused on sustainability and financial aspects (Loijos, 2013).

EMS also provide procurement tools fully integrated with suppliers. These procurement tools can be used to provide instant quotes. Modern EMS are also capable of supporting the top power management networking vendors such as Cisco and VMware. EMS provide support for virtualization, cloud computing, and demand response software. EMS can also provide corporate sustainability reports and customized internal reports with detailed analysis. These reports can be used for better transparency and to identify problems and recommend solutiions.

EMS also include data aggregation tools. These tools are capable of extracting disparate energy data from many sources. This data can belong to all energy types and operational data sources. These tools can then merge this data with financial data for analysis purposes. The same data tools are capable of importing from and exporting to data from a variety of enterprise system databases. EMS also provide information dashboards and kiosks that provide easy-to-view detailed graphs and tables based on complex energy data. Businesses can use these tables and graphs to easily understand specific breakdown of their energy consumption across their sites. Companies having different organizational structures can benefit from data sharing tools available with EMS.

Various levels of users can report and access EMS data. EMS are also capable of working with different systems and as such can integrate data from various sources. EMS makes available a large variety of metrics suitable for users having different goals such as social, environmental, and economic goals. Automated Key Performance Indicators (KPIs) are also available with many EMS. EMS allow collection of both qualitative and quantitative data. The data can be collected from a variety of sources on both flexible schedule and ad hoc basis.

Businesses can integrate EMS with enterprise software to better track the overall impact of energy management. EMS are available in both standalone and web-based versions. While stand-alone EMS can be accessed on a particular machine, a web-based EMS can be accessed anywhere. For companies tight on budget, web-based EMS offer an attractive option because companies can get a subscription service to use these EMS. In a subscription service, EMS is offered as software-as-a-service (SaaS) and businesses can choose a pay-as-you-go model.

5.3 Load Modelling and Forecasting (LMF) and Load Optimization

The two very important advanced aspects of EMS are load modelling and forecasting (LMF) and load optimization. The following sections will discuss the scientific relevance of these two aspects.

5.3.1 Load Modelling and Forecasting (LMF)

The issue of energy consumption forecasting has attracted great attention in power system research. Historically, this area was focused by electricity utilities but recently manufacturing organizations have also started showing interest in this issue. Since increased energy consumption leads to increased carbon emissions, the same issue has also attracted the attention of many global warming research scholars. As such we see that the subject of Load Modelling and Forecasting (LMF) will continue gain importance for industrial users.

For a long time, energy consumption and load forecasting has been applied on the user-side (Guntermann, 1982). First, to determine primary driver of energy consumption, we gathered information about energy consumption and plotted this against some variable (such as production activity). The relationship between energy consumption and the primary driver of energy consumption was then used as base information to provide a model of consumptions by simple interpolation of acquired data (Van Gorp, 2004). Benchmarking was used to compare the energy consumption of different buildings or equipment against each other. This benchmarking established why some buildings were more efficient than other. This subsequently was followed by certain actions to improve the less-efficient buildings.

According to Alfares & Nazeeruddin (2002), existing energy consumption forecasting methods can be divided into two broad categories: hard computing techniques and soft computing tech-

Table 2. Hard and soft computing techniques

Hard Computing Techniques	Soft Computing Techniques
Multiple linear regression (MLR), Exponential Smoothing (ES), Iterative Reweighted Least Squares (RLS), Mixed Models (Conejo, Plazas, Espinola, & Molina, 2005; Garcia-Martos, Rodriguez, & Sanchez, 2007) Autoregressive moving average with exogenous variable (ARMAX) (Contreras, Espinola, Nogales, & Conejo, 2003).	Bayesians nets, Genetic Algorithm Modelds (GAs), Fuzzy Logic, Fuzzy Neural Networks (Amjady, 2006) Neural networks (NNs) (Garcia-Martos, Rodriguez, & Sanchez, 2007) Expert Systems (Lora, Santos, Exposito, Ramos, & Santos, 2007)

niques. Aggarwal, Saini, & Kumar (2009) agreed with this classification. Table 2 provides examples of techniques in each category.

Another category of energy consumption forecasting methods has recently emerged, the hybrid model. The examples of hybrid model include methods of data pre-processing, techniques of modelling, evolutionary approaches in order to fix the internal parameters of models. (Wei, Ning, & Zhengang, 2009) combined Principal Component Analysis (PCA) and Support Vector Machine (SVM) for long-term energy consumption forecasting. Wei, Ning, & Zhengang (2009) used a combination of Principal Component Analysis and Support Vector Machine to perform long-term energy consumption forecasting. A variant of Support Vector Regression was proposed by (Hong, 2011) for electric energy consumption forecasting. A study of energy consumption forecasting in Spain used a combination of particle swarm optimization and Adaptive-Network based Fuzzy Inference System (Pousinho, Mendes, & Catalão, 2012).

Computer Intelligence (CI) technologies are the mainstream techniques used for load modeling and forecasting. Zhou et al. (2004) developed a linear regression model with nonnegative coefficients for electricity consumption forecasting. This model was based on Nonnegative Least Squares (NNLS) technique. However, this technique lacked appropriate selection of explanatory variables. The more recent models of energy forecast use feature selection (based on correlation analysis) to select explanatory variables(Zhang, Wang, Gao, & Zhou, 2012). The flexibility of CI tools has made them an attractive approach used to approach load modeling and fore acting. CI techniques can be mixed with other tools and techniques. Khotanzad, Davis, Abaye, & Maratukulam (1996) used an approach where they mixed NN and FL and get NFN approach. Catalao, Pousinho, & Mendes (2011) and Hong (2009) used evolutionary algorithms to get trained AN structure. Statistical and signal processing functions can highlight cyclical behavior of demand load profiles. Examples of these functions are wavelet transformations [(Catalao, Pousinho, & Mendes, 2011); (Hanmandlu & Chauhan, 2011)], integral or Kalman filters [(Hanmandlu & Chauhan, 2011); (Sargunaraj, Sen Gupta, & Devi, 1997)], correlation functions (Fan, Wu, Lee, & Lee, 2011), etc. The hybrid approach of combining CI tools with with statistical, signal processing functions has produced promising results in recent years (Hanmandlu & Chauhan, 2011; Wu, 2010; Wu, 2011).

5.3.2 Load Optimization

Common mathematical approaches are not sufficient to optimize an energy system having multiple objectives. Such a system presents a problem that has conflicting objectives with a set of Pareto optimal solutions. Many of the methods used to handle such problems carry their respective advantages and disadvantages. Many of these methods do not provide either efficient result or take long time to get the solution. (Zhou et al., 2011). Kienzle & Andersson (2009) applied mean-variance portfolio theory to a model of multiple forms of energy. Geidl & Andersson (2005) studied power flow and optimization approach for power systems including multiple energy carriers.

Geidl, M., & Andersson, G. (2005) implemented a classic economic dispatch method using price relation between inputs and outputs to optimize the multi-energy structure. Although the developed models could optimize the functionality of the system, they were sufficiently accurate for other optimal power flow applications.

6. ENERGY MANAGEMENT SYSTEM CONSIDERATIONS

Modern EMS are extraordinarily feature-rich and cost effective to deploy. They require nominal monthly fees and provide rapid return-on-investment in just 6 to 12 months. The dashboard touch-panel controls provide users an easy way to view and reduce their energy consumption. Users are able to reduce their monthly energy bills by 20% or more. Such advancements in EMS have generated enormous interest in EMS by businesses of all sizes who are looking to optimize their energy use while protecting the environment and improving their bottom lines. There are some key considerations for businesses that could help businesses in making informed decisions when selecting an EMS.

6.1. Focus on the HVAC System

Most EMS today are either too complex or too simple to achieve the desired goals. Companies need to take few additional steps in order maximize their energy savings. These steps include minimization of related costs, realization of fast payback, and an EMS that specifically focused on HVAC systems. This is because HVAC systems represent single largest controllable source of energy consumption. As such, they are the first target in producing significant savings in energy consumption. Today's EMS are scalable and flexible. These EMS also allow integration of additional sensors to the ESM in order monitor and control other sources of energy use (such as lighting, security, and office equipment).

6.2. On-Site and Remote Measurement and Control

Remote control of an EMS is always a desired feature. An EMS should be controllable remotely by either company personnel or even the manufacturer's support personnel. Through remote connectivity, EMS manufacturer can offer real-time monitoring of organizational energy consumption and make sure that various parameters are under the defined limits. With such real-time monitoring, companies can immediately know any consumption swings and other anomalies. Remote monitoring also ensures that EMS is being used to the best of its capabilities and is facilitating the expected energy and cost savings.

6.3. Quick-Install with Low Initial Fees and Quick ROI

The best commercial energy management systems on the market today can be delivered and installed for $2,000 or even less and with a monthly service contract below $50 (Ezcurra, 2009). For example, if a business pays monthly utility bill of $3,000, an ROI can be obtained in less than six months with an EMS that provides 15% energy saving. Some EMS manufacturers offer flexible payment mechanisms. In such mechanism, the initial payments are lower while the upfront payments are higher. These payment mechanisms are suitable for multiple unit operators. Companies also need to ensure that EMS can be installed quickly in order reduce the upfront costs.

6.4. Robust and User-Friendly Core Feature Set

A graphical interface that provides real-time and historical energy resource consumption is a common feature of all modern EMS. Besides this common feature, modern EMS also provide a robust feature set. This feature set include ability to set thermostat heating and cooling limits, set points for different times of the day, ability to block unauthorized access and changes to the settings, allow authorized users to view and alter the settings, provide information to the authorized users via the Web, smart phones, or the on-site display, and a 'peak shaving' option. Peak shaving is the process of reducing the amount of energy purchased from the utility company during peak hours when the charges are highest. Many businesses are paying for their electricity consumption on a time-of-use basis.

6.5. Established Track Record of Success

The EMS selected should be a product with significant number of successful real-world installations in the field. EMS manufacturer should also allow you to speak to the companies where this EMS is being used. During a comparison of competing EMS, companies should check references about the quality and user-friendliness of the product, ease of installation, the realized cost savings, time to realize cost savings, and quality of after-deployment support from the manufacturer.

These considerations will allow a business obtain a well-rounded and cost effective energy management system that will reduce their carbon footprint and energy consumption while delivering notable monetary gains.

7. FEATURE COMPARISON OF POPULAR ENERGY MANAGEMENT SYSTEMS

7.1. Energy Management Systems for Businesses

This section provides comparison of features of some popular energy management systems for businesses.

7.1.1. Utlities Direct

This EMS offers full online gas and electricity supplier switching. Users can access any reports regarding their energy use including a fully detailed real-time breakdown of energy consumption. The software is available as both standalone and web version. The software supports multiple sites spread across large geographical areas. Users can aggregate information from multiple sites for monitoring performance and meeting internal targets. The software is designed for large energy consumers. The focus is on consumption management. One great feature of this EMS is the procurement tool fully integrated with all suppliers and able to provide instant quotes. This EMS is capable of transforming complex information into an easy-to-view and detailed graphs and tables.

7.1.2. SAP Energy Management Software

SAP EMS is focused on sustainability with increased profits and streamlined processes. Users can view individual site energy use report or the energy usage report for entire enterprise. Data integration tools allow pulling disparate energy data of all types and merging it with financial data for analysis. Data integration can be run across all energy types and operational data sources and users can also share EMS data. The analysis of combined EMS and financial data can help companies reduce environmental impact while lowering costs. Multiple users can report and access data in SAP EMS. SAP EMS can work with different source systems and users can set reminders that inform them when a Key Performance Indicator (KPI) request is due. SAP EMS has a large library of metrics that users can choose from based on his/her specific goals. SAP EMS support collection of both qualitative and quantitative data from different sources on either a flexible schedule or ad hoc basis.

7.1.3. EnergyCAP

EnergyCAP energy management software is for utility bill & greenhouse gas tracking, analysis, reporting, auditing, benchmarking & ENERGY STAR ratings. There are different available versions of EnergyCAP such as Enterprise, Express, Professional versions. EnergyCAP Enterprise is a powerful enterprise energy management software for utility bill processing and energy reporting. Choose the features that meet your complex needs. This product is suitable for fortune 1000 organizations, colleges, universities, or large government entities. EnergyCAP Express is a fully online energy efficiency software for managing utility bills and energy data. It is suitable for colleges, towns, small counties, and businesses. It provides weather data, greenhouse gas tracking, benchmarking, reporting, energy dashboards, and automatic utility bill auditing. GreenQuest is a free website for home or business. Users can use this site to track utility bills for a single building through energy, cost and carbon footprint charts. While the site is free, users can use it to track one electric, gas, and water bill. It also includes ENERGY STAR rating for commercial buildings

EnergyCAP is a more expensive EMS. First reason is that EnergeyCAP provides regular updates to its EMS so that compliance standards and algorithms are always up-to-date. Second, EnergeyCAP EMS has many complex features. EnergyCAP only runs on Windows operating system. EnergyCAP Express uses cloud-based data storage, while EnergyCAP Enterprise uses traditional data storage. ENERGY STAR Portfolio Manager is an online tool you can use to measure and track energy and water consumption, as well as greenhouse gas emissions. Using EnergyCAP, businesses can submit data directly to the Energy Star Portfolio Management program. EnergyCAP provides a wide array of customizable reports and graphs. These graphs and reports can be exported to software such as Excel, Access, or Crystal Reports. EnergyCAP is flexible and provides top-of-the-line analyses.

7.1.4. Schneider Electric PowerLogic ION EEM

PowerLogic® ION® EEM enterprise energy management software unites business and energy strategies across entire enterprise while performing wide-area analysis of events and conditions. It's KPIs and advanced analytics helps manage energy in financial terms and gain unique insight into the impacts of power quality on your business and all energy assets. Stakeholders from management to operations get actionable energy intelligence to reveal opportunities, isolate problems and drive cost and risk reduction strategies. ION EEM focusses on financial aspects of a business and achievement of business environment goals. It uses a web portal that provides customizable dashboards of real-time data. Users can set up their portal to show the data that is most important to them. It automatically uploads and cleans up energy data and helps in eventually predicting, controlling, and reducing energy-related expenses. It provides one-way integration with enterprise software and data sharing with interested parties. Data is automatically acquired, cleansed and warehoused. Personalized, browser-based dashboards and innovative visualization and modeling tools help you accurately monitor, validate, predict and ultimately control all energy-related expenses and risks to reliability.

You can manage all utilities (electricity, gas, water, etc.) and emissions through a single, unified interface. You can benchmark and compare facility performance across an entire enterprise to identify energy inefficiencies or losses; measure and verify savings from energy conservation projects or performance contracts; reduce operational costs, improve processes, and prolong the life of energy assets; manage demand control schemes, load shedding, peak shaving, base loading or on-site generation; and enable participation in real-time pricing and load curtailment programs. You can also optimize procurement by forecasting and budgeting for energy needs and comparing utility rates. Utility billing errors can be identified and you can allocate and recover utilities costs from tenants, departments, processes, etc.

7.1.5. Credit360

Credit360 focuses on enterprise-wide energy tracking, environmental goals, and sustainability. It is used by Fortune 500 organizations. It provides forecasting, planning, and reporting. This helps businesses reduce their energy consumption, reduce company's environmental footprint, and provide reporting in line with different agencies to achieve energy and carbon emissions compliance. It has a built-in carbon-factor management system. One important issue of Credit360 is the lack of more complex data analysis. While it can import data from other enterprise reporting platforms, the capabilities of integration and tools offered are not comparable with other EMS such as SAP EMS and EnergyCAP EMS. Few strengths of Credit360 are its reporting capabilities, sophisticated estimating techniques, and ability to collect data on carbon emissions. It can import data on carbon emissions and create carbon-factor management goals as well as energy goals. Credit360 gathers real time information from meters and automatically tracks high impact areas.

7.1.6. eSightEnergy

This EMS comes in two versions: Web-based Software-as-a-Service and software that can be installed on site. It is a modular software that can be tailored to business requirements. It supports broad analysis of energy data using a wide range of techniques; energy analysis in terms of costs; management of supplier contracts; and creations and monitoring of budgets. Users can run a variety of pre-defined report templates against different companies, sites and meters. Data can be imported and exported in a large variety of formats. This EMS can generates automatic alarms for data outside of tolerated levels and notifies users by email. For data aggregation and plant efficiency, users can create virtual meters. You can calculate energy consumption for tenants and generate bills on a scheduled basis. You can also compare and verify bill values against collected meter data. This EMS can be used by both small and large businesses and web-based software can provide access from anywhere. A subscription service for the software is also available for the companies who are not ready to invest in an energy management system. This EMS can work with multiple currencies and local guidelines and users can import data from virtually any source data stream. Since the energy consumption data can be converted into monetary units, it becomes easier to understand data. One specific feature of this EMS is energy consumption contracts. Provided with financial data, this EMS can also validate invoices.

7.1.7. JouleX

JouleX EMS provides monitoring, analysis, and control of energy usage any network-connected device. JouleX doesn't require any client-side agent or hardware meter provides automatic remote power management. Depending on enterprise demand, JouleX can alter capacity. JouleX is focused on sustainability and financial aspects of the firm. This EMS is more suitable for organizations with high levels of energy consumption and support all major power management networking vendors such as Cisco and VMware. One very strong feature of this EMS is that it can measure dynamic energy consumption as well as the utilization of any device attached to the network. With this capability, energy use can be optimized and waste can be reduced. This EMS is optimized for both virtualization and cloud computing and provides customized reports with detailed analysis. These reports are useful in identifying major areas of power consumption and poor utilization and suggesting remedies to resolve them.

7.1.8. C3

The C3 Energy Company offers a range of EMS software products, namely, residential, commercial and enterprise software and analytics, customer experience software, customer segmentation and targeting software, EM&V, and customer channel analytics under the brand C3 Energy. These software products provide data aggregation, real-time analytics, and an intuitive and responsive user experience (BusinessWire.com, 2015). C3 EMS is intended for both small and large businesses. This EMS also has residential and mobile version. The mobile version is specifically tailored for the iPad.This EMS is capable of pulling data from varying data sets and developing a cohesive energy management strategy out of this data. C3 EMS focuses on both supply and demand side operations and provide separate analysis of each side of operations so that companies better understand their energy management. As a comprehensive demand response tool, C3 EMS utilizes smart grid analysis to calculate value of changing energy usage and its organizational impact. This EMS is user-friendly that can be used to provide accessible information and recommendation. This information and recommendations are useful for businesses to make decisions, save energy, and get energy rebates. C3 has a large library of rebate options. These options can be paired with suggestions made by C3 EMS. C3 EMS is also provided as Software-as-a-Service that can be run in most standard browsers. One drawback of this EMS is that it lack some sophisticated tools available with other EMS such as EnergyCAP EMS.

7.1.9. Hara Environmental and Energy Management (EEM)

Hara is a leading provider of energy and environmental management software. Hara has emerged as a leader in the energy and environmental management market through consistent product delivery, customer deployments and domain expertise. Deployed in over 100 countries, Hara EEM is a cloud-based solution that enables organizations to strategically manage and optimize their energy and natural resource usage and identify and execute efficiency strategies across the enterprise and value chain. Hara EEM can be used by organizations to track, verify and manage energy usage and emissions. Hara EEM gives organizations access to management of distributed energy and resource activities through one centralized system. Organizations get improved ability to actively track consumption, forecast future use, manage and monitor global environmental performance, and manage its resource efficiency on a wider enterprise basis. Hara EMS objective is to decrease operating costs, decrease energy and water consumption and improve organization's bottom line. Some notable customers of Hara include Hara customers include: Aerojet, Coca-Cola, Hasbro, Intuit, News Corporation, Safeway, US Bank and City of Las Vegas and City of Philadelphia. The Web-based version of HARA EEM provides a dashboard that organizations can use to monitor their water and fuel consumption and reduce their environmental impact. Besides electricity, Hara EEM can also monitor travel fuel consumption, GHG emissions, and water usage. The user-friendly interface of Hara EEM can also monitor the company's environmental projects. Hara EEM dashboards can be setup with RSS feed to stay updated with latest industry information. The generated reports are highly customizable and can be exported to Excel. Hara EEM is also capable of supporting company's efforts to gain rebates or meet green standards. The search functionality available helps companies find supplier and billing data easily.

7.1.10. CA Technologies

The EMS offered by CA Technologies has three versions: CA ecoGovernance, CA ecoMeter, and CA ecoDesktop. CA ecoGovernance is designed to enable organizations to accelerate energy and sustainability business performance with automated data management, analysis and reporting. It features project planning and management, scenario planning, portfolio management and supply chain management capabilities that help enable corporate energy and sustainability transformation. CA ecoGovernance allows companies to measure and analyze energy as well as carbon, water, and waste. It provides web-based questionnaires that businesses can use access facilities and suppliers. It integrates with a variety of third-party ERP's. CA ecoGovernance can process different types of data (such as invoice data).It can also provide automated data collection and real-time tracking of KPIs. CA ecoMeter is designed to capture real time data from datacenters and building infrastructure providing you with a clear and comprehensive picture of energy use and other environmental variables. With CA ecoDesktop you can manage power across a broad range of PCs and other desktop devices for remote forced adherence to power management policies. Organizations can use CA ecoDesktop to track desktop performance metrics and subsequently use the auditable data to assist with corporate carbon and environmental disclosure for reporting.

7.2. Feature Comparison of Popular Home Energy Management Systems

This section provides comparison of features of some popular home EMS.

7.2.1. Opower Software

This EMS for households collects and assesses data taken from the smart meter. This data is converted into a report that can be accessed by the customer through email, SMS, or Web.

7.2.2. Energy Lens

Energy Lens can be used to provide analysis of detailed energy-consumption data. It works with the existing smart meter to collect the required date. Energy Lens turns a meter's detailed records of energy consumption into useful charts and tables that will help to figure out energy usage, schedule of usage, recommendations to reduce energy consumption, and track progress of the energy reduction strategy. Energy Lens offers more sophisticated analysis than the mass-market tools and it is capable of working with Excel/CSV files of energy data. Energy Lens is a completely stand-alone system, which requires no connection to any hardware. The software has been developed to be an add-on to the Microsoft Excel package.

7.2.3. The Solent Electronic Home

This company offers different residential energy management solutions. Lighting Control System is a fully automated solution to control each individual lighting source to ensure device is only used when needed. The device can be activated by iPhone or iPad, via a PC or vocally. Heating Control System ensures that energy output is more effective by making sure heat is only available when needed and at the appropriate levels. The system supports multiple heating methods including wood burners and fossil fuel and solar energy.

7.2.4. GEO

The simple and accessible home energy management solution provided by this company can be used to serve basic to advanced needs of residential energy management. The EMS provide easy to understand display showing energy usage and carbon emissions. A basic EMS provides about energy consumption. A more advanced EMS (the 'connected home' model) provides energy consumption monitoring of appliances by attaching a sensor to the appliance.

8. ENERGY MANAGEMENT SYSTEMS STANDARDS

ISO 50001 is the latest international energy management standards published in 2011. ISO 50001 provides a framework for energy systems management. This framework help companies better plan to save energy and reduce pollution, reduce costs, resolve energy efficiency problems, and improve energy usage of energy-consuming assets, estimate environmental footprint, improve energy management and communication, set best practices for energy efficiency, and prioritize new energy-saving technology. According to ISO, the use of these standards can result in an estimated 60% savings in global energy consumption. ISO50001 is based on an energy management model, which applies the "Plan, Do, Check, and Action" cycle (PDCA) (ISO, 2015).

Broadly speaking, companies need to take few steps to get started on ISO 50001 implementation. These steps include securing support from top management; collecting, tracking, and analyzing energy data; identifying key energy uses; establishing a baseline; identifying energy-saving opportunities; and prioritizing opportunities.

There are a number of benefits of effectively implementing ISO 50001, including reduced operational and overhead costs that lead to increased profitability; reduced air emissions, such as GHGs; increased efficiency of energy sources; increased assurance of legal, internal compliance; identification of variables affecting energy use and consumption; and increased understanding of energy use and consumption via defined methods, processes of data collection.

9. ENERGY MANAGEMENT SYSTEMS: EMERGING TRENDS

The rapidly changing landscape of technology continues significantly influence the design and implementation of future EMS. The automation systems, both home and commercial, are increasingly become complex and continuing to provide several functions such as information acquisition, decision- making, and actuation. Future EMS would require full integration with automation systems, support control of a variety of appliances, enhanced interoperability and security, and provide advance features such as surveillance using sensors. The future EMS would need wireless communication with other devices and sensors through wireless protocols such as ZigBee. EMS mobility is an important issue and future EMS would provide various mobile functionalities for real-time communication with consumers about their energy consumption, carbon footprint, and current energy tariffs. Such communication using mobile devices can provide value-added services such as recommendation for an energy-efficient appliance and crowdsourcing user- collected information. With the help of such information, EMS can correlate user location and activity with user behavior and aggregate this on a large scale to gain insight into energy usage trends.

One of the fundamental goals of smart grid is demand-side energy management (Aggarwal, Saini, & Kumar, 2009). Future EMS can play an important role in achieving this goal by influencing consumers' energy usage behavior. However, this would require an insight into energy usage patterns, within homes and organizations, and their impact on overall consumption in the smart grid. The utility can send real-time notifications, regarding curtailment of energy use, to the EMS. Future EMS may have embedded intelligence that facilitate automatic decision making and control of appliances based on these notifications. This direct communication channel with EMS may also be utilized by other entities, besides utility, for communication with consumers. One such entity is a social network provider that can use this channel for communicating social and comparative messages.

Future EMS would require a better user interface. Weiser (Weiser, 2002) envisioned a pervasive future computing environment in which sensing, computing, and response occur in a non-intrusive way and without consumer intervention. Future EMS would need to conform to such standard. One possible way of doing this is to build EMS with embedded learning and intelligence capabilities. With such capabilities, EMS would be able to store user preferences, comfort level, and use this information to automate control of appliances. Medland, R. C., Foth, M., and Petkov, P. (2011) developed pervasive user interface guidelines that EMS developers can use.

Future EMS would need to integrate information about multiple consumptions besides consumption of electricity. These consumptions may include consumption of gas and water. Future EMS may be adopted by local utility companies. The local utility companies can offer these EMS to customers at subsidized prices, to make them more affordable, and to offer energy efficiency tips to the consumers. This could help utility companies achieve their long-term sustainability and energy efficiency goals and make their operations more reliable and cost effective.

Future EMS would require enhanced capability to perform what-if scenarios. In such scenarios, consumers can evaluate fine-grained effects of the impact of changes in their energy consumption behavior to the overall energy consumption. EMS would also need to provide a goal-setting framework for consumers. This framework would help consumers track their progress toward their self-specified goals related to behavior change.

Security and privacy will be a potential challenge for future EMS. It is especially important given that EMS increasingly use the collected data to monitor, infer, and devise user actions. This data is also shared among peers through social networking sites and other communication means. This trend warrants special measures against possible threats to data security and privacy.

The top three challenges faced by industrial EMS are accurate measurement of energy metrics, integration with disparate systems and data sources, and lack of organizational culture supporting energy management (Wed, 2013). Future industrial EMS would need to address these challenges by working closely with the industrial consumers.

10. FUTURE RESEARCH AREAS

In recent years, HEMS have changed considerably. EMS offerings have expanded and new research areas have emerged such as dynamic pricing, collaboration with non-utility vendors, and the integration of more grid-edge technologies. Research is needed to find ways to get customers used to the idea of dynamic (or variable) pricing, using starter schemes that are all carrot and no stick.

The concept of bring-your-own-thermostat is getting popular and few utilities are offering programs where those customers can enroll in peak-use rebate programs. In the United Kingdom, in-home displays will be mandatory as smart meters are rolled out, although it is unclear how much those devices will affect long-term energy efficiency. Smart meter data, however, could be the first step to get homeowners to think more seriously about efficiency measures. Research is needed to determine ways of delivering this data to the customers quickly and in a useful format. The automated communications standard for demand response, OpenADR, has been gaining ground for commercial applications. OpenADR is important because it allows utilities to verify the load they are shedding in real time. OpenADR 2.0 could very well become the de facto standard. Research is needed to determine how utilities can use this standard optimally for measurement and verification from residential demand response programs than they have had access to in the past.

One of the promises of smart meters was their help to average person better manage energy use. While that may be the case, vendors are also finding that the smart meter simply is not necessary for demand response, and many utilities are not using it to get into the home. Another concept "broadband to enter the home" instead of the meter is gaining ground. While there are about 30 million deployed smart meters, broadband is ubiquitous. Amongst other advantages, broadband access across rural population could help drive down the cost of connected home services to customers. Research is needed to determine how this service could be optimally offered.

Another trend is the use of behavioral psychology in the energy management space. A number of firms are applying machine-learning algorithms that help optimize software and hardware to change homeowners' behavior. Future research is needed on integrating weather forecasting into energy management, managing storage solutions and distributed energy such as rooftop solar, and complementing the use of micro-grids.

There are continued efforts to create new commercially viable ways to reduce overall energy consumption in existing buildings through improved energy management systems. Future research is needed in many areas including systems for identifying poor energy performance, diagnosing the reasons, visualizing the analysis, proposing intelligent solutions, and forecasting performance improvements. Another focus area is combining energy management systems with other smart capabilities valued by customers, such as measuring real-time energy use alongside space-use efficiency or productivity. Another research area is systems that overcome the uncertainty of achieving forecasted energy savings, for instance, through real-time monitoring and verification.

In industrial energy management systems, a future research area is how to cultivate a supportive energy management culture in the organizations. The answer to this question is not easy but very significant for the future success of industrial energy management systems.

11. CONCLUSION

An increasing trend of heterogeneity in both homes and industrial concerns pose several challenges to developing energy management approaches. While sharing a common goal of energy usage optimization, modern EMS are fundamentally different from their predecessors from design and operations point of view. Modern EMS can communicate and interact with both the users and the energy appliances to match energy consumption with energy supply. This chapter presents a survey of the existing landscape of EMS. Modern EMS is bringing a change in user's energy consumption behavior e.g. by providing real-time consumption monitoring with similar users. Future EMS need more focused attention on security and privacy issues. Currently, these issues are not adequately handled by EMS. Future EMS also need to utilize analytics and learning to provide intelligence that could be used to gain insights into consumer behavior and automate the EMS operation. Emerging trends point to a future scenario where we will have dynamic user communities. These communities, in contrast to the existing isolated consumers, will have a mutual goal of energy sustainability. These communities will perform smart and collective energy consumption management and share information among community members to achieve their energy-related goals.

REFERENCES

C3 Energy Featured as Top Vendor in Global Energy Management System Market Report. (2015, July 13). Business Wire. Retrieved from http://www.businesswire.com/news/home/20150713005254/en/C3-Energy-Featured-Top-Vendor-Global-Energy#.VaSjwLWLFf4

Aggarwal, S. K., Saini, L. M., & Kumar, A. (2009). Electricity price forecasting in deregulated markets: A review and evaluation. *International Journal of Electrical Power & Energy Systems*, *31*(1), 13–22. doi:10.1016/j.ijepes.2008.09.003

Alfares, H. K., & Nazeeruddin, M. (2002). Electric load forecasting: Literature survey and classification of methods. *International Journal of Systems Science*, *33*(1), 23–34. doi:10.1080/00207720110067421

Amer, M., Naaman, A., M'Sirdi, N. K., & El-Zonkoly, A. M. (2014). Smart home energy management systems survey. *Proceedings of the 2014 International Conference on Renewable Energies for Developing Countries REDEC* (pp. 167–173). IEEE. doi:10.1109/REDEC.2014.7038551

Amft, O., Medland, R., Foth, M., Petkov, P., Abreu, J., Pereira, F. C., & Paulos, E. et al. (2011). Smart Energy Systems. *IEEE Pervasive Computing / IEEE Computer Society [and] IEEE Communications Society*, *10*(1), 63–65. doi:10.1109/MPRV.2011.10

Amjady, N. (2006). Day-ahead price forecasting of electricity markets by a new fuzzy neural network. *IEEE Transactions on Power Systems*, *21*(2), 887–896. doi:10.1109/TPWRS.2006.873409

Bartram, L., Rodgers, J., & Muise, K. (2010). Chasing the Negawatt: Visualization for Sustainable Living. *IEEE Computer Graphics and Applications*, *30*(3), 8–14. doi:10.1109/MCG.2010.50 PMID:20650713

Bose, S., Wu, C., Wierman, A., & Mohsenian-Rad, H. (2014, January). Assessing Market Power in Deregulated Electricity Markets - IEEE Smart Grid. Retrieved from http://smartgrid.ieee.org/january-2014/1025-assessing-market-power-in-deregulated-electricity-markets

Brewer, R. S., & Johnson, P. M. (2010). WattDepot: An Open Source Software Ecosystem for Enterprise-Scale Energy Data Collection, Storage, Analysis, and Visualization. *Proceedings of the 2010 First IEEE International Conference on Smart Grid Communications (SmartGridComm)* (pp. 91–95). doi:10.1109/SMARTGRID.2010.5622023

Catalao, J. P. S., Pousinho, H. M. I., & Mendes, V. M. F. (2011). Hybrid Wavelet-PSO-ANFIS Approach for Short-Term Wind Power Forecasting in Portugal. *IEEE Transactions on Sustainable Energy*, *2*(1), 50–59. doi:10.1109/TSTE.2010.2076359

Cole, A., & Albicki, A. (2000). Nonintrusive identification of electrical loads in a three-phase environment based on harmonic content. *Proceedings of the 17th IEEE conference on Instrumentation and Measurement Technology Conference IMTC '00* (Vol. 1, pp. 24–29). IEEE. doi:10.1109/IMTC.2000.846806

Conejo, A., Plazas, M., Espinola, R., & Molina, A. (2005). Day-ahead electricity price forecasting using the wavelet transform and ARIMA models. *IEEE Transactions on Power Systems*, *20*(2), 1035–1042. doi:10.1109/TPWRS.2005.846054

Contreras, J., Espinola, R., Nogales, F. J., & Conejo, A. (2003). ARIMA models to predict next-day electricity prices. *IEEE Transactions on Power Systems*, *18*(3), 1014–1020. doi:10.1109/TPWRS.2002.804943

D'Alessandro, S., Tonello, A. M., Monacchi, A., & Elmenreich, W. (2014). Home energy management systems: Design guidelines for the communication infrastructure. *Proceedings of the 2014 IEEE International Energy Conference ENERGYCON* (pp. 805–812). IEEE. doi:10.1109/ENERGYCON.2014.6850518

Dong, Q., & De-jiang Zhang. (2010). The research and realization of energy management system in iron and steel enterprise. *Proceedings of the 2010 International Conference on Information Networking and Automation ICINA* (Vol. 1, pp. V1–448–V1–451). doi:10.1109/ICINA.2010.5636520

The What, Why, and How of Energy Management. (2015). Energylens.com. Retrieved from http://www.energylens.com/articles/energy-management

Ezcurra, G. (2009, November 9). Top Five Reasons Your Business Needs an Energy-Efficiency System. *Greentech Media*. Retrieved from http://www.greentechmedia.com/articles/read/top-five-reasons-your-business-needs-an-energy-efficiency-system

Fan, S., Wu, Y.-K., Lee, W.-J., & Lee, C.-Y. (2011). *Comparative study on load forecasting technologies for different geographical distributed loads*. IEEE Power and Energy Society General Meeting. Doi:10.1109/PES.2011.6039364

Farhangi, H. (2014). A Road Map to Integration: Perspectives on Smart Grid Development. *Power and Energy Magazine, IEEE*, *12*(3), 52–66. doi:10.1109/MPE.2014.2301515

Fischer, C. (2008). Feedback on household electricity consumption: A tool for saving energy? *Energy Efficiency*, *1*(1), 79–104. doi:10.1007/s12053-008-9009-7

Fitzpatrick, G., & Smith, G. (2009). Technology-Enabled Feedback on Domestic Energy Consumption: Articulating a Set of Design Concerns. *IEEE Pervasive Computing / IEEE Computer Society [and] IEEE Communications Society*, *8*(1), 37–44. doi:10.1109/MPRV.2009.17

Forth, B., & Tobin, T. (2002). Right power, right price [enterprise energy management systems]. *IEEE Computer Applications in Power*, *15*(2), 22–27. doi:10.1109/67.993756

Froehlich, J., Findlater, L., & Landay, J. (2010). The Design of Eco-feedback Technology. *Proceedings of the SIGCHI Conference on Human Factors in Computing Systems* (pp. 1999–2008). New York, NY, USA: ACM. doi:10.1145/1753326.1753629

Garcia-Martos, C., Rodriguez, J., & Sanchez, M. J. (2007). Mixed Models for Short-Run Forecasting of Electricity Prices: Application for the Spanish Market. *IEEE Transactions on Power Systems*, *22*(2), 544–552. doi:10.1109/TPWRS.2007.894857

Geidl, M., & Andersson, G. (2005). A modeling and optimization approach for multiple energy carrier power flow. In *Power Tech, 2005* (pp. 1–7). IEEE Russia; doi:10.1109/PTC.2005.4524640

Geidl, M., & Andersson, G. (2005, August). Optimal power dispatch and conversion in systems with multiple energy carriers. *Proc. 15th Power Systems Computation Conference PSCC.*

González, P. A., & Zamarreño, J. M. (2005). Prediction of hourly energy consumption in buildings based on a feedback artificial neural network. *Energy and Building*, *37*(6), 595–601. doi:10.1016/j.enbuild.2004.09.006

GTM Smart Grid. (2013, September 11). Greentechmedia. Retrieved from http://smart562.rssing.com/chan-8437336/all_p7.html

Guntermann, A. E. (1982). Are Energy Management Systems Cost Effective? *IEEE Transactions on Industry Applications*, *IA-18*(6), 616–625. doi:10.1109/TIA.1982.4504118

Hanmandlu, M., & Chauhan, B. K. (2011). Load Forecasting Using Hybrid Models. *IEEE Transactions on Power Systems*, *26*(1), 20–29. doi:10.1109/TPWRS.2010.2048585

Hart, G. W. (1992). Nonintrusive appliance load monitoring. *Proceedings of the IEEE*, *80*(12), 1870–1891. doi:10.1109/5.192069

Hong, W.-C. (2009). Hybrid evolutionary algorithms in a SVR-based electric load forecasting model. *International Journal of Electrical Power & Energy Systems*, *31*(7–8), 409–417. doi:10.1016/j.ijepes.2009.03.020

Hong, W.-C. (2011). Electric load forecasting by seasonal recurrent SVR (support vector regression) with chaotic artificial bee colony algorithm. *Energy*, *36*(9), 5568–5578. doi:10.1016/j.energy.2011.07.015

Hong, Y.-Y., & Chou, J.-H. (2012). Nonintrusive energy monitoring for microgrids using hybrid self-organizing feature-mapping networks. *Energies*, *5*(7), 2578–2593. doi:10.3390/en5072578

ISO. (2015). ISO 50001 - Energy management - ISO. Retrieved from http://www.iso.org/iso/home/standards/management-standards/iso50001.htm

Karnouskos, S. (2011). Crowdsourcing information via mobile devices as a migration enabler towards the SmartGrid. *Proceedings of the 2011 IEEE International Conference on Smart Grid Communications SmartGridComm* (pp. 67–72). doi:10.1109/SmartGridComm.2011.6102394

Khotanzad, A., Davis, M. H., Abaye, A., & Maratukulam, D. J. (1996). An artificial neural network hourly temperature forecaster with applications in load forecasting. *IEEE Transactions on Power Systems*, *11*(2), 870–876. doi:10.1109/59.496168

Kienzle, F., & Andersson, G. (2009). A greenfield approach to the future supply of multiple energy carriers. *Proceedings of the IEEE Power Energy Society General Meeting PES '09* (pp. 1–8). doi:10.1109/PES.2009.5275692

Kim, Y., Schmid, T., Charbiwala, Z. M., & Srivastava, M. B. (2009). ViridiScope: Design and Implementation of a Fine Grained Power Monitoring System for Homes. *Proceedings of the 11th International Conference on Ubiquitous Computing* (pp. 245–254). New York, NY, USA: ACM. doi:10.1145/1620545.1620582

Lazos, D., Sproul, A. B., & Kay, M. (2015). Development of hybrid numerical and statistical short term horizon weather prediction models for building energy management optimisation. *Building and Environment*, *90*, 82–95. doi:10.1016/j.buildenv.2015.03.025

Leeb, S. B. (1993). *A conjoint pattern recognition approach to nonintrusive load monitoring*. Massachusetts Institute of Technology.

Loijos, A. (2013, June 3). Review of Top 10 Energy Management Software (EMS) Comparison. LinkCycle. Retrieved from http://www.linkcycle.com/review-of-top-energy-management-software/

Lora, A., Santos, J. M. R., Exposito, A., Ramos, J. L. M., & Santos, J. C. R. (2007). Electricity Market Price Forecasting Based on Weighted Nearest Neighbors Techniques. *IEEE Transactions on Power Systems*, *22*(3), 1294–1301. doi:10.1109/TPWRS.2007.901670

Medland, R. C., Foth, M., & Petkov, P. (2011). Connecting people and resource consumption in real time. *IEEE Pervasive Computing / IEEE Computer Society [and] IEEE Communications Society*, *10*(1), 63–65.

Nastu, P. (2013, August 7). BEMS Market Will Grow To $5.6 Billion By 2020. *Energy Manager Today*. Retrieved from http://www.energymanagertoday.com/bems-market-will-grow-to-5-6-billion-by-2020-094340/

Paradiso, J., Dutta, P., Gellersen, H., & Schooler, E. (2011). Guest Editors' Introduction: Smart Energy Systems. *IEEE Pervasive Computing / IEEE Computer Society [and] IEEE Communications Society*, *10*(1), 11–12. doi:10.1109/MPRV.2011.4

Pousinho, H. M. I., Mendes, V. M. F., & Catalão, J. P. S. (2012). Short-term electricity prices forecasting in a competitive market by a hybrid PSO–ANFIS approach. *International Journal of Electrical Power & Energy Systems*, *39*(1), 29–35. doi:10.1016/j.ijepes.2012.01.001

Home Energy Management Systems: Vendors, Technologies and Opportunities. (2014, February 11). PRNewsire.com. Retrieved from http://www.prnewswire.com/news-releases/home-energy-management-systems-vendors-technologies-and-opportunities-2013-2017-244880121.html

Saini, N., & Prakash, P. (2014). Factors Affecting Adaptation of Energy Management & Information Systems in Organizations.

Sargunaraj, S., Sen Gupta, D. P., & Devi, S. (1997). Short-term load forecasting for demand side management. *IEE Proceedings on Generation, Transmission and Distribution*, *144*(1), 68–74. doi:10.1049/ip-gtd:19970599

Sundramoorthy, V., Cooper, G., Linge, N., & Liu, Q. (2011). Domesticating Energy-Monitoring Systems: Challenges and Design Concerns. *IEEE Pervasive Computing / IEEE Computer Society [and] IEEE Communications Society*, *10*(1), 20–27. doi:10.1109/MPRV.2010.73

Szablya, L. (2012, April 26). Home Energy Management with or without an Advanced Metering Infrastructure - Part 2. Intelligent Utility. Retrieved from http://www.intelligentutility.com/article/12/04/home-energy-management-or-without-advanced-metering-infrastructure-part-2

US Department of Housing. (2006, March). Public Housing Energy Conservation Clearinghouse New. Retrieved from https://portal.hud.gov/hudportal/documents/huddoc?id=DOC_10679.pdf

US Energy Department. (2005). *On the Road to Energy Security; Implementing a Comprehensive Energy Strategy: A Status Report*. Retrieved from http://energy.gov/sites/prod/files/edg/media/FINAL20DOE_20booklet20copy20sep.pdf

Van Gorp, J. C. (2004). Enterprising energy management. *IEEE Power and Energy Magazine*, *2*(1), 59–63. doi:10.1109/MPAE.2004.1263421

Wed, M. L. on. (2013, August 21). Empowering Shop-Floor Leaders to Optimize Energy Performance [DATA]. Retrieved from http://blog.lnsresearch.com/blog/bid/185439/Empowering-Shop-Floor-Leaders-to-Optimize-Energy-Performance-DATA

Wei, L., Ning, Y., & Zhengang, Z. (2009). Study on long-term load forecasting of MIXSVM based on principal component analysis. *Proceedings of the International Conference on Future BioMedical Information Engineering FBIE '09* (pp. 439–441). doi:10.1109/FBIE.2009.5405820

Weiser, M. (2002). The Computer for the 21st Century. *IEEE Pervasive Computing / IEEE Computer Society [and] IEEE Communications Society*, *1*(1), 19–25. doi:10.1109/MPRV.2002.993141

Weiss, M., Mattern, F., Graml, T., Staake, T., & Fleisch, E. (2009). Handy Feedback: Connecting Smart Meters with Mobile Phones. *Proceedings of the 8th International Conference on Mobile and Ubiquitous Multimedia* (pp. 15:1–15:4). New York, NY, USA: ACM. doi:10.1145/1658550.1658565

Wu, Q. (2010a). A hybrid-forecasting model based on Gaussian support vector machine and chaotic particle swarm optimization. *Expert Systems with Applications*, *37*(3), 2388–2394. doi:10.1016/j.eswa.2009.07.057

Wu, Q. (2010b). Power load forecasts based on hybrid PSO with Gaussian and adaptive mutation and Wv-SVM. *Expert Systems with Applications*, *37*(1), 194–201. doi:10.1016/j.eswa.2009.05.011

Wu, Q. (2011). Hybrid forecasting model based on support vector machine and particle swarm optimization with adaptive and Cauchy mutation. *Expert Systems with Applications*, *38*(8), 9070–9075. doi:10.1016/j.eswa.2010.11.093

Xu, Z., Jia, Q.-S., Guan, X., & Shen, J. (2013). Smart Management of Multiple Energy Systems in Automotive Painting Shop. *IEEE Transactions on Automation Science and Engineering*, *10*(3), 603–614. doi:10.1109/TASE.2012.2236554

Yang, G.-H., & Li, V. O.-K. (2010). Energy Management System and Pervasive Service-Oriented Networks. *Proceedings of the 2010 First IEEE International Conference on Smart Grid Communications (SmartGridComm)* (pp. 1–6). doi:10.1109/SMARTGRID.2010.5622001

Zhang, H., Wang, Z., Gao, F., & Zhou, D. (2012). A New Method for the Daily Electricity Consumption Forecast of Energy Intensive Enterprise. *Proceedings of the 2012 Asia-Pacific Power and Energy Engineering Conference APPEEC* (pp. 1–4). doi:10.1109/APPEEC.2012.6307482

Zhao, P., Suryanarayanan, S., & Simoes, M. G. (2013). An Energy Management System for Building Structures Using a Multi-Agent Decision-Making Control Methodology. *IEEE Transactions on Industry Applications*, *49*(1), 322–330. doi:10.1109/TIA.2012.2229682

Zhou, A., Qu, B.-Y., Li, H., Zhao, S.-Z., Suganthan, P. N., & Zhang, Q. (2011). Multiobjective evolutionary algorithms: A survey of the state of the art. *Swarm and Evolutionary Computation*, *1*(1), 32–49. doi:10.1016/j.swevo.2011.03.001

Zhou, D., Gao, F., Guan, X., Chen, Z., Li, S., & Lu, Q. (2004). Daily electricity consumption forecast for a steel corporation based on NNLS with feature selection. *Proceedings of the 2004 International Conference on Power System Technology PowerCon '04* (*Vol. 2*, pp. 1292–1297). doi:10.1109/ICPST.2004.1460201

Chapter 14
Energy Management in Cloud Computing

Ganesh Neelakanta Iyer
Progress Software Development, India

Pattabhi Mary Jyosthna
B.V. Raju Institute of Technology, India

Suman Jonnalagadda
B.V. Raju Institute of Technology, India

ABSTRACT

Due to the growing demand of businesses, enterprises are concentrating on performance improvement as well as investment reduction for their applications. Cloud Computing, an emerging technology which provides enormous services through internet is the best choice for the enterprises to improve their business performance and to reduce their capital budget. Cloud datacenters have thousands of servers to provide uninterrupted services to the consumers. These High Performance Computing systems consumes high energy, which leads to increase in Cloud Service Providers operational cost. Apart from the operational cost it also increases CO2 emission, which causes Global warming. So many researchers are evolving engineering techniques that are required to improve performance as well as optimization in power consumption. This chapter describes some of the techniques that are used to reduce operational cost, CO2 emission, SLA violation and able to maintain Quality of Service (QoS). It also describes about limitations on existing methods and further enhancement issues for better performance.

INTRODUCTION

Cloud Computing is an on-demand Internet service which provides everything as a service (XasS) on pay-as-you-go basis. Some of the services provided by cloud computing are Infrastructure as a Service (IaaS), Platform as a Service (PaaS), and Software as a Service (SaaS). Organizations are utilizing these services to improve their computing performance and also to reduce their maintenance investment. Consumers can communicate with Cloud Service Providers (CSPs) through internet to access these services without any human interaction. CSP provides shared pool of resources to

DOI: 10.4018/978-1-4666-9911-3.ch014

multiple consumers using a multi-tenant model with respect to consumer demand. Consumer can also extend his limit of resources utilization capability based on his requirement. Small, medium and large-size organizations are moving towards the cloud technology to reduce their maintenance cost - includes the required number of computing systems, storage and software - to run their applications. These organizations are also thinking about the energy consumption, because high energy consumption of computing systems can limit further performance growth and carbon dioxide footprints. Hence, the goal of the computing systems has been shifted from performance improvements to power and energy efficiency. Organizations can improve performance and reduce CO2 emission by adopting cloud technology. A recent research by Accenture shows that moving business applications to cloud can reduce carbon footprint of organizations. According to the report, if enterprises adopt cloud technology, small businesses can observe up to 90% reduction in carbon emission. Large and medium-size organizations can save at least 30-60 and 60-90 percent. So organizations are showing interest to adopt cloud technology because of its services and essential characteristics.

Cloud Services

Infrastructure as a Service (IaaS)

According to IETF (Internet Engineering Task Force), IaaS providers offer Physical or virtual machines, storage, networking, firewalls and datacenter space to the organizations. It is best option for the small scale organizations who cannot invest more on hardware equipment and maintaining own data centers. Clients can pay to the provider based on the amount of resources consumed by them. Some examples of IaaS providers include AWS and Rack space.

Platform as a Service (PaaS)

With PaaS cloud providers can provide the complete environment required to building and developing web based applications. This includes operating system, database, and web server. PaaS is useful when multiple members involve in developing a project or when the people want to interact with the development process and when developers wish to automate testing and deployment services. Some of the example PaaS providers are Google App Engine, Microsoft Azure Services

Software as a Service (SaaS)

Software as a Service is a software delivery method which provides the consumers or the organizations to access the software and its functions remotely as a Web-based service at a cost typically less then paying for licensed applications. Moreover it removes the need for organizations to handle the installation, set-up, daily upkeep and maintenance. Google Apps and Facebook games some commonly used SaaS applications (figure 1).

Cloud Computing Deployment Models

Cloud computing has four deployment models: Public Cloud, Private Cloud, Hybrid Cloud and Community Cloud. Business can chose any of them to deploy their applications.

Public Cloud

The Cloud Services which are rendered over a network that is open for public use is called as "Public Cloud". Public cloud services may be free or offered on a pay-per-usage model. Generally, public cloud providers like Amazon AWS, Microsoft and Google own and operate the infrastructure and offer access over the Internet.

Figure 1. Cloud computing services

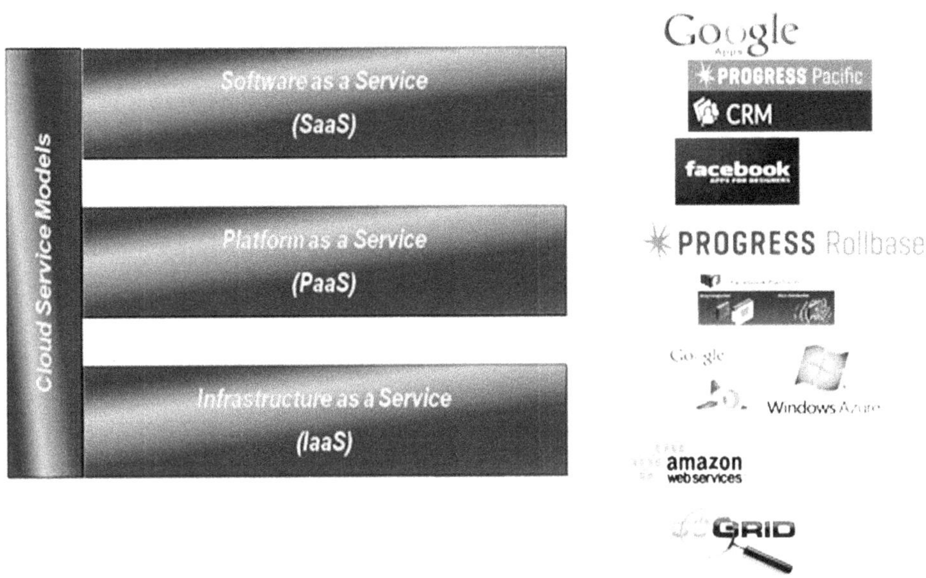

Private Cloud

Private Cloud is a Cloud Infrastructure which is dedicated and operated solely to a single or particular organization. This can be managed internally or by a third-party and hosted either internally or externally. Private clouds allow businesses to host applications in the cloud, while addressing concerns regarding data security and control, which is often lacking in a public cloud environment.

Hybrid Cloud

Hybrid cloud is a composition of two or more clouds (private, community or public) that remain distinct entities but are bound together, offering the benefits of multiple deployment models. Hybrid cloud can also mean the ability to connect collocation, managed and/or dedicated services with cloud resources.

Community Cloud

Community Cloud shares infrastructure between several organizations from a specific community with common concerns (security, compliance, jurisdiction, etc.), whether managed internally or by a third-party, and either hosted internally or externally. Community Clouds are a hybrid form of private clouds built and operated specifically for a targeted group. These communities have similar cloud requirements and their ultimate goal is to work together to achieve their business objectives. The goal of community clouds is to have participating organizations realize the benefits of a public cloud with the added level of privacy, security, and policy compliance usually associated with a private cloud. Community clouds can be either on-premise or off-premise (figure 2).

For the growing demand of businesses, Cloud providers need to maintain large data centers to offer cloud services. Each datacenter contain thousands of servers, cooling systems and high speed networking elements with distributed power supply to provide uninterrupted services to their clients. When the cloud data centers are growing, its power consumption also increases, which in turn increases the operational cost of cloud service provider and also increases carbon emission. To solve these problems cloud service providers have to consider the following.

Figure 2. Cloud computing deployment models (www.jmir.org/article/viewfile/1867/1/1)

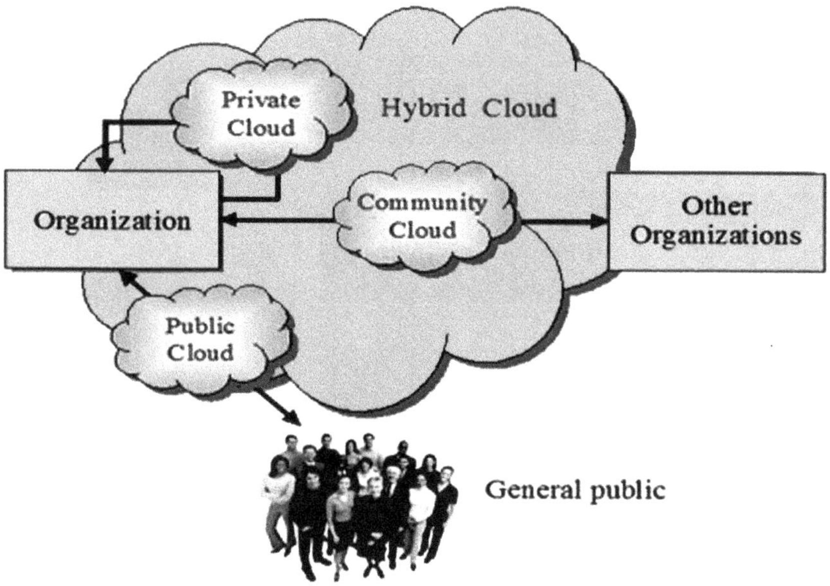

- Reduction in power consumption
- Performance improvement (Avoid SLA violation, maximize quality of service)

If cloud providers maintain the above two considerations, all corporations can show interest to adapt cloud technology. Now many researches are going on to reduce operational cost of service provider without degrading the performance. The objective of this chapter is to aware the researchers about the importance of energy management at cloud data centers. In view of this, rest of the chapter includes description about the reasons of considering energy efficiency in Background section, available techniques for reduction in energy consumption in related work section, and future research issues in Conclusion section.

BACKGROUND

Energy Consumption Statistics

Data centers consume 10 to 100 times more electricity than a city can consume. In 2007 Environmental Protection Agency (EPA) of USA reported that data centers in U.S consumed 1.5% of all power consumed in the U.S. which is equivalent to the cost $4.5 billion (United States Environmental Protection Agency, 2007). Gartner estimated that the environmental impact of these data centers was about 2% of the total global CO2 emissions (Gartner, Apr 2007). J Koomey reported that large-scale data centers consumed about 1.3% of all electricity use of the world, and 2% of all electricity usage for the United States in the year of 2010 (Jonathan Koomey, *August* 2010, 2011). According to (Datacenter Knowledge, 2011) Google's datacenter electricity use is about 0.01% of total worldwide electricity use and less than 1 percent of global datacenter electricity use in 2010. I.e. Google alone used about 2.26 million megawatt hours of electricity to run its data centers and generates a carbon footprint of 1.46 million metric tons of carbon dioxide. Face book's data centers consumed about 678m kWh of energy in 2012 – up from 509m kWh they consumed the year before (Datacenter dynamics, 2012). A typical datacenter can consume 10 Megawatt of power to operate 1000 racks. According to a report published by the European Union (Baer, Dec 2008), a decrease in emission volume of 15-30% is required before the year 2020 to keep the global temperature increase below 2°C. So the cloud vendors have started implementing new techniques to reduce the impact of increase in CO2 emission.

In fact, the cloud service providers like Google, Microsoft, Apple and Facebook are paying billions of dollars for just electricity usage every year, and 30-50 percent of operational expenses bill in data centers come from electricity. So, CSPs are required to pay more attention on maintaining data centers in terms of performance and energy usage. As per the energy consumption statistics at data centers, one-third of the total IT energy is consumed by communication links, switching, and aggregation elements, while the remaining two-thirds are allocated to computing servers. Other systems contributing to the datacenter energy consumption are cooling and power distribution systems that account for 45% and 15% of total energy consumption, respectively (High Performance Computer Architecture, 2003). L. A.Barroso and U.Holzle. have collected data from 5000 production servers and showed that servers operate only at 10-50% of their full capacity most of the time (L. A. Barroso and U. Holzle, 2007). Energy efficiency at data centers can be measured by Power Usage Effectiveness (PUE) metric. Where

PUE= IT power / Total building power

Data centers PUE metric has been reducing over the years. In 2003, the PUE metric for a typical datacenter was estimated to be about 2.6 (S. Ghemawat, H. Gobioff, and S.-T. Leung, 2003). In 2010, Koomey estimated that the average PUE was between 1.83 and 1.92 (Jonathan Koomey, *August* 2010, 2011). Some of the recent data centers built by Google, Facebook, and Microsoft have pushed PUEs under 1.2 or even 1.1 (S. Higginbotham, 2012), (T. Seubert, 2012). A datacenter be greenery when it maintains PUE with minimum 2. Due to that reason data centers are using many techniques to improve their PUE value, because USA EPA's rating will be given to the data centers based on this PUE. CSPs are reducing their non IT infrastructure (cooling and air-conditioning equipment) electrical energy consumption by moving their latest and largest data centers to cold places. Recently CSPs have designed a structure to maintain evaporative cooling, which keeps the datacenter cool by spraying water into incoming air (table 1).

Energy Inefficiency in Data Centers

In general, the power to the data centers is supplied from Power Grid and Renewable energy sources. As shown in the figure 3 power in the datacenter distribute to IT and Non-IT infrastructure. UPS is the storage element, which helps to give uninterrupted power supply to the IT equipment such as Servers, Storage devices and Network equipment (figure 3).

Fans and cooling systems are considered to be Non-IT equipment. Data centers can maintain advanced cooling systems such as Computer Room Air Conditioning (CRAC) units and air handling units with demand-driven, variable frequency drive (VFD) fans within heat exchangers to cool the returned hot air. VFDs are also available in

Table 1. Yearly statistics of energy consumption

Year	Data Centers Energy Consumption	Cost	CO2 Emission
2006-2008	1.5%	$4.5 billion	116.2 million metric tons
2009-10	between 1.7% and 2.2% of total electricity use in the U.S.	$11.5 billion	
2011-12		$7.4 billion	
2013	91 billion kilowatt-hours	$9.0 billion	97 Million Metric Tons

Figure 3. Typical Energy consumption architecture at data centers

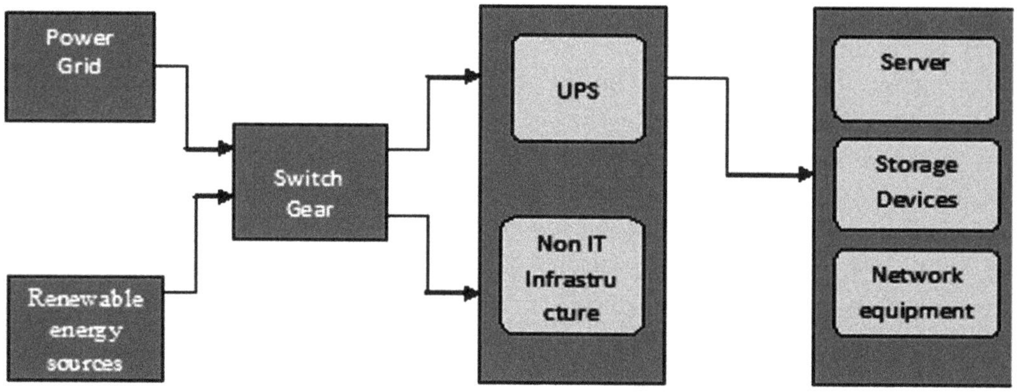

chilled water pumps, chillers, and cooling tower fans. By raising the outlet temperature of the CRAC units or by reducing the speed of the fans Hot spots can be avoided at the data centers.

At the traditional data centers, hundreds to thousands of servers are running to provide services to clients. According to the survey, most of the electricity is consuming by IT infrastructure, where servers are major elements which consumes high power than other elements. So inefficient servers results energy inefficiency at data centers, which in turn increases the operational cost. Data centers are inefficient due to the following reasons.

1. Servers' power consumption is not proportional to their load. Servers consume 80% of the peak power even at 20% utilization.
2. Datacenter is provisioning more servers to handle peak work load. It occurs infrequently. It results in underutilized server hardware.
3. Energy-Inefficient Legacy Server Hardware.
4. If the servers at data centers provide more performance than what is required for the requested application then it leads to poor power management.
5. Most UPS modules in a datacenter operate at 10%–40% of their full load capacity only. Unfortunately, at these low load levels, the UPS conversion efficiency (ac-SC-ac) is quite low.
6. Energy Cost of Cooling and Air Conditioning Units.

Techniques

As the demand for the cloud technology is increasing, the Energy management at the data centers is the major task to cloud service providers for providing services as mentioned in the SLA and to maintain Quality of Service (QoS). The optimization of Energy consumption can be achieved by changing the hardware design or the server functionality by using algorithms. Techniques for reducing the energy consumption at data centers are categorized and illustrated in Figure 4.

Figure 4. Hierarchy of energy consumption optimization techniques

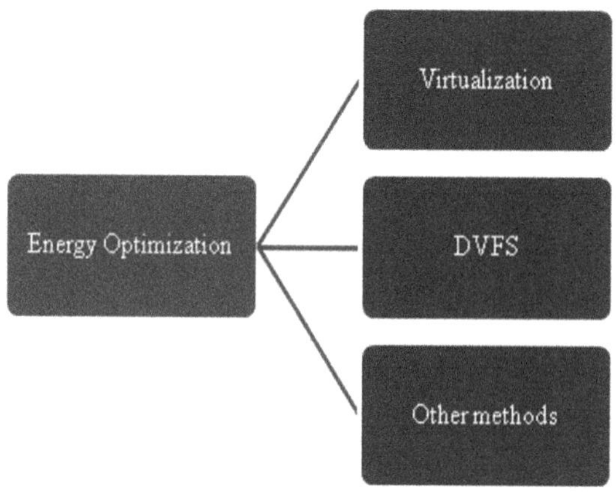

Virtualization

Virtualization means, creating of multiple virtual machine instances on a single physical machine to provide on-demand internet services to multiple tenants and it also helps to reduce the power consumption at cloud data centers. Most of the research work is carrying on Virtual Machines (VM) placement, consolidation and migration.

VM consolidation is used to place more number of VMs on minimum number of PMs and VM migration is used to move virtual machines of under loaded host to active host, which helps to switch off the idle servers. To reduce energy consumption, the minimum number of servers should be in active mode for the client applications. All the applications may not be homogeneous. So resources are virtualized and allocated to execute the applications. Two or more virtual machines are assigned to the physical server based on servers' capacity. The number of virtual machines to be placed on the physical server can be decided based on the CPU, Memory and Network bandwidth consumption of the application. In general, systems that are utilizing above 70% of CPU capacity can be treated as overloaded system. Some techniques are using upper and lower threshold values to decide whether the system is overloaded or under loaded. If the system is overloaded or under loaded then its virtual machines would be placed on separate active physical machine which has the sufficient capacity to accommodate it and switch off the idle physical server. The fixed threshold values cannot be suitable for heterogeneous applications.

As mentioned in Table 2, there are many techniques that utilize the virtualization to re-

Table 2. Different virtualization techniques

Reference	Primary Objective	Secondary Objective	Problem Formulation	Methodology
Beloglazov, A., & Buyya, R., 2010	Minimize power consumption	To maintain the level of SLA violation in the system at as low as 1%	Auto-adjustment of the utilization thresholds based on a statistical analysis of the historical data collected during the lifetime of VMs.	Dynamic Threshold (DT), MFBD
Beloglazov, A., Abawajy, J., & Buyya, R., 2012	To reduce operational costs	To provide required QoS	Decentralized architecture of the energy aware resource management system for Cloud data centers and introduced VM selection policies	System architecture
Takeda, S., & Takemura, T., 2010	Power Optimization	To avoid system performance degradation due to live migration	Unique server ranks are used to remap the VMS	First-Fit decreasing (FFD) Alg. Two threshold values R High, RLow
Kim, K. H., Beloglazov, A., & Buyya, R., 2011	To minimize power consumption.	–	It performs VM provisioning based on the User application request requirements and its power consumption.	Dynamic Voltage Frequency Scaling (DVFS) schemes.
Lee, E. K., Viswanathan, H., & Pompili, D., Dec 2012	To minimize power consumption	To maximize computing resource utilization	proposed a proactive cross-layer approach to autonomic datacenter management	proactive cross-layer approach
Beloglazov, A., & Buyya, R., 2013	To maximize the mean time between inter migration	QoS goal based	VM consolidation by maximizing the mean time between the migrations with the help of host overload detection	Optimal offline algorithm (OPT), Markov Chain model
LI Hongyou1, W. J., 2013	To consolidate heterogeneous workloads	To reduce power consumption		ESWCT and ELMWCT

duce power consumption at data centers. One of the techniques for heterogeneous applications is (Beloglazov, A., & Buyya, R, 2010) a dynamic consolidation of Virtual Machines (VMs) based on adaptive utilization thresholds for energy optimization. The objective of the approach is to maintain the level of SLA violation in the system at as low as 1%. In this method VMs have been selected for migration and then place them on the appropriate hosts that have enough resources to accommodate the VM. VMs can be selected for migration with the help of dynamic upper and lower utilization thresholds. These threshold values can be set by getting the data of VMs CPU utilization and sum of all VMs utilization at host during some time period. Here lower utilization threshold is same for all the hosts to determine the hosts that have lower utilization. With the modification of the Best Fit Decreasing Algorithm (MBFD) it sort all the VMs in decreasing order based on its current CPU utilizations and allocate each VM to a host that provides the least increase of the power consumption caused by the allocation. For this purpose select the most power-efficient ones first. It selects VMs for migration by considering three migration policies stated in (Beloglazov, A., Abawajy, J., & Buyya, R, 2012): (1) The number of VMs to be selected for migration should be as low as possible to minimize migration overhead (MM) (2) Get Highest Potential Growth (HPG) by migrating VMs that have the lowest usage of CPU (3) migrating the necessary number of VMs by picking them according to a uniformly distributed random variable. This consolidation leverages live migration of VMs. These policies are used to improve quality of service and to reduce SLA violation. To improve the quality of VM consolidation there is another algorithm (Beloglazov, A., & Buyya, R, 2013) dynamic consolidation of VM with the help of overloaded time fraction (OTF) value of each server. This OTF and CPU utilization threshold value are explicitly stated in the SLA document. The OTF value can be defined as follows:

$$OTF = \frac{\mathrm{To}}{Ta}$$

where *To* is the fraction of the time during which the host has been over loaded and *Ta* is the total active time of host. It uses Markov Chain model to detect overloaded hosts. According to this model a host is overloaded when its CPU utilization exceeds threshold value and then it initiates for dynamic VM consolidation. To improve the quality of VM consolidation it uses optimal offline algorithm (OPT) to maximizing the mean time between VM migrations. When the dynamic consolidation of VM is initiate by Markov chain model, OPT algorithm compares the host OTF with the maximum allowed OTF (M). If OTF of history $\leq$ M then it will not change the last migration time otherwise set the migration time. It also uses Multi size Sliding Window workload estimation approach for unknown non stationary workloads. These methods considered only CPU utilization to decide overloading systems.

Apart from the techniques used to detect overloaded hosts and migrating VMs of those hosts, there are other approaches (Takeda, S., & Takemura, T, 2010) that can do an autonomic VM remapping to optimize power consumption. It uses a manager server to remap VMs and to change the Physical machine (PM) power state while doing migration. It collects resource capacities about PM from the agent who keeps on monitoring the PM. He also uses unique server ranks to PMs for power saving and avoid hotspots by allotting the ranks from the side near the air conditioners. These server ranks can be changed dynamically during the operation to avoid damaging on PM machine. He used First-Fit decreasing (FFD) Algorithm for VMs remapping. Before remapping it decides resource allocation of VMs. Remapping is used to reduce the number of PMs running at the datacenter with FFD Algorithm. The heavy loaded VM will place on the highest rank PM. It uses two threshold values R_{High}, R_{Low}

for VM remapping and not to degrade system performance. The other algorithms that are used are used to consolidate heterogeneous workloads for energy optimization is (LI Hongyou1, W. J., 2013) Energy-aware Scheduling algorithm using Workload-aware Consolidation Technique (ESWCT) and the Energy-aware Live Migration algorithm using Work load-aware Consolidation Technique (ELMWCT) algorithms. ESWCT Compute every component (CPU, Memory and network bandwidth) capability of Physical Machine (PM) and get every component capability of the VM that is going to be placed. Calculate the imbalanced utilization value (IUVi*)* of sever *I* and assign the VM to that sever with the smallest value of IUV*i*. ELMWCT determine which VMs should be migrated, with node utilization threshold vector if the utilization of CPU, memory and network of a single physical machine is below the given node utilization threshold, all the VMs on the physical machine have to be migrated to other physical machines using the ESWCT algorithm. After all the VMs migrated to other machines, the physical machine has to be switched to the sleep mode to reduce power consumption.

Energy consumption at data centers can reduce by introducing framework for VM provisioning. As In (Kim, K. H., Beloglazov, A., & Buyya, R., 2011) the framework has two modules that are resource broker and VM provisioning. In this framework resource broker create a Real time Virtual machine (RT-VM) request Vi by analyzing information provided by user about real-time application. I.e. based on CPU utilization, MIPS rate and Deadline requirements for that application. User requests VMs by either Hard Real Time VM (HRT-VM) or Soft Real Time VM (SRT-VM) depending on deadline types. SRT-VM includes an additional parameter called penalty function. The penalty function indicates the diminished value of a service by executing a VM that has missed the deadline. Resource broker sends this request message for VM to the to the VM provisioning. Now VM provisioning finds the appropriate processing element that meets Vi requirements. Finally it maps the appropriate VM to the application. There are also (Lee, E. K., Viswanathan, H., & Pompili, D., Dec 2012) a proactive cross-layer approach to autonomic datacenter management, which is information centric and requires continuous processing and analysis of real-time feedback from multiple layers of abstraction. Heat-generation rate module takes information from the application layer to estimates heat generation at a server n hosting a set of VMs. Heat-extraction module uses information provided by the environment as well as hardware resource layers to find the rate of heat extraction at a server n. These two rates are used to formulate heat- imbalance model for calculating the increase or decrease in temperature and it helps to take management decisions such as VM placement, VM migration, and cooling system optimization

Dynamic Voltage and Frequency Scaling

The power aware computing idea lies in the roots of Dynamic Voltage and Frequency Scaling (DVFS) technique and Varying on Varying off (VOVO) method. DVFS is first utilized in mobile computing but nowadays many researchers utilize DVFS method to reduce energy consumption in clusters' scale along with a rapid increment of energy use. It is a technique through which the frequency of the CPU can be dynamically adjusted to reduce energy consumption to provision virtual machines in Cloud data centers. The DVFS scheme reduces the dynamic power consumption by decreasing the supplying voltage and frequency, which results in a slowdown of the CPU and increased execution time. To tradeoff between the number of servers and the frequency of each server for a given work load is a challenging problem. DVFS is a combination of Dynamic Voltage Scaling (DVS) and Dynamic Frequency Scaling (DFS). DVS is a power management technique in computer architecture, where the voltage used in a component is increased or decreased depending upon circumstances. DFS is a technique in computer architecture whereby the frequency of a microprocessor can be automatically adjusted "on the fly" either to conserve power or to reduce the amount of heat generated by the chip to save power. VOVO method is the traditional method preparing the number of servers to adjust a given workload in the scenario where servers are homogeneous.

Some existing methods that utilize the DVFS method to reduce power consumption at data centers are outlined in Table 3.

The primary objective of power aware computing is to improve power management and minimize power consumption using power aware ability of system devices, such as processors, disks, and communication links (Kim, K. H., Buyya, R., & Kim, J., May 2007). The two main reasons for need of power aware computing in cluster systems are operational cost and system reliability. Thus, efficient power management of cluster systems becomes important issue of data centers not only for reducing their operational cost but also for system reliability. A Dynamic Voltage and Frequency Scaling (DVFS) technique is introduced to make the CPUs working at proper frequencies (Kim, K. H., Beloglazov, A., & Buyya, R., Nov 2009). Another real- time service is proposed as a virtual machine request and to provision virtual machines of data centers using DVFS schemes. Power consumption at data centers is proportional to CPU frequency and voltage levels. By reducing the CPU frequencies, we can reduce the power consumption but it reduces the server's capacity and leads to increase in total number of servers. Therefore the server's capacity should be maintained by maintaining proper frequencies at each server. There are 2 algorithms such as Power Best Fit (PBF) and Load Balancing (LB) to solve this problem (Huai, W., Qian, Z., Li, X., & Lu, S., July 2013).

Other Methods

Methods that use neither DVFS nor Virtualization are described here. As specified in the Table.4 some of the techniques can reduce power consumption and electricity cost by reducing the number of active servers. Some other techniques use a frame work to offer cloud services based on

Table 3. Different methods using DVFS

Reference	Primary Objective	Secondary Objective	Problem Formulation	Methodology
Kim, K. H., Buyya, R., & Kim, J., May 2007	To minimize power consumption	To meet the deadlines specified by application users.	Providing the DVS scheduling algorithms for both time-shared and space-shared resource sharing policies.	DVS scheduling algorithms for bag-of-tasks applications with deadline constraints based on two different resource sharing policies: one for spaces sharing and the other for time-sharing is proposed.
Kim, K. H., Beloglazov, A., & Buyya, R., Nov 2009	To provide real-time Cloud service framework	To reduce power consumption	By providing a real time cloud service frame work	Investigate based on DVFS schemes.
Huai, W., Qian, Z., Li, X., & Lu, S., July 2013	To prove that the problem of dynamic resource allocation based on DVFS with the target of minimizing energy consumption is NP-Hard.	To reduce power consumption caused by increase in number of servers	The dynamic resource allocation is NP- Hard	Proposing two algorithms and comparing the results with FFD algorithm.

client request. All these techniques main aim is to reduce providers operational cost and to reduce CO2 emission.

A technique (Yamini, B & Selvi D. V, Nov 2010) is used for reducing the energy consumption by reducing number of servers that are connected to few nodes. They concerned about the reduction of number of servers that is connecting more number of nodes using clique star cover number theorem. In this clique star cover technique a server can be accomplished at a place where maximum number of nodes interconnected. According to clique star cover number minimum number of servers can connect maximum number of nodes. In (Kliazovich, D., Bouvry, P., & Khan, S. U., 2013) servers can be selected for the user requests using Data centers Energy-efficient Network – aware Scheduling (DENS) algorithm. It is a energy efficient and network aware scheduling algorithm to select servers for requested job. DENS methodology minimizes the total energy consumption of a datacenter by selecting the best-fit servers for job execution based on the load level and communication potential of datacenter components. DENS Algorithm computes DENS metric as a weighted sum of server-level, rack-level and module-level functions. Server-level function combines server load and its communication potential. Here rack load is sum of servers load and module load is sum of rack loads. Finally DENS selects a server having highest DENS metric for job processing. DENS approach can receive and analyze a run-time feedback from the datacenter switches and links as well as take decisions and actions based on the network feedback.

A strategy (Ghamkhari, M., & Mohsenian-Rad, H., 2013) can be used for profit maximization at data centers. It reduces power consumption by proportionate the Number of switched on servers at datacenter to the arrival of service requests. To maximize the profit they concentrate on rate of service requests arrival, and processing rate of requests at servers. First the running time of datacenter divide into sequence of time intervals. Based on the service requests, servers can update at the beginning of every time slot. At each time interval, if Service requests are processed within the SLA deadline cloud service providers can get revenue. Otherwise service provider pays the penalty. Sum of revenue of all time intervals is the total revenue of the datacenter. To process a new service request it must process all service requests that are already in queue (Q). I.e. it takes Q/μ seconds until all existing requests are removed from the queue. So it models the SLA-deadline (D) by a *finite-size queue* with the length μD. According to this strategy it handles a service request if $Q/\mu \leq D$. If $Q/\mu > D$, then the service request does not handle and it is dropped. Here μ is the datacenter's service rate. This strategy can make use of both power grid and renewable energy sources (table 4).

Different cloud architectures can also be used to maximize the profit. Architecture (Garg, S. K., Yeo, C. S., & Buyya, R., 2011) includes Green Offer Directory and carbon Emission Directory. Green Offer Directory contains offered cloud green services registered by cloud service providers and it can be accessed by green broker. The Carbon Emission Directory maintains data related to the energy efficiency of Cloud services. Green broker gets the current status of energy parameters of Cloud services from Carbon Emission Directory. It performs the operation layer by layer. First layer analyzes the service request and their QoS requirements. Its second layer calculates the cost and carbon footprint of leasing particular Cloud services based on information about various Cloud offerings and current CO2 emission factors obtained from Green Offer Directory and Carbon Emission Directory respectively. With these calculations, Green Policies make the decisions of leasing Cloud services. If no exact match is found for a request, alternate `Green Offers' are suggested to users by Cloud Request Services. Architecture (Tu, C. Y., Kuo, W. C., Teng, W. H., Wang, Y. T., & Shiau, S., Sep 2010) is a power-aware cloud architecture based on DRBL (Diskless Remote

Table 4. Other methods for minimizing energy consumption

References	Technique	Primary Objective	Secondary Objective	Problem Formulation	Methodology
Yamini, B., & Selvi, D. V., Nov 2010	Minimizing number of servers	Reducing energy consumption	–	Server can be placed where maximum number of nodes connected.	Clique star cover number technique
Kliazovich, D., Bouvry, P., & Khan, S. U. 2013	Network-aware scheduling for server selection	To minimize the amount of computing servers	To avoid hotspots in the datacenter network	Selects servers based on its load and communication potential	Equal CostMulti-Path (ECMP) routing,Top-of-Rack (ToR) switches
Garg, S. K., Yeo, C. S., & Buyya, R., 2011	Cloud service offerings	To reduce CO2 emission	To reduce operational cost	Green broker has three layers of analysis to provide services to cloud user.	Carbon Efficient Green Policy (CEGP)
Guo, Y., & Fang, Y., 2013	Energy Storage	To reduce electricity cost	Guaranteeing the QoS	Charge the battery when the electricity price is low while discharging it when the electricity price is high	Lyapunov optimization technique
Ghamkhari, M., & Mohsenian-Rad, H., 2013	Reducing number of switched on servers	To maximize profit,	Optimizing energy consumption	Service requests are processed within the time interval (the SLA dead line) to get revenue.	Interior point method (IPM)
Tu, C. Y., Kuo, W. C., Teng, W. H., Wang, Y. T., & Shiau, S., Sep 2010	Power-aware Cloud architecture	To minimize power consumption	–	Introduced a low-cost smart metering system based on open hardware Arduino board	Cloud architecture based on DRBL (Diskless Remote Boot in Linux), cpufreqd and xenpm

Boot in Linux), cpufreqd and xenpm. They also introduced a low-cost smart metering system based on open hardware Arduino board. It composes with existing technique, such as Dynamic Voltage Frequency Scaling (DVFS), ACPI, diskless design and RAM disk storage.

A different method (Guo, Y., & Fang, Y., 2013) can be used to reduce power consumption by joint load balancing, server configuration, and battery management scheme for multiple distributed data centers to reduce electricity cost under wholesale electricity markets with the utilization of energy storage capability. It modifies the Lyapunov optimization technique for determining the traffic distribution, number of servers at data centers (datacenter sizing), services provided by servers (service rate), and charging/discharging decisions. Usually multiple data centers can locate at different places and each datacenter has its own UPS system. Lyapunov optimization algorithm distributes the heavy work load to the data centers, where the electricity prices are low. At data centers power charging (P_i (t)) when electricity price is low while discharging when electricity price is high. It is an online algorithm.

Mathematical Techniques

There are a few Mathematical techniques which are commonly used in power optimization. They are as follows:

1. Markov Chain Model
2. Lyapunov optimization
3. Graph Theory

Markov Chain Model

A Markov chain model is a mathematical system which undergoes transitions from one state to another state on a state space in a random process and is usually characterized as memory less, in which the next state depends only on the current state and not on the sequence of events that preceded it. This specific kind of "memorylessness" is called the Markov property. This model is named after Andrey Markov, and is also known as Discrete-Time Markov Chain (DTMC). This model has many applications as statistical models of real-world processes. Markov chain refers to the sequence of random variables such a process moves through, with the Markov property defining serial dependence only between adjacent periods as in a "chain". Thus it can be used for describing systems that follow a chain of linked events, where what happens next depends only on the current state of the system. The changes of states in this process are called as transitions and the probabilities associated with various changes are known as transition probabilities. The process is characterized by a state space, which is a transition matrix describing the probabilities of particular transitions, and an initial state across the state space. Moreover we consider all the possible states and transitions have been included in the definition of the process by which there is always a next state, and the process does not get terminated.

A discrete-time random process involves a system which is in a certain state at each step, with the state changing randomly between steps which are often considered as moments in time, but they can equally be well referred to a physical distance or any other discrete measurement. Properly these steps are integers or natural numbers and the random process is a mapping of these two states. The Markov property states that the conditional probability distribution for the system at the next step (and in fact at all future steps) depends only on the current state of the system, and not additionally on the state of the system at previous steps.

Lyapunov Optimization

Lyapunov optimization refers to the use of a Lyapunov function to optimally control a dynamic system. Lyapunov functions are used extensively in control theory to ensure different forms of system stability. The state of a system at a particular time is often described by a multi-dimensional vector. A Lyapunov function is a nonnegative scalar measure of this multi-dimensional state. Typically, the function is defined to grow large when the system moves towards undesirable states. System stability is achieved by taking control actions that make the Lyapunov function drift in the negative direction towards zero. Lyapunov optimization used when we need to minimize the time average of a cost which depends only on the inputs (controlled or stochastic) and not the state, while keeping the system stable.

Graph Theory

Graph theory has a major role in solving computer applications. In the concept of energy optimization at cloud data centers, graph theory concepts can use to reduce number of active physical servers. In (Yamini, B., & Selvi, D. V, Nov 2010) the placement of number of servers can be decided by Clique star technique. A subset of vertex set $C \subseteq V$ of an undirected graph G (V, E) is called clique. Every two vertices in the clique are connected, that is, there is an edge between every two vertices of clique. Some times clique graph is refers to as a subgraph. A clique of largest possible size in a given graph is called maximum clique. The vertices in the maximum clique are called clique number ω (G) of graph G (figure 5). At cloud data centers servers can be placed where more number of node are connected. It uses to reduce power consumption at data centers.

The graph shown in Figure 5 has 23 1-vertex cliques (its vertices), 42 2-vertex cliques (its edges), 19 3-vertex cliques (the light and dark blue triangles), and 2 4-vertex cliques (dark blue areas). The six edges not associated with any triangle and the 11 light blue triangles form maximal cliques. The two dark blue 4-cliques are both maximum and maximal, and the clique number of the graph is 4.

FUTURE RESEARCH DIRECTIONS

With increasing adoption of Cloud by more and more organizations, we need to have efficient energy optimization approaches which do not violate SLA and should maintain QoS. There are some hardware based approaches and software based techniques for energy management in Cloud which uses Mathematical models, Graph theory and some Scheduling algorithms. Although the techniques were used for reducing energy consumption by minimizing the active servers or switch off the idle servers, the time taken by the physical machine to switch on when a new request arrive is considerable for completing the user request with in the time. Most of the above techniques are not considering time as a key factor to provide cloud services efficiently.

Recently some research works have included Game Theory for the optimization of energy consumption at cloud data centers. Such decision theory techniques need to be applied for better efficiency. Game Theory is used heavily for modeling problems involving multiple rational agents. It helps in modeling problems where mutually-conflicting interested parties need to cooperate for many things. Some of the techniques that has been using Game Theory are (Khani, H., Yazdani, N., & Mohammadi, S., Nov 2012),

1. Collocation game to share the cloud infrastructure expenses.
2. Nash Equilibrium game to model competing situation between cloud users to get resources of one cloud provider. This game determines the amount of resources that must be allocated to various users. However,

Figure 5. Clique Graph (wikipedia.org/wiki/Clique_ (graph theory))

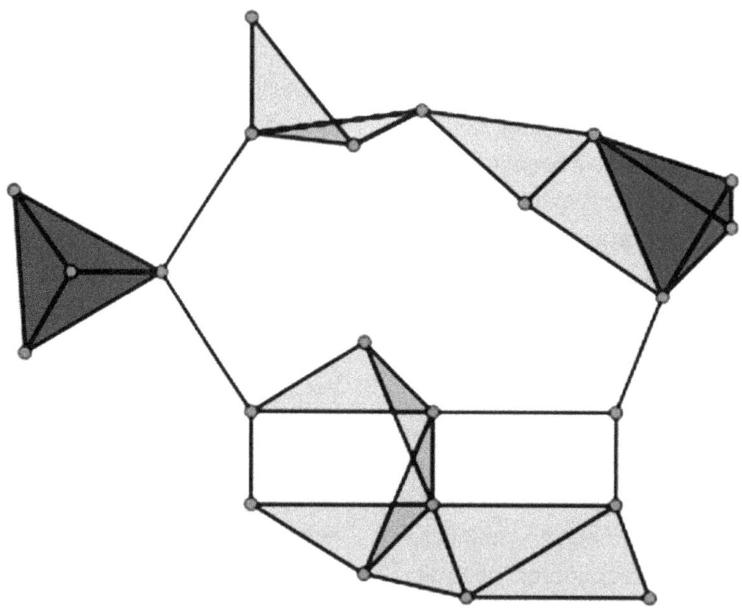

it does not determine which server must be allocated onto which virtual machine.

3. Horizontal Dynamic Cloud Federation platform has defined by a Game theory for cloud providers to dynamically exchange their resources in an open market. It follows price-based resource allocation strategy.
4. Power-Aware Game (PAG) is a distributed self-organized mechanism for minimizing power consumption and for resource management in cloud datacenter.

CONCLUSION

Cloud Service Providers are maintaining data centers – includes thousands of servers- to satisfy the needs of consumers. Power consumption at data centers is increased when number of active servers is increased. As we know, high energy consumption reduces cloud service provider's profit and it also increases CO_2 emission. So there is a lot of need to optimize energy consumption at data centers. There are different methods and algorithms to make cloud data centers as Green data centers. This chapter describes some of the mathematical approaches to tackle the problem of energy optimization in Cloud. But still it requires new approaches for better performance in terms of time, SLA and QoS. This chapter also describes some of the future research issues and some ways to tackle them.

REFERENCES

Baer, P. Exploring the 2020 global emissions mitigation gap. Retrieved from http://www.ippr.org/uploadedFiles/globalclimatenetwork/Exploring_the_Mitigation_Gap[1].pdf

Barroso, L. A., & Holzle, U. (2007). The case for energy-proportional computing. *Computer*, *40*(12), 33–37. doi:10.1109/MC.2007.443

Beloglazov, A., Abawajy, J., & Buyya, R. (2012). Energy-aware resource allocation heuristics for efficient management of data centers for cloud computing. *Future Generation Computer Systems*, *28*(5), 755–768. doi:10.1016/j.future.2011.04.017

Beloglazov, A., & Buyya, R. (2010, November). Adaptive threshold-based approach for energy-efficient consolidation of virtual machines in cloud data centers. *Proceedings of the 8th International Workshop on Middleware for Grids, Clouds and e-Science* (p. 4). ACM. doi:10.1145/1890799.1890803

Beloglazov, A., & Buyya, R. (2013). Managing overloaded hosts for dynamic consolidation of virtual machines in cloud data centers under quality of service constraints. *IEEE Transactions on* Parallel and Distributed Systems, *24*(7), 1366–1379.

Garg, S. K., Yeo, C. S., & Buyya, R. (2011). *Green cloud framework for improving carbon efficiency of clouds. Proceedings of Euro-Par 2011 Parallel Processing* (pp. 491–502). Springer Berlin Heidelberg.

Gartner Estimates ICT Industry Accounts for 2 Percent of Global CO2 Emissions. (n. d.). *Gartner*. Retrieved from http://www.gartner.com/it/page.jsp?id=503867

Ghamkhari, M., & Mohsenian-Rad, H. (2013). Energy and performance management of green data centers: A profit maximization approach. *IEEE Transactions on* Smart Grid, *4*(2), 1017–1025.

Ghemawat, S., Gobioff, H., & Leung, S.-T. (2003). The Google file system. *Proc. of the ACM Symp. Operating Syst. Principles* (pp. 29–43).

Guo, Y., & Fang, Y. (2013). Electricity cost saving strategy in data centers by using energy storage. *IEEE Transactions on* Parallel and Distributed Systems, *24*(6), 1149–1160.

Higginbotham, S. (2012, Mar.). Whose Data centers Are More Efficient? Facebook's or Google's? Retrieved from http://gigaom.com/cloud/whose-data-centers-are-more-efficient-facebooks-or-googles

Huai, W., Qian, Z., Li, X., & Lu, S. (2013, July). Towards Energy Efficient Data centers: A DVFS-Based Request Scheduling Perspective. *Proceedings of the 2013 Seventh International Conference on Innovative Mobile and Internet Services in Ubiquitous Computing IMIS* (pp. 299-306). IEEE.

Khani, H., Yazdani, N., & Mohammadi, S. (2012, November). Power-aware game for cloud computing: A distributed mechanism based on Game Theory for minimizing power consumption in cloud scale datacenter. *Proceedings of the 2012 Sixth International Symposium on Telecommunications IST* (pp. 598-601). IEEE. doi:10.1109/ISTEL.2012.6483058

Kim, K. H., Beloglazov, A., & Buyya, R. (2009, November). Power-aware provisioning of cloud resources for real-time services. *Proceedings of the 7th International Workshop on Middleware for Grids, Clouds and e-Science*. ACM. doi:10.1145/1657120.1657121

Kim, K. H., Beloglazov, A., & Buyya, R. (2011). Power-aware provisioning of virtual machines for real-time Cloud services. *Concurrency and Computation*, *23*(13), 1491–1505. doi:10.1002/cpe.1712

Kim, K. H., Buyya, R., & Kim, J. (2007, May). *Power Aware Scheduling of Bag-of-Tasks Applications with Deadline Constraints on DVS-enabled Clusters* (Vol. 7, pp. 541–548). CCGRID. doi:10.1109/CCGRID.2007.85

Kliazovich, D., Bouvry, P., & Khan, S. U. (2013). DENS: Datacenter energy-efficient network-aware scheduling. *Cluster Computing*, *16*(1), 65–75. doi:10.1007/s10586-011-0177-4

Koomey, J. (2011). Growth in datacenter electricity use 2005 to 2010. Oakland, CA: Analytics Press.

Kuo, A. M.-H. (2011). Opportunities and Challenges of Cloud Computing to Improve Health Care Services. *Journal of Medical Internet Research*, *13*(3), e67. Retrieved from www.jmir.org/article/viewfile/1867/1/1 doi:10.2196/jmir.1867 PMID:21937354

Lee, E. K., Viswanathan, H., & Pompili, D. (2012, December). Vmap: Proactive thermal-aware virtual machine allocation in HPC cloud data centers. *Proceedings of the 19th International Conference on High Performance Computing HiPC '12* (pp. 1-10). IEEE.

Li Hongyou, W.J. (2013). *Energy-Aware Scheduling Scheme Using Workload-Aware Consolidation Technique in Cloud Datacentres*. ZHANG Liangjie.

Miller, R. (2011). Report: Google Uses About 900,000 Servers. *Data Center Knowledge*. Retrieved from: http://www.datacenterknowledge.com/archives/2011/08/01/report-google-uses-about-900000-servers

Seubert, T. (2012, Feb. 27). Microsoft Builds New Datacenter in Dublin. Retrieved from http://facilitygateway.com/news/?p=1937

Shang, L., Peh, L.-S., Jha, K.N. (n. d.) Dynamic voltage scaling with links for power optimization of interconnection networks.

Sverdlik, Y. (2013). Facebook data centers' energy use up in 2012. *Data Center Dynamics*. Retrieved from http://www.datacenterdynamics.com/focus/archive/2013/06/facebook-data-centers-energy-use-2012

Takeda, S., & Takemura, T. (2010). A rank-based vm consolidation method for power saving in data centers. *IPSJ Online Transactions*, *3*, 88–96. doi:10.2197/ipsjtrans.3.88

Tu, C. Y., Kuo, W. C., Teng, W. H., Wang, Y. T., & Shiau, S. (2010, September). Power-aware cloud architecture with smart metering. *Proceedings of the 2010 39th International Conference on Parallel Processing Workshops ICPPW* (pp. 497-503). IEEE. doi:10.1109/ICPPW.2010.73

Report to congress on server and data center energy efficiency. (2007). United States Environmental Protection Agency. Retrieved from http://hightech.lbl.gov/documents/datacenters/epa-datacenters.pdf

Wikipedia. Clique (graph theory). Retrieved from: https://en.wikipedia.org/wiki/Clique_(graph_theory)

Yamini, B., & Selvi, D. V. (2010, November). Cloud virtualization: A potential way to reduce global warming. Proceedings of the 2010 conference on Recent Advances in Space Technology Services and Climate Change RSTSCC (pp. 55-57). IEEE. doi:10.1109/RSTSCC.2010.5712798

KEY TERMS AND DEFINITIONS

Cloud Computing: Cloud computing is an On-demand self service mechanism through which consumer can share pool of resources over the internet.

Cloud Service Providers: Vendors who provide various cloud services to consumers on pay-as -per –usage basis.

CO2 Emission: High power consumption by the infrastructure leads to substantial carbon dioxide (*CO*2) emissions contributing to the greenhouse effect.

Data Center Power Usage Effectiveness (PUE): Data center efficiency metric which is defined as the ratio of the total datacenter power consumption divided by the power consumption of the IT equipment.

Data Centers: Containing thousands of servers to store and process the enterprises Data.

Energy Optimization: Method to reduce energy consumption at data centers which leads to optimize the operational cost of service providers.

Global Warming: The increase in the overall temperature of the earth's atmosphere is attributed to the greenhouse effect which is caused by increased levels of CO2, CFCs and other pollutants.

High Performance Computing: High-performance computing (HPC) is the use of super computers and parallel processing techniques for solving complex computational problems.

Quality of Service: Providing cloud services to the clients without violating the conditions provided in the SLA document.

Service Level Agreement: A document which defines the relationship between service provider and client is called a service level agreement.

Chapter 15
Fuzzy Logic Based Approach for Power System Fault Section Analysis

Neeti Dugaya
Sagar Institute of Research, Technology & Science, India

Smita Shandilya
Sagar Institute of Research, Technology & Science, India

ABSTRACT

In this chapter, a fuzzy expert system is developed to assist the operators in fault detection. It requires much less memory to store the database (power system topology and the post fault status of circuit breakers and protective relays). The fuzzy expert system identifies two basic network section sets, $S_{healthy}$ *for the healthy sub network and* S_{island} *for the fault islands, using the post fault status of circuit breakers and relays. It then calculates membership function for each possible fault section. The objective of this calculation is to determine the likelihood of each candidate fault section as the actual fault section. Moreover membership functions provide a convenient means of ranking among possible (or candidate) fault sections, and are the most important factors in decision making. During decision making, the most possible fault section is determined by maximum selection method. In this method most possible fault section is the one which is having highest membership grade. MATLAB code for the proposed scheme is developed and the results obtained in four cases for a power- system network.*

1. INTRODUCTION

Power distribution reliability has a high impact on the service restoration cost when a fault occurs and on customer satisfaction and is therefore an important topic in an electric power industry. To reduce the outage time and to enhance service reliability, it is essential for operators to locate the fault sections as quickly as possible. Fault section location aims to identify fault components in a power system using the post fault status of protective relays and circuit breakers and diagnose the cause of a fault that has occurred. Accurately identifying a fault location and cause can significantly reduce the restoration cost. After a fault has been detected and located, it becomes easy

DOI: 10.4018/978-1-4666-9911-3.ch015

to restore the service. Today's in highly complex power systems Supervisory Control and Data Acquisition (SCADA) systems are in use for efficient operation of power transmission and distribution networks.

But, many false alarms are triggered in a SCADA system due to inaccurate operation or maloperations of the system's relays and circuit breakers in the post fault state (Srinivasan, Ruey, Young, Albert & Chwee, 2000). When a breaker fails to operate, the fault is then removed by back-up breakers. In such cases, the outage range is very large and it is difficult for the dispatchers to judge where the fault is located. However, the relay time settings in power transmission systems are much smaller. Consequently, difficulties may arise in setting the time delays with sufficient accuracy and failure would have serious implications. Moreover, occasionally multiple faults may take place, with many breakers being tripped within a short time. In these circumstances, so many alarm messages pour into the dispatch centre that it is impossible for the dispatchers to analyze the situation satisfactorily (Zhu, Yang, Hogg, Zhang, & Gao, 1994).

2. LITERATURE REVIEW

Over the last ten decades a larger number of papers have published in the area of fault section identification in power network. Here in this chapter, an attempt has been made review some of major contribution in this area.

(Zhu, Yang, Hogg, Zhang & Gao 1994) have described the development of an expert system for estimating fault section in control cEnter based on protective system simulation. They have presented as a principle for stable power supply, any fault sections in a power system must be disconnected quickly, to meet this principle, utilities are making effort to educate and train their operators for recovering power supply in proper manners. System operation and protective relay actions, but also accurate and extremely quick judgment based on the knowledge. Because of the importance of system restoration, development of an expert system efficiently supporting operators is eagerly awaited. Therefore, fault section judgment has been studied a some of the major themes in the field of artificial intelligence application to electric power systems .In many studies in this field, knowledge about cause and effect between the fault section and relay action is represented. There are many devices in power system and therefore, in order to describe a power system and its protective system accurately, a great numbers of consistent rules are required. This makes acquisition and verification of the knowledge difficult.

To overcome this problem, they proposed a method for estimating fault section including protective system simulation; the authors have developed a practical expert system which infers the protective coordination of relays on the bas is of the protection zone and time setting of the relays.

They proposed expert system features

- Relay operation functions are represented by protection zone and time setting with numerical figures, which service as database provided outside the expert system.
- The explanation facility of many other expert systems displays only the inference chain of rules applied, requiring the operators to comprehend the rules and their knowledge representation structure totally.
- Separating teleological knowledge and non-teleological knowledge
- Judgment for "single", "multiple", "phase to phase" and "phase to earth" as well as "malfunction" of relays, "failure to operate" of relays and circuit breakers.

(Zhu, Yang, Hogg, Zhang, & Gao 1994) have explained about An expert system for substation fault diagnosis and alarm processing developed by the Electric Power Systems Laboratory of the National Technical University of Athens. And they have implemented on a personal computer. They expanded it easily by adding misinformation to its knowledge base or by changing the existing information. The basic idea behind this expert system is to provide a non-expert with the necessary information and interaction in order to make a fault diagnosis through the substation alarms in a very friendly environment.

B. F. Wollenberg, & T. Sakaguchi (1987) have presented Intelligent operation of distribution network using control system operating in conjunction with a SCADA system with minimum supervision and interaction with a human operator. The application of model-based expert system results in an intelligent power distribution system with learning, self-organizing and diagnostic capabilities. The work deviates from previous approaches in two important aspects: the use of model-based reasoning allows many different Application knowledge bases (e.g. fault diagnosis, restoration, maintenance planning, etc.) to operate on the common model of the distribution network and it introduces for the first time the concept of intelligence in a power distribution system performance. The expert system has been tested extensively for operation with a real-time 22kV SCADA system covering a large area of 180 substations. The expert system gives accurate fault analysis for all the fault cases and proposes the best restoration plans in less than 15 seconds. In addition, it advises the control cEnter operators in cases of emergency & helps them in maintenance planning and can be used as a training simulator in off-line mode.

C.S.Chang, J.M.Chen, D.Srinivasan, F.S.Wen, A.C.Liew (1997) have described about using fuzzy sets to model the uncertainty in the fault location process of distribution networks. In this the heuristic knowledge of the control cEnter operators can be combined with the information obtained from the network database and SCADA system. However, the nature of the heuristic knowledge is inexact and uncertain. Also the information obtained from the remote control system contains uncertainty and may be incorrect, conflicting or inadequate. This paper proposes a method based on fuzzy set theory to deal with the uncertainty involved in the process of locating faults in distribution networks. The method is implemented in a prototype version of the distribution network operation support system.

Chang.C.S., Chen.J.M, Srinivasan.D. Wen.F.S. Liew.A.C. (1996) have proposed a novel fault section diagnosis method for power systems based on the fuzzy relation theory. Especially, the proposed method utilizes information on operating time sequences of the actuated relays and the tripped circuit breakers. The method proposed in this paper is intended to develop an expert system for fault-sectioned angels of power systems focusing on utilizing operation sequences of protective devices. In the case studies for real power system, it provided less fault section candidates than without considering the time sequences. As a result, the method reduced the uncertainty and ambiguity for diagnosing the final fault section.

L. Xu & M.-Y. Chow (2006) have presented an expert system for locating distribution system faults. They described about the advent of Artificial Intelligence (AI) in electrical power systems operation and control has been called as may be the next major enhancement for the functions of an energy management system. Among several potential applications of AI and in particular expert systems are contingency analysis, load shedding and system restoration. Contingency selection and alarm processing function using AI have already been reported. Other applications of expert system include power system planning, fault detection and transient stability analysis and more developments in this way are expected. They also described about a knowledge based system support used for data acquisition in power flow problem and an "online" diagnosis and advisor tool recently incorporated to a microcomputer based power network control cEnter simulator. Following a brief review of simulator capabilities, reasons that motivated the use of an expert system are stated together with a description of the added package and typical examples of application as well as the academic environment in which the expert system is used.

C. Fukui, & J. Kawakami (1986) have presented an expert system for diagnosis of power system fault allocation in real time (SIDUF-TR). First, the expert system structure is presented, including a description of the inference method used to determine the most probable failures places. Then the architecture for real-time operation of SIDUF-TR in a control cEnter is described. Finally, the result of the application of the system to a real disturbance is presented.

3. TYPES OF FAULTS

Faults in power systems are invariably due to short circuits. A fault is deemed to have occurred when there is a conducting path between points that are normally of different potential. The consequences of a fault may be accompanied by a noticeable change in voltage or current values, depending on the path resistance. Faults can generally be classified into balanced and unbalanced faults, as in figure 1

For a three-phase-to-ground fault, all three phases of a balanced system are shorted to ground, and only one phase of the system needs to be considered in computation. For unbalanced faults, the most common is the single-line-to-ground fault, when one phase of the system is shorted to ground. A line-to-line fault occurs if there is a short circuit between two phases, while a double-line-to-ground fault occurs when two phases are shorted to ground. The positive, negative, and zero sequence components of the power-system variables, i.e., current and voltage, are to be considered in the treatment of unbalanced faults.

4. THE PROTECTION SYSTEM

Protection systems in power networks consist mainly of relays and their associated circuit breakers. Protective relays detect faults in power systems through current and potential transformers, and subsequently activate the appropriate circuit breakers to isolate the faulty equipment from the system. They essentially act as switching mechanisms that disconnect the faulty regions quickly, so as to minimize the damage. The general philosophy of relay applications is to divide the power system into zones that can be protected adequately using fault recognition and removal, producing a disconnection of as few sections of the network as possible. The elements in a power system typically consist of generators, transformers, buses, transmission/distribution circuits and motors. Each of these elements is protected by circuit breakers, located in three different zones of protection: primary, secondary and tertiary. In the event that a fault occurs, the circuit breakers in

Figure 1. Types of faults

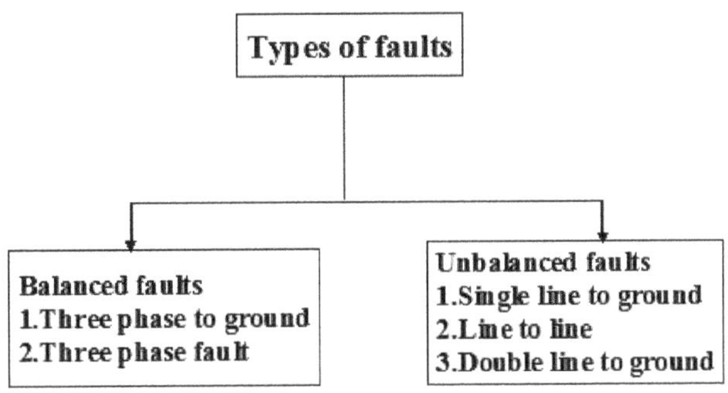

the primary zone are activated by the relays, and trip first. The circuit breakers in the secondary zone provide back-up protection to circuit breakers in the primary zone, in case they do not operate. Tertiary zone circuit breakers back-up the circuit breakers in the secondary zone.

5. FAULT DETECTION AND CIRCUIT BREAKER STATUS

Under normal conditions, the system is healthy, and the circuit breakers may be open or closed. The circuit breakers that are not activated are assumed to have a STATUS 0. When a fault occurs, the appropriate circuit breakers experience a change of state, and are activated to STATUS 1. By monitoring the status of these circuit breakers, faulty elements can be quickly detected and isolated from the system. The relationship between faults and circuit-breaker operations is the most crucial factor in fault detection. In a complex power system, a single fault may occur at a time in the power system or multiple faults can occur at different locations simultaneously. If all the relays and circuit breakers are in perfect working condition, they will be activated in the primary zone of protection. However, in many cases, malfunctions of the relays or circuit breakers in the primary zone cause circuit breakers in the secondary or even tertiary zones of protection to trip. Also, the overlapping protection zones of circuit breakers make fault identification difficult. In such scenarios, a deluge of information coming from the SCADA does not allow the power system control centre operators enough time to interpret the data and correctly identify the actual fault causing element that has tripped the circuit breakers. Moreover, many false alarms are triggered in the SCADA system due to inaccurate operation or malfunctioning of the system's relays and circuit breakers (Srinivasan, Ruey, Young, Albert & Chwee., 2000).

6. PROPOSED SCHEME

The main objective of this project is to help the power-system control cEnter operators in quick identification of all cases of fault and circuit breaker mal-operation. This automated incident-detection algorithm has the following objectives:

1. To determine faults along power sections using the states of circuit breakers and relays.
2. To have a high detection rate by accurately localizing the actual fault areas.
3. To have a low false alarm rate in misidentifying healthy areas as fault locations.

It is requisite that the average fault detection rate be as high as possible and the false-alarm rate as low as possible. Fault section estimation aims to identify fault components in a power system using the post fault status of protective relays and circuit breakers. For that a fuzzy expert system is developed which requires much less memory to store an active database. The fuzzy expert system identifies two basic network section sets, $S_{healthy}$ for the healthy sub network and S_{island} for the fault islands, using the post fault status of circuit breakers and relays. It then conducts inference to determine the most likely fault sections. The proposed fuzzy expert system structure is shown in Figure 2 (Chang, Chen, Srinivasan, Wen, & Liew, 1997). Its database contains the power system topology, and the post fault status of all circuit breakers and protective relays.

Knowledge base of the fuzzy expert system contains all the data of the protection system. The information is based on known statistics of protection performance used in the system. If this data is not available when a fault occurs, the fuzzy expert system asks the dispatcher to provide them and then saves them in the database for future use.

The fuzzy expert system is activated after circuit-breaker tripping to control each step of

Figure 2. Layout of fuzzy expert system

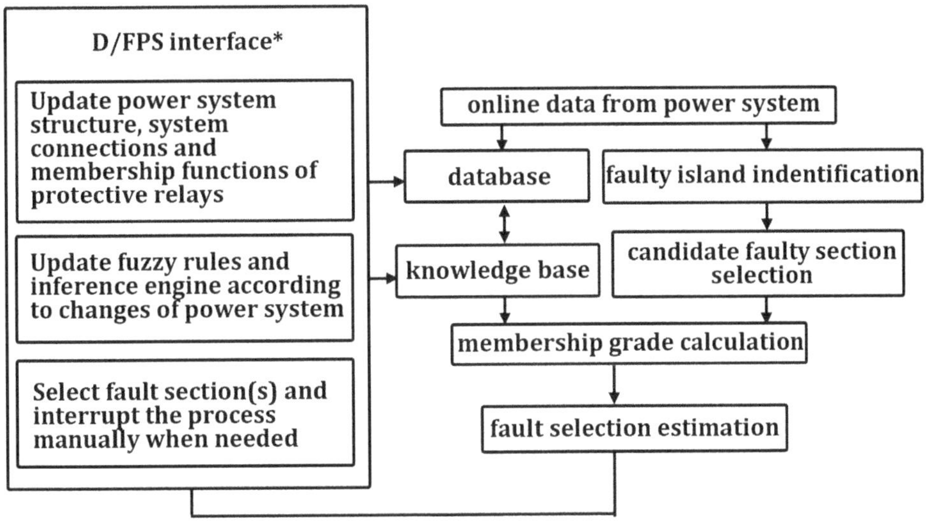

reasoning: to identify all the fault islands, to calculate membership grades for all candidate fault sections in each fault island, and to estimate the most likely fault sections.

The following three functional blocks are executed consecutively during fuzzy reasoning:

1. **Fault Island Identification and Candidate Fault Section Selection:** The fuzzy expert system keeps track of the post fault status of circuit breakers. The fault island identification algorithm (Prais 1989) & (Wen & Chang 1996) is executed to compare the prevail system configuration with the post fault system configuration, in order to identify the fault islands.
2. **Membership Grade Calculation for Candidate Fault Sections:** After the fault island identification, all the fault sections from each fault island are taken as the candidate fault sections, whose membership grades are calculated. The objective of this calculation is to determine the likelihood of each candidate fault section as the actual fault section.
3. **Fault Section Estimation:** On the list of membership grades, two parallel methods, the maximum selection method and the α-level selection method, are used to determine the most likely fault section. For convenience maximum selection method is used in this project.

Detail description of the above functional blocks are given below.

7. FAULT ISLAND IDENTIFICATION

When a fault occurs in a power system, the relays corresponding to the fault sections should trip the circuit breakers to isolate the fault sections. Thus the power system is separated into several parts named fault islands after operation of these protective relays and circuit breakers. A fuzzy expert system is developed to identify the fault islands by using the post fault status of circuit breakers and the real-time network topology determination method (Wen & Chang 1996). This efficient method is described as follows:

1. **Initializing the Network:** The fuzzy expert system identifies the power system pre fault status by using the real time network topology determination method. When a fault occurs, the power system status would be changed by operation of relays and circuit breakers.
2. **Healthy Sub Network Identification:** The next step is to identify the network topology of the healthy part of the post-fault power system by using the real time network topology determination method (Prais 1989). The healthy sub network is represented by a set $S_{healthy}$.
3. **Fault Island Identification:** By comparing the initial network topology with the healthy sub network topology, the differences between them are identified as the fault island. This sub network is called S_{island}.

8. FAULT SECTION IDENTIFICATION

When a fault occurs, the fuzzy expert system is activated shortly after opening of the first circuit breaker in a disturbance. A fault hypothesis F_i is formed as follows:

$$\begin{aligned} F_i &= F_i(CB) \cup F_i(RL) \\ &= \{C_i, \cup(C_i) \mid C_i \varepsilon S_{island}\} \\ P_{fault} &= \{F_i\} \end{aligned} \tag{1}$$

$$F_i(CB) = \{C_i, Ucb(C_i)\} \tag{2}$$

$$F_i(RL) = \{C_i, Url(C_i)\} \tag{3}$$

where C_i is one of the possible fault sections.

P_{fault} is the fuzzy set which contains all the possible fault sections and their membership functions

F_i (CB) is the fuzzy subset containing only the tripped circuit breakers.

F_i (RL) is the fuzzy subset containing only the operated relays.

$U(C_i)$ is the membership function that Ci belongs to the fuzzy set.

Ucb (C_i) is the membership function that Ci belongs to the fuzzy set by considering only the tripped circuit breakers.

Url (C_i) is the membership function that Ci belongs to the fuzzy set by considering only the operated relays.

Maloperated circuit breakers are determined after a fault hypothesis F_i is formed. The main protection should trip all the circuit breakers directly connected to possible fault section C_i to clear the fault. If some circuit breakers are not tripped, it means that these circuit breakers are jammed or maloperated.

Now a brief analysis of the above equations is given below.

8.1 Fault Section Identification Based on Circuit Breakers Information

The basic protection principles are:

1. Any fault equipment should be isolated by its own breaker.
2. If any associated relays or breakers fail to operate, the remote backup relays should trip the back-up breakers to clear the fault.
3. If both stage-1 and stage-2 breakers fail to operate, the fault is cleared by stage-3 back-up breakers.

When a fault occurs in a power system, the tripped circuit breakers are considered first. A fuzzy subset is formed by considering only the tripped circuit breakers associate with the possible fault section C_i.

F_i^k (CB) = {C_i, u^k (C_i)}

where k = 1, 2, 3 represents the first, second, third stage protection.

u^k (C_i) is the membership function to measure the extent to which C_i, belongs to fuzzy set P_{fault} by considering only the kth stage tripped circuit breakers.

The grade of membership function of a possible fault section measures its likelihood of becoming a fault section.

Analysis:

S1= total number of sections got faulted
$S_{k=}$ number of buses or lines or transformers got faulted
k=1 for buses
k=2 for lines
k=3 for transformer
n_k= total number of circuit breakers tripped
k=1 for buses
k=2 for lines
k=3 for transformers
b_k = number of circuit breakers tripped at each stage (primary, secondary, tertiary) for each section (figure 3)

where k=1, 2,3.

$$\text{sumy}_k = \sum_{j=1}^{b_k} y_1$$

y1=membership function for the circuit breaker to operate.

Membership function to measure the extent to which fault section belongs to fuzzy set by considering circuit breakers

$u^k = r_k \times (\text{sum } y_k / b_k)$ where $r_{k=}$weight coefficient for each stage

r_1=1
$r_{2=}$1-(sumy1/b_1)
r_3= (1-(sumy1/b_1)) × (1-(sumy2/b_2))

Considering all the tripped circuit breakers associated with a possible fault section **C_i**, F_i (CB) can be obtained by their union.

$$\begin{aligned} F_i(CB) &= F_i^1(CB) \cup F_i^2(CB) \cup F_i^3(CB) \\ &= \{C_i, Ucb(C_i)\} \end{aligned}$$

$$\begin{aligned} Ucb(C_i) &= (u^1) \cup (u^2) \cup (u^3) \\ &= u^1 + u^2 + u^3 - (u^1 \times u^2) - (u^2 \times u^3) \\ &\quad - (u^3 \times u^1) + (u^1 \times u^2 \times u^3) \end{aligned}$$

Figure 3. Linear fuzzy membership function for circuit breakers

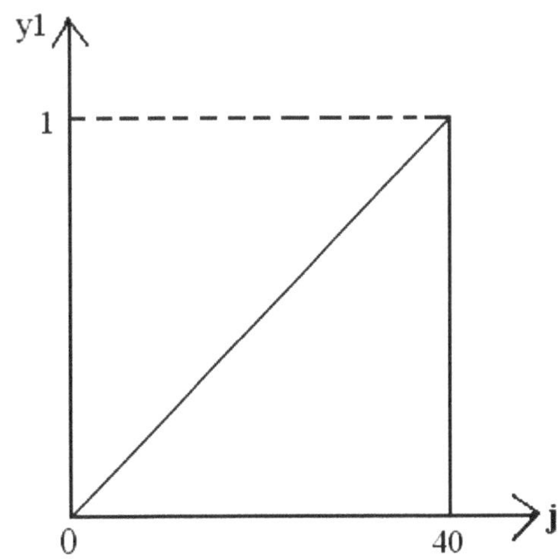

8.2 Fault Section Identification Based on Relays Information

Sometimes, circuit breakers maloperate even when the associated relays have operated correctly. Relay signals must therefore be considered. Similarly, the fuzzy subset and membership function associated with the possible fault section $\mathbf{C_i}$ are considered for protective relays as follows:

$$F_i^k (RL) = \{C_i, U^k (C_i)\}$$

where k = 1, 2, 3 represents the first, second, third stage protection.

$U^k (C_i)$ is the membership function to measure the extent to which C_i, belongs to fuzzy set P_{fault} by considering only the operated relays of the kth stage protection.

Analysis:

N_k= total number of relays tripped
k=1 for buses
k=2 for lines
k=3 for transformers
B_k=number of relays operated at each stage (primary, secondary, tertiary) for each section

where k=1, 2, 3

$$SUMY_k = \sum_{J=1}^{B_k} Y_1$$

Y1=membership function for the relay to operate.

Membership function to measure the extent to which fault section belongs to fuzzy set by considering relays

$$U^K = R_k \times (SUMY_k/B_k)$$

where R_k is weight coefficient

A fault set considering the operated protective relays is formed as below:

$$F_i (RL) = F_i^1 (RL) \cup F_i^2 (RL) \cup F_i^3 (RL)$$

$$= \{C_i, Url (C_i)\}$$

$$Url (C_i) = (U^1) \cup (U^2) \cup (U^3)$$

$$= U^1 + U^2 + U^3 - (U^1 \times U^2) - (U^2 \times U^3) - (U^3 \times U^1) + (U^1 \times U^2 \times U^3)$$

After calculating all the membership grades, the fault hypothesis is formed using eqn. 1

$$Fi = (Ficb) \cup (Firl)$$

$$U(C_i) = Ucb(C_i) + Url(C_i) - (Ucb(C_i) \times Url(C_i))$$

After calculating membership grades for all the sections, most possible fault section is determined by Maximum selection method.

Maximum selection: The most possible fault section is the one which is having highest membership grade.

9. ALGORITHM OF THE PROPOSED SCHEME AND FLOWCHART

STEP1: Identification of Fault Island.
STEP2: Identification of fault section from each fault island.
STEP3: Check whether the fault section is a bus or line or transformer.
STEP4: Get the tripped circuit breakers information for corresponding fault section.
STEP5: Get the tripped relays information for corresponding fault section.
STEP6: Calculate the membership function for each fault section based on the information of circuit breakers and relays.

STEP7: After calculating membership functions for all the fault sections; from the Maximum selection method the most likely fault section is the one which is having highest membership grade.

The operation of the algorithm is tested on IEEE 33KV distribution network system (figure 4).

10. THE TEST SYSTEM

The power-system network used as the test system is a 33 kV distribution network as shown in Figure 5 (Chang, Chen, Srinivasan, Wen, Liew, 1997). The network consists of 12 bus bars (B1-B12), eight line sections (L1-L8) and eight transformers (T1-T8). Each of the components is classified as an element and there are 28 elements altogether in this power network. These elements are protected by 84 protective relays (not shown in the figure 5) and 40 circuit breakers (CB1-CB40), with each circuit breaker providing primary, secondary and tertiary zones of protection to all the elements. Under normal operating conditions, the 40 circuit breakers may be open or closed and a change in a breaker's status is only activated by the corresponding relays (see tables 1-5).

11. CONCLUSION

This work has proposed a fuzzy expert system for fault detection in a power-system network. This system will consider only the post fault status of circuit breakers and operated relays of the fault model. So the database is relatively small compared with the full database that describes the

Table 1. Fault islands and tripped circuit breakers of case studies 1- 4

Test	Circuit Breakers Tripped	Island
1	CB4,CB5,CB7,CB9,CB12,CB27	B1,B2,L2,L4
2	CB19,CB20,CB29,CB30,CB32, CB33,CB34,CB35,CB36,CB37,CB39	B7,B8,L5,L6,L7,L8,T7,T8
3	CB4,CB5,CB6,CB7,CB8,CB9, CB10,CB11,CB12	B1,B2,L1,L2
4	CB7,CB8,CB11,CB12, CB29,CB30,CB39,CB40	L1,L2,L7,L8

Table 2. Results of Case Study 1

Test	Possible Fault Sections	Membership Grades	Circuit Breakers Tripped	Circuit Breakers Jammed	Estimated Fault Section
1	B1	0.8467	CB4,CB5,CB7,CB9, CB12,CB27	CB6,CB8,CB10	B1
	B2	0.8213	CB4,CB5,CB7,CB9, CB12,CB27	CB6,CB8,CB10	
	L2	0.7197	CB4,CB5,CB7,CB9, CB12,CB27	CB6,CB8,CB10	
	L4	0.7197	CB4,CB5,CB7,CB9, CB12,CB27	CB6,CB8,CB10	

Figure 4. System flowchart

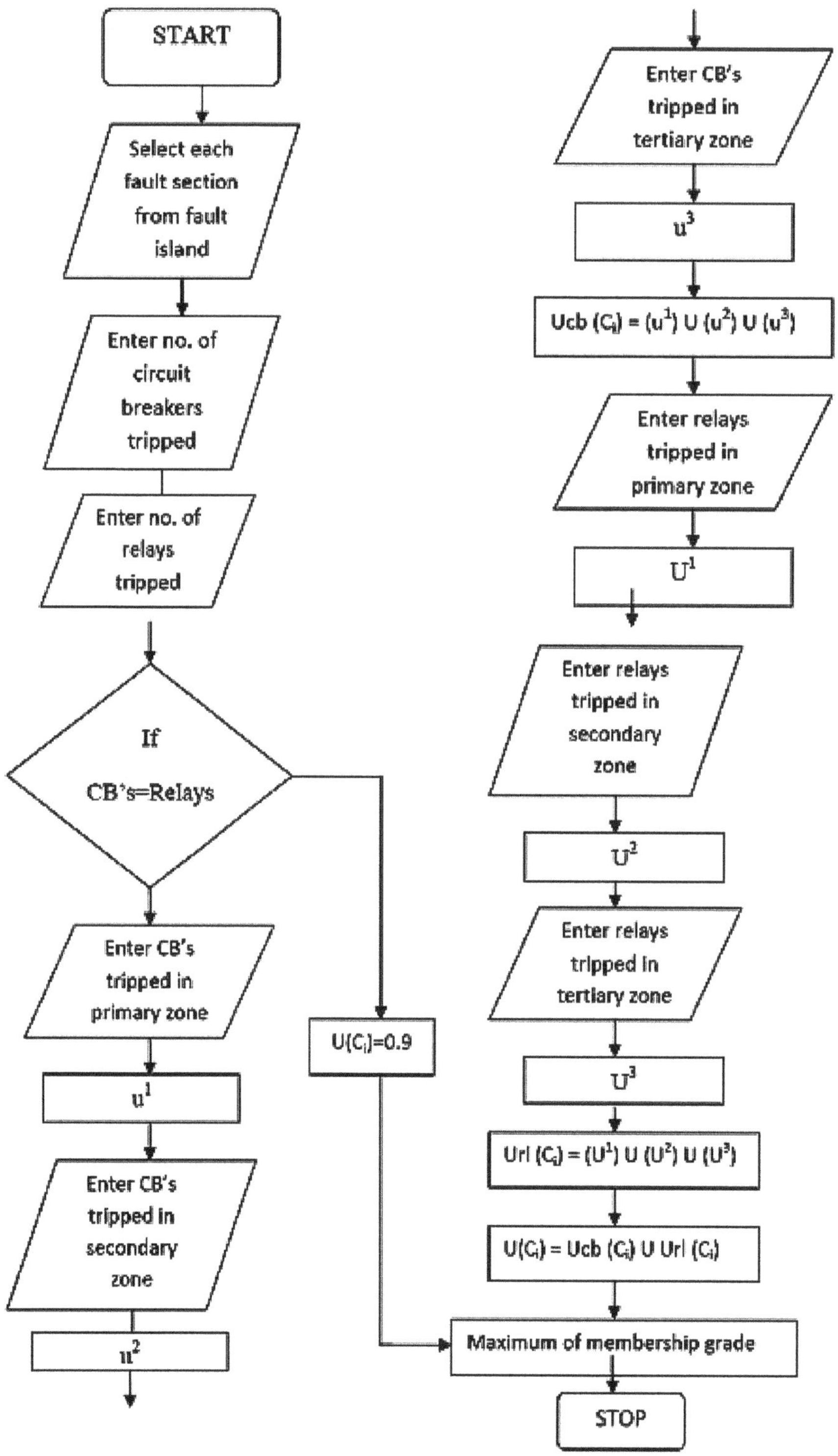

Figure 5. Circuit diagram of the test network

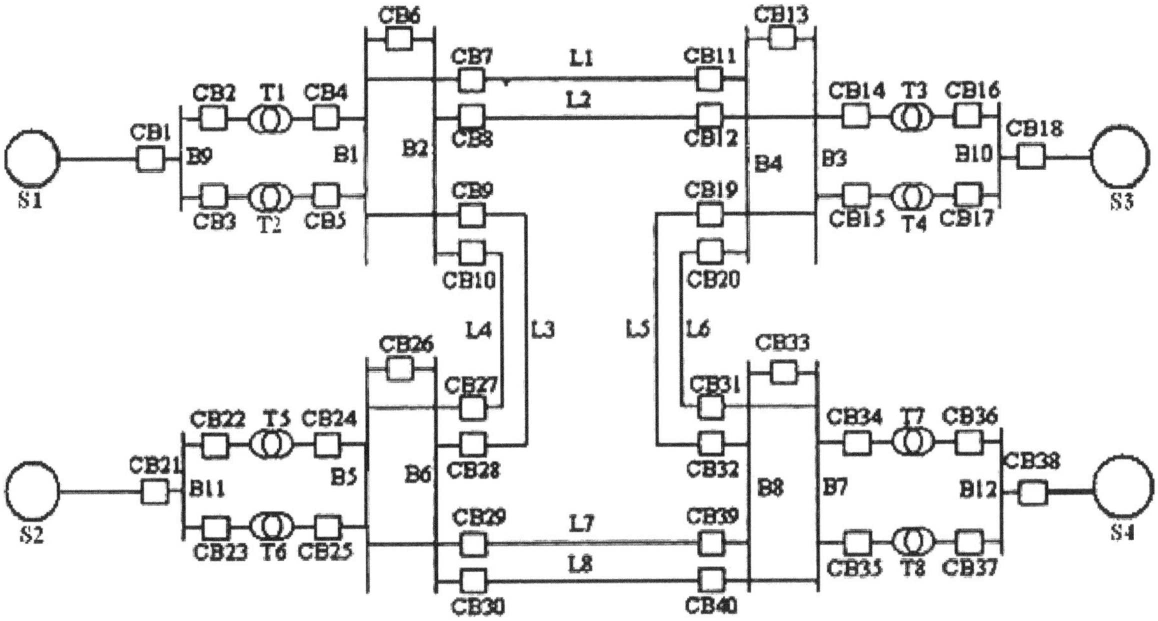

Table 3. Results of Case Study 2

Test	Possible Fault Sections	Membership Grades	Circuit Breakers Tripped	Circuit Breakers Jammed	Estimated Fault Section
2	B7	0.8156	CB20,CB30,CB33, CB34,CB35	CB31,CB40	
	B8	0.9	CB32,CB33,CB39	NO	B8
	L5	0.9	CB19,CB32	NO	L5
	L6	0.7305	CB20,CB30,CB33, CB34,CB35	CB31,CB40	
	L7	0.9	CB29,CB39	NO	L7
	L8	0.7305	CB20,CB30,CB33, CB34,CB35	CB31,CB40	
	T7	0.9	CB34,CB36	NO	T7
	T8	0.9	CB35,CB37	NO	T8

whole protection system. Fuzzy sets have been used to process inexact and uncertain information in fault diagnosis. The proposed approach has many advantages:

- Small memory space and computer time.
- Small number of rules.
- Flexibility of the system.
- Membership functions.

Table 4. Results of Case Study 3

Test	Possible fault sections	Membership Grades	Circuit Breakers Tripped	Circuit Breakers Jammed	Estimated Fault Section
3	B1	0.9	CB4,CB5,CB6, CB7,CB9,	NO	B1
	B2	0.9	CB6,CB8,CB10	NO	B2
	L1	0.9	CB7,CB11	NO	L1
	L2	0.9	CB8,CB12	NO	L2

Table 5. Results of Case Study 4

Test	Possible Fault Sections	Membership Grades	Circuit Breakers Tripped	Circuit Breakers Jammed	Estimated Fault Section
4	L1	0.9	CB7,CB11	NO	L1
	L2	0.9	CB8,CB12	NO	L2
	L7	0.9	CB29,CB39	NO	L7
	L8	0.9	CB30,CB40	NO	L8

Small memory space and computer time: Only the information on tripped breakers is needed and a few relay signals are added, when the fault cannot be determined by the status of the breakers only in the inference process. The knowledge base is also very small because there are only a few rules needed during inference. Computation time is small as most of inference algorithms used in this approach are made up of simple rules.

Small number of rules: Only a few fuzzy rules are needed in the fault section determination, and only one candidate fault section is assessed at a time.

Flexibility of the system: Membership functions representing breaker and relay performance can be changed dynamically according to the fault situation. The proposed approach can form a part of the overall architecture of a diagnosis system.

Membership functions: Membership grades would help the dispatcher to make faster, more definite and more effective estimation of the most likely fault sections. In actual system, membership functions of protective equipments are different for different voltage levels and different circumstances. Protective relays should operate with a very high reliability at the rate of 99% or even higher.

12. FUTURE SCOPE

To develop database & knowledge base for supporting fuzzy system and also it can adapt to Supervisory Control and Data Acquisition & distribution automation and applications of expert-fuzzy system include power system planning, fault detection and transient stability analysis and more developments in this way are expected. They also described about a knowledge based system support used for data acquisition in power flow problem and an "online" diagnosis and advisor tool recently incorporated to a microcomputer based power network control cEnter simulator. Following a brief review of simulator capabilities, reasons that motivated the use of an expert system are stated together with a description of the added package and typical examples of application as well as the academic environment in which the expert system is used. It can be solved more effectively by using neuro-fuzzy, neuro-genetic, fuzzy-genetic hybrids.

REFERENCES

Ayhan, B., Chow, M. Y., & Song, M.-H. (2005). *Monolith and Partition Schemes with LDA and Neural Networks as Detector Units for Induction Motor Broken Rotor Bar Fault Detection*. KIEE International Transactions on Electrical Machinery and Energy Conversion Systems.

Chang, C. S., Chen, J. M., Srinivasan, D., Wen, F. S., & Liew, A. C. (1996). *Power system fault diagnosis using fuzzy sets for uncertainties processing*. IEEE Transactions. doi:10.1109/ISAP.1996.501094

Chang, C. S., Chen, J. M., Srinivasan, D., Wen, F. S., & Liew, A. C. (1997). Fuzzy logic approach in power system fault section identification. *IEEE Proceedings-Part C, Generation Transmission and Distribution*, 144(5), 406-414. doi:10.1049/ip-gtd:19971278

Fukm, D., & Kawakame, J. (1986). An expert system for fault section estimation using mfonnatlon from protective relays and circuit breakers. IEEE Trans Power Delivery, 4, 83-90.

Fukui, C., & Kawakami, J. (1986). An expert system for fault section estimation using information from protective relays and circuit breakers. *IEEE Transactions on Power Delivery*, *1*(4), 83–90. doi:10.1109/TPWRD.1986.4308033

Fukui, C., & Kawakami, J. (1986). An expert system for fault section estimation using information from protective relays and circuit breakers. *IEEE Transactions on Power Delivery*, *1*(4), 83–90. doi:10.1109/TPWRD.1986.4308033

Khr, G. J., & Folger, T. A. (1988). *Fuzzy Sets, Uncertainty, and Information*. Prentice Hall.

Komal, K., Matsumoto, K., & Sakaguch, T. (1992). A diagnostic application speed expert systems shell for network fault diagnosis. *Electrical Power & Energy Systems*, 14, (2/3), 217-224.

Liu, C. C., & Damborg, M. J. (1988). *Development* of *Expert Systems as On-Line Power System Operational Aids* (Final Report). EPRI Research Project EL5635.

Prais, M. (1989). A topology processor that tracks network modifications over time'. *IEEE-PWRS*, *3*(3), 992–998.

Prince, W. R., Wollenberg, B. F., & Bertagnolli, D. B. (1989). Survey on excessive alarms. *IEEE Transactions on Power Systems*, *4*, 950–956. doi:10.1109/59.32584

Rodriguez, C., Rementeria, S., Martin, J. I., Lafuente, A., Muguerza, J., & Perez, J. (1996). Fault analysis with modular neural networks. *International Journal of Electrical Power & Energy Systems*, *18*(2), 99–110. doi:10.1016/0142-0615(95)00007-0

Srinivasan, D., Ruey Long Cheu, Young Peng Poh, Albert Kim, & Chwee Ng. (2000). *Automated fault detection in power distribution networks using a hybrid Fuzzy-genetic algorithm approach*.

Sundareswaran, K. (n. d.). A Learner's Guide to Fuzzy Logic Systems.

Wen, F., & Chang, C. S. (1996). A new approach to fault section estimation in power systems based on the set covering theory and the refined genetic algorithm. *Proceedings of 12th Power system computation conference PSCC '96*, Dresden (Vol 1, pp 358- 365).

Wen, F., & Han, Z. (1995). Fault section estimation in power systems using a genetic algorithm. *Electric Power Systems Research*, *34*(3), 165–172. doi:10.1016/0378-7796(95)00974-6

Wollenberg, B. F., & Sakaguchi, T. (1987). Artificial intelligence inpower system operations. *Proceedings of the IEEE*, 12, 1678-1685.

Xu, L., & Chow, M.-Y. (2006). A classification approach for power distribution systems fault cause identification. *IEEE Transactions on Power Systems*, *21*(1), 53–60.

Xu, L., Chow, M.-Y., & Taylor, L. S. (2007). Power Distribution Fault Cause Identification with Imbalanced Data Using the Data Mining Based Fuzzy Classification E-Algorithm. *IEEE Transactions on Power Systems*, *22*(1), 164–171. doi:10.1109/TPWRS.2006.888990

Zhu, Y. L., Yang, Y. H., Hogg, B. W., Zhang, W. Q., & Gao, S. (1994). An expert system for power systems fault analysis. *IEEE Transactions on Power Systems*, *9*(1), 503–509.

Chapter 16
Parameter Estimation of an High Performance I.M. Drive for Sensorless Operation

Monika Dixit
Sagar Institute of Research, Technology & Science, India

Smita Shandilya
Sagar Institute of Research, Technology & Science, India

ABSTRACT

A modern AC motor drive is a very intelligent system which covers a wide range of different electro technical apparatus and a wide scope of electrical engineering skills. A modern AC motor drive consists of four closely acting main parts: the AC machine, the power electronics, the motor control algorithm and the control hardware, i.e. the signal electronics. The advances in semiconductors and microelectronics have made the rapid development of AC motor drives possible. Semiconductors used in the switching converters provide the electric energy processing capability and microcontrollers and digital signal processors provide the data processing power for complex control algorithms.

1. INTRODUCTION

In previous days 1950s nearly DC motors were used in all industrial applications because of its easy speed control. The use of induction motor is very limited normally these motors have a constant speed depending on the frequency of the supply and the no of windings. In the past it was not possible to control the speed of the induction motors according to the need. That's why their use was limited and despite having many a motors they advantages over dc motors they could not be used because of this disadvantage. But at the field of drivers have improved due to the availability of thyristor or SCRthyristors, power transistors, IGBTs and GTOs the variable speed induction motor drives have been invented. Though the cost of these drivers are more than dc driver, but still the use of induction motors are increasing and they are replacing dc motors because of their advantages. Now days with increase in technology modern AC motor drives performance is increases

DOI: 10.4018/978-1-4666-9911-3.ch016

comparable to DC motors. This improvement in performance of AC motor drive is achieved due to the advancement of power electronics and microprocessors.

Takahashi and Nagouchi have reported a new type of control technique for ac motor drives. This technique controls both the stator flux linkage and the torque directly, and is different from the vector control technique in which, these variables are controlled via the stator currents. Takahashi and Nagouchi have named this new control technique as direct torque control (DTC). Almost simultaneously, Depenbrock has also proposed the same type of control technique and named it as direct self-control (DSC). The DTC method was developed as an alternative and fast control method for an induction motor drive.

In a direct torque control induction motor drive, the basic concept is to control both stator flux and electromagnetic torque of the machine simultaneously with the help of torque control loop, flux control loop and speed control loop. In comparison with the conventional Vector-controlled drives, in a DTC-based Drive the only requirement is the knowledge of stator resistance, and thus the sensitivity associated to parameter variations is reduced. Moreover, the DTC-based drives do not require fulfilling the coordinate transformation between stationary frame and synchronous frame. Thus the complexities of the algorithms involved in a Field Oriented control are minimized. DTC algorithm is more efficient and easiest technique among all available techniques.

For the implementation of DTC algorithm, requires an independent control of electromagnetic torque and flux with two independent control loops. Hence the parameters which are to be measured or estimated for the implementation of DTC algorithm are,

1. Stator flux (estimated)
2. Electromagnetic torque (estimated/measured)
3. Rotor speed (estimated/measured)
4. Motor phase currents (measured)
5. Stator phase voltage (estimated/measured)
6. DC link voltage (measured)

There is a strong interdependency among these variables. For example the flux of an induction motor drive can be estimated with the help of motor back emf, which in turn depends upon stator phase voltage, phase current and stator resistance. The mathematical expression for stator flux in stationary reference frame can be given by (1)–(3)

$$\lambda_{s\alpha} = \frac{1}{s}\left(E_{s\alpha}\right) \text{and } \lambda_{s\beta} = \frac{1}{s}\left(E_{s\beta}\right) \tag{1}$$

where

$$E_{s\alpha} = v_{s\alpha} - i_{s\alpha}R \tag{2}$$

$$E_{s\beta} = v_{s\beta} - i_{s\beta}R \tag{3}$$

Thus for the estimation of stator flux in the required variables are phase voltage, phase current and stator resistance.

Similarly, the electromagnetic torque of an induction motor drive can be either sensed or can be estimated. Estimation of electromagnetic torque rather than its measurement by any torque transducer makes a DTC drive more cost effective and less complex. The electromagnetic torque in stator reference frame is given by (4)

$$T_e = \frac{3}{2} P * \left(\lambda_{s\alpha} * i_{s\beta} - \lambda_{s\beta} * i_{s\alpha}\right) \tag{4}$$

And the interdependency among the different parameters can again be seen from the torque equation, which says that electromagnetic torque can be estimated with the help of orthogonal stator flux components ($\lambda_{s\alpha}$, $\lambda_{s\beta}$) and stator phase currents.

In a sensor less drive application, estimation of the rotor speed can be achieved with the help of mathematical expression, thus avoiding the installation of shaft mounted speed transducers, which may again may affect the reliability and cost effectiveness of any drive. Parameter estimation is done to reduce the number of measuring devices and provide sensorless operation of drive. The use of large number of devices increases the cost of the drive and reduces the reliability of the system. For sensor less drives, the estimation of speed is based upon the electrical frequency, and is given by

$$w_r = \frac{We}{P} - w_{slip} \tag{5}$$

where

$$w_e = \frac{E_{s\beta}\lambda_{s\alpha} - E_{s\alpha}\lambda_{s\beta}}{|\lambda_s|^2} \tag{6}$$

Thus once again it can be verified that for speed estimation the variables that are to be known are stator phase voltage, motor parameters and stator flux.

A voltage source inverter is the heart of any variable frequency drive, and it is a well-established fact that the output voltage of any voltage source inverter, which feeds the induction motor drive, can be easily estimated with the help of DC link voltage and duty cycle of switching signals. Thus a further reduction in sensor count with increased reliability and cost effectiveness can be achieved by estimated the inverter output voltage rather than its actual measurement.

This chapter describes the basic principle of a conventional DTC drive with a discussion on following important aspects pertaining to DTC drives.

1. Implementation of DTC algorithm
2. Discussion on flux control loop and its influence on the performance of a DTC drive
3. The torque control loop and its analysis
4. Implementation of different speed control algorithm for closed loop speed control applications.

2. LITERATURE REVIEW

Bodkhe & Aware (2009) proposed "A new control strategy for three-phase induction motor which includes speed and torque control loops and the current regulation thereby overcoming the limitation of volts per hertz control drives. For close-loop control, the feedback signals including the rotor speed, flux and torque are not measured directly but are estimated by means of an algorithm. The inputs to this algorithm are the reconstructed waveforms of stator currents and voltages obtained from the dc link and not measured directly on stator side. The proposed drive thus requires only one sensor in the dc link to implement the close-loop speed and torque control of a three-phase induction motor. Unlike the conventional flux and speed observers, the proposed estimation algorithm needs less computation and is less dependent on machine parameters. The proposed scheme was simulated using Matlab/Simulink software and was tested on a 2.2 kW drive for various steady-state and dynamic operating conditions. The results show fast dynamic response and good agreement between the actual values and the estimated values of torque and speed. Replacement of the open-loop control strategy of existing v/f drive by the proposed close-loop strategy appears to be possible without requiring any additional power components and sensors."

Shin, Hyun, Cho, & Choe (2000) proposed "A programmable low pass filter (LPF) to estimate stator flux for speed sensorless stator flux orientation control of induction motors. The programmable LPF is developed to solve the dc drift problem associated with a pure integrator and a LPF. The pole of the programmable LPF is located far from the origin in order to decrease the time constant with the increasing speed. In addition, the programmable LPF has the phase/gain compensator to estimate exactly stator flux in a wide speed range. Consequently, the drift problem is much improved and the stator flux is exactly estimated in the wide speed range."

Rumzi, Idris, Halim, & Yatim (2002) proposed "An improved stator flux estimation technique based on a voltage model with some form of low-pass (LP) filtering. In voltage-model-based stator flux estimation, an LP filter is normally used instead of a pure integrator to avoid integration drift problem due to dc offset, noise, or measurement error present in the back electromotive force. In steady-state condition, the LP filter estimator will degrade the performance and efficiency of the direct torque control (DTC) drive system since it introduced magnitude and phase errors, thus resulting in an incorrect voltage vector selection. The stator flux steady-state error between the LP filter and a pure integrator estimator technique is derived and its effect on the steady-state DTC drive performance is analyzed. A simple method is proposed to compensate for this error which results in a significant improvement in the steady-state drive performance."

Luukko, Niemelä, & Pyrhönen (2001) proposed "An improved integration method for the estimation of the stator flux linkage for speed- and position-sensor less direct-torque-controlled ac machine drives. The method is based on monitoring the scalar product of the estimated stator flux linkage and the measured stator current. The ac part of the scalar product is extracted using filtering and the correction for the estimated stator flux linkage is formed from that part. Adequate performance is obtained by using simple low-pass filtering. By using adaptive filtering in the extraction, the performance of the drive is excellent."

Cirrincione, Pucci, Cirrincione, & Capolino (2004) proposed "A simple adaptive notch filter for the elimination of the dc component in the integration of signals used for the flux estimation in high performance ac drives. This integration method is composed of two identical adaptive noise cancellers using a linear neural network with just one bias weight. Its behaviour has been investigated in simulation as applied to electrical drives and compared with other four traditional integration algorithms."

Wang & Deng (2012) proposed "An integration algorithm for stator flux estimation of direct torque control (DTC). This algorithm only contains a fifth-order low-pass filter (LPF), a high-pass (HP) filter, and a simple logical calculation part, where the fifth-order LPF and the HP filter can effectively filter out HF harmonics and the dc drift in the back EMF, respectively. The α- and β-axis back EMFs passing through the two filters will obtain β-axis stator flux linkage and opposite polarity α-axis stator flux linkage, respectively, and then by the logical calculation part can finally achieve α- andβ-axis stator flux linkages. At any synchronous angular frequency, the amplitude- and phase-frequency characteristics of this algorithm are same as those of the pure integrator, and meanwhile its dc gain is zero. The proposed algorithm with high estimation accuracy behaves well in both steady state and dynamic performance when used in a DTC-based electrical excitation flux-switching generator dc power system."

Wang & Deng (2012) proposed "A direct torque linear control (DTLC) scheme for voltage regulation of a parallel HESF (PHESF) generator. This scheme can achieve direct and smooth control of the instantaneous torque to improve the system steady-state and dynamic performance. However, the stator flux estimation accuracy will affect torque control quality. Therefore, a practical stator flux estimation method is presented for both direct torque control and the DTLC schemes of the PHESF generator. It contains a third-order low-pass (LP) filter, a high-pass (HP) filter, and two simple coordinate transformation modules. The third-order LP filter and the HP filter can effectively filter out high-frequency harmonics and the dc drift in the back-EMF, respectively. Then, the produced leading and the lagging phase angles can be compensated by the two coordinate transformation modules. Hence, this method has same amplitude-frequency and phase-frequency characteristics as those of the pure integrator. Meanwhile, its dc gain is zero. The DTLC scheme for the PHESF generator dc power systems using the improved estimation method exhibits good steady-state and dynamic performance."

Lascu & Andreescu (2006) proposed and compared "Two flux observers for wide speed range direct torque control (DTC) of sensor less induction-motor drives. The first one is a full-order sliding-mode observer with proportional plus integral (PI) compensation, without rotor speed adaptation. The second one is based on a zero phase delay- improved integrator of the voltage model, which uses only a PI flux amplitude control with stator-flux reference magnitude in the correction loop. In both cases, an estimated dc offset is built up and memorized by the PI integral component and this totally compensates for all dc offsets and drifts originated in the acquisition channels. Two feasible solutions for on-line stator resistance identification are proposed. Simulation and experimental results prove the accuracy, robustness, and high-dynamic performance of both observers when employed in sensor less DTC drives. The effectiveness of state estimation is confirmed by a steady state and transient sensor less operation at very low speed with rated load torque and step-speed reversal."

Rehman & Xu (2011) proposed "Indirect field-oriented control, direct field oriented control, and direct torque control techniques for the drive system of alternative energy vehicles (AEVs). Three candidate alternating-current machines are evaluated and investigate the suitability of different control techniques for the drive systems of battery-powered electric vehicles, hybrid electric vehicles, and integrated starter alternators. Accurate flux and torque estimation is the core of any opted control technique for all types of AEV drive systems. Therefore, an accurate closed-loop voltage model flux and torque estimator that is insensitive to stator resistance variation has been designed. In this scheme, loss model control and search control (SC) techniques for the drive system's efficiency optimization is also analyzed and an offline SC efficiency optimization technique is proposed."

Seyoum, Rahman & Grantham (2003) proposed "A low pass filter based scheme to eliminate the integrator output errors. Integration error comprises of a drift produced by the integrator, somehow related to initial condition, and the drift produced by error present in the back electromotive force (emf) due to measurement offset error in the terminal voltage or/and phase current. The performance of the induction machine control mainly depends on the accuracy of the estimated flux. The proposed method compensates the error produced by the inherent problem in the pure integrator and measurement error. This method can be easily applied in a DSP based induction machine control to estimate the flux. The execution time required to implement the proposed system is small that there will not be much software computation burden."

Bodkhe & Aware (2010) proposed "A new algorithm for the reconstruction of stator currents of an inverter-fed, three-phase induction motor drive. Unlike the classical and improved methods available in literature, the proposed method is neither based on pulse width modulation pattern modification nor is a machine model based observer and therefore is independent of demerits of existing methods like higher switching loss and increased sensitivity to parameter variations. This makes it a robust solution. The proposed algorithm makes use of simple logic gates and if else statements and as a result it generates stable three-phase current signals even during short periods of measurement. It also carries out auto-calibration of dc link current to remove the unwanted dc-offsets that are caused due to asymmetry in switching of semiconductor devices. The proposed algorithm is tested on a 2.2 kW, current-regulated, field oriented control, induction motor drive by using MATLAB/Simulink software."

Tripura, Babu, & Tagore (2011) proposed "Space Vector Pulse Width Modulation (SVPWM) Scheme. In this scheme, three SVPWM schemes, called 7-segment space vector modulation (SVM), 7-segment SVM with even-order harmonic elimination and 5-segment (discontinuous) SVM are studied in detail. SVPWM method is an advanced, computation intensive PWM method and possibly the best among all the PWM techniques for variable frequency drive applications. The SVPWM is an alternative method for the determination of switching pulse width and their position. The major advantage of SVPWM stem from the fact that, there is a degree of freedom of space vector placement in a switching cycle. This feature improves the harmonic performance of this method. This method has been finding widespread application in recent years because of the easier digital realization and better dc bus utilization."

Stojic (2005) proposed "A new method for the induction motor stator flux estimation used in the sensor less IM drive applications. Proposed algorithm advantageously solves problems associated with the pure integration, commonly used for the stator flux estimation. An observer based structure is proposed based on the stator flux vector stationary state, in order to eliminate the undesired DC offset component present in the integrator based stator flux estimates. By using a set of simulation runs it is shown that the proposed algorithm enables the DC-offset free stator flux estimated for both low and high stator frequency induction motor operation."

Pillay & Levin (1995) proposed "Different mathematical models have been used over the years to examine different problems associated with induction motors. These range from the simple equivalent circuit models to more complex d,q models and abc models which allow the inclusion of various forms of impedance and/or voltage unbalance. Recently, hybrid models have been developed which allow the inclusion of supply side unbalance but with the computational economy of the d,q models. This paper presents these models with typical results and provides guidelines for their use."

Holtz & Quan (2002) proposed "An approach to overcome the problems of poor performance of induction motor drive at low speed. In this algorithm author proposed implementation of pure integrator for stator flux estimation. The time variable DC offset is estimated from the flux drift in a parallel stator model and used to eliminate the offset by feedforward control. Residual high-frequency disturbances are compensated by the feedback flux amplitude control. A linearization of the PWM inverter transfer function and an improved stator resistance estimation scheme further enhance the system performance."

Boussak & Jarray (2006) proposed "A new method for the implementation of a sensorless indirect stator-flux-oriented control (ISFOC) of induction motor drives with stator resistance tuning is proposed in this paper. The proposed method for the estimation of speed and stator resistance is based only on measurement of stator currents. The error of the measured q-axis current from its reference value feeds the proportional plus integral (PI) controller, the output of which is the estimated slip frequency. It is subtracted from the synchronous angular frequency, which is obtained from the output integral plus proportional (IP) rotor speed controller, to have the estimated rotor speed. For current regulation, this paper proposes a conventional PI controller with feedforward compensation terms in the synchronous frame. Owing to its advantages, an IP controller is used for rotor speed regulation. Stator resistance updating is based on the measured and reference d-axis stator current of an induction motor on $d-q$ frame synchronously rotating with the stator flux vector."

Jemli, Azza, Boussak & Gossa (2009) proposed "A method of rotor speed estimation based on the measurement of the main and auxiliary windings stator currents and that of a reference q-axis current generated by the control algorithm. The error of the measured q-axis current from its reference value feeds the proportional plus integral controller, the output of which is the estimated slip angular frequency."

Raghu, Rao, & Chandrasekhar (2012) proposed "An Adaptive Pseudoreducedorder Flux Observer (APFO) for speed control of the Induction Motor drive (IM) without a speed sensor. In this scheme an Adaptive Pseudo reduced-order Flux Observer (APFO) is used instead of the Adaptive Full-order Flux Observer (AFFO). In comparison with the AFFO, this method consumes less computational time, and provides a better speed estimation at very low speeds. APFO is used for estimate the IM rotor speed and stator resistance, and these are used as feedback signals for the Field Oriented Control (FOC), which is a widely used control method for Induction Motor drive (IM)."

Jouili, Jarray, Koubaa & Boussak (2009) proposed "The implementation of a sensorless indirect rotor-flux oriented control (IRFOC) of induction motor (IM) drives. The proposed method for speed estimation is based on Luenberger state observer (LSO). The adaptive rotor speed is actually derived from Lyapunov stability theory by using measured and estimated stator currents and estimated rotor flux. The proposed control scheme, as a result, achieves a good performance with computational complexity reduction obtained by using the analytical relation to determine the LSO gain matrix. Again, the observer is simple and robust, when compared with the previously developed observers, and suitable for online implementation."

Schauder (1992) proposed "A model-reference adaptive system (MRAS) for the estimation of induction motor speed from measured terminal voltages and currents. The estimated speed is used as feedback in a vector control system, thus achieving moderate bandwidth speed control without the use of shaft-mounted transducers. This technique is less complex and more stable than previous MRAS tacholess drives"

Kubota, Matsuse, & Takayoshi (1993) proposed "A new method of estimating the speed of an induction motor. This method is based on adaptive control theory;"

Kubota, & Matsuse (2010) proposed "A method of estimating simultaneously the motor speed and the rotor resistance of an induction motor by superimposing ac components on the field current command."

Versèle, Deblecker & Lobry (2008) proposed "A project in which the parameter determination and set up of a vector control structure of a voltage-fed inverter induction motor (IM) drive using a dSPACE platform and Matlab/Simulink environment is developed."

Lascu, Boldea, & Blaabjerg (2006) proposed "A new direct torque and flux control based on space-vector modulation (DTC-SVM) for induction motor sensorless drives. It is able to reduce the acoustical noise, the torque, flux, current, and speed pulsations during steady state. The DTC transient merits are preserved, while better quality steady-state performance is produced in sensorless implementation for a wide speed range. The flux and torque estimator is presented and an improved voltage–current model speed observer is introduced."

Haron, Rumzi, & Idris (2006) proposed "A study of the performance of the rotor flux based MRAS (RF-MRAS) and back e.m.f based MRAS (BEMF-MRAS) for estimating the rotor speed. Model Reference Adaptive System (MRAS) based techniques are one of the best methods to estimate the rotor speed due to its performance and straightforward stability approach. These techniques use two different models (the reference model and the adjustable model) which have made the speed estimation a reliable scheme especially when the motor parameters are poorly known or having large variations. The scheme uses the error vector from the comparison of both models as the feedback for speed estimation. Depending on the type of tuning signal driving the adaptation mechanism, there could be a number of schemes available such as rotor flux based MRAS, back e.m.f based MRAS, reactive power based MRAS and artificial neural network based MRAS. Both schemes use the stator equation and rotor equation as the reference model and the adjustable model respectively. The output error from both models is tuned using a PI controller yielding the estimated rotor speed."

Bojoi, Guglielmi, & Pellegrino (2006) proposed "A sensor less direct rotor field-oriented control (SDRFOC) scheme of three-phase induction motors for low-cost applications. The SDRFOC algorithm is based on a sensor less closed-loop rotor flux observer whose main advantages are simplicity and robustness to motor parameter detuning. The whole algorithm has been developed and implemented on a low-cost fixed-point digital signal processor controller."

Duran, Duran, Perez, & Fernandez (2004) proposed "Sensor less drive control has been widely studied in recent years due to the numerous advantages regarding potential failures of position sensors, especially in applications such as automotive or aerospace. Among vector-control drives, indirect rotor-flux-oriented control (IRFOC) type is one of the most popular and tested options. However, it is still a challenging field since several aspects can be improved, such as the low-speed behaviour, parameter detuning, and current control. The present scheme includes temperature estimation to correct the deviation in steady state, a new control scheme with skin-effect estimation to improve the transient accuracy, and over current protection to be able to control the stator currents while allowing a good performance. The parameter estimation is carried out using lumped-parameter models, the control scheme is modified and is able to account for static friction, and the over current protection improves the performance allowing transient currents over the rated value. The validity and usefulness of the proposed scheme is experimentally tested on a TMS320C31 digital signal processor (DSP) from the Simulink/Matlab environment."

Kim & Jahns (2006) proposed "Three-phase current measurement using only a single current sensor in the inverter dc link is appealing for ac motor drives because it minimizes the number of current sensors, thereby reducing sensor cost, weight, and volume. However, the basic dc link single current sensor technique poses special challenges because the duration of the active voltage vectors must be long enough to measure the dc link current reliably during every PWM switching interval. A new single current sensor algorithm for reconstructing the phase currents called the measurement vector insertion method (MVIM) is presented that overcomes this problem using active voltage vectors that are applied for brief measurement intervals only when needed during each fundamental frequency cycle. The insertion of measurement vectors makes it possible to reconstruct the output phase currents with minimum impact on the machine's performance and operating envelope."

Hongrae & Thomas (2006) proposed "A new algorithm using a single dc-link current sensor to reconstruct all three inverter phase currents has been proposed that overcomes problems caused by the presence of regions in the voltage vector plane where the phase currents are not conveniently measurable. The measurement vector insertion method (MVIM) achieves this objective by applying additional active voltage vectors for brief intervals only when needed during each fundamental frequency cycle. It addresses techniques for incorporating the MVIM algorithm into a closed-loop current regulator designed to provide high-quality current control using a single dc-link current sensor. Virtual active resistance state feedback is introduced into the synchronous-frame current regulator in order to improve the controller's ability to reject disturbances caused by the MVIM measurement vectors."

Vaclavek & Blaha (2009) proposed "A method based on deterministic evaluation of measurement using the state observer based on the Lyapunov function. AC induction motors have become very popular for motion-control applications due to their simple and reliable construction. Control of drives based on ac induction motors is a quite complex task. Provided the vector-control algorithm is used, not only the rotor speed but also the position of the magnetic flux inside the motor during the control process should be known. In most applications, the flux sensors are omitted and the magnetic-flux phasor position has to be calculated. However, there are also applications in which even speed sensors should be omitted. In such a situation, the task of state reconstruction can be solved only from voltage and current measurements. The method has been proven in testing on a real ac induction machine."

3. DIRECT TORQUE CONTROLLED (DTC) IM DRIVE

Direct torque control (DTC) is one method used in variable frequency drives to control the torque (and thus finally the speed) of three-phase AC electric motors. This involves calculating an estimate of the motor's magnetic flux and torque based on the measured voltage and current of the motor.

Stator flux linkage is estimated by integrating the stator voltages. Torque is estimated as a cross product of estimated stator flux linkage vector and measured motor current vector. The estimated flux magnitude and torque are then compared with their reference values. If either the estimated flux or torque deviates from the reference more than allowed tolerance, the transistors of the variable frequency drive are turned off and on in such a way that the flux and torque errors will return in their tolerant bands as fast as possible. Thus direct torque control is one form of the hysteresis or bang-bang control.

The properties of DTC can be characterized as follows:

- Torque and flux can be changed very fast by changing the references.
- High efficiency & low losses - switching losses are minimized because the transistors are switched only when it is needed to keep torque and flux within their hysteresis bands.
- The step response has no overshoot.
- No coordinate transforms are needed, all calculations are done in stationary coordinate system.
- No separate modulator is needed, the hysteresis control defines the switch control signals directly.
- There are no PI current controllers. Thus no tuning of the control is required.
- The switching frequency of the transistors is not constant. However, by controlling the width of the tolerance bands the average switching frequency can be kept roughly at its reference value. This also keeps the current and torque ripple small. Thus the torque and current ripple are of the same magnitude than with vector controlled drives with the same switching frequency.
- Due to the hysteresis control the switching process is random by nature. Thus there are no peaks in the current spectrum. This further means that the audible noise of the machine is low.
- The intermediate DC circuit's voltage variation is automatically taken into account in the algorithm (in voltage integration). Thus no problems exist due to dc voltage ripple (aliasing) or dc voltage transients.
- Synchronization to rotating machine is straightforward due to the fast control; Just

make the torque reference zero and start the inverter. The flux will be identified by the first current pulse.

- Digital control equipment has to be very fast in order to be able to prevent the flux and torque from deviating far from the tolerance bands. Typically the control algorithm has to be performed with 10-30 microseconds or shorter intervals. However, the amount of calculations required is small due to the simplicity of the algorithm.
- The current measuring devices have to be high quality ones without noise because spikes in the measured signals easily cause erroneous control actions. Further complication is that no low-pass filtering can be used to remove noise because filtering causes delays in the resulting actual values that ruins the hysteresis control.
- The stator voltage measurements should have as low offset error as possible in order to keep the flux estimation error down. For this reason the stator voltages are usually estimated from the measured DC intermediate circuit voltage and the transistor control signals
- In higher speeds the method is not sensitive to any motor parameters. However, at low speeds the error in stator resistance used in stator flux estimation becomes critical.

4. OPERATING PRINCIPLE OF DIRECT TORQUE CONTROLLED IM DRIVE

According to the DTC principle, an independent control of torque and flux can be achieved by the application of appropriate voltage vectors in such a way that the error between the estimated torque and flux with their respective reference values remain within the limits of hysteresis comparators. The desired voltage vectors to compensate the errors are selected based on the output of the torque and flux hysteresis comparator as well as the locus of stator flux vector.

From the basic equation governing induction motor operation stator flux λs is given by

$$\lambda s=\int(Vs-RsIs)dt \tag{7}$$

Neglecting the drop in stator resistance,

$$\Delta\lambda s=Vs\Delta t, \tag{8}$$

where Δt is the time interval of application of the applied voltage vector .Electromagnetic torque in an induction motor is given by

$$Te = P\frac{Lm}{\sigma LsLr}|\lambda s||\lambda r|\sin(\delta) \tag{9}$$

where

$$\sigma = 1-\frac{L_m^2}{L_{sL_r}} \tag{10}$$

Due to the large time constant of rotor flux, it cannot be changed instantly, but the stator flux vector can change by the application of proper voltage vector. Hence from (9) and Figure 1 it can be concluded that an increment in torque can be achieved by increasing the angle between stator and rotor flux vector. Splitting the vector $\Delta\lambda_s$ into horizontal and orthogonal components it can be concluded that orthogonal component of $\Delta\lambda_s$ is responsible for torque control and the horizontal component controls the flux as shown in Figure 1. Thus depending upon the torque and flux requirement, feeding the induction motor with appropriate voltage vectors, helps to achieve an independent control of torque and flux.

A typical two level feed voltage source inverter is shown in Figure 2 below which has six switches S1,S2,S3,S4,S5,S6. According to the working of

Figure 1. Flux and torque control in a DTC drive

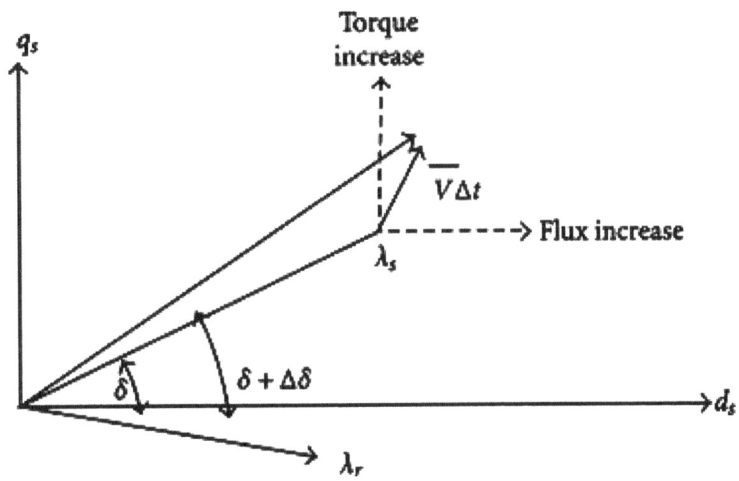

Figure 2. DTC drive fed from two level VSI

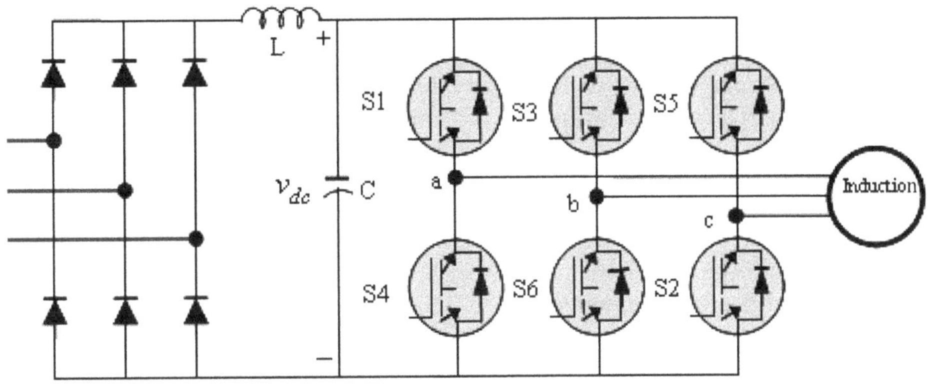

voltage source inverter it has a combination of three switching positions. Based on these three positions of switches inverter have six voltage vectors. According to this six voltage vectors the value of flux and torque is managed.

Different voltage vectors can be generated from a typical voltage source inverter shown in Figure 2, which depends upon the switching status of the force commutated switches (S1-S6).

Considering the combinations of status of phases a, b and c the inverter has six active (non-zero) voltage vectors V1 (100), V2 (110), V3 (010), V4 (011), V5 (001), V6 (101) and two null (zero) voltage vectors V0 (000) and V7 (111). During zero voltage vectors, the motor terminals get short-circuited. For example a voltage vector V2 can be generated by the combination of switching signal Sa=1, Sb=1 and Sc=0 (110) as shown in Figure 3.

In a classical DTC drive decoupled control of the torque and stator-flux can be achieved by the action of the radial and tangential components of the stator-voltage space vector, and thus they can be controlled by the appropriate inverter switching. Depending upon the location of the stator flux, torque and flux can be increased or decreased by the selection of appropriate voltage vectors.

Figure 3. Voltage vector V2 generated by switching signal (110)

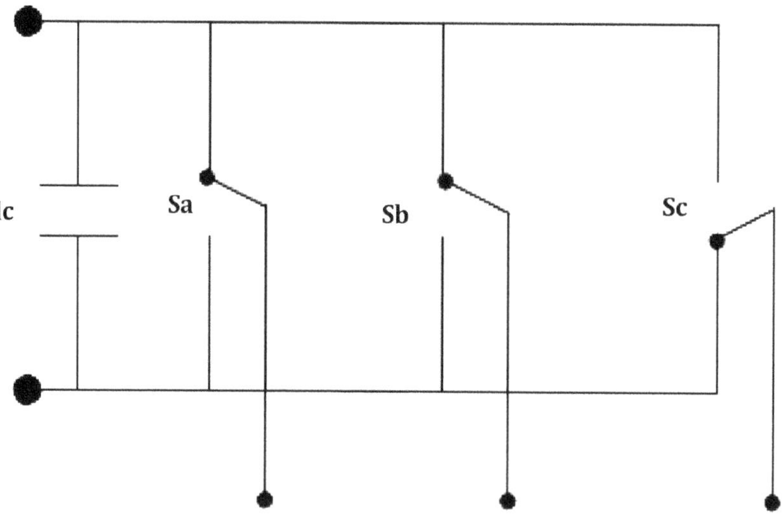

The voltage vector selection in a classical DTC drive can be explained from Figure 4. The stator flux plane is divided into six sectors of 60 degrees. For example if the stator flux vector lies in the first sector (−30° to 30°), then an increment in torque and flux can be achieved by voltage vector V2, and for an increment in flux and decrement in torque the selected voltage vector is V6. A torque increment and flux decrement is achieved by vector V3 and the vice versa by vector V5. By the application of any zero voltage vector (V0 or V7), the stator flux does not change due to the termination of integration operation. However the rotor flux vector, due to its large time constant, still continues to rotate in clockwise direction resulting in a decrement in angular separation between stator and rotor flux vector. This causes a decrease of the motor torque. However, this kind of torque reduction is only good for the middle and high speeds operations. In low speed range the rotor flux motion is too slow to achieve rapid torque reduction.

In general, if the stator flux linkage vector lies in the Kth sector, its magnitude can be increased by using switching vectors Vk-1 (for clockwise rotation) or Vk+1 (for anticlockwise rotation), and can be decreased by applying voltage vectors Vk-2 (for clockwise rotation) or Vk+2 (for anticlockwise rotation).

An increment or decrement in torque is decided on the basis of magnitude of torque error. This torque error is the error between the reference torque command and actual estimated torque. Similarly an increment or decrement in flux vector is decided by the magnitude of torque error between reference flux and actual estimated flux. Thus a comparison has to be carried out between reference torque and actual torque by a torque hysteresis comparator and a similar comparator is required for flux comparison .The action of the torque and flux hysteresis comparator is based on equations (11) – (13).

$$\text{If } \Delta T_e \geq Ht, \text{ then } Htorque = 1 \tag{11}$$

$$\text{If } -Ht < \Delta T_e < Ht, \text{ then } Htorque = 2 \tag{12}$$

$$\text{If } \Delta T_e \leq -Ht, \text{ then } Htorque = 3 \tag{13}$$

Figure 4. Stator flux veztor in sector IV and corresponding voltage vectors for torque and flux increment

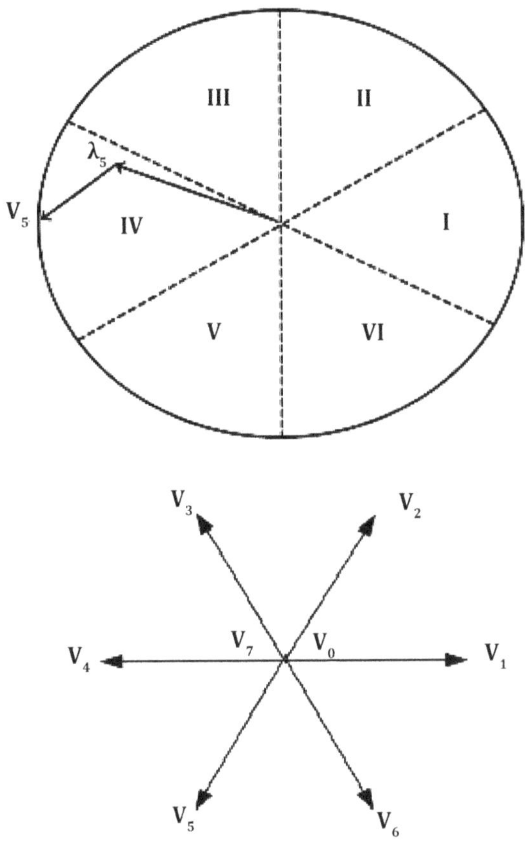

where ΔT_e is the magnitude of torque error, Ht is the bandwidth of torque hysteresis comparator and Htorque is the control output of the torque hysteresis comparator which is shown in Figure 5.

Similarly for flux comparison the following equations are used

$$\text{Hflux}=1, \text{ for } \Delta\lambda_{se} \geq H\phi \tag{14}$$

$$\text{Hflux}=2, \text{ for } \Delta\lambda_{s} \leq -H\phi \tag{15}$$

where $\Delta\lambda_{se}$ the magnitude of stator is flux error and $H\phi$ is the bandwidth of flux comparator. Since voltage vector selection depends upon the magnitude of torque error, flux error and the location of the stator flux vector in a particular sector. Thc locus of the stator flux vector in a particular sector depends on angle 'θ' given by (16).

$$\theta = \tan^{-1}\frac{\lambda_{\beta s}}{\lambda_{\alpha s}} \tag{16}$$

The location of stator flux vector in a particular sector is then determined as follows.

If $\theta \in (-30°$ to $30°)$, then sector 1(S1) is the location of stator flux vector. Where $\lambda_{\beta s}$ and $\lambda_{\alpha s}$ are the orthogonal stator flux vectors in stationary reference frame.

Based on the voltage vector selection strategy shown in Figure 4, finally the decoupled control of torque and flux can be achieved by the selection of the voltage vectors with the help of the classical look up table proposed by Hisao Kubota, Kouki Matsuse, & Takayoshi Nakmo (1993), which is shown below as Table 1.

According to this look up table the position of flux and torque is maintained on the basis of six sections and position of voltage vectors. There are six sections of 60^0 intervals and eight voltage vectors (V0 to V7) positions. As the values

Figure 5. Torque hysteresis comparator

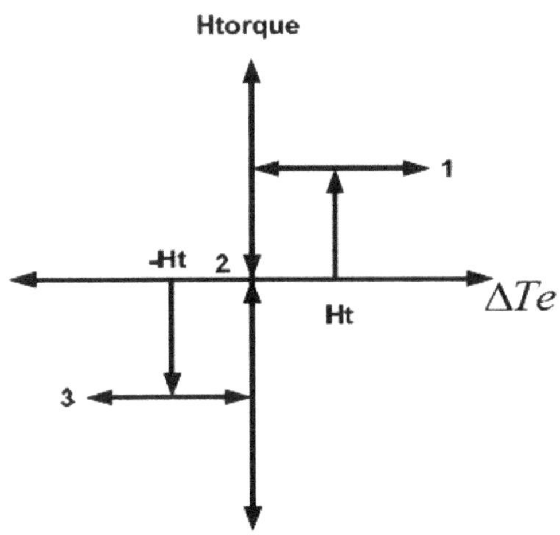

Table 1. Classical Look-up Table

Hflux	Htorque	Sec1	Sec2	Sec3	Sec4	Sec5	Sec6
1	1	V2	V3	V4	V5	V6	V7
	2	V0	V7	V0	V7	V0	V7
	3	V6	V1	V2	V3	V4	V5
2	1	V3	V4	V5	V6	V1	V2
	2	V7	V0	V7	V0	V7	V0
	3	V5	V6	V1	V2	V3	V4

of flux and torque is varies any increment and decrement shown in the flux and torque position, according to that requirement the voltage vector section is selected.

5. MODELING OF SPEED CONTROLLERS

A DTC drive can be operated in two distinct control modes. There can be a torque control mode irrespective of its operating speed and a speed control mode. The closed loop speed control in a DTC drive is achieved indirectly from the torque control loop. Measured or estimated speed is compared with reference speed and the error signal is fed to a controller which generates the reference torque command. This reference torque command is in turn compared with actual estimated torque. Appropriate voltage vectors are selected on the basis of the torque and flux error to compensate the respective errors. Thus a decoupled control of torque/speed and flux is achieved. Since in a direct torque control drive, closed loop speed control is achieved by the torque control loop. Hence it can be concluded that ripples in torque leads to speed ripple, thus deteriorating the steady state performance of the drive. Although such ripples are primarily filtered by the inertia of the motor and the coupled load at high speeds. But at low speeds this torque and speed ripples during no load condition may not be damped and are the cause of acoustic noise and mechanical vibrations. The speed control loop has to be designed carefully for an efficient dynamic response of the drive. The inherent nonlinearities of an induction motor drive along with parameter variations demand a robust and efficient speed control algorithm. The different commonly adopted control algorithms for induction motor drive are,

- Conventional PI Control
- Fuzzy logic control (Hongrae & Thomas, 2006)
- Sliding mode control (Duran, Duran, Perez, & Fernandez, 2004)
- Neural network based speed control algorithm (Bodkhe & Aware, 2010)
- Adaptive control (Vaclavek & Blaha, 2009)

The most simple control algorithm to implement is the PI control. But it has certain limitations such as it requires fine tuning and is not immune to parameter variations. The fuzzy logic control can compensate the system non linearity's through human expertise. The limitations of fuzzy logic control are that it relies too much on expertise and experience of the designer. A robust controller to cope up with the system nonlinearities and parameter variation is an adaptive controller, which can adjust its parameters according to the system dynamics, but its control algorithm is complex in implementation.

In this chapter the feasibility of implementation of the following speed control algorithms has been studied for the speed control of the proposed DTC drive.

1. Conventional PI Control
2. Fuzzy Logic Control
3. Sliding Mode Control
4. Hybrid Speed Controller

The above mentioned control algorithm are implemented through numerical simulations and a few of them have been validated on the experimental drive.

6. CONVENTIONAL PI CONTROL

The most simple control algorithm to implement is the conventional PI control. The schematic block diagram of the proposed controller is shown in Figure 6.

The output of the PI controller is the reference torque command T(k) which is passed through a limiter which saturates at $\pm T_e^{max}$, where T_e^{max} is the rated torque of the induction motor drive. A low pass filter with appropriate cutoff frequency is employed to block the high frequency components in the speed feedback signal. Furthermore the sampling frequency of the sensed speed feedback signal is kept low by introducing a zero order hold in the feedback loop to avoid the oscillations in rotor speed due to the fast acting speed controller when implemented on a DSP based controller.

The mathematical expression of the proposed PI speed controller in discrete form can be given by (17) - (20).

$$N_r^*(k) - N_r(k) = E(k) \tag{17}$$

$$E(k) * \left(K_p + \frac{Ki}{s} \right) = T(k) \tag{18}$$

$$E(k) * (K_p = Ki) = sT(k) \tag{19}$$

$$K_p(E(k) - E(k-1)) + K_i \Delta t E(k) \tag{20}$$

Finally the output expression of the controller is given by (21)

$$T(k) = T(k-1) + K_p(E(k) - E(k-1)) + K_i \Delta t E(k) \tag{21}$$

7. FUZZY LOGIC CONTROL

The schematic block diagram of the proposed fuzzy logic speed controller (FLC) is shown in Figure 7. The basic consideration in the design

Figure 6. Conventional PI controller

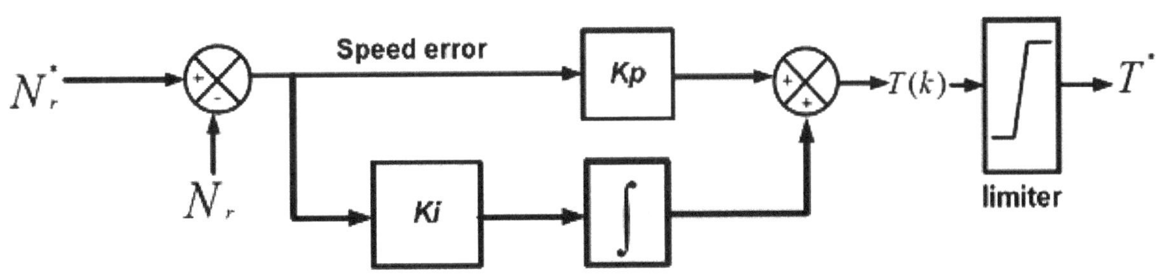

Figure 7. Schematic block diagram of fuzzy logic based speed controller

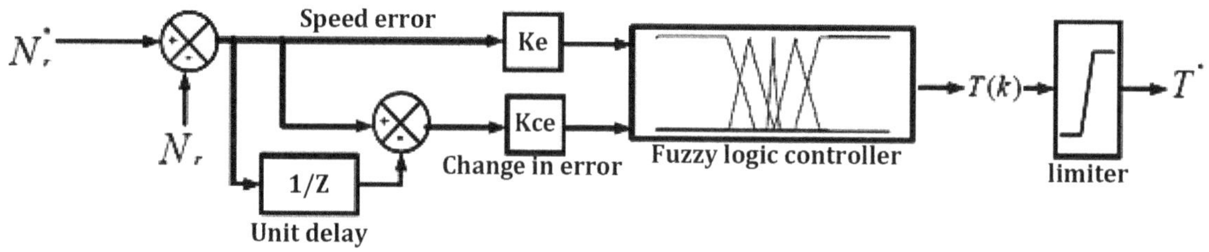

of the control rules is aimed at a faster dynamic response of the drive while maintaining the simplicity of control. The two inputs to the FLC are speed error and change in speed error, which are expressed by (22) and (23) in discrete form for implementation on a DSP based controller.

$$\Delta E(k) = N_r^*(k) - N_r(k) \tag{22}$$

$$\Delta E(k) = (\Delta E(k) - \Delta E(k-1)) \tag{23}$$

8. SLIDING MODE CONTROL

A sliding mode (SM) speed controller is proposed for obtaining a robust dynamic performance of the DTC drive. The schematic diagram of the speed controller is given in Figure 8 .The equations for modeling the SM speed controller are given by equations (24) – (26).

The rotor speed error can be given as

$$X_1 = N_r^*(k) - N_r(k) \tag{24}$$

where $N_r^*(k)$ is the reference speed at k sampling instant and $N_r(k)$ is the actual rotor speed.

The rate of change of speed error in discrete form is given by (25)

$$X_2 = \frac{X1(k) - X1(k-1)}{\Delta t} \tag{25}$$

The output of the sliding mode controller is given by (26)

$$T(k) = C_3 X_1(k) Y_1(k) + C_4 X_2(k) Y_2(k) \tag{26}$$

Reference torque command T*(k) is obtained after limiting the output of the sliding mode controller. In the eqn. (26), C3 and C4 are gains of sliding mode controller and Y(k) can be given as:

Y1 (k) = +1 if ZX1 (k) =0

Figure 8. Schematic block diagram of sliding mode based speed controller

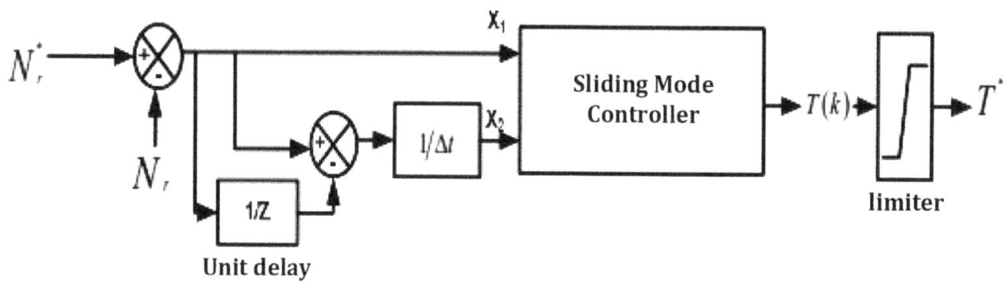

Y1 (k) = -1 if ZX1 (k) <0

And, Y2 (k) = +1 if ZX2 (k) =0

Y2 (k) = -1 if ZX2 (k) <0

Z is switching line on which the sliding mode action occurs and is given by (27)

$$Z=C1\ X1(k) + C2\ X2(k) \tag{27}$$

where C1 and C_2 are adjustable parameters.

9. MODELING OF TORQUE, FLUX AND SECTOR ESTIMATOR

Implementation of DTC algorithm requires an independent control of electromagnetic torque and flux with two independent control loops. Hence the parameters which are to be measured or estimated for the effective implementation of DTC algorithm are

1. Stator flux (estimated)
2. Electromagnetic torque (estimated/measured)
3. Rotor speed (estimated/measured)
4. Motor phase currents (measured)
5. Stator phase voltage (estimated/measured)
6. DC link voltage (measured)

There is a strong interdependency among these variables. For example the flux of an induction motor drive can be estimated with the help of motor back emf, which in turn depends upon stator phase voltage, phase current and stator resistance. The mathematical expression in stationary reference frame can be given by (28) – (30).

$$\lambda_{s\alpha} = \frac{1}{S}\left(E_{s\alpha}\right) \text{ and } \lambda_{s\beta} = \frac{1}{S}\left(E_{s\beta}\right) \tag{28}$$

where

$$E_{s\alpha} = V_{s\alpha} - i_{s\alpha}R \tag{29}$$

$$E_{s\beta} = V_{s\beta} - i_{s\beta}R \tag{30}$$

Thus for the estimation of stator flux the required variables are phase voltage, phase current and stator resistance.

Similarly the electromagnetic torque of an induction motor drive can be either sensed or can be estimated. Estimation of electromagnetic torque rather than its measurement by any torque transducer makes a DTC drive more cost effective and less complex. The electromagnetic torque in stator reference frame is given by (31)

$$T_e = \frac{3}{2}P * \left(\lambda_{s\alpha} * i_{s\beta} - \lambda_{s\beta} * i_{s\alpha}\right) \tag{31}$$

The interdependency among the different parameters can again be seen from the torque equation, which says that electromagnetic torque can be estimated with the help of orthogonal stator flux components ($\lambda_{s\alpha,}$ $\lambda s_{\beta)}$ and stator phase currents.

A voltage source inverter is the heart of any variable frequency drive, and it is a well- established fact that the output voltage of any voltage source inverter, which feeds the induction motor drive, can be easily estimated with the help of DC link voltage and duty cycle of switching signals. Thus a further reduction in sensor count with increased reliability and cost effectiveness can be achieved by estimated the inverter output voltage rather than its actual measurement.

Thus following conclusions can be drawn for the effective implementation of DTC algorithm.

- For the implementation of DTC algorithm, flux, torque and speed are to be measured/estimated.

- Stator flux can be estimated in stationary reference frame with the help of stator phase voltage and phase current.
- Electromagnetic torque can be estimated with the help of stator flux components and stator phase currents in stationary reference frame.
- Estimation of rotor speed can be achieved with the help of machine parameters and stator flux.
- Finally the stator phase voltage which is required for the estimation of flux, electromagnetic torque and rotor speed can be estimated with the help of DC link voltage and switching signals.
- Thus the variables that need to be estimated are stator flux, electromagnetic torque and stator voltage.

Hence for the implementation of DTC algorithm with reduced complexity in control, mathematical expressions for the estimation of the above mentioned variables are developed in stationary reference frame, which are discussed in the following sections.

10. TORQUE AND STATOR FLUX ESTIMATOR

The Simulink model for flux, torque and sector estimation can be developed from the basic equations in stationary reference frame given by (32)–(34). The stator flux components are computed by integration of the voltages in the stationary reference frame given by the equation (32)

$$\lambda_s(k) = \Delta t_s E_s(k) + \lambda_s(k-1) \quad (32)$$

Stator flux magnitude and its phase angle for sector determination can be given by (33) while electromagnetic torque is given by (34)

$$\left|\lambda_s\right| = \sqrt{\lambda_{s\alpha}^2 + \lambda_{s\beta}^2}, \theta = \tan^{-1}\frac{\lambda_{\beta s}}{\lambda_{\alpha s}} \quad (33)$$

$$T_e = \frac{3}{2}P * \left(\lambda_{s\alpha} * i_{s\beta} - \lambda_{s\beta} * i_{s\alpha}\right) \quad (34)$$

11. RESULTS AND DISCUSSIONS

The classical DTC scheme has been implemented with a modification in the look up Table 1. For implementation of DTC algorithm the voltage vectors have been given the new nomenclature such that any voltage vector Vk is designated as V k+1 as shown below. The verification of the control strategy can be done from experimental results shown in Figure 9. The experimental results shown in this thesis have been captured with the help of control desk software of dSPACE. The real time signals are converted into mat. files with the help of control desk, and subsequently these captured signals can be either plotted or utilized for any further mathematical computations.

In Figure 10 a variation of electromagnetic torque with respect to applied voltage vector along with the variation of the different control variables is shown. The criterion for voltage vector selection is in accordance with the modified switching table for a stator flux location in sector 1.

Following observations can be made from Figure 10

- For a positive torque error (when actual estimated torque is less than reference torque) between the sampling instants 1.073 – 1.0735, then the output of torque hysteresis comparator (Htorque) is equal to 1.
- The value of Hflux which is the control signal generated by flux comparator, during the same time period changes between 1 and 2.

Figure 9. Simulink Model of Flux and Torque Estimation Subsystem

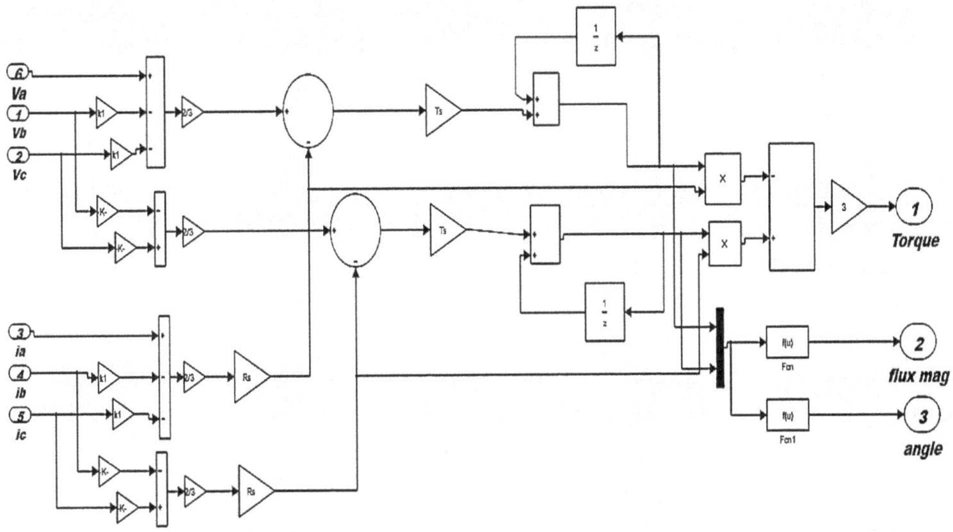

Figure 10. Experimental results verifying the classical DTC algorithm

Hflux	Htorque	Sec 1
1	1	V_{2+1}
	2	V_{0+1}
	3	V_{6+1}
2	1	V_{3+1}
	2	V_{7+1}
	3	V_{5+1}

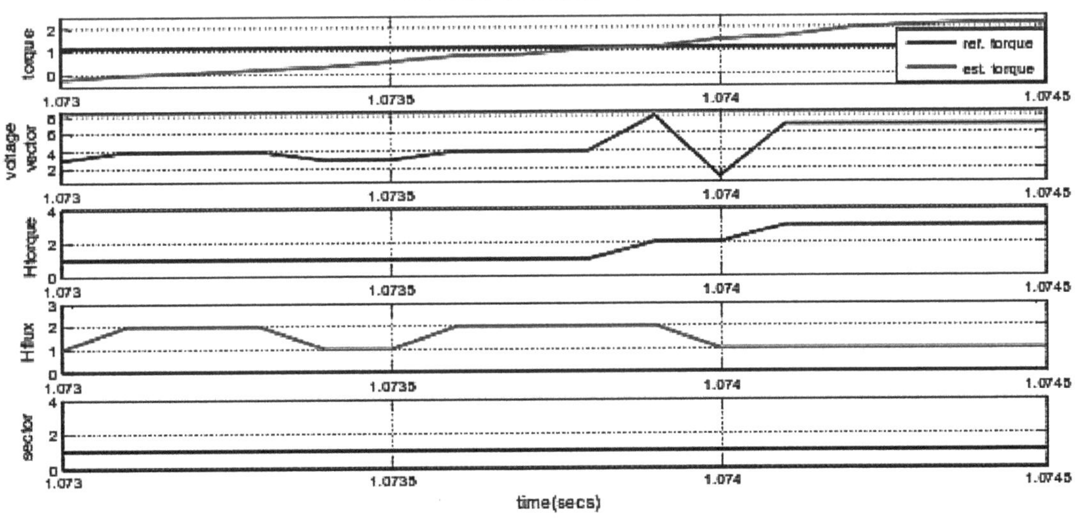

- Stator flux is located in sector 1.
- Hence according to the modified look up table, the applied voltage vector should be either V3 or V_4(shown highlighted), and it can be verified that the same vectors are selected.
- Due to the application of these vectors the torque error reduces, which confirms that selected voltage vectors results in an increment in torque.
- At the sampling instant 1.074 sec, when torque error is zero and approaching towards a negative value, then Htorque attains a value of 2 and zero vectors (V_1 and V_2) are selected for the compensation of torque error.
- After 1.074 sec, as the torque continues its excursion in the positive direction and exceeds reference torque, then to compensate the negative torque error, vector V_7 is selected.

12. CONCLUSION

An analysis of classical DTC algorithm has been carried out from the perspective of its effective implementation for achieving a decoupled control of torque and flux, with closed loop speed control. The following conclusions can be drawn from the findings based on these simulation and experimental results.

For the effective implementation of DTC algorithm with a decoupled torque and flux control with independent control loops, the identified key factors are

- Flux estimation and its control
- Torque control with minimized ripple
- Voltage vector selection algorithm
- Speed control algorithm

Thus a comprehensive study on conventional DTC was carried with the emphasis on flux control, torque control and speed control and the critical parameters for the effective implementation of these control loops were identified.

REFERENCES

Bodkhe, S. B., & Aware, M. V. (2009), " A variable-speed, sensorless, induction motor drive using DC link measurements", 1 Industrial Electronics and Applications, 4th IEEE Conference on pp. 359, 3596.

Bodkhe, S. B., & Aware, M. V. (2010). Robust method for stator current reconstruction from DC link. *International Journal of Engineering Science and Technology*, *2*(10), 87–99.

Bojoi, R., Guglielmi, P., & Pellegrino, G. (2006). Sensorless Direct Field Oriented Control of Three-Phase Induction Motor Drives for Low Cost Applications. *Proceedings of the IEEE Conference on Industry Applications* (Vol. 2, pp. 866 – 872). doi:10.1109/IAS.2006.256627

Boussak, M., & Jarray, K. (2006). A High-Performance Sensorless Indirect Stator Flux Orientation Control of Induction Motor Drive. *IEEE Transactions on Industrial Electronics*, *53*(1), 41–49. doi:10.1109/TIE.2005.862319

Cirrincione, M., Pucci, M., Cirrincione, G., & Capolino, G. A. (2004). A Improved Adaptive Integration Methodology for Estimating Flux In Induction Machine Drives. *IEEE Transactions on Power Electronics*, *19*(1), 25–34. doi:10.1109/TPEL.2003.820565

Stojic, D.M. (2005). A New Induction Motor Drive Based on the Flux Vector Acceleration Method. IEEE Transactions on Energy Conversion, 20(1), 173-180.

Duran, M.J., Duran, J.L., Perez, F., & Fernandez, J. (2004). A novel induction motor IRFOC with on-line parameter estimation taking into account thermal and deep-bar effects. *Proceedings of the 2004 IEEE conference on Industrial Electronics* (Vol. 1, pp. 435–440).

Haron, A.R., & Idris, N.R.N. (2006). Simulation of MRAS-based Speed Sensorless Estimation of Induction Motor Drives using MATLAB/SIMULINK. *Proceedings of the IEEE International Power and Energy Conference* (pp. 411 – 415).

Holtz, J., & Quan, J. (2002). Sensorless Vector Control of Induction Motors at Very Low Speed Using a Nonlinear Inverter Model and Parameter Identification. *IEEE Transactions on Industry Applications*, *38*(4), 1087–1095. doi:10.1109/TIA.2002.800779

Jemli, M., Azza, H., B., , Boussak, M., & Gossa, M. (2009). Sensorless Indirect Stator Field Orientation Speed Control for Single-Phase Induction Motor Drive. *IEEE Transactions on Power Electronics*, *24*(6), 1618–1627. doi:10.1109/TPEL.2009.2014867

Jouili, M., Jarray, K., Koubaa, Y., & Boussak, M. (2009). Sensorless Indirect Stator Field Orientation Speed Control for Single-Phase Induction Motor Drive. *IEEE Transactions on Power Electronics*, 24(6).

Kim, H., & Jahns, T.M. (2006). Integration of the Measurement Vector Insertion Method (MVIM) with Discontinuous PWM for Enhanced Single Current Sensor Operation. Proceedings of the Industry Applications Conference (Vol. 5, pp. 2459-2465).

Kubota, H., & Matsuse, K. (2010). Speed sensorless field-oriented control of induction motor with rotor resistance adaptation. *IEEE Transactions on Industry Applications*, 30(5), 1219–1224

Kubota, H., Matsuse, K., & Nakmo, T. (1993). DSP-Based Speed Adaptive Flux Observer of Induction Motor. *IEEE Transactions on Industry Applications*, *29*(2).

Lascu, C., Boldea, I., & Blaabjerg, F. (2006). DTFC-SVM motion-sensorless control of a PM-assisted reluctance synchronous machine as starter alternator for hybrid electric vehicles. *IEEE Transactions on Power Electronics*, 21(3), 711–719.

Lascu, C., & Andreescu, G. (2006). Sliding-Mode Observer and Improved Integrator With DC-Offset Compensation for Flux Estimation in Sensorless-Controlled Induction Motors. IEEE transactions on industrial electronics, 53(3).

Luukko, J., Niemelä, M., & Pyrhönen, J. (2003). Estimation of the flux linkage in a direct-torque-controlled drive. *IEEE Transactions on Industrial Electronics*, *50*(2), 283–287. doi:10.1109/TIE.2003.809392

Myoung-Ho Shin, Dong-Seok Hyun, Soon-Bong Cho, & Song-Yul Choe (2000). An improved stator flux estimation for speed sensorless stator flux orientation control of induction motors. IEEE Transactions on Power Electronics, 15(2), 312-318.

Nik, R. N. I., & Abdul, H. M. Y. (2002). An Improved Stator Flux Estimation in Steady-StateOperation for Direct Torque Control of Induction Machines. *IEEE Transactions on Industry Applications*, *38*(1).

Pillay, P., & Levin, V. (1995). Mathematical Models for Induction Machines. *Proceedings of the Industry Applications Conference* (*Vol. 1*).

Raghu, T., Srinivas Rao, J., & Chandra Sekhar, S. (2012). Simulation of Sensorless Speed Control of Induction Motor Using APFO Technique. *International Journal of Computer and Electrical Engineering*, 4(4).

Rehman, H., & Xu, L. (2011). Alternative Energy Vehicles Drive System: Control, Flux and Torque Estimation, and Efficiency Optimization. IEEE Transactions on Vehicular Technology, 60(8).

Schauder, C. (1992). Adaptive Speed Identification for Vector Control of Induction Motors without Rotational Transducers. *IEEE Transactions on Industry Applications*, *28*(5), 1054–1061. doi:10.1109/28.158829

Seyoum, D., Rahman, F., & Grantham, C. (2003). Simplified flux estimation for control application in induction machines. Proceedings of the IEEE International Electric Machines and Drives Conference (Vol. 2, pp.691-695). doi:10.1109/IEMDC.2003.1210311

Tripura, P., Kishore Babu, Y.S., & Tagore, Y.R. (2011). Space Vector Pulse Width Modulation Schemes for Two-Level Voltage Source Inverter. *Int. J. on Control System and Instrumentation*, 2(3).

Vaclavek, P., & Blaha, P. (2009). Lyapunov function based design of PMSM state observer for sensorless control. *Industrial Electronics & Applications (Vol. 1*, pp. 331–336).

Versèle, C., Deblecker, O., & Lobry, J. (2008). Implementation of a vector control scheme using dSPACE material for teaching induction motor drive and parameters identification. *Proceedings of the 18th International Conference on Electrical Machines* (pp. 1-6). doi:10.1109/ICELMACH.2008.4799951

Wang, Y., & Deng, Z. (2012). Improved Stator Flux Estimation Method for Direct Torque Linear Control of Parallel Hybrid Excitation Switched-Flux Generator. IEEE transactions on energy conversion, 27(3).

Wang, Y., & Deng, Z. (2012). An Integration Algorithm for Stator Flux Estimation of a Direct-Torque-Controlled Electrical Excitation Flux-Switching Generator. IEEE transactions on energy conversion, 27(2).

Chapter 17
Restructuring Power Industry in India

Kashika Dubey Pathak
Bansal Institute of Research and Technology, India

ABSTRACT

Since the beginning of the nineties decade, many nations have undergone a change in the method of operations of the power network. The power sector was switched to open access market from vertically integrated market. The reasons for this change are many and they differ from nation to nation. The developing countries faced the problem of high demand growth along with irrational tariff policies and lack of efficiency in managing the power system. On the contrary, the developed countries restructured the power sector to provide electricity at lower prices and give the customers a number of operators to choose from. In short, the goal of restructuring is to encourage competition in the market along with providing consumers with more choices in purchasing economic energy and economic benefits. This might seem to be less complex but involves several issues which are discussed further. This chapter describes this restructuring of power sector with special reference to Indian scenario.

1. INTRODUCTION

The need to restructure the electric power industry was felt to give economic access of power to the consumers as well as facilitate its easy generation (Khaparde 2004). It was realized by the investors that the competitive market and the reduced government regulatory costs would result in increased power supplies at low prices. Thus, sufficient competition amongst the generation sector has developed, which also gives consumer a number of choices to choose from.

The electricity is usually generated at the central station by rotating a magnet inside of a coil of wire. This electricity, is then, transmitted from the station to long distance substations (that deliver the power to consumers) over high voltage transmission lines. At a generating plant, electric power is "stepped up" to several thousand volts by a transformer and delivered to the transmission line. At numerous substations on the transmission system, transformers step down the power to a lower voltage and deliver it to distribution lines. The type of transmission structures used for any

DOI: 10.4018/978-1-4666-9911-3.ch017

project is determined by the characteristics of the transmission line's route, including terrain and existing infrastructure. One of the chief concerns of restructuring involves ensuring fair and reliable market for generating electricity along with ensuring the equal access to transmission grids.

2. POWER INDUSTRY IN INDIA

Electricity is one of the major components of infrastructure that plays a crucial role in the development of the nation. India stands 5th in position in terms of power generation. About 65% of the electricity consumed in India is generated by thermal power plants, 22% by hydroelectric power plants, 3% by nuclear power plants and rest by 10% from other alternate sources like solar, wind, biomass etc (The Electricity Act, 2003). 53.7% of India's commercial energy demand is met through the country's vast coal reserves. In India, sources for power generation range from commercial sources such as coal, lignite, natural gas, oil, hydro and nuclear power to other viable non-conventional sources such as wind, solar, and agriculture and domestic waste. With development of the nation at a fast pace and an increase in population every year, the demand for electricity is also growing at a rapid rate and is also expected to increase further in the years to come. It is expected that by 2030-35, energy demand in India will be the highest among all the nations. Government of India has also started several projects like the Jawaharlal Nehru National Solar Mission to attract investors to India.

Power Sector in India has undergone severe changes since the enforcement of the Electricity Act, 2003 (The Electricity Act, 2003). It has paved the way for competitive and open access market. Basically, the Power sector has three foundations to its support: Generation, Transmission and Distribution. The Generation of Power is divided amongst Central, State and Private Sectors. The main focus in Electricity generation is mainly concentrated on generation that involves low cost, optimization of capacity utilization, upgradation of technology and complete utilization of non-conventional resources.

Transmission of electricity involves bulk transfer of power over a long distance at high voltages. The entire country has been divided into five regions for transmission systems, namely, Northern Region, North Eastern Region, Eastern Region, Southern Region and Western Region. The Interconnected transmission system within each region is also called the regional grid.

In India, many of the state governments provide electricity at subsidized rates or even free to some sections. This includes for use in agriculture and for consumption by backward classes. The subsidies are mainly as cross-subsidization, with the other users such as industries and private consumers paying the deficit caused by the subsidized charges collected. One of the major disadvantages of providing subsidized electricity has resulted in the financial losses to the state governments.

3. ISSUES IN INDIAN POWER SECTOR

India being a developing nation faces a lot of problems in meeting the rapidly growing demand of power consumption. Some of them are:

1. *Limited Fuel Resource*: In India, electricity is mainly produced at the thermal power stations where the main source of electricity production is coal. As per a study, the Indian power sector used about 400 million tonnes of coal in 2011–2012. However, the production of coal is insufficient to meet the growing demand of power consumption. This shortage is being addressed by importing coal from other nations (Figure 1).
2. *Shortage of Core Components*: There is inadequate of availability of some core equipments needed in power production such as

Generators, Boilers, Turbines, coal-handling plants, ash-handling plants etc.

3. *Losses During Transmission and Distribution of Electricity*: This is one of the most severe problems faced by Indian Power Industry. In India, average T & D (Transmission & Distribution) losses have been officially indicated as 23 percent of the electricity generated. However, as per sample studies carried out by independent agencies including TERI, these losses have been estimated to be as high as 50 percent in some states. Controlling the T&D losses would result in minimizing the demand supply gap.
4. *Inefficient Power Plants and Transmission Network*: Since many of the existing power plants in India were established immediately after Independence in year 1947, they have become old and are inefficient in handling the growing power demands. This fact also contributes to the low growth and transmission rate of electricity production and transmission. These power plants need to be improvised as per the modern techniques of electricity production and transmission to meet its fast pace.
5. *Delay in Project Construction and Approval:* In India, the starting up of new projects is facing a delay due to lack of financial support and slow flow of money from various departments involved in the project.
6. *Less Usage of Renewable Resources:* India is a nation with abundant amount of renewable resources available with all the states. However, this availability varies from state to state. Some states have more than sufficient renewable resources available with them whereas some lack in the availability of these resources. But still, the inclination of Indian Power Sector towards the renewable resources is very less and most of them remain unexplored.

4. RESTRUCTURING IN INDIA

Overview

The first obvious question is "What actually restructuring is?" As the name suggests, restructuring means creating a new framework for the power industry. The term restructuring can be used synonymously with the term 'deregulation'.

Power sector across the globe is undergoing restructuring; India is no exception to this. At the time of Independence, i.e. around 1947, the nation

Figure 1. Electricity Production in India

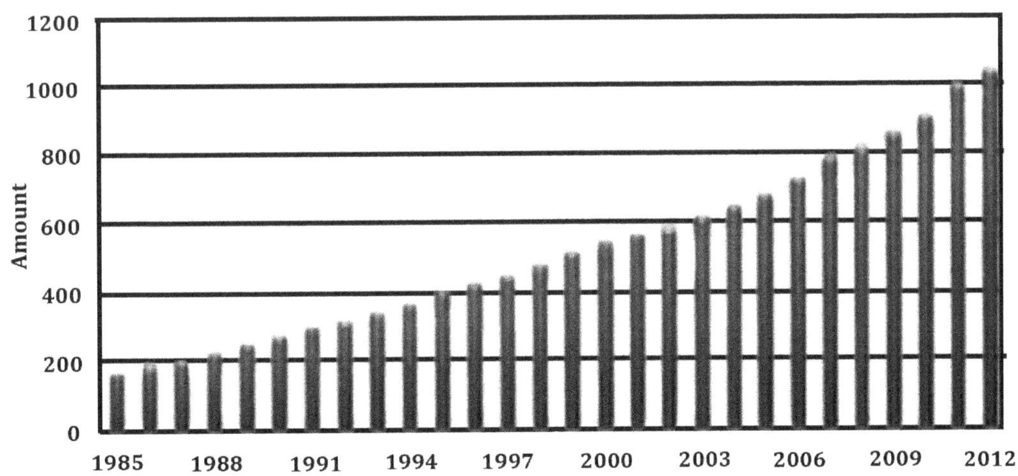

was not much developed and the Indian power sector was mainly concentrated to few urban areas only to cater to entire needs. As the time passed by, several projects were started in merely few decades to provide power to as much population as possible and also to serve the needs of irrigation for agriculture sector and flood control. Sixties saw the further development in the Power Sector in India calling for the integration and evolution of power grids. But, attempt to join these power grids to form the regional grids was successful during the seventies and eighties, with an increase in unit size from 210 to 500 MW and in transmission voltage from 220 to 400 KV.

The need for restructuring the Power Industry in India was felt due to lack of financial resources with the Central and State Governments as well as improving its commercial efficiency. On 24th July, 1998, CERC (Central Electricity Regulatory Commission) was formed with an aim to [1] rationalize and regulate electricity tariffs of Power Generating Companies owned or controlled by Govt. of India; [2] undertake APDRP (Accelerated Power Development and Reform Programme) with an objective of improvement in the quality of supply.

Proposed Model for Restructuring

Basically two models prevail in the areas across the globe where separation of generation, transmission and distribution has taken place. The two models are ISO (Independent System Operator) model and TSO (Transmission System Operator) model. The ISO has responsibility for controlling the access to and use of the transmission grid by competing generators and retailers. Transmission assets may still be owned by a single generation company or retailer, but real time control of their operation is vested in the ISO, to ensure that the ownership of transmission assets does not facilitate market foreclosure or other anti-competitive behavior by integrated generators or retailers. TSO owns the transmission wires as well as controls their operation. A common objective of both the ISO and the TSO model is that they seek to facilitate increased trading within the area over which they have control.

To form an efficient market, neutrality is an important aspect that TSO addresses and due to the same reason it becomes more suitable for restructuring electrical market in India over ISO. Some of the Developed nations are also diverging from ISO model and are converging towards the TSO Model.

Benefits of Restructuring

Since restructuring paved the way for open market access, the competitive market provides a wide range of benefits to both the private companies and the customers. Some of these benefits include:

1. **Choice of Retailer:** The customer now has a provision to choose amongst various retailers. The retailers will compete on the price and various other facilities provided to their customers. This competition will ultimately lead to the benefit of the customer.
2. **Lowering Down of Electricity Rates:** It is a proven fact that in a competitive market, the prices tend to go down as the retailers will try to sell the product at minimum marginal cost to attract more number of customers.
3. **Customer-Centric Services:** The retailers would provide better facilities and services than that provided by the monopolists.

5. SALIENT FEATURES OF THE ELECTRICITY ACT, 2003

The salient features of this act are summarized as follows in Table 1:

Table 1.

S. No.	Category	Key features of Act
1.	Objectives	• Encouraging autonomous regulation with the separation of policy regulation and operational aspects • Rationalizing tariff and lowering the cross-subsidization levels • Creating competition in the industry • Ensuring supply of electricity to all areas • Protecting consumer interests
2.	Policy	National electricity plan shall be prepared in accordance with National Electricity Policy every 5 years National policy on stands alone systems for rural areas and Nonconventional energy systems National policy on electrification and local distribution in rural areas
3.	Restructuring	• Vertical integration instead of horizontal unbundling of State Electricity Boards (SEBs) to make them financially strong • State governments will have the freedom to decide the sequence and phases of restructuring, and also retain the integrated structure of the SEB for a limited period • Introduction of the concept of power trading as a distinct activity, and the introduction of a spot market for bulk electricity
4.	Generation	• Removal of captive power plants from the ambit of licensing and other permissions • Generators can contract directly with DISCOMs (Distribution Companies) • DISCOMs can have embedded generation • Captive generation allowed freely—can supply to associates • Elimination of Licensing requirement and techno-economic clearances for generation projects except hydel projects
5.	Transmission and Distribution	• Provision for Private participation in distribution • Surcharge for open access to meet current cross-subsidy burden (except for CPP's) • Dedicated transmission lines allowed (not regulated) • Central and State transmission Utilities (CTU and STUs) not permitted to trade • Transmission licensees allowed • Multi Year Tariff (MYT) recommended • Bidding allowed

6. OPEN ACCESS TO TRANSMISSION NETWORK

Prior to the enforcement of Electricity Act 2003, the Distributed Companies (DISCOMs) and the consumers were not provided with any legal framework so as to choose the electricity suppliers. The Act enabled the DISCOMs and the consumers to form a choice of suppliers of electricity. On the same grounds, the generator also has the choice to choose amongst the various DISCOMS of its choice. The Act specifies the provisions for non-discriminatory use of transmission lines or distribution system or associated facilities with such lines or system by any licensee or consumer or a person engaged in generation. This is known as 'open access.'

In view of effective implementation of 'open access' in Indian Market, a concept paper was prepared and published by CERC in 2003, which was circulated among different stakeholders and interested persons with a view to considering their suggestions before formulating the regulations envisaged in the Act. In this paper, various pricing methods like contract path method, postage stamp method, MW-mile method and congestion pricing method have been discussed.

Contract path method is presently applied for determination of wheeling charges for State Electricity Board (SEB) system used for inter-state transmission. National Thermal Power Corporation (NTPC), Western Regional Electricity Board (WREB) supported this method due to its simplicity and distance sensitivity. Central Electricity

Authority (CEA) and Eastern Regional Electricity Board (EREB) opposed this method since it was felt that the actual path may be different from contract path and that it would not reflect true cost in many cases such as transmission taking place in displacement mode.

Incremental Postage Stamp method considers the country to be parted into squares of 100 km x 100 km and the charges payable by the open access customer are determined by counting the squares vertically and horizontally from the source to sink of the transaction. The method makes the rate sensitive to distances exceeding 100 km. Grid Corporation of Orissa Ltd. (GRIDCO), West Bengal State Electricity Board (WBSEB), Railways came up in favor of this method. Madhya Pradesh Electricity Regulatory Commission (MPERC) recommended application of this method in cases where the line involved in power flow cannot be determined. The Energy and Resources Institute (TERI) was against accepting this method arguing that the sharing of stamps between State Transmission Utility (STU) and Central Transmission Utility (CTU) would add complexity.

In Flow Based Pricing or Megawatt-mile method, the transmission rates explicitly highlight the cost of transmission, based on both the power flow and the network usage between the receipt and delivery points. MW-mile is a sophisticated analytical method that involves load flow analysis to model power flows on the transmission network to determine charges and hence requires complex data. Southern Regional Electricity Board (SERB) and Industrial Development Bank of India (IDBI) pointed out that MW-mile method is most scientific but too complex to be applied in present conditions.

CEA has come out with an improvised version of incremental postage stamp method, whereby the country is divided into 14 zones, each represented by one or more states. Stamps between various zones are counted not only by the physical distance but also by considering the mode of transmission (i.e. actual flow or displacement). CEA has also suggested a matrix of stamps between various zones which are further based on notional distances. The notional distance captures the existing flow pattern and impact of incremental flow due to open access transaction. Thus, this method tries to replicate results obtained from MW-mile method, without going into complexities of the later.

Some different thoughts like 'Single National Postage Stamp' Methods and 'Free Wheeling' were also proposed by some of the participants. The former was ruled out because of the regional basis of the electricity grid and lack of capacity and capability to transfer freely substantive amount of power across the region. The later was opposed by the state beneficiaries.

The issues other than transmission pricing, like customer categorization were also mentioned in the concept paper. The Commission categorized the open access customers into two specific categories: (i) Short Term Customers, who wish to use the transmission service up to one year and (ii) Long Term Customers, who wish to use the transmission access for 5 years or more. Allotment priority of long term open access customers shall be higher than the short term customers.

The Commission even ordered that the nodal agency for long term access shall be the Central Transmission Utility (CTU), which is POWERGRID at present. The nodal agency for short term access shall be the Regional Load Despatch Center (RLDC) of the region in which the point of drawal is located.

The Availability Based Tariff (ABT) has been implemented by Central Commission in all the five regions of the country which has built-in mechanism for energy accounting in respect of any deviations from the declared schedules. The Commission has recommended that same methodology and procedure for energy accounting should also be implemented at the State level also.

7. CONCLUSION

Restructuring of power industry is being fast adopted by many countries. The introduction of competition in the electricity industry is a complex process and is only solution for survival of same. For a developing nation like India, the prospectus and the economic growth are basically dependent on the private sectors, especially when the concern is the power sector. It is also important to create a healthy competition among the players who participate in power generation without violating the policies in a deregulated electricity market. India is still facing major challenges in restructuring its power sector such as providing electricity access to all the households, improving reliability and quality of power supply. Its power systems are struggling to overcome power shortages and poor power quality. The major constraint in achieving the target is shortage of capital resources.

Considering all these problems some if the proposed solutions are mentioned in this study. Some of them include optimum utilization of existing resources, reducing transmission and distribution losses, demand side management through greater conservation of electrical energy, policy changes in pricing mechanism, shift and emphasis on renewable energy sources for power generation, total energy systems, etc.. Steps which may help large scale integration of renewable power with conventional power generation are also enumerated.

REFERENCES

Animesh, P. (2013). Power Sector in India: Growth, Policies and Challenges. *International Journal of Emerging Technology and Advanced Engineering*, 3(Special Issue 3), 527–536.

Khaparde, S. A. (2004). Power Sector Reforms and Restructuring in India. *Proceedings of the IEEE Power Engineering Society General Meeting* (Vol. 2, pp. 2328–2335).

Official Website of Central Electricity Regulatory Commission (CERD). (n. d.). Retrieved from http://www.cercind.gov.in/

Srivastava, A., & Shahidehpour, M. (2002). M. Restructuring Choices for the Indian Power Sector. *IEEE Power Engineering Review*, *22*(11), 25–29. doi:10.1109/MPER.2002.1045559

Impact of Restructuring of SEBs (Study Report). (n. d.). *Ministry of Power, Govt. of India*. Retrieved from http://powermin.nic.in/whats_new/pdf/IIPA.pdf

The Electricity Act, 2003. (n. d.). *Central Electricity Regulatory Commission (CERD)*. Retrieved from http://www.cercind.gov.in/08022007/Act-with-amendment.pdf

Tripta, T. (2005). Impact Assessment of the Electricity Act 2003 on the Indian Power Sector. *Energy Policy*, *33*(9), 1187–1198. doi:10.1016/j.enpol.2003.11.016

Chapter 18
Application of Soft Computing Techniques in Power Engineering

Vivek Venkobarao
IEEE Senior Member, India

ABSTRACT

A discharge estimator is designed and analyzed in this chapter to estimate the discharge with voltage, frequency and current without prior knowledge of head. A control of discharge in submersible pump is achieved by varying the motor speed. In VSI fed induction motor the speed is varied by varying the switching frequency. The ratio of frequency to the applied voltage is maintained constant, to keep the air gap flux at a desired value. The development of discharge estimator, PID and Fuzzy controllers are discussed in this chapter.

1. INTRODUCTION

The discharge and head of submersible pump are difficult to monitor and record under real-time conditions. So a discharge estimator, which will estimate the discharge close to actual value, is needed. This is achieved by designing a discharge estimator using ANN. In the proposed estimator, discharge is considered as a function of motor input voltage, stator current and frequency. Figure 1 shows the block diagram of the proposed estimator.

The ANN techniques available in MATLAB are used in implementation of the estimator as shown in Figure 2. The training data for the neural

Figure 1. Block diagram of the estimator

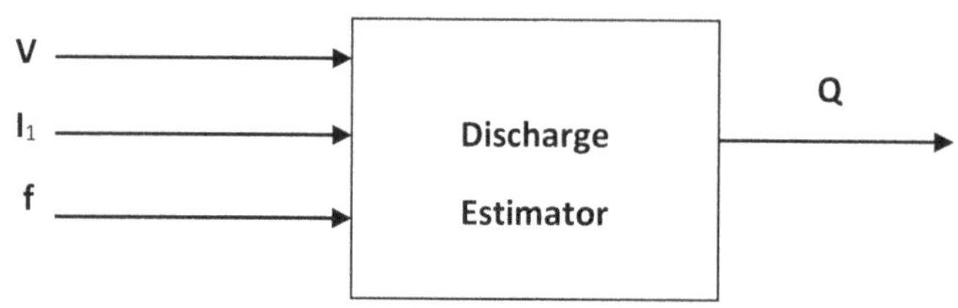

DOI: 10.4018/978-1-4666-9911-3.ch018

Figure 2. ANN Model of Discharge Estimator

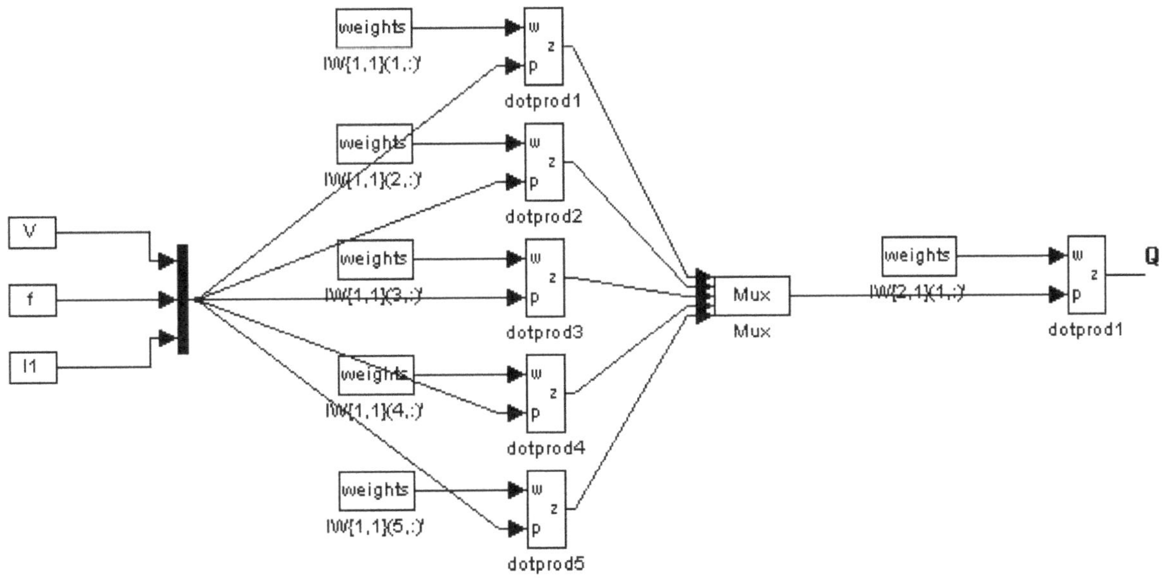

network is obtained from the VSI fed induction motor driven submersible pump. The input and output training sets are normalized and the ANN is trained with this normalized data. The estimator developed is tested for various realistic values of inputs and the discharge thus obtained is compared with the simulated results as shown in Figure 3. It can be seen that the error is less than 2%.

Hence this estimator can be used to sense discharge effectively within ± 2% error.

2. DISCHARGE CONTROLLERS

The function of discharge estimator with the controller is to control discharge without moni-

Figure 3. Percentage Error Plot

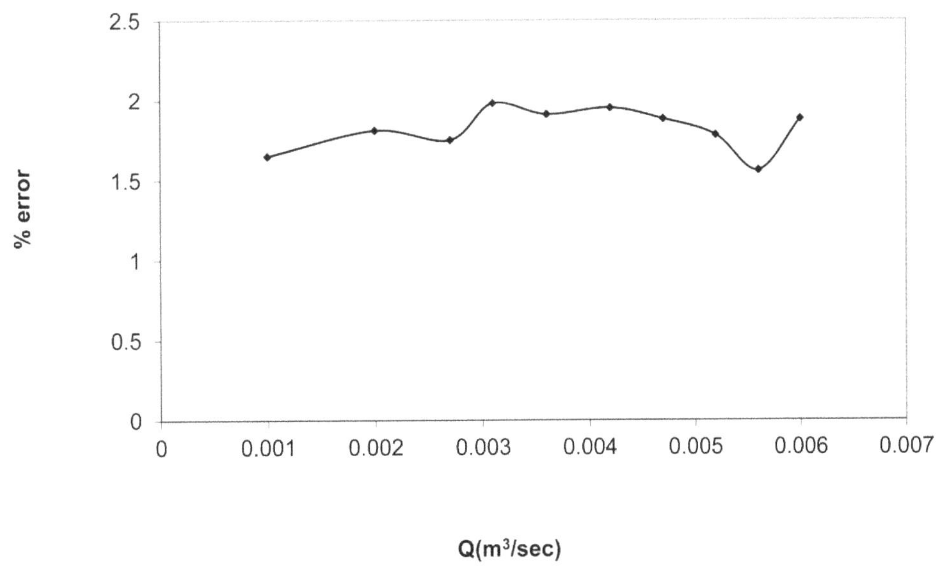

toring it. The discharge control is achieved using simple closed loop V/f control. The power circuit consists of uncontrolled rectifier with three-phase ac supply, filter and a PWM inverter. The speed command corresponding to the required discharge is predicted using the block diagram given in Section 4 and is compared with the actual speed command, which is equivalent to the actual discharge. The error is fed to the PID controller and which generates a control signal. The control signal is added with the actual speed signal to generate a frequency command. The frequency command also generates the voltage command through a V/f map. The frequency command and the voltage command are fed to the PWM inverter fed induction motor in order to achieve the required discharge.

Modeling of discharge estimator and PWM inverter as sub blocks for closed loop control to implement maximum efficiency control and constant flow control are given in Figures 4 and 5 respectively. The block diagram includes reference signal generator and current limiter, which are discussed in next section.

3. PID CONTROL SCHEME

PID controller is designed for the induction motor driven submersible pump, as this is widely used in the industry.

4. DESIGN OF REFERENCE SIGNAL ESTIMATOR

The reference signal estimator generates reference discharge, which can be either for maximum efficiency or constant discharge operations based on the application.

In maximum efficiency operation the reference discharge is generated by tracing the locus

Figure 4. Closed Loop Operation for maximum efficiency control with constant V/f

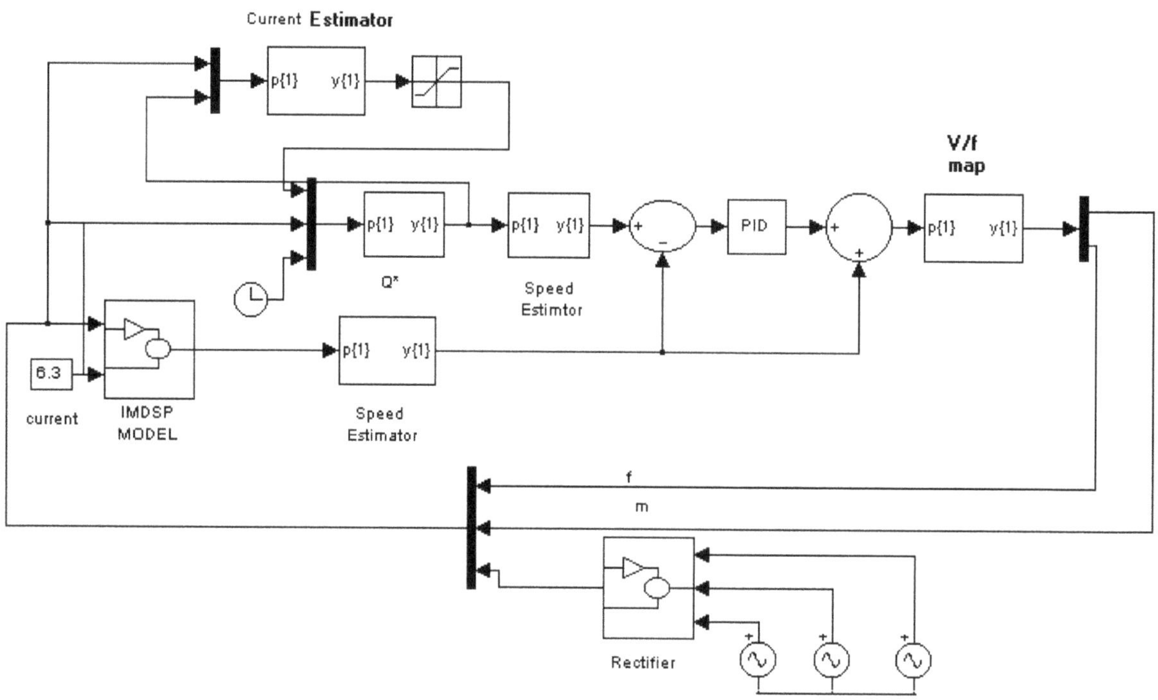

Figure 5. Closed Loop Operation constant flow control with constant V/f

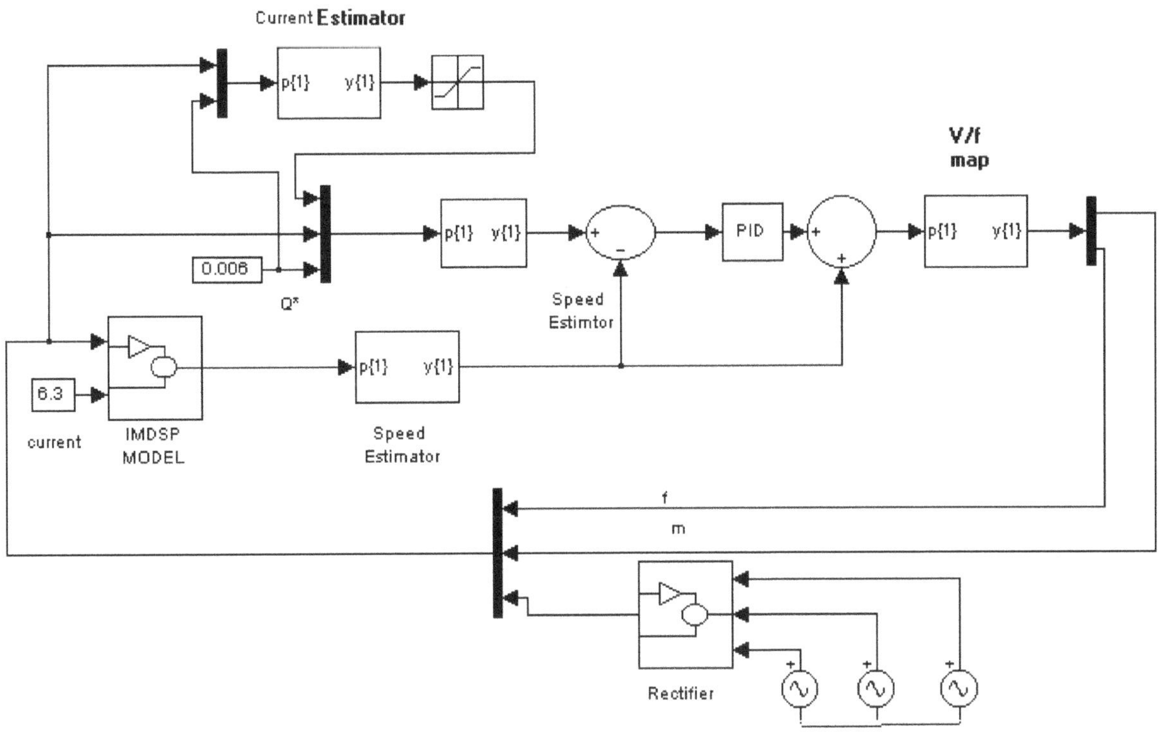

of maximum efficiency at a particular line current, whereas in constant flow operation required discharge is set based on the requirement.

The block diagram and Simulink model for generating the reference discharge for maximum efficiency operation are given in Figures 6 and 7 respectively.

Block-diagram and Simulink model of speed estimator are shown in Figures 8 and 9 respectively. Speed estimator is used for both the control strategies.

1.2 Design of Current Limiter

The current in the motor should not exceed the rated value for wide range of heads. A limiter takes care of this requirement. The line current of the induction motor driven submersible pump is highly non-linear. ANN technique is used to model this non-linearity.

The ANN model is developed with inputs as motor voltage, frequency and discharge and output as current .The Simulink model of the same is shown in Figure 10.

Figure 6. Block Diagram of Reference Discharge Estimator under Maximum Efficiency Operation

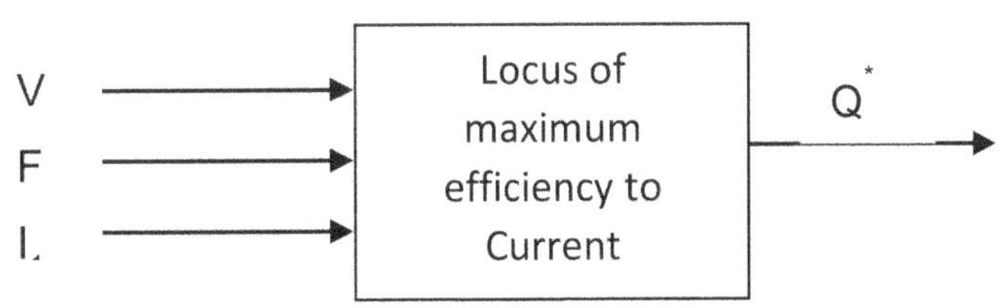

Figure 7. Simulink Model of Reference Discharge Estimator under Maximum Efficiency Operation

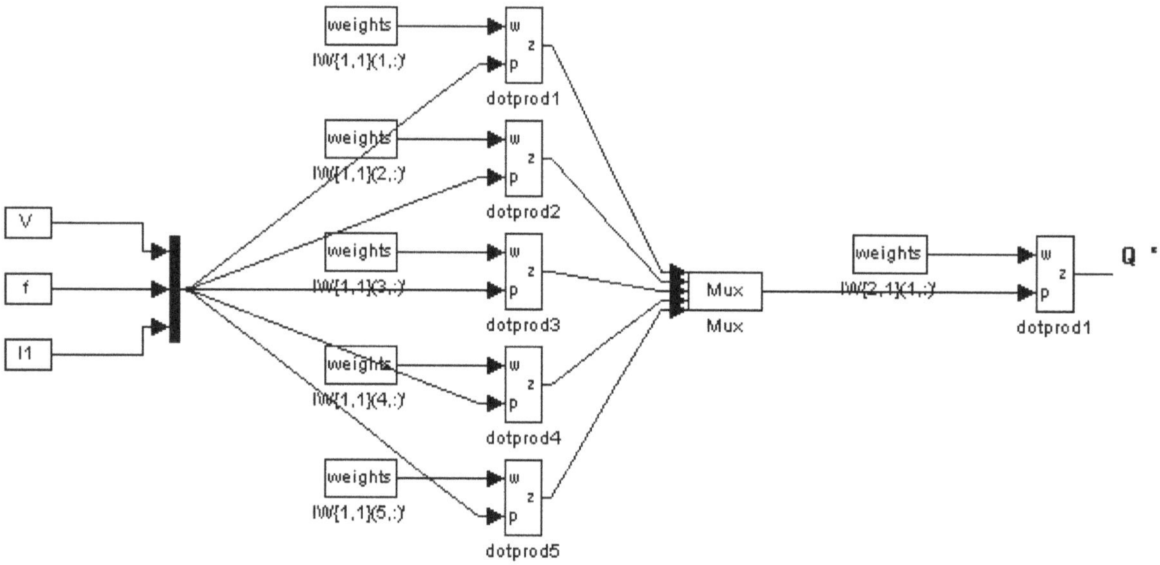

The optimal parameters of the controller can be determined by many methods. In the present work, Ziegler Nichols tuning is used. The tuned values of proportional gain (K_p) integral time constant (T_i) and derivative time constant (T_d) are given in Table 1.

The drive system is simulated with PID type discharge controller. Energy consumed by the pump under various heads is used for analysis.

1.3 Performance of the Drive Under Maximum Efficiency Operation

The initial current is set as 6.3 amps as it corresponds to the rated head. An increase of 0.1 amps for every hour is incorporated in the model. It is seen that from simulation of the drive under maximum efficiency water flow rate decreases from 0.006 m^3/sec to 0.0012 m^3/sec with a speed variation of 1360 rpm to 1450 rpm for pumping out 100 liters of water at rated head.

A speed range of 1020 rpm to 1470 rpm is achieved for wide varying heads without exceeding the current limit.

1.4 Performance of the Drive Under Constant Flow Operation

The maximum discharge of 0.006 m^3/sec is set as the reference signal. It is seen that from simulation the efficiency of the unit decreases from 68.8% to 63.8% during pumping out 100 liters of water with a speed range of 1350 rpm to 1420 rpm at rated head.

Figure 8. Block diagram of speed estimator for a given operating condition

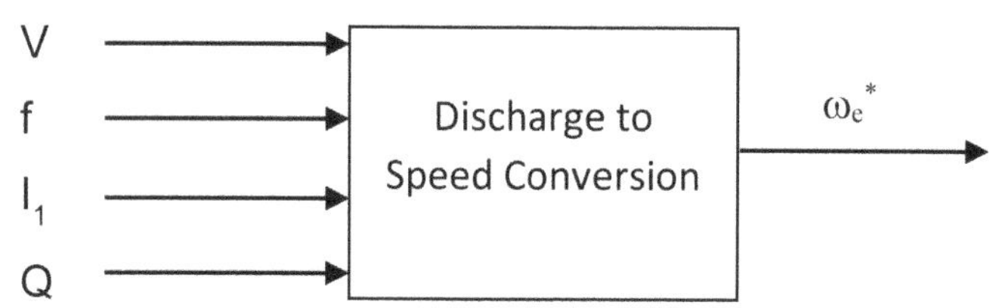

Figure 9. Simulink Model of Speed Estimator

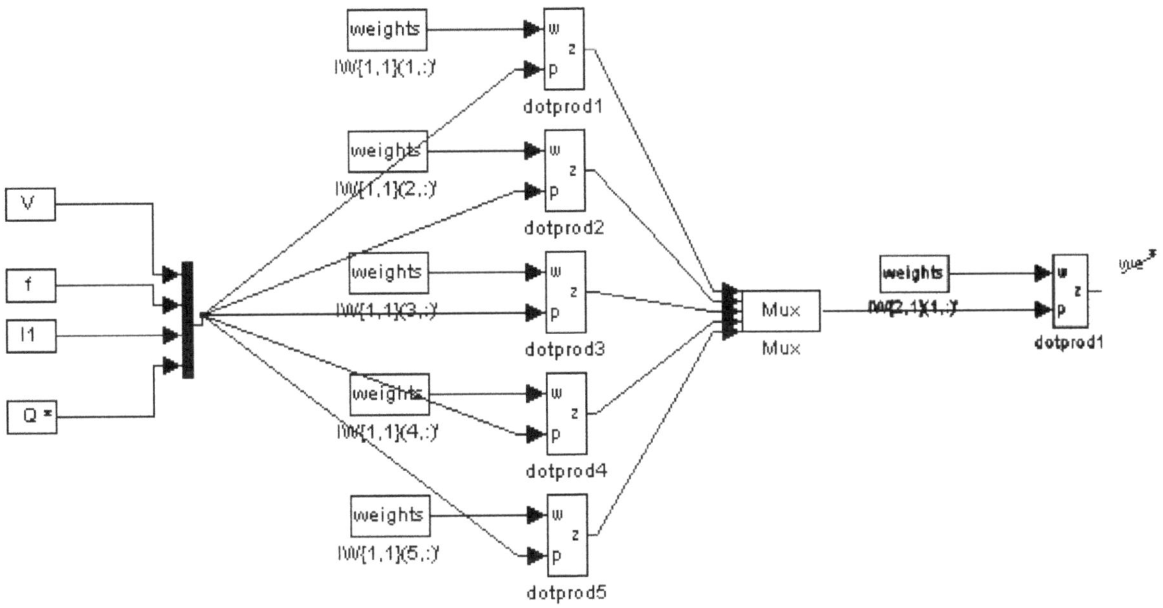

Figure 10. Simulink Model of the Current Limiter 4.2 PID Controller

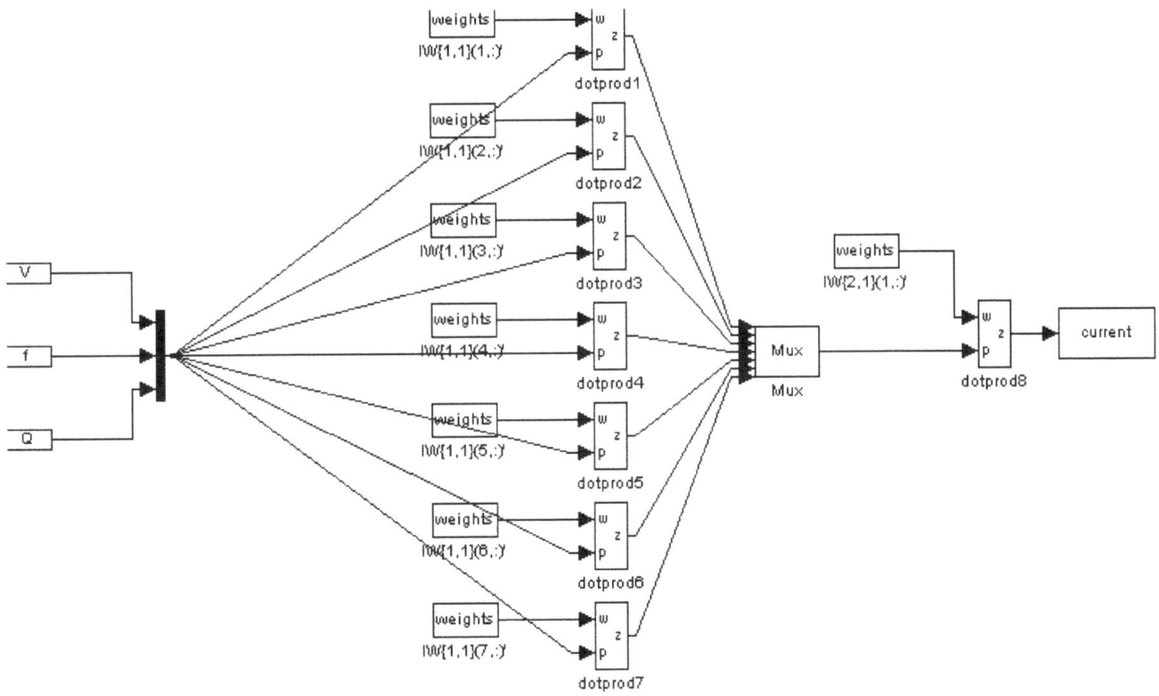

Table 1. Settings for PID Controller

Gain	Value
Proportional Gain (K_p)	1.21
Integral Time (T_i)	0.025 sec
Derivative Time (T_d)	0.09 ec

For a given discharge and wide varying heads a speed range of 1020 rpm to 1430 rpm is observed in the simulation without exceeding the current limit.

Table 2 shows the comparison of the performance for the maximum efficiency and constant discharge operation for pumping 100 liters of water at rated head.

In order to obtain the energy consumed in pumping water over a year, the variation of head in a year should be known in advance.

Figure 11 shows a typical distribution of head in hours in a year.

Figure 12 shows the energy consumed for pumping 0.75 tmc of water for various heads.

1.5. Fuzzy Logic Controller

Fuzzy Logic Controller (FLC) uses fuzzy logic as a design methodology, which can be utilized for developing linear and non-linear systems. A fuzzy logic controller does not require an exact mathematical model. It however requires knowledge-based set of heuristic rules. These rules are impressive and are expressed in terms of linguistic

Table 2. Comparison of Performance of Maximum Efficiency and Constant Discharge Operation at Rated Head

Performance Parameter	Maximum Efficiency Operation	Constant Discharge Operation
Speed Variation	1360-1450 rpm	1350-1420 rpm
Efficiency Range	71%-70%	68.8%-63.8%
Energy Consumed	0.35 Kwhr	0.46 Kwhr
Time taken to pump 100 liters of water	1.3 min	48 sec

Figure 11. Typical variation of head during an year

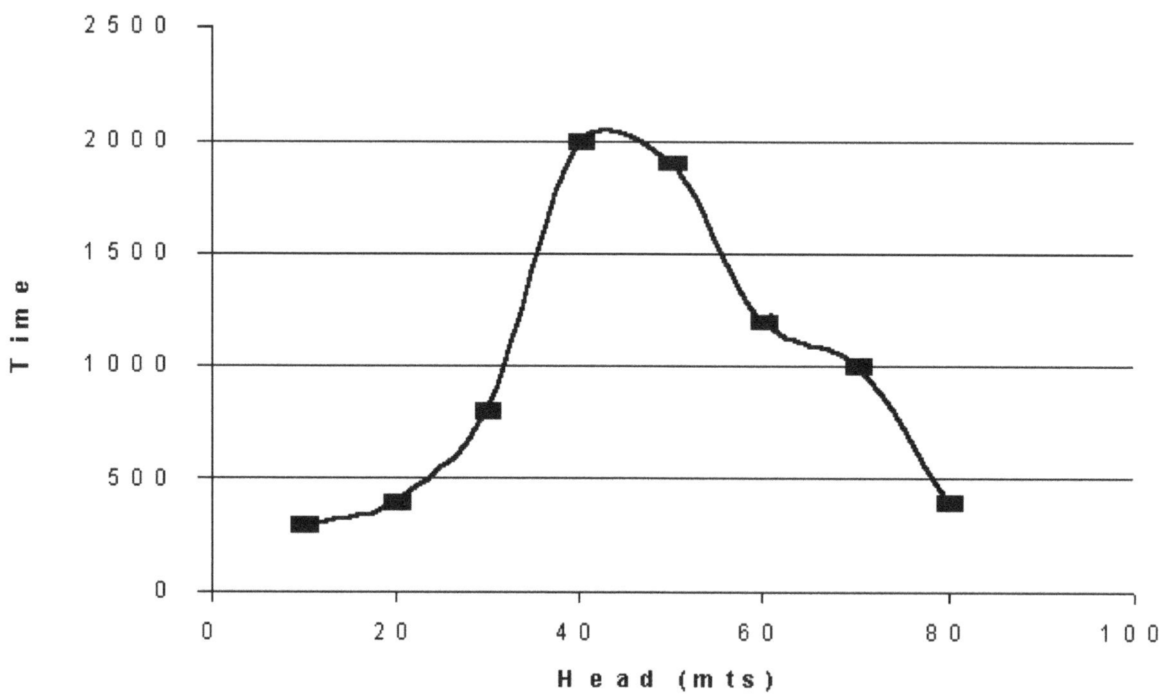

Figure 12. The energy consumed by the submersible pump in pumping 0.75 mc of water

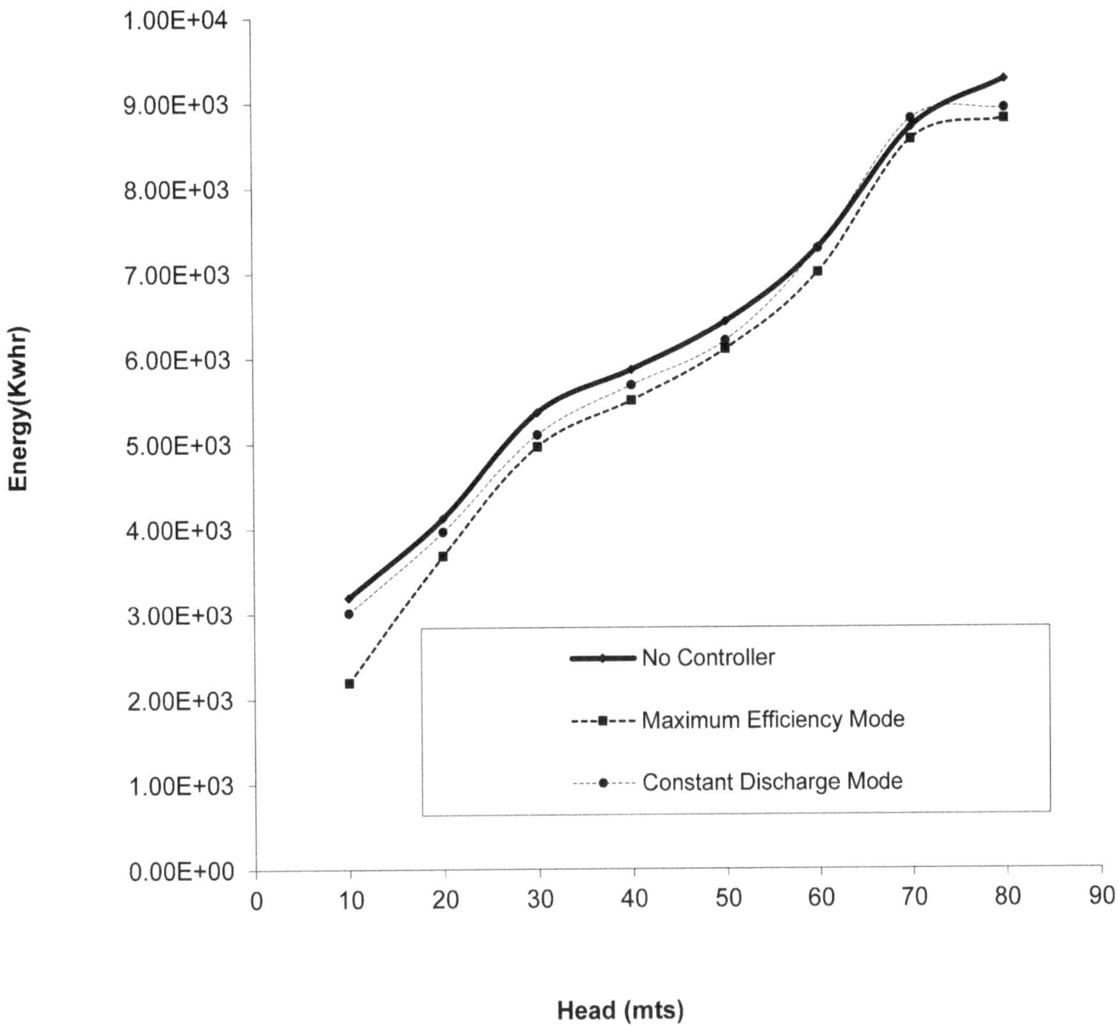

variables. The block diagram of a fuzzy logic controller is shown in Figure 13. The steps involved in the design of fuzzy controller are fuzzification, rule identification and defuzzification.

Fuzzification is the process of transforming crisp control variables to corresponding fuzzy variables. The error e and derivative of error Δe have been selected as inputs and cf as output in the present work. Linguistic variables used are LP, MP, SP, VS, SN, MN, LN which stand for large positive, medium positive, small positive, very small, small negative, medium negative and large negative respectively. Linear triangular membership function is used.

Fuzzy rules are used to formulate the conditional statements that relate the input output fuzzy set. The system with two input variables and seven labels has a 7x7 decision table as given in Table 3. Every entry in the table represents a rule. As an example, if e is LN and Δe is LP then cf is VS.

Decisions are based on the testing of all the rules in the FIS and the rules must be combined

Figure 13. Block Diagram of Fuzzy Logic Controller

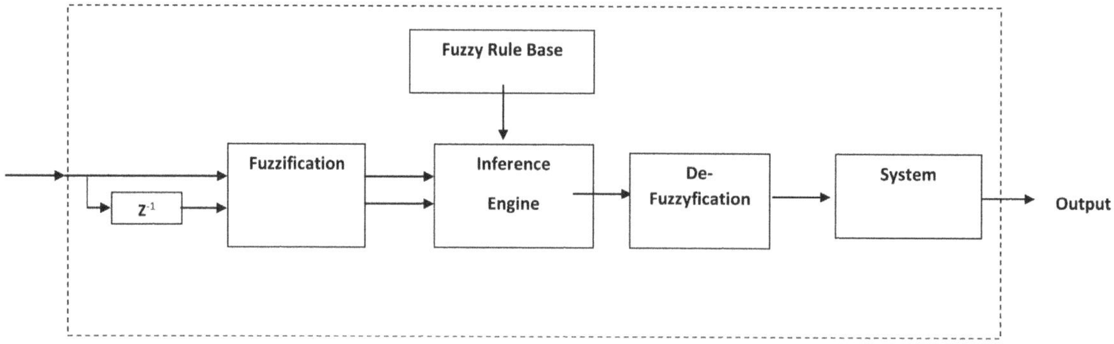

Table 3. Rule Table for FLC

e \ Δe	LN	MN	SN	VS	SP	MP	LP
LP	VS	SP	MP	LP	LP	LP	LP
MP	SN	VS	SP	MP	MP	LP	LP
SP	MN	SN	VS	SP	SP	MP	LP
VS	MN	MN	SN	VS	SP	MP	MP
SN	LN	MN	SN	SN	VS	SP	MP
MN	LN	LN	MN	MN	SN	VS	SP
LN	LN	LN	LN	LN	MN	SN	VS

in some manner in order to make a decision, which is known as aggregation. In this work, built-in-function Max available in the fuzzy toll box of MATLAB is used for aggregation. The output of the aggregation process is the input for the defuzzification. Centroid calculation, which returns the center of area under the curve, is used for defuzzification. Tables 4 and 5 show the design parameters and the crisp values of linguistic variables of FLC.

Table 4. Design Parameters of FLC

Number of fuzzy rules	49
Defuzzification	Centroid
Aggregation	Max
And Method	Prod

The maximum efficiency and constant discharge operation is simulated by replacing the PID control with the FLC. The energy consumed in pumping out 0.75 tmc of water is shown in Figure 14.

It is seen from the simulation that FLC and PID controller give almost the same result. Hence PID controller can be replaced by FLC, as they require complicated mathematical models. FLC techniques need less development time and can be easily used in embedded control.

Table 5. Values of Linguistic Variables

LN	MN	SN	VS	SP	MP	LP
-0.5	-0.28	-0.14	-0.05	0.12	0.38	0.45

Figure 14. The energy consumed by the submersible pump in pumping 0.75 tmc of water using FLC

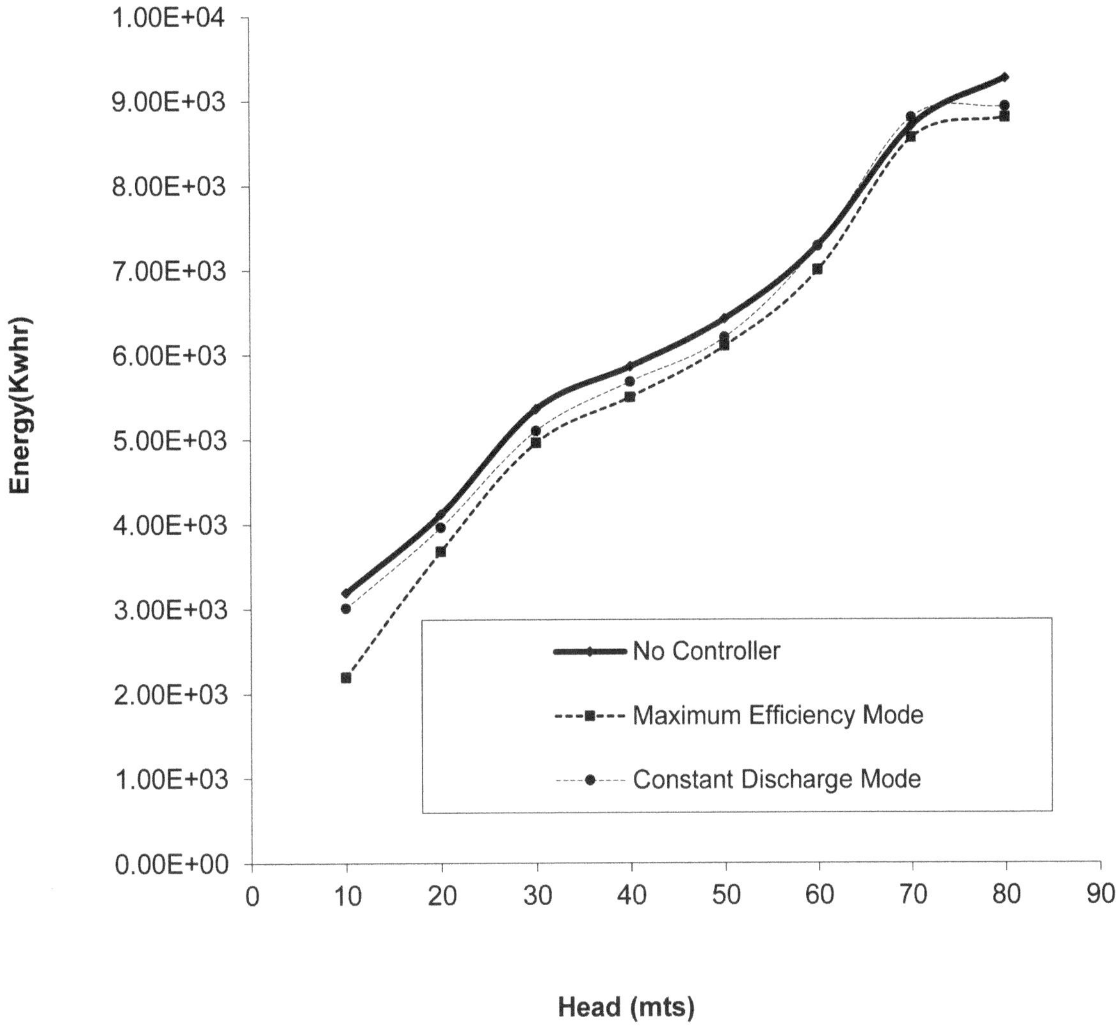

1.6. CONCLUSION

The pump can be operated at either constant flow control or maximum efficiency method up to the heads of 65 mts. Above this head, current limit becomes active and constant power control is the only choice.

The energy saved by the maximum efficiency scheme is higher than the constant discharge. The energy saved in the constant flow control is 12% whereas it is 15.8% at maximum efficiency control at rated head. An efficient planning is needed for the selection of proper scheme based on production and maintenance schedule.

1.7. REFERENCES

ABB. (1999). ACS 600 Single drive Operation and Maintenance: course G300. Student binder (pp. 15-18).

ABB. (1999). *Mandatory installation instruction according to EMC directive for type ACS400 frequency converters.*

Boldea I., & Naser S.A. (1986). *Electrical machine dynamics.* New York: McMillan publishing company.

Brogliato, P., & Orhant, P. (1998). Contact Stability analysis of a One degree of freedom Robot. *Journal of Dynamics and Control*, 8, 37-53.

Byoung-Kuk Lee, , & Ehsami, M. (2001). A simplified functional simulation model for three phase voltage source inverter using switching function concept. *IEEE Transactions on Industrial Electronics*, *48*(2), 309–321. doi:10.1109/41.915410

Gault, T., & Rucker, K. (1986). Reduction of electrical consumption in paper mill pumping systems. *Tappi Journal Pumps/Energy*, 7, 78-85.

Blackburn, J.F., Rethof, G., & Shearer J.L. (1960). Fluid power control. Massachusetts, USA: The M.I.T press.

Khaligh, A., & Emadi, A. (2007). Suitability of the pulse adjustment technique to control single DC/DC choppers feeding vehicular constant power loads in parallel with conventional loads. *Int. J. Electric and Hybrid Vehicles*, *1*(1), 20–45. doi:10.1504/IJEHV.2007.014445

Kuwabara, T., Shibuya, A., & Furuta, H. (1996). Design and dynamic response characteristics of 400MW Adjustable speed pumped storage unit for ohkawachi power station. IEEE Trans on Energy conversion (Vol. 2, pp. 376-384).

Shridhar, L., Singh, B., Jha, C. S., Singh, B. P., & Murthy, S. S. (1994). Design of energy efficient motor for irrigation pumps operating under realistic conditions. *IEE Proceedings. Electric Power Applications*, *141*(6), 269–273. doi:10.1049/ip-epa:19941472

Thorsen, O. V., & Dalva, M. (2001). Combined electrical and mechanical model of electric submersible pumps. *IEEE Transactions on Industry Applications*, *37*(2), 541–547. doi:10.1109/28.913720

Vivek, V., & Chellamuthu, C. (2010). Sliding mode controller for internal combustion engine for energy-efficient operation of hybrid electric vehicles. *Int. J. Electric and Hybrid Vehicles*, *2*(4), 282–297. doi:10.1504/IJEHV.2010.034981

Vivek, V., Kumudini Devi, R.P., & Chellamuthu, C. (n. d.). Design and development of energy efficient discharge controllers for Induction motor driven pump for Pulp and Paper Industry. *International Journal of Modeling and Simulation*, 45-51.

Chapter 19
Smart Grid and Cloud Computing

Girish Neelakanta Iyer
Aryanet Institute of Technology, India

Ganesh Neelakanta Iyer
Progress Software Development, India

ABSTRACT

The network is created through a mixture of IT, communication technology and electrical power systems. The smart grid means it is "smarter" in the sense it has its own intelligence hence a smarter power grid. The generated resources are being decentralized by the smart power grids. The major challenges with the smart grid include huge number of smart meters. Due to the large traffic and limited bandwidth, scheduling is very important. Traditional algorithms will maximize the throughput or minimize the average delay may not be valid in smart grid. Power grid plays an important role in datacenters which forms the backbone for the cloud computing infrastructure. The cloud computing and smart grid interacts through the datacenters. The load balancing and robustness can be achieved through this datacenter and cloud system. Many cloud computing features help to achieve this load balancing.

INTRODUCTION

Cloud computing is an emerged expression used to describe different kinds of computing concepts that involve a large number of computers that are connected through a real-time communication network such as internet (Mariana Carroll, 2012). Cloud computing is a synonym for distributed computing or parallel computing over a network and means the ability to run a program on many connected computers at the same time. Its use in marketing to sell hosted services in the sense of application service provisioning that run client server software on a remote location makes the cloud more popular. In simple words Cloud computing is the current century's technology which provides various kinds of services (wikipedia.org/wiki/CloudComputing) to users through internet on cost of pay-as-you go basis. Cloud is simply any-time-anywhere model provided a constant internet should be there.

Cloud computing always requires searching for a cloud provider. Cloud can be public, private, or hybrid. It will have the feature of elasticity, scal-

DOI: 10.4018/978-1-4666-9911-3.ch019

ability, self-service provisioning, standardized interfaces, and pay per usage (Peter Mell, 2011). These are the important characteristic features of cloud paradigm.

Elasticity and scalability: The cloud is elastic, which means the resource allocation can get bigger or smaller depending on demand. Elasticity entitles scalability, means that the cloud can have the ability to scale upward for high demand and downward for low demand. Scalability also refers to an application that can scale when adding users and when there is a change in application requirements.

Self-service provisioning: Cloud customers can provision cloud services without going through a lengthy process. An automatic provisioning is made possible at the completion of using up of these kind of resources.

Standardized interfaces: Cloud services would have standardized Application Program Interfaces, which provide instructions on how the communication between two application or data sources can be made possible in successful manner. A standardized interface helps the customer in an easier manner to link the cloud services together.

Billing and service usage metering: The billing of resources is based on how much the user is using. This pay-as-you-go model indicates that the usage is metered and user needs to pay only for what he/she consumes.

The cloud computing is having several advantages (Mell, 2011), which includes:

- **Cost Efficient:** For use, upgrade and maintain, Cloud computing is probably the most cost efficient methodology at present. The licensing fees for multiple users can prove to be very expensive for the establishment concerned. The availability of cloud for the users is high and is in a cheaper rate, which will significantly lower the company's IT expenses.
- **Unlimited Storage:** Storing information in the cloud gives you almost unlimited storage capacity. Hence no need to bother about the running out of storage space or increasing your current storage space availability.
- **Backup and Recovery:** Since all our data is stored in the cloud, backing up of data and restoring the same is relatively much easier than storing the same on a physical device. Most of the cloud service providers are usually competent for handling recovery of information, thereby making the entire process of backup and recovery in a simple manner than other traditional methods of data storage.
- **Automatic Software Integration:** Automatic occurrence means that the user does not need to take additional efforts for customization and integration of the applications as per users' preferences.
- **Easy Access to Information:** Once a user get registered the cloud, accessing of information from anywhere, where there is an Internet connection is possible. This feature helps the user to move beyond the time zone and geographic location issues.
- **Quick Deployment:** If the user opts for quick deployment, the whole system can be fully functional in a few minutes. The time taken will be depending on the exact kind of technology that is needed for business.

Major disadvantage with the cloud paradigm is described as below:

- **Technical Issues:** Though it is true that information and data on the cloud can be accessed anytime and from anywhere. User should be aware of the fact that this technology is always prone to outages and other technical issues. Rather than keeping up high standards of maintenance, the cloud service providers run into these kinds of problems. User always needs a very good constant Internet connection with high speed and should be logged onto the server at all times. User will invariably be stuck in case of network and connectivity problems.

- **Security in the Cloud:** The other major issue while in the cloud is that of security issues such that the user should aware that he will be surrendering all your company's sensitive information to a third-party cloud service provider. Great risk will occur in this situation. Hence, the user needs to make absolutely sure that the chosen provider is the most reliable service provider, for ensuring that all the information are totally secure.
- **Prone to Attack:** Keeping information in the cloud could make the company vulnerable to external hack attacks and threats.

BACKGROUND

The network is created through a mixture of IT, communication technology and electrical power systems. The smart grid means it is "smarter" in the sense it has its own intelligence hence a smarter power grid. The generated resources are being decentralized by the smart power grids.

The major challenges with the smart grid include huge number of smart meters. Due to the large traffic and limited bandwidth, scheduling is very important. The time requirement for the data transmission is very high as the power grid is a highly dynamic system. Traditional algorithms will maximize the throughput or minimize the average delay may not be valid in smart grid.

The U.S. Department of Energy defines smart grid as "an automated, distributed energy delivery network in wide manner incorporating the benefits of distributed computing and communications for delivering real-time information and enabling the near-instantaneous balance of supply and demand is specified in the article (Nayab, n. d.). The philosophies, concepts, and technologies are brought by the smart grid that enables the Internet to the utility and energy grid. The smart grids are very expensive duc to its installation costs.

Cloud computing is the new paradigm where enterprise relay on for the day-to-day activities. Its major advantages such as on-demand self-service, resource pooling, pay-as-you-go model etc helps to increase the efficiency and reliability with less cost. Power grid plays an important role in datacenters which forms the backbone for the cloud computing infrastructure. The cloud computing and smart grid interacts through the datacenters. The load balancing and robustness can be achieved through this datacenter and cloud system. Many cloud computing features help to achieve this load balancing in power grid.

In this chapter, we discuss about the applications of cloud paradigm for smart grid. Datacenter always need enormous amount of electricity. The enormous amounts of power consumption in datacenters are expected to have major impact on power grid by increasing the load at location where they built. Increasing the size of the data centers will result in huge increase in the cost of electricity. So the consumption of power will be a major issue in the data center. So the power grid is getting smarter for reducing the power consumption thereby the name smart grid.

Power applications for cloud includes the time varying computational needs, which includes the applications that are used in the planning and operation of power systems will have the computational needs which vary in time can be because of the usage in periodic manner or occasional manner. But the computation capacity still remains the same. In power markets, electricity prices are determined by the market operators. The price will change for different uses and the usage will change day by day. Hence, if pay per use model is implemented in this case the pricing can be reduced to minimum which will save the money in huge amount. Another important application of power grid in cloud is the data storage. Cloud based storage provide reliable storage facility where the user may get access anytime anywhere with on demand capacity is getting, which the charges is applicable only to the amount of data stored in the cloud and the amount of data downloaded from the cloud.

Ease of access is another important application of cloud that can be applicable to power grid. The cloud based storage helps for accessing the data in a convenient manner and the security constraint s comparatively low. There are a few scheduling and energy efficient approaches for making the power grid more efficient and cost effective with the help of cloud paradigm. The security of supply, environmental challenges and public scrutiny are key concerns for leaders in the utility industry. Smart Grid technology transforms the utility infrastructure into a distributed network of systems such as clouds that senses and reports abnormal activities and automatically performs self-optimization. Hence the detection and responding to potential faults before problems impact customers can be made and thereby with the help of cloud computing makes the power grid smarter to get a cloud grid approach for the power management.

Cloud computing helps to balance the load on the data centers which helps to minimize the energy consumption in the data center. The power Minimization helps to improve the efficiency of smart grid. Various algorithms are there for the load balancing feature. This chapter will detail out the overview of cloud computing and smart grid and demand response in contrast with cloud computing and smart grid.

CLOUD COMPUTING AND SMART GRID

The cloud resources can be exploited to gain massive speedups without locking the solution to any specific cloud vendor (Wilde, 2011). The interactions between cloud computing systems and smart grid through massive, energy-hungry, and geographically distributed data centers is shown in figure 1. The key elements of the cloud computing is the Data center (Armbrust, 2009). They are also expected to have major impact on the Internet and affect routing and congestion control algorithms (Shimonishi, 2010). Also, because of enormous energy consumption, datacenters are expected to have major impact on the electric grid by increasing the load at locations they are built. Moreover, as data centers grow in size, the cost of electricity is dominating all other cost aspects in cloud computing.

In (Ketan, 2013), the cloud computing techniques for Smart Power Grid Design Using Parallel Scripting has been evaluated and also analyzed the cloud side of a network-intensive problem characterized by wide-area data collection and processing. A representative parallel scripting paradigm has been used in this method. The properties of multiple cloud systems as applied to their problem space have been applied. A major limitation of each of the environments is that the fault tolerance techniques are not much effective and seamless assurance of data availability in the event of failure. Not only computational but performance bandwidth resources are needed for achieving desired application performance. Apart from a basic application execution, in a complex and networked environment, additional requirements are foreseen. These requirements include high assurance, dynamic configuration, fault tolerance, transparent connection migration, distributed data repository, and overall task

Figure 1. Flow diagram

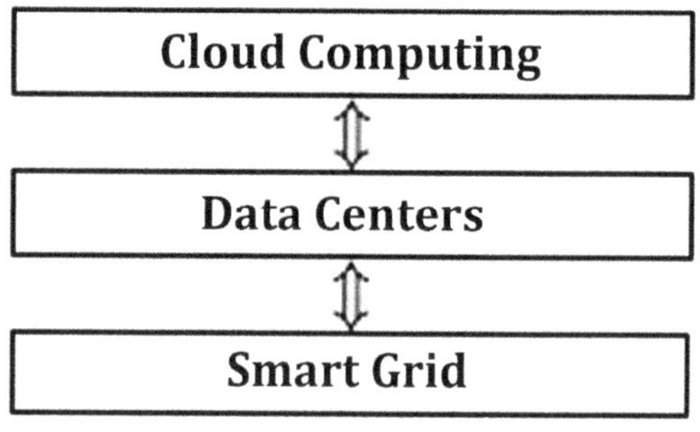

coordination and orchestration of computation. Not all requirements are addressed in this work. However, the resource provision model of the implementation forms is a strong basis to address all requirements. The inter-cloud bandwidth analysis is useful for large scale task placement. Each independent instance of pipelines could be placed on nodes showing high affinity in terms of bandwidth.

In (Moselehi, 2010), a reliability perspective of smart grid has been provided. In a smart grid, real-time prices enable wider voluntary participation by consumers through either automatic or manual response to price signals, or through a bidding process based on direct communication between the consumer and the market/system operator or through aggregators and/or local utilities (Figure 2). In addition to capability to flatten the load profile, demand response can serve as an ancillary resource to help reliability.

The realization of the smart grid requires meeting the ever-increasing reliability challenges by attaching modern communication and information technologies to enable an IT infrastructure that provides grid wide coordinated monitoring and control capabilities. Such IT infrastructure, while expediting the utilization of modern cyber security should be capable of providing fail proof and nearly quick bidirectional communications among all devices ranging from individual loads to the grid wide control centers including all important equipment at the distribution and transmission levels. This involves processing a vast number of data transactions for automation and analysis. Managing the communication burden and resulting that the data latency is the primary concern for efficient analysis and fast control responses and calls for distribution of intelligence throughout the infrastructure, since centralized systems are too slow for this purpose. A distributed system enables local data processing and minimizes the need for massive data exchanges (e.g., bad data detection at substation level, feeder level forecasts aggregated at substation level). A distributed system can enable the high performance needed for preventing or containing rapidly evolving adverse events.

In (Birmen, 2012), gives the cloud computing in relation to future electric power system needs. The authors note that scalability is a major need in future at which, a kind that only cloud computing can offer. There will be power system requirements

Figure 2. Communication for Demand Response

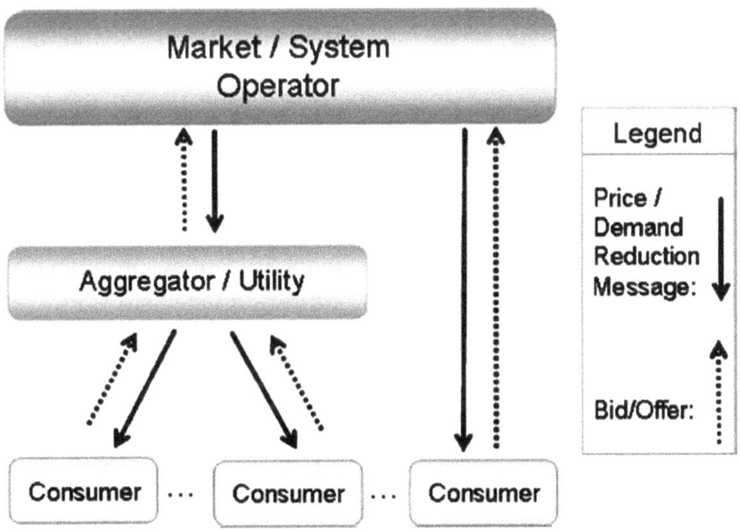

(real-time, consistency, privacy, security, etc.) that cloud computing cannot currently support and that many of these needs, which are specific to the expected future electric power paradigm, will not filled by the cloud industry in a stretch. These problems were addressed by them.

DEMAND RESPONSE IN CLOUD COMPUTING AND SMART GRID

In (Kim, 2011), a cloud based demand response for smart grid has been proposed which uses distributed algorithms. This architecture was in contrast to master/slave based demand response where each participant will directly interact with the utility using host address-centric communication. Cloud based demand response leverages data-centric communication; publisher/subscriber and topic-based group communication to make demand response secure, scalable and reliable. Using this implementation framework, a two market-based distributed algorithm has been proposed.

The most demand response systems proposed till are based on master/slave architecture. Utility's energy management system (EMS) interacts with customers' EMS individually. In general, the master/slave architecture is based on host-address centric communication and is good for a small scale network due to its simplicity. However, from system protection view, the master/slave architecture, for demand response, has several drawbacks. When there is an occurrence of communication on IP-enabled networks, it is possible that meters and home EMSes can be compromised by cyber attackers because these systems are usually installed outside the physical security perimeter of the utility. In master/slave based demand response systems, all slave nodes (e.g., meters, home EMS) must know the IP addresses of the designated master nodes in utility sides. Therefore from a security perspective, the traditional approach applied for demand response causes extensive exposure to cyber attacks such as distributed denial of services (DDoS) from compromised slave nodes.

The proposed system in Mell (2011) is attractive to the utility because the utility is spared from the specific details of implementation, which is performed as a cloud service. In realizing CDR, the demand response optimization problem is solved in a decentralized manner, and proposed the bisection and Illinois methods.

In the technical report (Simmhan, 2011), the authors examined a real-world smart grid project to characterize the challenges posed for informatics researchers by the emerging application domain and exploring the opportunities for research into topics such as stream processing, semantic complex event processing and Cloud computing systems, They have proposed a new software architecture that combines these technologies to address demand response optimization in smart grids.

The characterization of smart grid is by a large scale heterogeneous communication network in which the data flows through various media, wired and wireless. There are enormous security concerns with any large scale communication system as it is susceptible to security attacks on the various media through which the data travels. At the same time the vulnerabilities in the power grid occurrence is very high because most of the current devices were built specifically with the focus on securing the devices from physical tampering and with less focus on securing the communication. Data and cyber security issues in the smart grid domain deal with securing the transportation channel carrying smart meter and other utility and consumer data between different entities in the smart grid. It also aims to prevent disruption of the power grid through introducing unauthorized code or "viruses", or attempts by consumers to modify the power usage data reported to the utility. It is extremely important to integrate with the legacy system in a secure manner in such a way that the vulnerabilities in the existing system do not put the entire smart grid at risk. A number of unique security concerns occur at smart grid due to the close interaction between the cyber and physical infrastructure.

In (Xi Fang,2012), many cloud computing and smart grid applications are mentioned which includes, User Data Management and Application Service, Advanced Trading Information Service, Recommending, Grid Operation Information Analysis Service and Consulting and Advertising Service.

SCALABILITY USING DISTRIBUTED ALGORITHMS

In (Kim, 2011), a new architecture named SeDAX (SEcure Data-centric Application eXtensible) platform for Smart Grid applications has been proposed. SeDAX implements scalable, resilient which is elastic and secure data delivery and data sharing in a wide area network which can be used in smart grid environments. The platform can scale and handle high volumes of data generated by both applications and sensors. The SeDAX architecture has as its basis a Delaunay Triangulation (DT) network. The properties of the DT graph are leveraged to scalably support secure data-centric (or information-centric) group communication. The primary goals of this platform were to support communication elasticity and data availability. The key functional blocks of the SeDAX platform were, a geographic hash forwarding algorithm that operates over the DT graph (DT-GHF), and a DT-based data replication scheme (Kim, 2012). The forwarding and replication schemes are scalable and cost effective in terms of communication overhead and memory and the hash functions employed can be designed to provide increased data availability. In addition to Smart Grid systems spanning across electricity consumers and utilities, the SeDAX platform can be employed for wide-area measurement and control in continental-scale transmission grids as well.

LOAD BALANCING IN SMART GRIDS USING CLOUD TECHNIQUES

In (Wang, 2014), due to the enormous energy consumption, the data centers are expected to have major impact on the electric power grid. It can be achieved by significantly increasing the load at locations where they are built. The dynamic energy pricing policies in the recently proposed smart grid technology can incentivize the cloud computing controller to shift their computation load towards data centers in regions with cheaper electricity. Also, the distributed data centers also provide opportunities to help the smart grid for improving the load balancing and robustness with the help of cloud techniques. In this, a nested two stage game based formulation has been proposed based on the location-dependent real-time pricing policy of the smart grid. The leading player in this game is the smart grid controller that announces the relationship between the electricity price at each power bus and the total load demand at that bus. In the second stage, the cloud computing controller performs resource allocation as response to the pricing functions, whereas the other load devices perform demand side management. The smart grid controller is having the objective for maximizing its own profit and performs load balancing among power buses, whereas the objective of the cloud computing controller is to maximize its own profit with respect to the location-dependent pricing functions. The optimal strategies are derived based on the backward induction principle for the smart grid controller, the cloud computing controller, and the other load devices, using convex optimization and heuristic search.

In the article (Hao Wang, 2013), the interactions between power grid and data centers as a two-stage problem, where the power grid operator aims to balance the electric power load in the first stage, and the data centers seek to minimize their total energy cost in the second stage. We show that this two-stage problem is a bilevel program with an indefinite quadratic objective function, which cannot be solved efficiently using standard convex optimization algorithms. Therefore, we reformulate this bilevel optimization problem as a linear program with additional finite complementarily slackness conditions, and propose a branch and bound algorithm to attain the globally optimal solution.

CONCLUSION

We addressed current state of the art demand response techniques, scalability and load balancing techniques for smart grid using cloud computing. We categorized existing techniques into multiple categories and explained them in detail.

REFERENCES

Birmen, K. (2012). Running smart grid control software on cloud computing architectures. *Proceedings of the Computational Needs for the Next Generation Electric Grid Workshop*

Wang, H., Huang, J., Lin, X., & Mohsenian-Rad, H. (2013). Exploring Smart Grid and Data Center Interactions for Electric Power Load Balancing. *ACM SIGMETRICS Performance Evaluation Review*, 41(3), 89-94.

Shimonishi, H., Higuchi, J., Yoshikawa, T., & Iwata, A. (2010). A Congestion Control Algorithm for Data Center Area Communications. *Proc. of IEEE International Communications Quality and Reliability Workshop, Vancouver, Canada.*

Kim, H., Kim, Y.-J., Yang, K., & Thottan, M. (2011). Cloud-based Demand Response for Smart Grid: Architecture and Distributed Algorithms. *Proceedings of the IEEE International Conference on Smart Grid Communications* (pp. 398-403).

Maheshwari, K., Birman, K., Wozniakz, J. M., & Van Zandt, D. (2013). Evaluating Cloud Computing Techniques for Smart Power Grid Design Using Parallel Scripting. *Proceedings of the International Symposium on Cluster, Cloud and Grid Computing (CCGrid)*, IEEE/ACM. doi:10.1109/CCGrid.2013.26

Moslehi, K., & Kumar, R. (2010). A Reliability Perspective of the Smart Grid. IEEE transactions on Smart Grid, 1(1), 57-64.

Mariana, C., Kotzé, P., & van der Merwe, A. (2012) Securing Virtual and Cloud Environments. In Cloud Computing and Services Science, Service Science, Springer Science & Business Media, LLC. 4 doi:10.1007/978-1-4614-2326-3

Armbrust, M., Fox, A., Griffith, R., Joseph, A.D., Katz, R.H., …, Zaharia, M. (2009). *Above the clouds: A Berkeley view of cloud computing* (Research Report). University of California at Berkeley.

Wilde, M., Hategan, M., Wozniak, J. M., Clifford, B., Katz, D. S., & Foster, I. (2011). Swift: A language for distributed parallel scripting. *Parallel Computing, Emerging Programming Paradigms for Large-Scale Scientific Computing, 39*(9), 633–652.

Nayab, N. (2010). Advantages of Smart Grid technology. *Brighthub*. Retrieved from http://www.brighthub.com/environment/renewable-energy/articles/74984.aspx#

Mell, P., & Grance, T. (2011). The NIST Definition of Cloud Computing (Special Publication 800-145 Cloud Computing). *NIST.* Retrieved from http://csrc.nist.gov/publications/nistpubs/800-145/SP800-145.pdf

Fang, X., Misra, S, Guoliang Xue, & Dejun Yang (2012). Managing Smart Grid Information in the Cloud: Opportunities, Model, and applications. *IEEE Network*, 26(4), 32-38.

Yanzhi Wang, Xue Lin, Pedram, M. (2014). Coordination of the Smart Grid and Distributed Data Centers: A Nested Game-Based Optimization Framework. *Proceedings of the 5th IEEE Annual Power & Energy Society Innovative Smart Grid Technologies Conference to Showcase Smart Grid Investment Grant Projects.*

Simmhan, Y., Aman, S., Cao, B., Giakkoupis, M., Kumbhare, A., Zhou, Q., & Prasanna, V. et al. (2011). *An Informatics Approach to Demand Response Optimization in Smart Grids* (Technical Report). Department of Computer Science in the University of California.

Kim, Y.-J., Lee, J., Atkinson, G., Kim, H., & Thottan, M. (2012). SeDAX: A Scalable, Resilient, and Secure Platform for Smart Grid Communications. *IEEE Journal on Selected Areas in Communications, 30*(6), 1119–1136. doi:10.1109/JSAC.2012.120710

Kim, Y.-J., Kolesnikov, V., Kim, H., & Thottan, M. (2011). SSTP: a Scalable and Secure Transport Protocol for Smart Grid Data Collection. *IEEE Smart Grid Communication, Communication networks for smart grid.*

Compilation of References

A.A. Edris, (1987). Controllable VAR Compensation: A Potential Solution to Loadability Problems of Low Capacity Power Systems. *IEEE Trans. on Power Systems*, 2(3).

ABB. (1999). ACS 600 Single drive Operation and Maintenance: course G300. Student binder (pp. 15-18).

ABB. (1999). *Mandatory installation instruction according to EMC directive for type ACS400 frequency converters*.

ABB. (2010). *Technical note: Starting methods for AC motors*. ABB.

Abdel-Halim, M. A., Almarshoud, A. F., & Alolah, A. I. (2001). Control of grid connected induction generator using naturally commutated AC voltage controller. *Proceedings of the Canadian Conference on Electrical and Computer Engineering* (Vol. 2, pp. 839-843). IEEE.

Abdelmoumene, A. (2010). Etude de la redondance sur le système de protection du transformateur de puissance [Thèse de magistère]. Boumerdes University.

Abdelmoumene, A., & Benatrzi, H. (2012). Simulation du transformateur de puissance en régime de fonctionnement perturbé. *Proceedings of the deuxième Conférence Internationale sur la Maintenance, la Gestion, la Logistique et l'Electrotechnique CIMGLE'2012*, ENSET Oran, Algeria.

Abdelmoumene, A., Benatrzi, H., & Ouahdi, D. (2012, June 13-15). Power transformer protective system reliability assessment. *Proceedings of the 5th International conference on reliable engineering computing*, Brno, Czech Republic.

Abdelmoumene, A. (2010). *Amélioration de la sureté de fonctionnement du transformateur de puissance. Journées d'Etude sur les Technologies Contrôle Commande JETCC'10*. Algeria: IAP Boumerdes.

Abdelmoumene, A., & Benatrzi, H. (2012). Study of the Influence of Power Fluctuation on Power Transformers. *International Journal of Industrial Engineering & Technology*, 2(2), 31–37.

Abdelmoumene, A., Ouahdi, D., & Benatrzi, H. (2011). *Contribution à la réduction des risques de défaillances des transformateurs de puissance. Proceedings of the séminaire Euro-Méditerranéen sur l'Environnement et la Sécurité Industrielle SEMESI'2011*. Algeria: ENSET Oran.

Aboytes, F., & Arroys, G. (1987). Security Assessment in the Operation of Longitudinal Power Systems. *IEEE Transactions on Power Systems*, 1.

Abu-Elhaija, W.S., & Muetze, A. (2013). Self-Excitation and Stability at Speed Transients of Self-Excited Single-Phase Reluctance Generators. *IEEE Transactions on Sustainable Energy*, 4(1), 136-144.

AC motors starting and protection systems. (2010Schneider Electric.

Aggarwal, S. K., Saini, L. M., & Kumar, A. (2009). Electricity price forecasting in deregulated markets: A review and evaluation. *International Journal of Electrical Power & Energy Systems*, *31*(1), 13–22. doi:10.1016/j.ijepes.2008.09.003

Aghamohammadi, M. R., & Tabandeh, S. M. (2012, 2-5 Dec. 2012). *Online coherency identification based on correlation characteristics of generator rotor angles.* Paper presented at the 2012 IEEE International Conference on Power and Energy PECon.

Aghamohammadi, M. R., & Shahmohammadi, A. (2012). Intentional islanding using a new algorithm based on ant search mechanism. *International Journal of Electrical Power & Energy Systems*, *35*(1), 138–147. doi:10.1016/j.ijepes.2011.10.006

Ahuja, R. K., Murthy, S. S., & Chaudhary, J. K. (2012). Steady state analysis of three phase self excited induction generator using MATLAB GUI to feed single phase loads with three capacitor configuration. *Proceedings of the 2012 IEEE 5th India International Conference on Power Electronics IICPE* (pp. 1–6). doi:10.1109/IICPE.2012.6450481

Ahvar, S., & Mahdavi, M. (2011). EEQR: An Energy Efficient Query-Based Routing Protocol for Wireless Sensor Networks. *Journal of Advances in Computer Research*, *2*(3), 25–38.

Akagi, H. (1996). New trends in active filters for power conditioning. *IEEE Transactions on Industry Applications*, *32*(6), 1312–1322. doi:10.1109/28.556633

Akagi, H., Kanazawa, Y., & Nabae, A. (1984). Instantaneous reactive power compensation comprising switching devices without energy components. *IEEE Transactions on Industry Applications*, *20*(3), 625–630. doi:10.1109/TIA.1984.4504460

Akaike, H. (1973). Maximum likelihood identification of Gaussian auto-regressive moving-average models. *Biometrika*, *60*, 255–266.

Akkaya, K., & Younis, M. (2003). "An Energy-Aware QoS Routing Protocol for Wireless Sensor Networks. *Proceedings of the IEEE Workshop on Mobile and Wireless Networks MWN '03, Providence, Rhode Island.* doi:10.1109/ICDCSW.2003.1203636

Al-Ali, A. R., El-Hag, A. H., Dhaouadi, R., & Zainaldain, A. (2011, April). Smart home gateway for smart grid. *Proceedings of the 2011 International Conference on Innovations in Information Technology IIT* (pp. 90-93). IEEE. doi:10.1109/INNOVATIONS.2011.5893876

Alfares, H. K., & Nazeeruddin, M. (2002). Electric load forecasting: Literature survey and classification of methods. *International Journal of Systems Science*, *33*(1), 23–34. doi:10.1080/00207720110067421

Al-Karaki, J. N. Ul-Mustafa v & Kamal A. E. (2004). Data Aggregation in Wireless Sensor Networks - Exact and Approximate Algorithms. *Proceedings of IEEE Workshop on High Performance Switching and Routing (HPSR), Phoenix, Arizona, USA* (pp. 241-245).

Al-Karaki, J. N., & Kamal, A. E. (2004). *Routing Techniques in Wireless Sensor Networks: A Survey* (pp. 6–28). IEEE Wireless Communications.

Al-Saffar, M., Nho, E. C., & Lip, T. (1998). Controlled Shunt Capacitor Self Excited Induction Generator. *Proceedings of the 33rd IAS Annual meeting on Industry applications conference* (Vol.2, pp. 1486 – 1490).

Amer, M., Naaman, A., M'Sirdi, N. K., & El-Zonkoly, A. M. (2014). Smart home energy management systems survey. *Proceedings of the 2014 International Conference on Renewable Energies for Developing Countries REDEC* (pp. 167–173). IEEE. doi:10.1109/REDEC.2014.7038551

Amft, O., Medland, R., Foth, M., Petkov, P., Abreu, J., Pereira, F. C., & Paulos, E. et al. (2011). Smart Energy Systems. *IEEE Pervasive Computing / IEEE Computer Society [and] IEEE Communications Society*, *10*(1), 63–65. doi:10.1109/MPRV.2011.10

Amjady, N. (2006). Day-ahead price forecasting of electricity markets by a new fuzzy neural network. *IEEE Transactions on Power Systems*, *21*(2), 887–896. doi:10.1109/TPWRS.2006.873409

Anderson, R.N., Boulanger, A., Powell, W.B., & Warren, S. (2010). Adaptive Stochastic Control for the Smart Grid. *Proceedings of the IEEE*, 99(6), 1098-1115.

Anderson, C. D., & Anderson, J. (2010). Electric and Hybrid Cars. *History (Historical Association (Great Britain))*, 2010, 189.

Anderson, M. D., & Carr, D. S. (1993). Battery energy storage technology. *Proceedings of the IEEE*, *81*(3), 475–479. doi:10.1109/5.241482

Andersson, G., Donalek, P., Farmer, R., Hatziargyriou, N., Kamwa, I., Kundur, P., & Sanchez-Gasca, J. et al. (2005). Causes of the 2003 major grid blackouts in North America and Europe, and recommended means to improve system dynamic performance. *Power Systems. IEEE Transactions on, 20*(4), 1922–1928.

Animesh, P. (2013). Power Sector in India: Growth, Policies and Challenges. *International Journal of Emerging Technology and Advanced Engineering,* 3(Special Issue 3),527–536.

Anwar, A. (2012). *Distributed Generation Planning for loss and cost minimisation in Power Distribution Systems* [Master's Thesis]. The University of New South Wales, Australia.

Anwar, A., & Mahmood, A. N. (2014). Swarm intelligence based multi-phase OPF for peak power loss reduction in a smart grid. *IEEE Power and Energy Society General Meeting,* USA.

Anwar, A., & Pota, H. R. (2012a). Optimum capacity allocation of DG units based on unbalanced three-phase optimal power flow. *IEEE Power and Energy Society General Meeting,* USA.

Anwar, A., & Pota, H. R. (2012b). Optimum allocation and sizing of DG unit for efficiency enhancement of distribution system. In *IEEE International Power Engineering and Optimization Conference (PEOCO)*, Malaysia. doi:10.1109/PEOCO.2012.6230854

Aoki, K., Nara, K., Itoh, M., Satoh, T., & Kuwabara, H. (1989). A New Algorithm for Restoration in Distribution Systems. *IEEE Transactions on Power Delivery, 4*(3), 1832–1839. doi:10.1109/61.32679

Arbib, M. (Ed.). (2003). *The Handbook of Brain Theory and Neural Networks.* MIT Press.

Armbrust, M., Fox, A., Griffith, R., Joseph, A.D., Katz, R.H., …, Zaharia, M. (2009). *Above the clouds: A Berkeley view of cloud computing* (Research Report). University of California at Berkeley.

Asynchronous motors. (2010). Electrical installation guide (p. 45). Schneider Electric.

Atwa, Y. M., El-Saadany, E. F., Salama, M. M. A., & Seethapathy, R. (2010). Optimal renewable resources mix for distribution system energy loss minimization. *IEEE Transactions on Power Systems, 25*(1), 360–370. doi:10.1109/TPWRS.2009.2030276

Awad, A., Sommer, C., German, R., & Dressler, F. (2008). Virtual Cord Protocol (VCP): A flexible DHT-like routing service for sensor networks. *Proceedings of the5th IEEE International Conference on Mobile Ad Hoc and Sensor Systems* (pp. 133-142). doi:10.1109/MAHSS.2008.4660079

Ayhan, B., Chow, M. Y., & Song, M.-H. (2005). *Monolith and Partition Schemes with LDA and Neural Networks as Detector Units for Induction Motor Broken Rotor Bar Fault Detection.* KIEE International Transactions on Electrical Machinery and Energy Conversion Systems.

B. Atakan, O. B. Akan & T. Tugcu, (2009). Bio-inspired Communications in Wireless sensor networks. In S.C. Misra, I. Woungang, & S. Misra (Eds.), *Guide to Wireless Sensor Networks* (pp. 659-687). Springer-Verlag London.

Baer, P. Exploring the 2020 global emissions mitigation gap. Retrieved from http://www.ippr.org/uploadedFiles/globalclimatenetwork/Exploring_the_Mitigation_Gap[1].pdf

Bains, V., & Sharma, K. (2012). Ant colony based routing in wireless sensor networks. *International Journal of Electronics and Computer Science Engineering, 1*(4), 2516–2524.

Ball, G. & Hall, D. (1965). ISODATA: A novel method of data analysis and pattern classification. Stanford Research Institute. Menlo Park.

Baran, M.E., & Wu, F.F. (1989). Network Reconfiguration in Distribution Systems for Loss Reduction and Load Balancing. *IEEE Transactions on Power Delivery, 4*(2), 1401–1407. doi:10.1109/61.25627

Barroso, L. A., & Holzle, U. (2007). The case for energy-proportional computing. *Computer, 40*(12), 33–37. doi:10.1109/MC.2007.443

Bartram, L., Rodgers, J., & Muise, K. (2010). Chasing the Negawatt: Visualization for Sustainable Living. *IEEE Computer Graphics and Applications*, *30*(3), 8–14. doi:10.1109/MCG.2010.50 PMID:20650713

Basics of fuel cell. (2003, August). U.S. Department of Energy, Office of Energy Efficiency and Renewable Energy.

Bassett, E.D. & Potter, F.M. (1935). Capacitive excitation for induction generators. *Electrical Engineering*, 54(5), 540-545. doi:10.1109/EE.1935.6540048

Batra N., Jain A. & Dhiman S., (2011). An Optimized Energy Efficient Routing Algorithm for Wireless Sensor Network. *International Journal of Innovative Technology &Creative Engineering, 1(5)*, 41-45.

BE Electra. Retrieved from http://electra.afrikblog.com

Beloglazov, A., Abawajy, J., & Buyya, R. (2012). Energy-aware resource allocation heuristics for efficient management of data centers for cloud computing. *Future Generation Computer Systems*, *28*(5), 755–768. doi:10.1016/j.future.2011.04.017

Beloglazov, A., & Buyya, R. (2010, November). Adaptive threshold-based approach for energy-efficient consolidation of virtual machines in cloud data centers. *Proceedings of the 8th International Workshop on Middleware for Grids, Clouds and e-Science* (p. 4). ACM. doi:10.1145/1890799.1890803

Beloglazov, A., & Buyya, R. (2013). Managing overloaded hosts for dynamic consolidation of virtual machines in cloud data centers under quality of service constraints. *IEEE Transactions on* Parallel and Distributed Systems, *24*(7), 1366–1379.

Benavides, J. (n. d.). *Precios, inversión y economía política de la energía eléctrica.* Retrieved August 22, 2012, from http://www.scielo.org.co

Bester, J., Robertson, I., & Groenewald, B. (2009). Wireless Control Network for Building Automation using IEEE 802.15.4.*Proceedings ICUE International Conference, Theme: Energy Management*, Cape Town, South Africa.

Bhatt, J. G., & Verma, H. K. (2010). RS-485/MODBUS based intelligent building automation system using LabVIEW.

Bhatt, J.G. (2009, November). *Design and Implementation of Building Automation Systems.* M.Tech. Project Report, Department of Electrical Engineering, Indian Institute of Technology Roorkee, Roorkee, India.

Bhatt, J.G. (2013, December). Literature Survey Report for Doctoral (Ph.D.) Research [Doctoral Thesis]. Pandit Deendayal Petroleum University, Gandhinagar, India.

Bhatt, J., Shah, V., & Jani, O. (2014). An instrumentation engineer's review on smart grid: Critical applications and parameters. *Renewable & Sustainable Energy Reviews*, *40*, 1217–1239. doi:10.1016/j.rser.2014.07.187

Bhatt, J., & Verma, H. K. (2015). *Design and Development of Wired Building Automation Systems.* Energy and Buildings.

Bian, X., Xu, Q., Li, B., & Xu, L. (2007). Equipment Fault Forecasting Based on a Two-level Hierarchical Model. *Proceedings of theIEEE International Conference on Automation and Logistics* (pp. 2095-2099). doi:10.1109/ICAL.2007.4338921

Birmen, K. (2012). Running smart grid control software on cloud computing architectures. *Proceedings of the Computational Needs for the Next Generation Electric Grid Workshop*

Bishop, C. M. (2006). *Pattern Recognition and Machine Learning (Information Science and Statistics).* New York: Springer-Verlag.

Blackburn, J.F., Rethof, G., & Shearer J.L. (1960). Fluid power control. Massachusetts, USA: The M.I.T press.

Bodkhe, S. B., & Aware, M. V. (2009), " A variable-speed, sensorless, induction motor drive using DC link measurements", 1 Industrial Electronics and Applications, 4th IEEE Conference on pp. 359, 3596.

Bodkhe, S. B., & Aware, M. V. (2010). Robust method for stator current reconstruction from DC link. *International Journal of Engineering Science and Technology*, *2*(10), 87–99.

Bojoi, R., Guglielmi, P., & Pellegrino, G. (2006). Sensorless Direct Field Oriented Control of Three-Phase Induction Motor Drives for Low Cost Applications. *Proceedings of the IEEE Conference on Industry Applications* (Vol. 2, pp. 866 – 872). doi:10.1109/IAS.2006.256627

Boldea I., & Naser S.A. (1986). *Electrical machine dynamics*. New York: McMillan publishing company.

Boldea, I., & Naser, S.A. (1986). Electrical machine dynamics. New York: McMillan Publishing.

Bollen, M.H.J., Zhong, J., Zavoda, F., Meyer, J., McEachern, A., & López, F.C. (2010). Power quality aspects of Smart Grids. *Proceedings of the International Conference on Renewable Energies and Power Quality ICREPQ '10*, Granada, Spain.

Booth, A., Demirdoven, N., & Tal, H. (2010). The smart grid opportunity for solution providers. *McKinsey & Company*.

Borozan, V., Rajicic, D., & Ackovski, R. (1997). Minimum Loss Reconfiguration of Unbalanced Distribution Networks. *IEEE Transactions on Power Delivery*, *12*(1), 435–442. doi:10.1109/61.568268

Bose, S., Wu, C., Wierman, A., & Mohsenian-Rad, H. (2014, January). Assessing Market Power in Deregulated Electricity Markets - IEEE Smart Grid. Retrieved from http://smartgrid.ieee.org/january-2014/1025-assessing-market-power-in-deregulated-electricity-markets

Boussak, M., & Jarray, K. (2006). A High-Performance Sensorless Indirect Stator Flux Orientation Control of Induction Motor Drive. *IEEE Transactions on Industrial Electronics*, *53*(1), 41–49. doi:10.1109/TIE.2005.862319

Braginsky, D., & Estrin, D. (2002). Rumor routing algorithm for sensor networks.*Proceedings of the First Workshop on Sensor Networks and Applications (WSNA), Atlanta*.

Brewer, R. S., & Johnson, P. M. (2010). WattDepot: An Open Source Software Ecosystem for Enterprise-Scale Energy Data Collection, Storage, Analysis, and Visualization. *Proceedings of the2010 First IEEE International Conference on Smart Grid Communications (SmartGridComm)* (pp. 91–95). doi:10.1109/SMARTGRID.2010.5622023

Broadwater, R. P., Khan, A. H., Shaalan, H. P., & Lee, R. E. (1993). Time Varying Load Analysis to Reduce Distribution losses through Reconfiguration. *IEEE Transactions on Power Delivery*, *8*(1), 294–300. doi:10.1109/61.180349

Brogliato, P., & Orhant, P. (1998). Contact Stability analysis of a One degree of freedom Robot. *Journal of Dynamics and Control*, 1998, 37-53.

Brogliato, P., & Orhant, P. (1998). Contact Stability analysis of a One degree of freedom Robot. *Journal of Dynamics and Control*, 8, 37-53.

Bruno, S., Lamonaca, S., Rotondo, G., Stecchi, U., & Scala, M. L. (2011). Unbalanced three-phase optimal power flow for smart grids. *IEEE Transactions on Industrial Electronics*, *58*(10), 4504–4513. doi:10.1109/TIE.2011.2106099

Buildings Energy Data Book. (2010). U.S. Department of Energy. Retrieved from http://buildingsdatabook.eren.doe.gov/DataBooks.aspx

Buso, S., Fasolo, S., Malesani, L., & Mattavelli, P. (2000). A Dead Beat Adaptive Hysteresis Current Control. *IEEE Transactions on Industry Applications*, *36*(4), 1174–1180. doi:10.1109/28.855976

Byoung-Kuk Lee, , & Ehsami, M. (2001). A simplified functional simulation model for three phase voltage source inverter using switching function concept. *IEEE Transactions on Industrial Electronics*, *48*(2), 309–321. doi:10.1109/41.915410

C3 Energy Featured as Top Vendor in Global Energy Management System Market Report. (2015, July 13). Business Wire. Retrieved from http://www.businesswire.com/news/home/20150713005254/en/C3-Energy-Featured-Top-Vendor-Global-Energy#.VaSjwLWLFf4

Canizares, C.A., De Souza, A.Z., & Quintana, V.H. (1996). Comparison of performance indices for detection of proximity of voltage collapse. IEEE Transactions on Power Systems, 11(3), 1441-1450.

Casteleiro-Roca, J., Quintián, H., Calvo-Rolle, J., Corchado, E., & del Carmen Meizoso-López, M. (2013). Intelligent Model to Obtain Initial and Final Conduction Angle of a Diode in a Half Wave Rectifier with a Capacitor Filter. In Á. Herrero, B. Baruque, F. Klett, A. Abraham, V. Snášel, A.C.P.L.F. de Carvalho, … E. Corchado (Eds.), *Proceedings of the International Joint Conference SOCO '13-CISIS'13-ICEUTE'13*. Springer International Publishing.

Catalao, J. P. S., Pousinho, H. M. I., & Mendes, V. M. F. (2011). Hybrid Wavelet-PSO-ANFIS Approach for Short-Term Wind Power Forecasting in Portugal. *IEEE Transactions on Sustainable Energy*, *2*(1), 50–59. doi:10.1109/TSTE.2010.2076359

Ceraolo, M. (2000). New dynamic models of lead-acid Batteries. *IEEE Transactions on Power Systems*, *15*(4), 1184–1190. doi:10.1109/59.898088

Chakrabarti Abhinandan De, A., Kothari, D.P., & Mukhopadhyay, A.K. (2010). *An Introduction to Reactive Power Control and Voltage Stability in Power Transmission System*. PHI publishers.

Chakrabarti, A., & Mukhopadhyay, A.K. (1989). Operating Problem in Longitudinal Power System.*Proc. IEEE International TENCON Conference*, India (pp. 932-935). doi:10.1109/TENCON.1989.177084

Chang C. S., Wang Zhaoxia, Yang Fan, & Tan W. W. (2012, March). Hierarchical Fuzzy Logic System for Implementing Maintenance Schedules of Offshore Power Systems. *IEEE Transactions Smart Grid*, 3(1).

Chang, C. S., Chen, J. M., Srinivasan, D., Wen, F. S., & Liew, A. C. (1997). Fuzzy logic approach in power system fault section identification. *IEEE Proceedings-Part C, Generation Transmission and Distribution*, 144(5), 406-414. doi:10.1049/ip-gtd:19971278

Chang, C. S., Chen, J. M., Srinivasan, D., Wen, F. S., & Liew, A. C. (1996). *Power system fault diagnosis using fuzzy sets for uncertainties processing*. IEEE Transactions. doi:10.1109/ISAP.1996.501094

Chang, J. H., & Tassiulas, L. (2004). Maximum lifetime routing in wireless sensor networks. *IEEE/ACM Transactions on Networking*, *12*(4), 609–619.

Chen, H., Cong, T. N., Yang, W., Tan, C., Li, Y., & Ding, Y. (2009). Progress in electrical energy storage system: A critical review. *Progress in Natural Science*, *19*(3), 291–312. doi:10.1016/j.pnsc.2008.07.014

Chen, Y.L. (1996). Weak Bus-Oriented optimal Multi-objective VAR Planning. *IEEE Transactions on Power Systems*, *11*(4), 1885–1890. doi:10.1109/59.544659

Chiang, S. J., Huang, S. C., & Liaw, C. M. (1995). Three-phase multifunctional battery energy storage system. *Electric Power Applications, IEE Proceedings*, *142*, 275–284. doi:10.1049/ip-epa:19951856

Chiang, D., & Rene, J. J. (1992). Optical Network Reconfiguration in Distribution System: Part2: Solution Algorithms and Numerical Results. *IEEE Transactions on Power Delivery*, *5*(3), 1568–1574. doi:10.1109/61.58002

Chiang, H. D., Wang, C. S., & Flueck, A. S. (1997). Look-ahead Voltage and Load Margin Contingency Selection Functions for Large Scale Power Systems. *IEEE Transactions on Power Systems*, *12*(1), 173–180. doi:10.1109/59.574937

Chiang, H.D., & Jumeau, R.J. (1990). Optimal network reconfigurations in distribution systems: part 1: a new formulation and solution methodology. *IEEE Transactions on Power Delivery*, *5*(4), 1902–1909. doi:10.1109/61.103687

Ching-Tzong Su, Chung-Fu Chang, Morrison, G.K., Gao, B., & Kundur, P. (2005). Voltage stability analysis using static and dynamic approaches. *IEEE Transactions on Power Systems*.

Chow, J. (1996). A nonlinear model reduction formulation for power system slow coherency and aggregation. In H. Khalil, J. Chow & P. Ioannou (Eds.), *Proceedings of Workshop on Advances in Control and its Applications* (Vol. 208, pp. 282-298). Springer Berlin Heidelberg. doi:10.1007/BFb0027710

Chow, J. H., & Date, R. (1988, 7-9 June 1988). *A nodal aggregation algorithm for linearized two-time-scale power networks*. Paper presented at the IEEE International Symposium on Circuits and Systems.

Chow, J. H., & Cheung, K. W. (1992). A toolbox for power system dynamics and control engineering education and research. *IEEE Transactions on* Power Systems, *7*(4), 1559–1564.

Chow, J. H., Galarza, R., Accari, P., & Price, W. W. (1995). Inertial and slow coherency aggregation algorithms for power system dynamic model reduction. *IEEE Transactions on* Power Systems, *10*(2), 680–685. doi:10.1109/59.387903

Cirrincione, M., Pucci, M., Cirrincione, G., & Capolino, G. A. (2004). A Improved Adaptive Integration Methodology for Estimating Flux In Induction Machine Drives. *IEEE Transactions on Power Electronics*, *19*(1), 25–34. doi:10.1109/TPEL.2003.820565

Civanlar, S., Grainger, J. J., Yin, H., & Lee, S. S. H. (1998). Distribution Feeder Reconfiguration for Loss Reduction. *IEEE Transactions on Power Delivery*, *3*(3), 1217–1223. doi:10.1109/61.193906

Código Técnico de la Edificación. (2010). Ministerio de Fomento, Gobierno de España. Retrieved from: http://www.codigotecnico.org

Cole, A., & Albicki, A. (2000). Nonintrusive identification of electrical loads in a three-phase environment based on harmonic content. *Proceedings of the 17th IEEEconference onInstrumentation and Measurement Technology Conference IMTC '00* (Vol. 1, pp. 24–29). IEEE. doi:10.1109/IMTC.2000.846806

Conejo, A., Plazas, M., Espinola, R., & Molina, A. (2005). Day-ahead electricity price forecasting using the wavelet transform and ARIMA models. *IEEE Transactions on Power Systems*, *20*(2), 1035–1042. doi:10.1109/TPWRS.2005.846054

Contreras, J., Espinola, R., Nogales, F. J., & Conejo, A. (2003). ARIMA models to predict next-day electricity prices. *IEEE Transactions on Power Systems*, *18*(3), 1014–1020. doi:10.1109/TPWRS.2002.804943

Corchado, E., & Baruque, B. (2012). WeVoS-ViSOM: An ensemble summarization algorithm for enhanced data visualization. *Neurocomputing*, *75*(1), 171–184. doi:10.1016/j.neucom.2011.01.027

Cybenko, G. (1989). Approximations by superpositions of sigmoidal functions. *Mathematics of Control, Signals, and Systems*, *2*(4), 303–314. doi:10.1007/BF02551274

D'Alessandro, S., Tonello, A. M., Monacchi, A., & Elmenreich, W. (2014). Home energy management systems: Design guidelines for the communication infrastructure. *Proceedings of the2014 IEEE InternationalEnergy Conference ENERGYCON* (pp. 805–812). IEEE. doi:10.1109/ENERGYCON.2014.6850518

Daniels, T. L., & Daniels, K. (2003). *The Environmental Planning Handbook for Sustainable Communities and Regions* (p. 524). American Planning Association.

Date, R. A., & Chow, J. H. (1991). Aggregation properties of linearized two-time-scale power networks. *IEEE Transactions on* Circuits and Systems, *38*(7), 720–730. doi:10.1109/31.135744

Davidson, D.R. (1975). Long Term Dynamic Response of Power Systems: An Analysis of Major Disturbances. *IEEE Trans. on Power Systems*, 94(3).

de Boer, P., & Jillis, R. (2007). *Flow Battery (Briefing Paper). Leonardo energy*. KEMA.

Deb, K. (2004). *Optimization for engineering design: algorithms and examples*. India: Prentice-Hall.

Dent, C. J., Ochoa, L. F., & Harrison, G. P. (2010b). Network distributed generation capacity analysis using OPF with voltage step constraints. *IEEE Transactions on Power Systems*, *25*(1), 296–304. doi:10.1109/TPWRS.2009.2030424

Dent, C. J., Ochoa, L. F., Harrison, G. P., & Bialek, J. W. (2010a). Efficient secure AC OPF for network generation capacity assessment. *IEEE Transactions on Power Systems*, *25*(1), 575–583. doi:10.1109/TPWRS.2009.2036809

Dickmann, G. (2011, April). DigitalSTROM: A centralized PLC topology for home automation and energy management. *Proceedings of the 2011 IEEE International Symposium on Power Line Communications and Its Applications ISPLC* (pp. 352-357). IEEE.

Ding, L., Gonzalez-Longatt, F. M., Wall, P., & Terzija, V. (2013). Two step spectral clustering controlled islanding algorithm. *IEEE Transactions on Power Systems*, *28*(1), 75–84. doi:10.1109/TPWRS.2012.2197640

Ding, L., Wall, P., & Terzija, V. (2014). Constrained spectral clustering based controlled islanding. *International Journal of Electrical Power & Energy Systems*, *63*, 687–694. doi:10.1016/j.ijepes.2014.06.016

Distribution system analysis subcommittee radial test feeders. (1991). *IEEE*. Retrieved from the http://ewh.ieee.org/soc/pes/dsacom/testfeeders/index.html

Divya, K. C., & Østergaard, J. (2009). Battery energy storage technology for power systems—An overview. *Electric Power Systems Research*, *79*(4), 511–520. doi:10.1016/j.epsr.2008.09.017

Diwakar, M., & Kumar, S. (2012). An Energy Efficient Level Based Clustering Routing Protocol For Wireless Sensor Networks. *International Journal of Advanced Smart Sensor Network Systems*, *2*(2), 55–65. doi:10.5121/ijassn.2012.2206

Dong, Q., & De-jiang Zhang. (2010). The research and realization of energy management system in iron and steel enterprise. *Proceedings of the2010 International Conference on Information Networking and Automation ICINA* (Vol. 1, pp. V1–448–V1–451). doi:10.1109/ICINA.2010.5636520

Dugan, R., & Kersting, W. (2006). Induction machine test case for the 34-bus test feeder- description. *Proceedings of the IEEE Power Engineering Society General Meeting PESGM '11*, USA.

Dugan, R., & McDermott, T. (2011). An open source platform for collaborating on smart grid research. *Proceedings of the IEEE Power and Energy Society General Meeting PESGM '11*, USA. doi:10.1109/PES.2011.6039829

Dugan, R., McDermott, T., & Ball, G. (2001). Planning for distributed generation. *IEEE Industry Applications Magazine*, *7*(2), 80–88. doi:10.1109/2943.911193

Duran, M.J., Duran, J.L., Perez, F., & Fernandez, J. (2004). A novel induction motor IRFOC with on-line parameter estimation taking into account thermal and deep-bar effects. *Proceedings of the 2004 IEEE conference on Industrial Electronics* (Vol. 1, pp. 435–440).

Egido Cortés, I., Rouco Rodríguez, L., Alonso Perez, J., Porras Muñoz, E., & Ruiz Mendoza, J. L. (n.d.). Impacto en regulación frecuencia-potencia de los cambios de programa en escalón de las unidades generadoras. *Revista de La Asociación de Ingenieros Del ICAI.*

Ekici, B. B., & Aksoy, U. T. (2009). Prediction of building energy consumption by using artificial neural networks. *Advances in Engineering Software*, *40*(5), 356–362. doi:10.1016/j.advengsoft.2008.05.003

El-Khattam, W., Bhattacharya, K., Hegazy, Y., & Salama, M. (2004). Optimal investment planning for distributed generation in a competitive electricity market. *IEEE Transactions on Power Systems*, *19*(3), 1674–1684. doi:10.1109/TPWRS.2004.831699

El-said S. A., Hassanien A. E. (2013). Artificial Eye Vision Using Wireless Sensor Networks. In *Wireless Sensor Networks: Theory and Applications*. CRC Press.

Energía y sociedad. (n. d.-a). *El proceso de liberalizacion de los sectores energéticos.* Retrieved from http://www.energiaysociedad.es

Energía y sociedad. (n. d.-b). *Energía y sociedad.*

Energía y sociedad. (n. d.-c). *Energías renovables: tecnología, economía, evolución e integración en el sistema eléctrico.*

Energy Performance Certification of Buildings. (2013). International Energy Agency. Retrieved from http://www.iea.org/publications/freepublications/publication/buildings_certification.pdf

Engelbrecht, A. (2007). *Computational intelligence: an introduction.* John Wiley & Sons. doi:10.1002/9780470512517

Erol-Kantarci, M., & Mouftah, H. T. (2011). Wireless sensor networks for cost-efficient residential energy management in the smart grid. *IEEE Transactions on* Smart Grid, *2*(2), 314–325.

Etemadi, A., & Fotuhi-Firuzabad, M. (2008). Distribution system reliability enhancement using optimal capacitor placement. *IET Generation. Transmission Distribution*, *2*(5), 621–631. doi:10.1049/iet-gtd:20070515

European Commission. (2010). Energy 2020: A strategy for competitive, sustainable and secure energy. Brussels. Retrieved from http://ec.europa.eu/energy/strategies/2010/2020_en.htm

Ezcurra, G. (2009, November 9). Top Five Reasons Your Business Needs an Energy-Efficiency System. *Greentech Media*. Retrieved from http://www.greentechmedia.com/articles/read/top-five-reasons-your-business-needs-an-energy-efficiency-system

Fabra, N. (2007). Ajustes regulatorios en el sector eléctrico español, La electricidad: mercado, inversiones y garantía de suministro. *ENEconomía Industrial*, 364.

Fang, X., Misra, S, Guoliang Xue, & Dejun Yang (2012). Managing Smart Grid Information in the Cloud: Opportunities, Model, and applications. *IEEE Network*, 26(4), 32-38.

Fang, X., Misra, S., Xue, G., & Yang, D. (2012). Smart grid the new and improved power grid: A survey. *IEEE Communications Surveys and Tutorials*, *14*(4), 944–980. doi:10.1109/SURV.2011.101911.00087

Fan, S., Wu, Y.-K., Lee, W.-J., & Lee, C.-Y. (2011). *Comparative study on load forecasting technologies for different geographical distributed loads*. IEEE Power and Energy Society General Meeting. Doi:10.1109/PES.2011.6039364

Fanxing, L. (2010, September). Smart Transmission Grid: Vision and Framework. *IEEE Transactions on Smart Grid*, *1*(2), 168–177. doi:10.1109/TSG.2010.2053726

Fardanesh, B., Shperling, B., Uzunovic, E., & Zelingher, S. (2000). Multi-converter FACTS Devices: The generalized unified power flow controller (GUPFC). *Proc. IEEE Power Eng. Soc. Summer Meeting* (Vol. 4, pp. 2511–2517). doi:10.1109/PESS.2000.867513

Farhangi, H. (2014). A Road Map to Integration: Perspectives on Smart Grid Development. *Power and Energy Magazine, IEEE*, *12*(3), 52–66. doi:10.1109/MPE.2014.2301515

Farmer, R. G., Schwalb, A. L., & Eli, K. (1977). Navajo project report on subsynchronous resonance analysis and solutions. *IEEE Transactions on* Power Apparatus and Systems, *96*(4), 1226–1232.

Ferguson, C. D. (2011). *Nuclear Energy: What Everyone Needs to Know* (p. 240). USA: Oxford University Press.

Fernández-Serantes, L. A., Estrada Vázquez, R., Casteleiro-Roca, J., Calvo-Rolle, J., & Corchado, E. (2014). Hybrid Intelligent Model to Predict the SOC of a LFP Power Cell Type. In M. Polycarpou, A. P. L. F. de Carvalho, J.-S. Pan, M. Wozniak, H. Quintian, & E. Corchado (Eds.), *Hybrid Artificial Intelligence Systems*. Springer International Publishing. doi:10.1007/978-3-319-07617-1_49

Fischer, C. (2008). Feedback on household electricity consumption: A tool for saving energy? *Energy Efficiency*, *1*(1), 79–104. doi:10.1007/s12053-008-9009-7

Fitzpatrick, G., & Smith, G. (2009). Technology-Enabled Feedback on Domestic Energy Consumption: Articulating a Set of Design Concerns. *IEEE Pervasive Computing / IEEE Computer Society [and] IEEE Communications Society*, *8*(1), 37–44. doi:10.1109/MPRV.2009.17

Forth, B., & Tobin, T. (2002). Right power, right price[enterprise energy management systems]. *IEEE Computer Applications in Power*, *15*(2), 22–27. doi:10.1109/67.993756

Foucquier, A., Robert, S., Suard, F., Stéphan, L., & Jay, A. (2013). State of the art in building modeling and energy performances prediction: A review. *Renewable & Sustainable Energy Reviews*, *23*, 272–288. doi:10.1016/j.rser.2013.03.004

Froehlich, J., Findlater, L., & Landay, J. (2010). The Design of Eco-feedback Technology.*Proceedings of the SIGCHI Conference on Human Factors in Computing Systems* (pp. 1999–2008). New York, NY, USA: ACM. doi:10.1145/1753326.1753629

Fukm, D., & Kawakame, J. (1986). An expert system for fault section estimation using mfonnatlon from protective relays and circuit breakers. IEEE Trans Power Delivery, 4, 83-90.

Fukui, C., & Kawakami, J. (1986). An expert system for fault section estimation using information from protective relays and circuit breakers. *IEEE Transactions on Power Delivery*, *1*(4), 83–90. doi:10.1109/TPWRD.1986.4308033

GaleanoGonzález, D. A., & Botero Castro, H. A. (n. d.). Representación mediante grafcet del accionamiento de generadores sincrónicos en una central hidroeléctrica modernizada. *Dyna, Revista de La Facultad de Minas - Universidad de Colombia*.

Ganesh, V., Sivanagaraju, S., & Thiruveedula, R. (2009). Feeder Reconfiguration for Loss Reduction in Unbalanced Distribution System Using Genetic Algorithm. *International Journal of Electrical and Electronics Engineering*, 3(12), 744-762.

Gao, B., Morison, G., & Kundur, P. (1992). Voltage Stability Evaluation Using Modal Analysis. *IEEE Transactions on Power Systems*, *7*(4), 1529–1542. doi:10.1109/59.207377

Garcia-Martos, C., Rodriguez, J., & Sanchez, M. J. (2007). Mixed Models for Short-Run Forecasting of Electricity Prices: Application for the Spanish Market. *IEEE Transactions on Power Systems*, *22*(2), 544–552. doi:10.1109/TPWRS.2007.894857

Garg, S. K., Yeo, C. S., & Buyya, R. (2011). *Green cloud framework for improving carbon efficiency of clouds. Proceedings of Euro-Par 2011 Parallel Processing* (pp. 491–502). Springer Berlin Heidelberg.

Gartner Estimates ICT Industry Accounts for 2 Percent of Global CO2 Emissions. (n. d.). *Gartner*. Retrieved from http://www.gartner.com/it/page.jsp?id=503867

Gault, T., & Rucker, K. (1986). Reduction of electrical consumption in paper mill pumping systems. *Tappi Journal Pumps/Energy*, 7, 78-85.

Geddada, N., Karanki, S. B., & Mishra, M. K. (2012). Synchronous reference frame based current controller with SPWM switching strategy for DSTATCOM applications. *Proc. of IEEE Conf. PEDES* (pp. 1-6). doi:10.1109/PEDES.2012.6484393

Geidl, M., & Andersson, G. (2005). A modeling and optimization approach for multiple energy carrier power flow. In *PowerTech, 2005* (pp. 1–7). IEEE Russia; doi:10.1109/PTC.2005.4524640

Geidl, M., & Andersson, G. (2005, August). Optimal power dispatch and conversion in systems with multiple energy carriers.*Proc. 15th Power Systems Computation Conference PSCC.*

Gellings, C. W. (1985). The concept of demand-side management for electric utilities. *Proceedings of the IEEE*, *73*(10), 1468–1470. doi:10.1109/PROC.1985.13318

Ghamkhari, M., & Mohsenian-Rad, H. (2013). Energy and performance management of green data centers: A profit maximization approach. *IEEE Transactions on* Smart Grid, *4*(2), 1017–1025.

Ghemawat, S., Gobioff, H., & Leung, S.-T. (2003). The Google file system. *Proc. of the ACM Symp. Operating Syst. Principles* (pp. 29–43).

Gil Gonzalez-Madroño, E. (2007). *Tecnologías de acumulación de energía para el uso en la compañía eléctrica. Baterías de sodio-azufre*. Universidad Carlos III.

Gil, H. A., & Joos, G. (2008). Models for quantifying the economic benefits of distributed Generation. *IEEE Transactions on Power Systems*, *23*(2), 327–335. doi:10.1109/TPWRS.2008.920718

Godoy Simoes, M., & Farret, F. A. (2004). *Renewable Energy Systems: Design and Analysis with Induction generators*. Boca Raton, FL: CRC Press.

Golari, M., Fan, N., & Wang, J. (2014). Two-stage stochastic optimal islanding operations under severe multiple contingencies in power grids. *Electric Power Systems Research*, *114*, 68–77. doi:10.1016/j.epsr.2014.04.007

Gomes, F.V., Carneiro, Jr, S., Pereira, J.L.R., & Vinagre, M.P. (2006). A new distribution system reconfiguration approach using optimum power flow and sensitivity analysis for loss reduction. IEEE Transaction on Power Systems, 21(4), 1616-1623.

Gomez-Gonzalez, M., Lopez, A., & Jurado, F. (2012). Optimization of distributed generation systems using a new discrete PSO and OPF. *Electric Power Systems Research*, *84*(1), 174–180.

González, P. A., & Zamarreño, J. M. (2005). Prediction of hourly energy consumption in buildings based on a feedback artificial neural network. *Energy and Building*, *37*(6), 595–601. doi:10.1016/j.enbuild.2004.09.006

Goswami, S. K., & Basu, S. K. (1992). A New Algorithm for the Reconfiguration of Distribution Feeders for Loss Minimization. *IEEE Transactions on Power Delivery*, 7(3), 1484–1491. doi:10.1109/61.141868

Grainger, J.J. & Stevenson, W.D. (1993). Power System Analysis (T.M.H. Edition).

Granelli, G., Montagna, M., Zanellini, F., Bresestib, P., Vailati, R., & Innorta, M. (2006). Optimal network reconfiguration for congestion management by deterministic and genetic algorithms. *Electric Power Systems Research*, 76, 549–556.

Granzer, W., Kastner, W., Neugschwandtner, G., & Praus, F. (2006). *A modular architecture for building automation systems.*

Greene, S., Dobson, I., & Alvarado, F. L. (1997). Sensitivity of the loading margin to voltage collapse with respect to arbitrary parameters. *IEEE Transactions on Power Systems*, *12*(1), 262–272. doi:10.1109/59.574947

GTM Smart Grid. (2013, September 11). Greentechmedia. Retrieved from http://smart562.rssing.com/chan-8437336/all_p7.html

Guerrero, J. M., Chandorkar, M., Lee, T., & Loh, P. C. (2013). Advanced control architectures for intelligent microgrids, part I: Decentralized and hierarchical control. *IEEE Transactions on Industrial Electronics*, *60*(4), 1254–1262. doi:10.1109/TIE.2012.2194969

Guntermann, A. E. (1982). Are Energy Management Systems Cost Effective? *IEEE Transactions on Industry Applications*, *IA-18*(6), 616–625. doi:10.1109/TIA.1982.4504118

Guowei, C., Junfeng, Z., Yang, D., Chan, K. W., & Long, X. (2010, 24-28 Oct. 2010). *The identification of coherent generator groups via EMD and SSI.* Paper presented at the 2010 International Conference on Power System Technology POWERCON.

Guo, Y., & Fang, Y. (2013). Electricity cost saving strategy in data centers by using energy storage. *IEEE Transactions on* Parallel and Distributed Systems, *24*(6), 1149–1160.

Gupta, D., & Sharma, A. K. (2011). Investigations on Energy Efficiency for WSN Routing protocols for Realistic Radio Models. *International Journal of Future Generation Communication and Networking*, *4*(3), 61–72.

Gyugyi, L. (1994). Dynamic Compensation of AC Transmission Lines by Solid-state Synchronous Voltage Sources. *IEEE Transactions on Power Delivery*, *9*(2), 904–911. doi:10.1109/61.296273

HadiSaadat. (1998). *Power Systems Analysis.* Mg-H College.

Han, D. M., & Lim, J. H. (2010). Smart home energy management system using IEEE 802.15. 4 and zigbee. *IEEE Transactions on* Consumer Electronics, *56*(3), 1403–1410.

Hanmandlu, M., & Chauhan, B. K. (2011). Load Forecasting Using Hybrid Models. *IEEE Transactions on Power Systems*, *26*(1), 20–29. doi:10.1109/TPWRS.2010.2048585

Haque, M. H. (2006). Damping improvement using facts devices. *Electric Power Systems Research*, *76*(9-10). doi:10.1016/j.epsr.2005.11.001

Haron, A.R., & Idris, N.R.N. (2006). Simulation of MRAS-based Speed Sensorless Estimation of Induction Motor Drives using MATLAB/SIMULINK. *Proceedings of the IEEE International Power and Energy Conference* (pp. 411 – 415).

Harrison, G. P., & Wallace, A. R. (2005). Optimal power flow evaluation of distribution network capacity for the connection of distributed generation. *IEE Proc.–Generation, Transmission and Distribution, 152*(1), 115–122

Harrison, G. P., Piccolo, A., Siano, P., & Wallacen, A. R. (2008). Hybrid GA and OPF evaluation of network capacity for distributed generation connections. *Electric Power Systems Research*, *78*(3), 392–398. doi:10.1016/j.epsr.2007.03.008

Harris, R., & Sollis, R. (2003). *Applied Time Series Modeling and Forecasting*. Hoboken, NJ: John Wiley & Sons Inc.

Hart, G. W. (1992). Nonintrusive appliance load monitoring. *Proceedings of the IEEE*, *80*(12), 1870–1891. doi:10.1109/5.192069

Havells. (2013). *Energy Efficient: Electric Motors*. Havells.

Haykin, S. (1999). *Neural Networks – A Comprehensive Foundation* (2nd ed.). New Delhi: Pearson.

Haykin, S. (2009). *Neural Networks and Learning Machines*. Prentice Hall.

Hazarika, D., Bhuyan, S., & Chowdhury, S. P. (2006). Line outage contingency analysis including the system islanding scenario. *International Journal of Electrical Power & Energy Systems*, *28*(4), 232–243. doi:10.1016/j.ijepes.2005.11.018

Heidemann, J., Estrin, D., & Xu, Y. (2001). Geography-informed Energy Conservation for Ad-hoc Routing. *Proceedings of the Seventh Annual ACM/IEEE International Conference on Mobile Computing and Networking*, Rome, Italy (pp. 70-84).

Heikalabad, S. R., Navin, A. H., Mirnia, M., Ebadi, S., & Golesorkhtabar, M. (2010). Energy balancing and dynamic hierarchical routing algorithm for wireless Sensor networks. *IECIE Electronics Express*, *7*(15), 1112–1118. doi:10.1587/elex.7.1112

Heinzelman, W., Chandrakasan, A., & Balakrishnan, H. (2000). Energy efficient communication protocol for wireless sensor networks. *Proceeding of the Hawaii International Conference System Sciences,* Hawaii.

Heinzelman, W., Kulik, J., & Balakrishnan, H. (1999). Adaptive protocols for information dissemination in wireless sensor networks.*Proceedings of the 5th Annual ACM/IEEE International Conference on Mobile Computing and Networking*, Seattle, WA (pp. 174-185). doi:10.1145/313451.313529

Higginbotham, S. (2012, Mar.). Whose Data centers Are More Efficient? Facebook's or Google's? Retrieved from http://gigaom.com/cloud/whose-data-centers-are-more-efficient-facebooks-or-googles

Hingorani, N. G., & Gyugyi, L. (1999). *Understanding FACTS: concepts and technology of flexible ac transmission systems*. NY: IEEE Press. doi:10.1109/9780470546802

Hingorani, N. G., & Gyugyi, L. (2011). *Understanding FACTS*. Delhi: IEEE Press.

Holtz, J., & Quan, J. (2002). Sensorless Vector Control of Induction Motors at Very Low Speed Using a Nonlinear Inverter Model and Parameter Identification. *IEEE Transactions on Industry Applications*, *38*(4), 1087–1095. doi:10.1109/TIA.2002.800779

Home Energy Management Systems: Vendors, Technologies and Opportunities. (2014, February 11). PRNewsire.com. Retrieved from http://www.prnewswire.com/news-releases/home-energy-management-systems-vendors-technologies-and-opportunities-2013-2017-244880121.html

Hong, H. (1998). An efficient point estimate method for probabilistic analysis. *Reliability Engineering & System Safety*, *59*(3), 261–267. doi:10.1016/S0951-8320(97)00071-9

Hong, W.-C. (2009). Hybrid evolutionary algorithms in a SVR-based electric load forecasting model. *International Journal of Electrical Power & Energy Systems*, *31*(7–8), 409–417. doi:10.1016/j.ijepes.2009.03.020

Hong, W.-C. (2011). Electric load forecasting by seasonal recurrent SVR (support vector regression) with chaotic artificial bee colony algorithm. *Energy*, *36*(9), 5568–5578. doi:10.1016/j.energy.2011.07.015

Hong, Y.-Y., & Chou, J.-H. (2012). Nonintrusive energy monitoring for microgrids using hybrid self-organizing feature-mapping networks. *Energies*, *5*(7), 2578–2593. doi:10.3390/en5072578

Hoyong Kim, Yunseok Ko, & Kyung-Hee Jung (1993). Artificial neural-network based feeder reconfiguration for loss reduction in distribution systems. *IEEE Transactions on Power Delivery*, 8(3), 1356 – 1366.

Hsiao, Y. T., & Chiem, C. Y. (2000). Enhancement of restoration service in distribution systems using a Combination Fuzzy-GA Method. *IEEE Transactions on Power Systems*, *15*(4), 1394–1400. doi:10.1109/59.898118

Huai, W., Qian, Z., Li, X., & Lu, S. (2013, July). Towards Energy Efficient Data centers: A DVFS-Based Request Scheduling Perspective. *Proceedings of the 2013 Seventh International Conference on Innovative Mobile and Internet Services in Ubiquitous Computing IMIS* (pp. 299-306). IEEE.

Huang, G., Zhu, Q. & Siew, C. (2006). Extreme learning machine: Theory and applications. *Neurocomputing*, *70*, 489–501.

Huggins, R. A. (2010). *Energy Storage* (p. 406). Springer. doi:10.1007/978-1-4419-1024-0

Huneault, M., Rosu, C., Manoliu, R., & Galiana, F. D. (1994). A study of knowledge engineering tools in power engineering applications. *IEEE Transactions on Power Systems*, *9*(4), 1825–1832. doi:10.1109/59.331437

Hurvich, C. M., & Tsai, C. L. (1989). Regression and time series model selection in small samples. *Biometrika*, *76*(2), 297–307. doi:10.1093/biomet/76.2.297

Hussain, S., Matin, A. W., & Islam, O. (2007). Genetic Algorithm for Energy Efficient Clusters in Wireless Sensor Networks. *Proceedings of theFourth International Conference on Information Technology* (pp. 147 –154). doi:10.1109/ITNG.2007.97

Ibrahim, I. (2011). *Source Node Expansion Algorithm for Coherency Based Islanding of Power Systems*.

Impact of Restructuring of SEBs (Study Report). (n. d.). *Ministry of Power, Govt. of India*. Retrieved from http://powermin.nic.in/whats_new/pdf/IIPA.pdf

Inam, M., Ikram, A. A., & Ahmad, M. (2013). Hierarchical Energy Efficient Routing Protocol (HEERP) for Wireless Sensor Networks (WSN). *International Journal of Computer Science and Telecommunications*, *4*(10).

Intanagonwiwat, C., Govindan, R., & Estrin, D. (2000). Directed diffusion: a scalable and robust communication paradigm for sensor networks. *Proceedings of the 6th annual international conference on Mobile computing and networking* (pp. 56-67).

ISO. (2015). ISO 50001 - Energy management - ISO. Retrieved from http://www.iso.org/iso/home/standards/management-standards/iso50001.htm

J., Joy,, Jasmin, E.A., John V.R. (2013, March). Challenges of Smart Grid. *International Journal of Advanced Research in Electrical, Electronics and Instrumentation Engineering*, *2*(3).

Jabbari, F., & Mohammadi-ivatloo, B. (2014). Static Voltage Stability Assessment Using Probabilistic Power Flow to Determine the Critical PQ Buses. *Majlesi Journal of Electrical Engineering*, *8*(4), 17–25.

Jasmon, G. B., & Lee, L. H. C. C. (1993). New contingency ranking technique incorporating a voltage stability criterion. *IEE Proceedings. Generation, Transmission and Distribution*, *140*(2), 87–90. doi:10.1049/ip-c.1993.0012

Jemli, M., Azza, H., B., , Boussak, M., & Gossa, M. (2009). Sensorless Indirect Stator Field Orientation Speed Control for Single-Phase Induction Motor Drive. *IEEE Transactions on Power Electronics*, *24*(6), 1618–1627. doi:10.1109/TPEL.2009.2014867

Jiang, Z., & Dougal, R. A. (2008, July). Hierarchical microgrid paradigm for integration of distributed energy resources. *Proceedings of the IEEE Power and Energy Society 2008 General Meeting: Conversion and Delivery of Electrical Energy in the 21st Century* (pp. 1–8). doi:10.1109/PES.2008.4596185

Jindal, A. K., Ghosh, A., & Joshi, A. (2007). Interline unified power quality conditioner. *IEEE Transactions on Power Delivery*, *22*(1), 364–372. doi:10.1109/TPWRD.2006.881581

Johnson, W. A. (1980). EHV Operating Problems Associated with Reactive Control (Paper no 80 SM 513-2). IEEE PES Summer Meeting. doi:10.1109/TPAS.1981.316611

Joshi, D., Sandhu, K. S., & Soni, M. K. (2006). Constant voltage constant frequency operation for self excited induction generators. *IEEE Transactions on Energy Conversion*, *21*(1), 228–234. doi:10.1109/TEC.2005.858074

Jouili, M., Jarray, K., Koubaa, Y., & Boussak, M. (2009). Sensorless Indirect Stator Field Orientation Speed Control for Single-Phase Induction Motor Drive. *IEEE Transactions on Power Electronics*, 24(6).

Jove, E., Aláiz-Moretón, H., Casteleiro-Roca, J., Corchado, E., & Calvo-Rolle, J. (2014). Modeling of Bicomponent Mixing System Used in the Manufacture of Wind Generator Blades. In E. Corchado, J. Lozano, H. Quintián, & H. Yin (Eds.), *Intelligent Data Engineering and Automated Learning-IDEAL*. Springer International Publishing; doi:10.1007/978-3-319-10840-7_34

Juan, L., Chen, S., & Chao, Z. (2007). Ant System Based Any cast Routing in Wireless Sensor Networks. *Proceedings of the International Conference on Wireless Communications, Networking and Mobile Computing* (pp. 2420-2423).

Jung, K.-H., Kim, H., & Ko, Y. (1993). Network Reconfiguration algorithm for Automated Distribution Systems based on Artificial Intelligence Approach. *IEEE Transactions on Power Delivery*, *8*(4), 1933–1941. doi:10.1109/61.248305

Kahrobaeian, A; Ibrahim Mohamed, Y.A.-R. (2015). Networked-Based Hybrid Distributed Power Sharing and Control for Islanded Microgrid Systems. *IEEE Transactions on Power Electronics*, 30(2), 603 – 617.

Kai, S., Da-Zhong, Z., & Qiang, L. (2003). Splitting strategies for islanding operation of large-scale power systems using OBDD-based methods. *IEEE Transactions on* Power Systems, *18*(2), 912–923. doi:10.1109/TPWRS.2003.810995

Kaltschmitt, M., Streicher, W., & Wiese, A. (2007). *Renewable Energy: Technology, Economics and Environment* (p. 596). Springer.

Karnouskos, S. (2011). Crowdsourcing information via mobile devices as a migration enabler towards the SmartGrid. *Proceedings of the2011 IEEE International Conference on Smart Grid Communications SmartGridComm* (pp. 67–72). doi:10.1109/SmartGridComm.2011.6102394

Kastner, W., Neugschwandtner, G., Soucek, S., & Newman, H. M. (2005). Communication systems for building automation and control. *Proceedings of the IEEE*, *93*(6), 1178–1203. doi:10.1109/JPROC.2005.849726

Kawade, S., & Nekovee, M. (2009, December). Wireless options for high data-rate indoor users: Cognitive access to TV white space. *Proceedings of the 2009 First UK-India International Workshop on Cognitive Wireless Systems UKIWCWS* (pp. 1-6). IEEE.

Kay, S. M. (1988). *Modern Spectral Estimation: Theory and Application*. Englewood Cliffs, NJ: Prentice Hall.

Keane, A., & O'Malley, M. (2005). Optimal allocation of embedded generation on distribution networks. *IEEE Transactions on Power Systems*, *20*(3), 1640–1646. doi:10.1109/TPWRS.2005.852115

Kessel, P., & Glavitsch, H. (1986). Estimating the voltage stability of a power system. *IEEE Transactions on Power Delivery*, *PWRD-1*(3), 346–354219. doi:10.1109/TPWRD.1986.4308013

Khaligh, A., & Emadi, A.Alireza Khaligh Ali Emadi. (2007). Suitability of the pulse adjustment technique to control single DC/DC choppers feeding vehicular constant power loads in parallel with conventional loads. *Int. J. Electric and Hybrid Vehicles*, *1*(1), 20–45. doi:10.1504/IJEHV.2007.014445

Khani, H., Yazdani, N., & Mohammadi, S. (2012, November). Power-aware game for cloud computing: A distributed mechanism based on Game Theory for minimizing power consumption in cloud scale datacenter. *Proceedings of the 2012 Sixth International Symposium on Telecommunications IST* (pp. 598-601). IEEE. doi:10.1109/ISTEL.2012.6483058

Khaparde, S. A. (2004). Power Sector Reforms and Restructuring in India. *Proceedings of the IEEE Power Engineering Society General Meeting* (Vol. 2, pp. 2328–2335).

Khotanzad, A., Davis, M. H., Abaye, A., & Maratukulam, D. J. (1996). An artificial neural network hourly temperature forecaster with applications in load forecasting. *IEEE Transactions on Power Systems*, *11*(2), 870–876. doi:10.1109/59.496168

Khr, G. J., & Folger, T. A. (1988). *Fuzzy Sets, Uncertainty, and Information*. Prentice Hall.

Kienzle, F., & Andersson, G. (2009). A greenfield approach to the future supply of multiple energy carriers. *Proceedings of the IEEE Power Energy Society General Meeting PES '09* (pp. 1–8). doi:10.1109/PES.2009.5275692

Kim, H., & Jahns, T.M. (2006). Integration of the Measurement Vector Insertion Method (MVIM) with Discontinuous PWM for Enhanced Single Current Sensor Operation. Proceedings of the Industry Applications Conference (Vol. 5, pp. 2459-2465).

Kim, H., Kim, Y.-J., Yang, K., & Thottan, M. (2011). Cloud-based Demand Response for Smart Grid: Architecture and Distributed Algorithms. *Proceedings of theIEEE International Conference on Smart Grid Communications* (pp. 398-403).

Kim, M., Kim, S., Kim, J., & Yoo, Y. (2011, August). Design and implementation of MAC protocol for SmartGrid HAN environment. *Proceedings of the 2011 IEEE 11th International Conference on Computer and Information Technology CIT* (pp. 212-217). IEEE. doi:10.1109/CIT.2011.78

Kim, Y.-J., Kolesnikov, V., Kim, H., & Thottan, M. (2011). SSTP: a Scalable and Secure Transport Protocol for Smart Grid Data Collection. *IEEE Smart Grid Communication, Communication networks for smart grid.*

Kim, H., Ko, Y., & Jung, K.-H. (1993). Artificial neural network based feeder reconfiguration for loss reduction in distribution systems. *IEEE Transactions on Power Delivery*, *8*(3), 1356–1366. doi:10.1109/61.252662

Kim, K. H., Beloglazov, A., & Buyya, R. (2009, November). Power-aware provisioning of cloud resources for real-time services.*Proceedings of the 7th International Workshop on Middleware for Grids, Clouds and e-Science*. ACM. doi:10.1145/1657120.1657121

Kim, K. H., Beloglazov, A., & Buyya, R. (2011). Power-aware provisioning of virtual machines for real-time Cloud services. *Concurrency and Computation*, *23*(13), 1491–1505. doi:10.1002/cpe.1712

Kim, K. H., Buyya, R., & Kim, J. (2007, May). *Power Aware Scheduling of Bag-of-Tasks Applications with Deadline Constraints on DVS-enabled Clusters* (Vol. 7, pp. 541–548). CCGRID. doi:10.1109/CCGRID.2007.85

Kim, Y.-J., Lee, J., Atkinson, G., Kim, H., & Thottan, M. (2012). SeDAX: A Scalable, Resilient, and Secure Platform for Smart Grid Communications. *IEEE Journal on Selected Areas in Communications*, *30*(6), 1119–1136. doi:10.1109/JSAC.2012.120710

Kim, Y., Schmid, T., Charbiwala, Z. M., & Srivastava, M. B. (2009). ViridiScope: Design and Implementation of a Fine Grained Power Monitoring System for Homes. *Proceedings of the 11th International Conference on Ubiquitous Computing* (pp. 245–254). New York, NY, USA: ACM. doi:10.1145/1620545.1620582

Kliazovich, D., Bouvry, P., & Khan, S. U. (2013). DENS: Datacenter energy-efficient network-aware scheduling. *Cluster Computing*, *16*(1), 65–75. doi:10.1007/s10586-011-0177-4

Komal, K., Matsumoto, K., & Sakaguch, T. (1992). A diagnostic application speed expert systems shell for network fault diagnosis. *Electrical Power & Energy Systems*, 14, (2/3), 217-224.

Koomey, J. (2011). Growth in datacenter electricity use 2005 to 2010. Oakland, CA: Analytics Press.

Krings, A. W., & Ma, Z. S. (2009). Bio-Inspired Computing and Communication in Wireless Ad Hoc and Sensor Networks. *Ad Hoc Networks, Elsevier*, *7*(4), 742–755. doi:10.1016/j.adhoc.2008.03.003

Kroese, D. P., Taimre, T., & Botev, Z. I. (2011). *Handbook of Monte Carlo Methods* (Vol. 706). John Wiley & Sons. doi:10.1002/9781118014967

Kubota, H., & Matsuse, K. (2010). Speed sensorless field-oriented control of induction motor with rotor resistance adaptation. *IEEE Transactions on Industry Applications*, 30(5), 1219–1224

Kubota, H., Matsuse, K., & Nakmo, T. (1993). DSP-Based Speed Adaptive Flux Observer of Induction Motor. *IEEE Transactions on Industry Applications*, *29*(2).

Kundur, P. (1994). *Power System Stability and Control* (1st ed.). New York: MacGraw-Hill.

Kundur, P., Balu, N. J., & Lauby, M. G. (1994). *Power system stability and control* (Vol. 7). McGraw-hill New York.

Kuo, S.-C., & Wang, L. (2001). Analysis of voltage control for a self-excited induction generator using a current-controlled voltage source inverter (CC-VSI). *Generation, Transmission and Distribution, IEE Proceedings*, 148(5), 431-438. doi:10.1049/ip-gtd:20010477

Kuo, A. M.-H. (2011). Opportunities and Challenges of Cloud Computing to Improve Health Care Services. *Journal of Medical Internet Research*, *13*(3), e67. Retrieved from www.jmir.org/article/viewfile/1867/1/1 doi:10.2196/jmir.1867 PMID:21937354

Kusiak, A., Li, M., & Tang, F. (2010). Modeling and optimization of *HVAC* energy consumption. *Applied Energy*, *87*(10), 3092–3102. doi:10.1016/j.apenergy.2010.04.008

Kuwabara, T., Shibuya, A., & Furuta, H. (1996). Design and dynamic response characteristics of 400MW Adjustable speed pumped storage unit for ohkawachi power station. IEEE Trans on Energy conversion (Vol. 2, pp. 376-384).

Lai, Y. X., Rodrigues, J. J., Huang, Y. M., Wang, H. G., & Lai, C. F. (2012). An intercommunication home energy management system with appliance recognition in home network. *Mobile Networks and Applications*, *17*(1), 132–142. doi:10.1007/s11036-011-0345-z

Langhammer, N., & Kays, R. (2012). Performance evaluation of wireless home automation networks in indoor scenarios. *IEEE Transactions on* Smart Grid, *3*(4), 2252–2261.

Larsen, E., Miller, N., Nilsson, S., & Lindgren, S. R. (1992). Benefits of GTO-Based Compensation Systems For Electric Utility Applications. *IEEE Transactions on Power Delivery*, *7*(4), 2056–2062. doi:10.1109/61.157009

Lascu, C., & Andreescu, G. (2006). Sliding-Mode Observer and Improved Integrator With DC-Offset Compensation for Flux Estimation in Sensorless-Controlled Induction Motors. IEEE transactions on industrial electronics, 53(3).

Lascu, C., Boldea, I., & Blaabjerg, F. (2006). DTFC-SVM motion-sensorless control of a PM-assisted reluctance synchronous machine as starter alternator for hybrid electric vehicles. *IEEE Transactions on Power Electronics*, 21(3), 711–719.

Laskar, S. H. Mohibullah, (2012, September). Power Quality Issues and Need of Intelligent PQ Monitoring in the Smart Grid Environment. *International Journal of Emerging Technology and Advanced Engineering*, 2(9).

Lazos, D., Sproul, A. B., & Kay, M. (2015). Development of hybrid numerical and statistical short term horizon weather prediction models for building energy management optimisation. *Building and Environment*, *90*, 82–95. doi:10.1016/j.buildenv.2015.03.025

Lee, E. K., Viswanathan, H., & Pompili, D. (2012, December). Vmap: Proactive thermal-aware virtual machine allocation in HPC cloud data centers. *Proceedings of the 19th International Conference on High Performance Computing HiPC '12* (pp. 1-10). IEEE.

Leeb, S. B. (1993). *A conjoint pattern recognition approach to nonintrusive load monitoring*. Massachusetts Institute of Technology.

Lee, S.-J., Lim, S.-I., & Ahn, B.-S. (1998). Service Restoration of Primary Distribution Systems Based on Fuzzy Evaluation of Multi-Criteria. *IEEE Transactions on Power Systems*, *13*(3), 1156–1163. doi:10.1109/59.709114

Leung, P. C. M. & Lee, E.W.M. (2013). Estimation of electrical power consumption in subway station design by intelligent approach. *Applied Energy* 101 C 634-643.

Lewis, F. L. (2004). *Smart Environments: Technologies, Protocols, and Applications. In Wireless Sensor Networks* (pp. 11–46). John Wiley.

Li Hongyou, W.J. (2013). *Energy-Aware Scheduling Scheme Using Workload-Aware Consolidation Technique in Cloud Datacentres.* ZHANG Liangjie.

Li Wang & Chang-Min Cheng. (2000). Excitation capacitance required for an isolated three-phase induction generator supplying a single-phase load. *Proceedings of the Power Engineering Society Winter Meeting 2000* (Vol. 1, pp. 299-303).

Liao, Z., & Dexter, A. L. (2004). A simplified physical model for estimating the average air temperature in multi-zone heating systems. *Building and Environment*, *39*(9), 1013–1022. doi:10.1016/j.buildenv.2004.01.034

Lien, C. H., Bai, Y. W., Chen, H. C., & Hung, C. H. (2009, January). Home appliance energy monitoring and controlling based on power line communication. *Proceedings of the Digest of Technical Papers International Conference on Consumer Electronics ICCE '09* (pp. 1-2). IEEE. doi:10.1109/ICCE.2009.5012315

Lien, C. H., Chen, H. C., Bai, Y. W., & Lin, M. B. (2008, May). Power monitoring and control for electric home appliances based on power line communication. *Proceedings of the Instrumentation and Measurement Technology Conference Proceedings IMTC '08* (pp. 2179-2184). IEEE. doi:10.1109/IMTC.2008.4547409

Lindesy, S., & Raghavendra, C. (2002). PEGASIS: Power-Efficient Gathering in Sensor Information System. *Proceedings of the Aerospace Conference*, *Big Sky, Montana*, (Vol. 3, pp. 9-16). IEEE. doi:10.1109/AERO.2002.1035242

Lin, W. M., & Chin, H. C. (1997). Preventive and Corrective Switching for Feeder Contingencies in Distribution Systems with Fuzzy Set Algorithm. *IEEE Transactions on Power Delivery*, *12*(4), 1711–1716. doi:10.1109/61.634195

Li, Q., Aslam, J., & Rus, D. (2011). Hierarchical Power-aware Routing in Sensor Networks. *Proceedings of the DIMACS Workshop on Pervasive Networking*.

Liu, C. C., & Damborg, M. J. (1988). *Development* of *Expert Systems as On-Line Power System Operational Aids* (Final Report). EPRI Research Project EL5635.

Liu, L., Liu, W., Cartes, D. A., & Chung, I.-Y. (2009). Slow coherency and Angle Modulated Particle Swarm Optimization based islanding of large-scale power systems. *Advanced Engineering Informatics*, *23*(1), 45–56. doi:10.1016/j.aei.2008.06.009

Llaria, A., Curea, O., Jiménez, J., & Camblong, H. (2011). Survey on microgrids: Unplanned islanding and related inverter control techniques. *Renewable Energy*, *36*(8), 2052–2061. doi:10.1016/j.renene.2011.01.010

Lloyd, S. P. (1982). Least squares quantization in PCM. *IEEE Transactions on Information Theory*, *28*(2), 129–137. doi:10.1109/TIT.1982.1056489

Loijos, A. (2013, June 3). Review of Top 10 Energy Management Software (EMS) Comparison. LinkCycle. Retrieved from http://www.linkcycle.com/review-of-top-energy-management-software/

Lopes, J. A. P., Hatziargyriou, N., Mutale, J., Djapic, P., & Jenkins, N. (2007). Integrating distributed generation into electric power systems: A review of drivers, challenges and opportunities. *Electric Power Systems Research*, *77*(9), 1189–1203. doi:10.1016/j.epsr.2006.08.016

López García, H., & Machón González, I. (2004). Self-organizing map and clustering for wastewater treatment monitoring. *Engineering Applications of Artificial Intelligence*, *17*(3), 215–225. doi:10.1016/j.engappai.2004.03.004

Lora, A., Santos, J. M. R., Exposito, A., Ramos, J. L. M., & Santos, J. C. R. (2007). Electricity Market Price Forecasting Based on Weighted Nearest Neighbors Techniques. *IEEE Transactions on Power Systems*, *22*(3), 1294–1301. doi:10.1109/TPWRS.2007.901670

Lucia López, J., & Meneu Ferrer, V. (2005). Los precios en los mercados reestructurados de electricidad: algunas lecciones básicas para la negociación derivada. *CICE Nº 69*.

Luenberger, D. G. (1984). *Linear and Nonlinear Programming*. Addison-Wesley.

Luiz, A. C. Lopes & Almeida, R.G. (2006). Wind–Driven Self–Excited Induction Generator With Voltage and Frequency Regulated by a Reduced- Rating Voltage Source Inverter. IEEE Trans. on Energy conversion, 21(2).

Luukko, J., Niemelä, M., & Pyrhönen, J. (2003). Estimation of the flux linkage in a direct-torque-controlled drive. *IEEE Transactions on Industrial Electronics*, *50*(2), 283–287. doi:10.1109/TIE.2003.809392

M. A. Kashem et al. (2000). Network Reconfiguration for Enhancement of Voltage stability in Distribution Network. *IEEE*, 147(3), 171 – 175.

M. Arun & P. Aravindha Babu (2010). Fuzzy based reconfiguration algorithm for voltage stability enhancement of distribution systems. *Expert Systems with Applications: An International Journal archive*, 37(10), 6974-6978.

MacKay, D. J. C. (2009). *Sustainable Energy: Without the Hot Air* (p. 366). Uit Cambridge Limited.

Macqueen, J. (1967). Some methods for classification and analysis of multivariate observations. *Proceedings of the5th Berkeley Symposium on Mathematical Statistics and Probability* (pp. 281-297).

Maheshwari, K., Birman, K., Wozniakz, J. M., & Van Zandt, D. (2013). Evaluating Cloud Computing Techniques for Smart Power Grid Design Using Parallel Scripting. *Proceedings of theInternational Symposium on Cluster, Cloud and Grid Computing (CCGrid)*, IEEE/ACM. doi:10.1109/CCGrid.2013.26

Mahmood, H., & Jiang, J. (2012). Modeling and control system design of a grid connected VSC considering the effect of the interface transformer type. *IEEE Transactions on Smart Grid*, *3*(1), 122–134. doi:10.1109/TSG.2011.2166412

Mahmud, M. A., Hossain, M. J., & Pota, H. R. (2014). Voltage Variation on Distribution Networks With Distributed Generation: Worst Case Scenario. *IEEE Systems Journal*, *8*(4), 1096–1103. doi:10.1109/JSYST.2013.2265176

Malakhov, A., Kopyriulin, P., Petrovski, S., & Petrovski, A. (2012) Adaptation of smart grid technologies. Proceedings of the 2012 IEEE International Conference on Fuzzy Systems (FUZZ-IEEE) (pp. 1 – 6).

Manjeshwar, A., & Agrawal, D. P. (2002). APTEEN: a hybrid protocol for efficient routing and comprehensive information retrieval in wireless sensor networks. *Proceedings of the 2nd International Workshop on Parallel and Distributed Computing Issues in Wireless Networks and Mobile computing, Ft. Lauderdale, FL.*

Manjeshwar, A., & Agrawal, D. (2001). TEEN: a Routing Protocol for Enhanced Efficient in Wireless Sensor Networks. *Proceedings of the 15th International Parallel and Distributed Processing Symposium, San Francisco* (pp. 2009-2015). doi:10.1109/IPDPS.2001.925197

Mansour, Y., Xu, W., Alverado, F., & Rinzin, C. (1994). SVC Placement Using Critical Modes of Voltage Instability. *IEEE Transactions on Power Systems*, *9*(2), 757–763. doi:10.1109/59.317664

Mariana, C., Kotzé, P., & van der Merwe, A. (2012) Securing Virtual and Cloud Environments. In Cloud Computing and Services Science, Service Science, Springer Science & Business Media, LLC. 4 doi:10.1007/978-1-4614-2326-3

Marsh, W. M. (2005). *Landscape Planning: Environmental Applications* (I. Ed, Ed.). 4th ed.). John Wiley & Sons.

Massoud, A. S. (2011). Smart Grid: Overview, Issues and Opportunities. Advances and Challenges in Sensing, Modeling, Simulation, Optimization and Control. *European Journal of Control*, *17*(5-6), 547–567. doi:10.3166/ejc.17.547-567

Math, H. J. Bollen, (2000). Understanding Power Quality Problems. IEEE Press Series on Power Engineering (pp. 6-124).

Ma, Y., Yu, J. Q., Yang, C. Y., & Wang, L. (2010). Study on power energy consumption model for large-scale public building.*Proceedings of the 2nd international workshop on intelligent systems and applications* (pp. 1-4). doi:10.1109/IWISA.2010.5473608

McBee, K.D., & Simões, M.G. (2012, June). Utilizing a Smart Grid Monitoring System to Improve Voltage Quality of Customers. *IEEE Transactions on smart grid*, 3(2), 738-743.

Medland, R. C., Foth, M., & Petkov, P. (2011). Connecting people and resource consumption in real time. *IEEE Pervasive Computing / IEEE Computer Society [and] IEEE Communications Society*, *10*(1), 63–65.

Medrea, N., Demean, A. (2008). Induction motor starting in a power system using edsa paladin design base software (Report No. 32). *Annals of the University of Craiova, Electrical Engineering series.*

Mell, P., & Grance, T. (2011). The NIST Definition of Cloud Computing (Special Publication 800-145 Cloud Computing). *NIST.* Retrieved from http://csrc.nist.gov/publications/nistpubs/800-145/SP800-145.pdf

Mendosa, J., Lopez, R., Morales, D., Lopez, E., Dessante, P., & Moraga, R. (2006). Minimal loss reconfiguration using genetic algorithms with restricted population and addressed operators: Real application. IEEE transactions on power systems, 21(2), 948-954.

Mendoza, J., Lopez, R., Morales, D., Lopez, E., Dessante, P., & Moraga, R. (2006). Minimal Loss Reconfiguration using Genetic Algorithms with Restricted Population and Addressed Operators: Real Application. *IEEE Transactions on Power Systems*, *21*(2), 948–954. doi:10.1109/TPWRS.2006.873124

Menniti, D., Sorrentino, N., Pinnarelli, A., Burgio, A., Brusco, G., & Power, B.G. (2014). In the future Smart Cities: Coordination of micro Smart Grids in a Virtual Energy District. Proceedings of the International Symposium on Electronics, Electrical Drives, Automation and Motion SPEEDAM '14 (pp. 676 – 682).

Metke, A. R., & Ekl, R. L. (2010). Security technology for smart grid networks. *IEEE Transactions on Smart Grid*, *1*(1), 99–107. doi:10.1109/TSG.2010.2046347

Miller, R. (2011). Report: Google Uses About 900,000 Servers. *Data Center Knowledge.* Retrieved from: http://www.datacenterknowledge.com/archives/2011/08/01/report-google-uses-about-900000-servers

Miller, T. J. E. (Ed.). (1982). *Reactive Power Control in Electric Power System.* John Wiley& Sons.

Miranda, V., Ranito, J. V., & Proenca, L. M. (1994). Genetic Algorithms in Optimal Multistage Distribution Network Planning. *IEEE Transactions on Power Systems*, *9*(4), 1927–1933. doi:10.1109/59.331452

Misra S., Woungang I., & Misra S. C. (2009). Guide to Wireless Sensor Networks. *Springer-Verlag London.*

Missaoui, R., Warkozek, G., Bacha, S., & Ploix, S. (2012, October). Real time validation of an optimization building energy management strategy based on power-hardware-in-the-loop tool. *Proceedings of the 2012 3rd IEEE PES International Conference and Exhibition on Innovative Smart Grid Technologies ISGT Europe* (pp. 1-7). IEEE. doi:10.1109/ISGTEurope.2012.6465791

Mohammad-Hossein, Shariatkhah, Mahmoud-Reza Haghifam, JavadSalehi, & Moser, A. (2012). Duration based reconfiguration of electric distribution networks using dynamic programming and harmony search algorithm. *Electrical Power and Energy Systems*, 41, 1–10.

Morrison, G.K., Gao, B., & Kundur, P. (1993). Voltage stability analysis using static and dynamic approaches. IEEE Transactions on Power Systems, 8(3), 1159-1171.

Morton, A. B., & Mareels, I. M. Y. (2000). Efficient Brute-Force solution to Network Reconfiguration problem. *IEEE Transactions on Power Delivery*, *15*(3), 996–1000. doi:10.1109/61.871365

Moslehi, K., & Kumar, R. (2010). A Reliability Perspective of the Smart Grid. IEEE transactions on Smart Grid, 1(1), 57-64.

Murthy, S.S., & Pinto, A.J.P. (2005, September). A Generalized Dynamic and Steady State Analysis of Self Excited Induction Generator (SEIG) Based on MATLAB. *Proceedings of the Eighth International Conference on Electrical Machines and Systems ICEMS '05* (Vol. 3, pp. 1933-1938).

Murthy, S.S., Jha, C.S., & Rao, P.S.N. (1990). Analysis of grid connected induction generators driven by hydro/wind turbines under realistic system constraints. *IEEE Transactions on Energy Conversion*, 5(1), 1-7. doi:10.1109/60.50804

Myoung-Ho Shin, Dong-Seok Hyun, Soon-Bong Cho, & Song-Yul Choe (2000). An improved stator flux estimation for speed sensorless stator flux orientation control of induction motors. IEEE Transactions on Power Electronics, 15(2), 312-318.

Mytech-info. Retrieved from http://www.mytech-info.com

N.C. Sahoo & K. Prasad, (2006). A fuzzy genetic approach for network reconfiguration to enhance voltage stability in radial distribution systems. *Energy Conversion Management*, 47(18-19), 3288-3306.

Najafi, S., Abedi, M., & Hosseinian, S. H. (2008, 1-3 Dec. 2008). *A strategy for frequency stability of islanded power systems.* Paper presented at the Power and Energy Conference, 2008. PECon 2008. IEEE 2nd International.

Najafi, S., Hosseinian, S. H., & Abedi, M. (2010). A Novel Strategy for Stability Evaluation of Islanded Power Systems. *IEEJ Iranian Journal of Electrical and Computer Engineering, 9*(1), 43-51.

Najafi, S. (2013). An Early Warning Mechanism to Encounter Interconnected Power Systems Catastrophic Failures. *International Journal on Technical and Physical Problems of Engineering*, *5*(17), 105–112.

Najafi, S., Hosseinian, S. H., & Abedi, M. (2010). Proper Splitting of Interconnected Power Systems. *IEEJ Transactions on Electrical and Electronic Engineering*, *5*(2), 211–220. doi:10.1002/tee.20519

Nara, K., Shiose, A., Kitagaowaand, M., & Ishihara, T. (1992). Implementation of Genetic Algorithm for Distribution System Loss Minimum Reconfiguration. *IEEE Transactions on Power Systems*, *7*(3), 1044–1051. doi:10.1109/59.207317

Nastu, P. (2013, August 7). BEMS Market Will Grow To $5.6 Billion By 2020. *Energy Manager Today*. Retrieved from http://www.energymanagertoday.com/bems-market-will-grow-to-5-6-billion-by-2020-094340/

Nayab, N. (2010). Advantages of Smart Grid technology. *Brighthub*. Retrieved from http://www.brighthub.com/environment/renewable-energy/articles/74984.aspx#

Nelles, O. (2001). *Nonlinear System Identification. From Classical Approaches to Neural Networks and Fuzzy Models*. Springer.

Newsham, G. R., & Birt, B. J. (2010). Building-level occupancy data to improve ARIMA-based electricity use forecasts. *Proceedings of the 2nd ACM workshop on embedded sensing systems for energy-efficiency in building* (pp. 13-18). New York, USA. doi:10.1145/1878431.1878435

Nik, R. N. I., & Abdul, H. M. Y. (2002). An Improved Stator Flux Estimation in Steady-StateOperation for Direct Torque Control of Induction Machines. *IEEE Transactions on Industry Applications*, *38*(1).

Ochoa, L. F., Dent, C. J., & Harrison, G. P. (2010). Distribution network capacity assessment: Variable DG and active networks. *IEEE Transactions on Power Systems*, *25*(1), 87–95. doi:10.1109/TPWRS.2009.2031223

Official Website of Central Electricity Regulatory Commission (CERD). (n. d.). Retrieved from http://www.cercind.gov.in/

Padiyar, K. R. (2007). *FACTS Controllers in Transmission and Distribution*. New Delhi: New Age International.

Padiyar, K. R., & Devi, A. L. (1994). Control and simulation of static condenser.*Proc. of the 9th Annual Applied Power Electronics Conf. Expo.*

Pahwa, S., Youssef, M., Schumm, P., Scoglio, C., & Schulz, N. (2013). Optimal intentional islanding to enhance the robustness of power grid networks. *Physica A: Statistical Mechanics and its Applications, 392*(17), 3741-3754.

Paliwal, M., & Kumar, U. A. (2009). Neural networks and statistical techniques: A review of applications. *Expert Systems with Applications*, *36*(1), 2–17. doi:10.1016/j.eswa.2007.10.005

Paradiso, J., Dutta, P., Gellersen, H., & Schooler, E. (2011). Guest Editors' Introduction: Smart Energy Systems. *IEEE Pervasive Computing / IEEE Computer Society [and] IEEE Communications Society*, *10*(1), 11–12. doi:10.1109/MPRV.2011.4

Patel R., Pariyani S. & Ukani V., (2011). Energy and Throughput Analysis of Hierarchical Routing Protocol (LEACH) for Wireless Sensor Network. *International Journal of Computer Applications, 20(4).*

Pedrasa, M. A. A., Spooner, T. D., & MacGill, I. F. (2010). Coordinated scheduling of residential distributed energy resources to optimize smart home energy services. *IEEE Transactions on Smart Grid*, *1*(2), 134–143. doi:10.1109/TSG.2010.2053053

Peponis, G. J., Papadopoulos, M. P., & Hatziargyriou, N. D. (1995). Distribution Network Reconfiguration to Minimize Resistive Line Losses. *IEEE Transactions on Power Delivery*, *10*(3), 1338–1342. doi:10.1109/61.400914

Pérez Lombard, L., Ortiz, J., & Pout, J. (2008). A review on buildings energy consumption information. *Energy and Building*, *40*(3), 394–398. doi:10.1016/j.enbuild.2007.03.007

Pillay, P., & Levin, V. (1995). Mathematical Models for Induction Machines. *Proceedings of theIndustry Applications Conference (Vol. 1).*

Pousinho, H. M. I., Mendes, V. M. F., & Catalão, J. P. S. (2012). Short-term electricity prices forecasting in a competitive market by a hybrid PSO–ANFIS approach. *International Journal of Electrical Power & Energy Systems*, *39*(1), 29–35. doi:10.1016/j.ijepes.2012.01.001

Prais, M. (1989). A topology processor that tracks network modifications over time'. *IEEE-PWRS*, *3*(3), 992–998.

Prince, W. R., Wollenberg, B. F., & Bertagnolli, D. B. (1989). Survey on excessive alarms. *IEEE Transactions on Power Systems*, *4*, 950–956. doi:10.1109/59.32584

Priolkar Jayeshkumar, G., & Shet Vinayak, N. (2013, April). A Review on Protection Issues in Microgrid. *International Journal of Emerging Trends in Electrical and Electronics*, *2*(1).

Quintián, H., Calvo-Rolle, J. L., & Corchado, E. (2014). A Hybrid Regression System Based on Local Models for Solar Energy Prediction. *Informatica*, *25*(2), 265–282. doi:10.15388/Informatica.2014.14

Raghu, T., Srinivas Rao, J., & Chandra Sekhar, S. (2012). Simulation of Sensorless Speed Control of Induction Motor Using APFO Technique. *International Journal of Computer and Electrical Engineering*, 4(4).

Rahul, C., & Rabaey, J. (2002). Energy Aware Routing for Low Energy Ad Hoc Sensor Networks. *Proceedings of the Wireless Communications and Networking Conference* (pp. 350-355). *IEEE.*

Ramirez, J.M., & Torres, M.E. (2007). An electronic load controller for self excited induction generators. *IEEE Trans. Energy Conversion,* 21(2), 546-548.

Rao, K. R. (2011). *Energy and Power Generation Handbook: Established and Emerging Technologies* (p. 700). American Society of Mechanical Engineers. doi:10.1115/1.859551

Rao, P., Crow, M. L., & Yang, Z. (2000). STATCOM Control for Power System Voltage Control Applications. *IEEE Transactions on Power Delivery*, *16*, 18–23.

Rashed Md, G., Kabir, M. H., Rahim, M. S., & Ullah, S. E. (2010). Cluster Based Hierarchical Routing Protocol For Wireless Sensor Network. *International Journal of Computer and Network Security*, *2*(5), 128–131.

Rashid, M.H. (2003). *Power electronics Circuits, Devices and Applications* (3rd ed.).

Rashid, M. H. (2007). *Power Electronics Handbook: Devices, Circuits and Applications* (2nd ed.). Academic Press.

Rau, N. S., & Wan, Y. H. (1994). Optimum location of resources in distributed planning. *IEEE Transactions on Power Systems*, *9*(4), 2014–2020. doi:10.1109/59.331463

Razibul, I.M., & Gabbar Hossam, A. (2012, February). Study of Micro Grid Safety & Protection Strategies with Control System Infrastructures. *Smart Grid and Renewable Energy*, 3, 1-9. Retrieved from http://www.SciRP.org/journal/sgre

Rehman, H., & Xu, L. (2011). Alternative Energy Vehicles Drive System: Control, Flux and Torque Estimation, and Efficiency Optimization. IEEE Transactions on Vehicular Technology, 60(8).

Report to congress on server and data center energy efficiency. (2007). United States Environmental Protection Agency. Retrieved from http://hightech.lbl.gov/documents/datacenters/epa-datacenters.pdf

River, C. (2005). *Primer on Demand-Side Management with an Emphasis on Price-Responsive Programs*. Tech. Rep. prepared for The World Bank by Charles River Associates.

Rodoplu, V., & Ming, T. H. (1999). Minimum energy mobile wireless networks. *IEEE Journal on Selected Areas in Communications*, *17*(8), 1333–1344. doi:10.1109/49.779917

Rodriguez, C., Rementeria, S., Martin, J. I., Lafuente, A., Muguerza, J., & Perez, J. (1996). Fault analysis with modular neural networks. *International Journal of Electrical Power & Energy Systems*, *18*(2), 99–110. doi:10.1016/0142-0615(95)00007-0

Rousseeuw, P. J. (1987). Silhouettes: A graphical aid to the interpretation and validation of cluster analysis. *Journal of Computational and Applied Mathematics*, *20*(0), 53–65. doi:10.1016/0377-0427(87)90125-7

Saeedifard, M., Nikkhajoei, H., & Iravani, R. (2007). A space vector modulated STATCOM based on a three level neutral point clamped converter. *IEEE Transactions on Power Delivery*, *22*(2), 1029–1039. doi:10.1109/TPWRD.2007.893448

Saini, N., & Prakash, P. (2014). Factors Affecting Adaptation of Energy Management & Information Systems in Organizations.

Salmerón, J. (2010). *Las renovables en el mercado de la electricidad: perspectivas de futuro.*

Santoso, S. (2011). *Fundamentals of Electric Power Quality* (p. 273). Surya Santoso.

Sargunaraj, S., Sen Gupta, D. P., & Devi, S. (1997). Short-term load forecasting for demand side management. *IEE Proceedings on Generation, Transmission and Distribution, 144*(1), 68–74. doi:10.1049/ip-gtd:19970599

Schauder, C. (1992). Adaptive Speed Identification for Vector Control of Induction Motors without Rotational Transducers. *IEEE Transactions on Industry Applications*, *28*(5), 1054–1061. doi:10.1109/28.158829

Schmidt, H.P., Ida, N., Kagan, N., & Guaraldo, J.C. (2005). Fast Reconfiguration of Distribution Systems Considering Loss Minimization. *IEEE Transactions on Power Systems*, *20*(3), 1311–1319. doi:10.1109/TPWRS.2005.846180

Selvakennedy, S., Sinnappan, S., & Shang, Y. (2007). A biologically-inspired clustering protocol for wireless sensor networks. *Computer Communications, Elsevier*, *30*(14/15), 2786–2801. doi:10.1016/j.comcom.2007.05.010

Seubert, T. (2012, Feb. 27). Microsoft Builds New Datacenter in Dublin. Retrieved from http://facilitygateway.com/news/?p=1937

Seyoum, D., Rahman, F., & Grantham, C. (2003). Simplified flux estimation for control application in induction machines. Proceedings of the IEEE International Electric Machines and Drives Conference (Vol. 2, pp.691-695). doi:10.1109/IEMDC.2003.1210311

Shang, L., Peh, L.-S., Jha, K.N. (n. d.) Dynamic voltage scaling with links for power optimization of interconnection networks.

Shareef, H., Ibrahim, A.A., Salman, N., Mohamed, A., & Ling Ai, W. (2014). Power quality and reliability enhancement in distribution systems via optimum network reconfiguration by using quantum firefly algorithm. *Electrical Power and Energy Systems*, 58, 160–169.

Shimonishi, H., Higuchi, J., Yoshikawa, T., & Iwata, A. (2010). A Congestion Control Algorithm for Data Center Area Communications. *Proc. of IEEE International Communications Quality and Reliability Workshop, Vancouver, Canada.*

Shirmohammadi, D., & Hong, H. W. (1989). Reconfiguration of Electric Distribution Networks for Resistive Line Loss Reduction. *IEEE Transactions on Power Delivery*, *4*(2), 1492–1498. doi:10.1109/61.25637

Shridhar, L., Singh, B., Jha, C. S., Singh, B. P., & Murthy, S. S. (1994). Design of energy efficient motor for irrigation pumps operating under realistic conditions. *IEE Proceedings. Electric Power Applications*, *141*(6), 269–273. doi:10.1049/ip-epa:19941472

Siew, Z.W., Wong, C.H., Chin, C.S., & Kiring, A., & Kin-Teo K.T. (2012). Cluster Heads Distribution of Wireless Sensor Networks via Adaptive Particle Swarm Optimization. *Proceedings of theFourth International Conference on Computational Intelligence, Communication Systems and Network*. doi:10.1109/CICSyN.2012.25

Simmhan, Y., Aman, S., Cao, B., Giakkoupis, M., Kumbhare, A., Zhou, Q., & Prasanna, V. et al. (2011). *An Informatics Approach to Demand Response Optimization in Smart Grids* (Technical Report). Department of Computer Science in the University of California.

Singh, B., & Kasal, G.K. (2006). Solid state voltage regulator for isolated asynchronous Generators. *Proc. of India International Conference on Power Electronics.*

Singh, B., & Kasal, G.K. (2008). Solid State voltage and frequency Controller for a Stand Alone Wind Power Generating System. *IEEE Trans. on Power Electronics*, 23(3).

Singh, B., & Murthy, S. S., Sushma Gupta (2006). STATCOM-Based Voltage Regulator for Self-Excited Induction Generator Feeding Nonlinear Loads. IEEE Transactions on Industrial Electronics, 53(5).

Singh, B., Murthy, S.S., Gupta, S. (2004). Analysis and Design of STAT-COM Based Voltage Regulator for Self Excited Induction Generator. *IEEE Trans. on Energy conversion, 19(4).*

Singh, B., Al-Haddad, K., & Chandra, A. (1998). Harmonic elimination, reactive power compensation and load balancing in three phase four wire electric distribution system supplying non linear loads. *Electric Power Systems Research*, *44*(2), 93–100. doi:10.1016/S0378-7796(97)01187-5

Singh, B., Murthy, S. S., & Gupta, S. (2005). Transient analysis of self excited induction generator with electronic load controller supplying static and dynamic loads. *IEEE Transactions on Industry Applications*, *41*(5), 1194–1204. doi:10.1109/TIA.2005.855047

Singh, B., Murthy, S. S., & Gupta, S. (2006). Analysis and design of electronic load controller for self excited induction generator. *IEEE Transactions on Energy Conversion*, *21*(1), 285–293. doi:10.1109/TEC.2005.847950

Singh, B., & Shilpakar, L. B. (1998). Analysis of a novel solid state voltage regulator for a self excited induction generator. *IEE Proceedings on Generation, Transmission and Distribution*, *145*(6), 647–655. doi:10.1049/ip-gtd:19982357

Singh, G. K. (2004). Self excited induction generator research-a survey. *Electric Power Systems Research*, *69*(2-3), 107–114. doi:10.1016/j.epsr.2003.08.004

Singh, T., & Vara, P.K. (2009, July). Smart metering the clouds. *Proceedings of the 2009 18th IEEE International Workshops on Enabling Technologies* (pp. 66–71).

Sirisukprasert, S. (2014). Power electronics-based energy storages: A key component for Smart Grid technology. *Proceedings of the International 2014 Electrical Engineering Congress iEECON* (pp. 1 – 7).

Smith, J., Dugan, R., & Sunderman, W. (2011). Distribution modelling and analysis of high penetration PV. *Proceedings of the IEEE Power and Energy Society General Meeting PESGM '11*, USA.

Sohrabi, K., & Pottie, J. (2000). Protocols for self-organization of a wireless sensor network. *Personal Communications, 7*(5), 16–27. doi:10.1109/98.878532

Soliman, S.A. & Al-Kandari A.M. (2010). Electric Load Modeling for Long-Term Forecasting. *Electrical Load Forecasting* (pp. 353-406).

Son, Y. S., Pulkkinen, T., Moon, K. D., & Kim, C. (2010). Home energy management system based on power line communication. *IEEE Transactions on* Consumer Electronics, *56*(3), 1380–1386.

Srinivasan, D., Ruey Long Cheu, Young Peng Poh, Albert Kim, & Chwee Ng. (2000). *Automated fault detection in power distribution networks using a hybrid Fuzzy-genetic algorithm approach.*

Srivastava, A., & Shahidehpour, M. (2002). M. Restructuring Choices for the Indian Power Sector. *IEEE Power Engineering Review, 22*(11), 25–29. doi:10.1109/MPER.2002.1045559

Stojic, D.M. (2005). A New Induction Motor Drive Based on the Flux Vector Acceleration Method. IEEE Transactions on Energy Conversion, 20(1), 173-180.

Su, C.-L. (2005). A new probabilistic load flow method. *Paper presented at thePower Engineering Society General Meeting.*

Su, C.-L. (2005). Probabilistic load-flow computation using point estimate method. *IEEE Transactions on* Power Systems, *20*(4), 1843–1851.

Sullivan, R. L. (1977). *Power system planning* (p. 324). McGraw-Hill.

Sundareswaran, K. (n. d.). A Learner's Guide to Fuzzy Logic Systems.

Sundramoorthy, V., Cooper, G., Linge, N., & Liu, Q. (2011). Domesticating Energy-Monitoring Systems: Challenges and Design Concerns. *IEEE Pervasive Computing / IEEE Computer Society [and] IEEE Communications Society, 10*(1), 20–27. doi:10.1109/MPRV.2010.73

Sun, K., Zheng, D.-Z., & Lu, Q. (2005). A simulation study of OBDD-based proper splitting strategies for power systems under consideration of transient stability. *IEEE Transactions on* Power Systems, *20*(1), 389–399.

Sverdlik, Y. (2013). Facebook data centers' energy use up in 2012. *Data Center Dynamics*. Retrieved from http://www.datacenterdynamics.com/focus/archive/2013/06/facebook-data-centers-energy-use-2012

Szablya, L. (2012, April 26). Home Energy Management with or without an Advanced Metering Infrastructure - Part 2. Intelligent Utility. Retrieved from http://www.intelligentutility.com/article/12/04/home-energy-management-or-without-advanced-metering-infrastructure-part-2

Takeda, S., & Takemura, T. (2010). A rank-based vm consolidation method for power saving in data centers. *IPSJ Online Transactions, 3*, 88–96. doi:10.2197/ipsjtrans.3.88

Tandon, A. K., Murthy, S. S., & Berg, C. J. (1984). Steady State Analysis of capacitor self-excited induction generators. *IEEE Transactions on Power Apparatus and Systems, 103*(3), 612–617. doi:10.1109/TPAS.1984.318748

Taylor, C.W. (1994). Power System Voltage Stability. New York, USA: McGraw Hill.

Taylor, T., & Lubkeman, D. (1990). Implementation on Heuristic Search Strategy for Distribution Feeder Reconfiguration. *IEEE Transactions on Power Delivery, 5*(1), 239–246. doi:10.1109/61.107279

Technical Note: Power Quality. (2007). Schneider Electric.

Teleke, M. E., Baran, M. E., Huang, A. Q., Bhattacharya, S., & Anderson, L. (2009). Control strategies for battery energy storage for wind farm dispatching. *IEEE Transactions on Energy Conversion, 24*(3), 725–732. doi:10.1109/TEC.2009.2016000

The Electricity Act, 2003. (n. d.). *Central Electricity Regulatory Commission (CERD)*. Retrieved from http://www.cercind.gov.in/08022007/Act-with-amendment.pdf

The official website of Denmark. (n. d.). *Proceedings of the Climate Conference COP '15*. Retrieved from http://denmark.dk

The What, Why, and How of Energy Management. (2015). Energylens.com. Retrieved from http://www.energylens.com/articles/energy-management

Thorsen, O. V., & Dalva, M. (2001). Combined electrical and mechanical model of electric submersible pumps. *IEEE Transactions on Industry Applications*, *37*(2), 541–547. doi:10.1109/28.913720

Tian, Z., Miao, W., Hongjie, J., & Zhou, J. (2012). Multilevel partitioning algorithm for power system splitting control. *Paper presented at the Innovative Smart Grid Technologies Asia ISGT-Asia.*

Tibshirani, R. (1996). Regression Shrinkage and Selection Via the Lasso. *Journal of the Royal Statistical Society. Series B. Methodological*, *58*, 267–288.

Tripta, T. (2005). Impact Assessment of the Electricity Act 2003 on the Indian Power Sector. *Energy Policy*, *33*(9), 1187–1198. doi:10.1016/j.enpol.2003.11.016

Tripura, P., Kishore Babu, Y.S., & Tagore, Y.R. (2011). Space Vector Pulse Width Modulation Schemes for Two-Level Voltage Source Inverter. *Int. J. on Control System and Instrumentation*, 2(3).

Trodden, P., Bukhsh, W., Grothey, A., & McKinnon, K. (2013). MILP formulation for controlled islanding of power networks. *International Journal of Electrical Power & Energy Systems*, *45*(1), 501–508. doi:10.1016/j.ijepes.2012.09.018

Tu, C. Y., Kuo, W. C., Teng, W. H., Wang, Y. T., & Shiau, S. (2010, September). Power-aware cloud architecture with smart metering. *Proceedings of the 2010 39th International Conference on Parallel Processing Workshops ICPPW* (pp. 497-503). IEEE. doi:10.1109/ICPPW.2010.73

Tushar W., Saad W., Poor, H.V., & Smith D. (2012). Economics of electric vehicle charging: a game theoretic approach. *IEEE Transactions on Smart Grid*, 3(4), 1767–1778

United Nations. (1998). *Kyoto Protocol on the Framework Convention of the United Nations about the Climatic Change*.

United Nations. (2005). *Proceedings of theMontreal Climate Change Conference*.

United Nations. (2007). *Bali Climate Change Conference*. Retrieved from http://unfccc.int

US Department of Housing. (2006, March). Public Housing Energy Conservation Clearinghouse New. Retrieved from https://portal.hud.gov/hudportal/documents/huddoc?id=DOC_10679.pdf

US Energy Department. (2005). *On the Road to Energy Security; Implementing a Comprehensive Energy Strategy: A Status Report*. Retrieved from http://energy.gov/sites/prod/files/edg/media/FINAL20DOE_20booklet20copy20sep.pdf

Utray Fabra, J. (n. d.). *El mercado de la electricidad: Algunas reflexiones y algunas propuestas.*

Vaclavek, P., & Blaha, P. (2009). Lyapunov function based design of PMSM state observer for sensorless control. *Industrial Electronics & Applications (Vol. 1*, pp. 331–336).

Vaimann T., Niitsoo J., Kivipold T., & Lehtla T. (2012). Power Quality Issues in Dispersed Generation and Smart Grids. *Elektronika Ir Elektrotechnika*, 18(8).

Van Custem, T., & Mailhot, R. (1997). Validation of a Fast Voltage Stability Analysis Method on The Hydro-quebec System. *IEEE Trans. on Power Systems*, 12(1), 282-288.

Van Cutsem, T., & Vournas, C. (1998). *Voltage Stability of Electric Power systems*. Boston, USA: Kluwer Academic Publishers. doi:10.1007/978-0-387-75536-6

Van Gorp, J. C. (2004). Enterprising energy management. *IEEE Power and Energy Magazine*, *2*(1), 59–63. doi:10.1109/MPAE.2004.1263421

VAR Management Task Force of IEEE. (1984). VAR Management –Problem Recognition and Control. *IEEE Trans. on Power Systems*, 103(8).

Vellido, A., Lisboa, P. J. G., & Vaughan, J. (1999). Neural networks in business: A survey of applications (1992–1998). *Expert Systems with Applications*, *17*(1), 51–70. doi:10.1016/S0957-4174(99)00016-0

Venkobarao, V., & Chinnagouder, C. (2010). Sliding mode controller for internal combustion engine for energy-efficient operation of hybrid electric vehicles. *Int. J. Electric and Hybrid Vehicles*, *2*(4), 282–297. doi:10.1504/IJEHV.2010.034981

Verbic, G., & Cañizares, C. A. (2006). Probabilistic optimal power flow in electricity markets based on a two-point estimate method. *IEEE Transactions on* Power Systems, *21*(4), 1883–1893.

Versèle, C., Deblecker, O., & Lobry, J. (2008). Implementation of a vector control scheme using dSPACE material for teaching induction motor drive and parameters identification. *Proceedings of the18th International Conference onElectrical Machines* (pp. 1-6). doi:10.1109/ICELMACH.2008.4799951

Vivek, V., Kumudini Devi, R.P., & Chellamuthu, C. (n. d.). Design and development of energy efficient discharge controllers for Induction motor driven pump for Pulp and Paper Industry. *International Journal of Modeling and Simulation*, 45-51.

Vovos, P. N., & Bialek, J. W. (2005). Direct incorporation of fault level constraints in optimal power flow as a tool for network capacity analysis. *IEEE Transactions on Power Systems*, *20*(4), 2125–2134. doi:10.1109/TPWRS.2005.856975

Vovos, P. N., Harrison, G. P., Wallace, A. R., & Bialek, J. W. (2005). Optimal power flow as a tool for fault level constrained network capacity analysis. *IEEE Transactions on Power Systems*, *20*(2), 734–741. doi:10.1109/TPWRS.2005.846070

Vovos, P. N., Kiprakis, A. E., Wallace, A. R., & Harrison, G. P. (2007). Centralized and distributed voltage control: Impact on distributed generation penetration. *IEEE Transactions on Power Systems*, *22*(1), 476–483. doi:10.1109/TPWRS.2006.888982

Wagner, T. P., Chikhani, A. Y., & Hackam, R. (1991). Feeder Reconfiguration for Loss Reduction: An Application of Distribution Automation. *IEEE Transactions on Power Delivery*, *6*(4), 1922–1931. doi:10.1109/61.97741

Wang, H., Huang, J., Lin, X., & Mohsenian-Rad, H. (2013). Exploring Smart Grid and Data Center Interactions for Electric Power Load Balancing. *ACM SIGMETRICS Performance Evaluation Review*, 41(3), 89-94.

Wang, J., Sun, Z., Zhou, Y., & Dai, J. (2012, May). Optimal dispatching model of smart home energy management system. Proceedings of the 2012 IEEE Conference on Innovative Smart Grid Technologies-Asia ISGT Asia (pp. 1-5). IEEE.

Wang, X., & Vittal, V. (2004). System islanding using minimal cutsets with minimum net flow. *Paper presented at the Power Systems Conference and Exposition.*

Wang, X., Shao, W., & Vittal, V. (2005). Adaptive corrective control strategies for preventing power system blackouts. *Proceedings of the 15th Power Systems Computation Conference*, Liège, Belgium.

Wang, Y., & Deng, Z. (2012). An Integration Algorithm for Stator Flux Estimation of a Direct-Torque-Controlled Electrical Excitation Flux-Switching Generator. IEEE transactions on energy conversion, 27(2).

Wang, Y., & Deng, Z. (2012). Improved Stator Flux Estimation Method for Direct Torque Linear Control of Parallel Hybrid Excitation Switched-Flux Generator. IEEE transactions on energy conversion, 27(3).

Wang, C., & Lin, Q. (2008). Swarm intelligence optimization based routing algorithm for Wireless Sensor Networks. *Proceedings of International Conference on Neural Networks and Signal Processing* (pp. 136-141). doi:10.1109/ICNNSP.2008.4590326

Warner, M. (2011). *The Electric Vehicle Conversion Handbook* (p. 176). Penguin Group US.

Warwick, K., & Ekwue, A. (Eds.). (1997). *Artificial Intelligence Techniques in Power Systems*. K. Warwick Arthur Ekwue Raj Aggarwal, Institution of Electrical Engineers. doi:10.1049/PBPO022E

Webb, A. R., & Copsey, K. D. (2011). *Statistical Pattern Recognition*. Wiley. doi:10.1002/9781119952954

Wed, M. L. on. (2013, August 21). Empowering Shop-Floor Leaders to Optimize Energy Performance [DATA]. Retrieved from http://blog.lnsresearch.com/blog/bid/185439/Empowering-Shop-Floor-Leaders-to-Optimize-Energy-Performance-DATA

Weedy, B.M., & Cox, B.R. (1968). Voltage Stability of Radial Power Links.*Proc. of IEEE (Vol. 115).*

Wei, L., Ning, Y., & Zhengang, Z. (2009). Study on long-term load forecasting of MIXSVM based on principal component analysis. *Proceedings of theInternational Conference on FutureBioMedical Information Engineering FBIE '09* (pp. 439–441). doi:10.1109/FBIE.2009.5405820

Weigend, A. S., & Gershenfeld, N. A. (Eds.). (1993). *Time Series Prediction: Forecasting the Future and Understanding the Past*. Addison-Wesley.

Weisberg, S. (2005). *Applied Linear Regression*. Hoboken, NJ: Wiley. doi:10.1002/0471704091

Weiser, M. (2002). The Computer for the 21st Century. *IEEE Pervasive Computing / IEEE Computer Society [and] IEEE Communications Society*, *1*(1), 19–25. doi:10.1109/MPRV.2002.993141

Weiss, M., Mattern, F., Graml, T., Staake, T., & Fleisch, E. (2009). Handy Feedback: Connecting Smart Meters with Mobile Phones.*Proceedings of the 8th International Conference on Mobile and Ubiquitous Multimedia* (pp. 15:1–15:4). New York, NY, USA: ACM. doi:10.1145/1658550.1658565

Weixiao Meng, Ruofei Ma, & Hsiao-Hwa Chen (2014). Smart grid neighbourhood area networks: a survey. Network, 28(1), 24 – 32.

Wen, F., & Chang, C. S. (1996). A new approach to fault section estimation in power systems based on the set covering theory and the refined genetic algorithm.*Proceedings of 12th Power system computation conference PSCC '96*, Dresden (Vol 1, pp 358- 365).

Wen, F., & Han, Z. (1995). Fault section estimation in power systems using a genetic algorithm. *Electric Power Systems Research*, *34*(3), 165–172. doi:10.1016/0378-7796(95)00974-6

Wengenmayr, R., Bührke, T., & Brewer, W. D. (2013). *Renewable Energy: Sustainable Energy Concepts for the Energy Change* (p. 170). Wiley.

Wikipedia. Clique (graph theory). Retrieved from: https://en.wikipedia.org/wiki/Clique_(graph_theory)

Wilde, M., Hategan, M., Wozniak, J. M., Clifford, B., Katz, D. S., & Foster, I. (2011). Swift: A language for distributed parallel scripting. *Parallel Computing, Emerging Programming Paradigms for Large-Scale Scientific Computing*, *39*(9), 633–652.

Wollenberg, B. F., & Sakaguchi, T. (1987). Artificial intelligence inpower system operations. *Proceedings of the IEEE*, 12, 1678-1685.

Wong, S. L., Wan, K. K. W., & Lam, T. N. T. (2010). Artificial neural networks for energy analysis of office buildings with daylighting. *Applied Energy*, *87*(2), 551–557. doi:10.1016/j.apenergy.2009.06.028

Wu, J.S., Tomsovic, K.L., & Chen, C.S. (1991). A Heuristic Search Approach to Feeders Switching Operation for Overload, faults, unbalanced flow and maintenance. IEEE Transactions on Power Delivery, 6(4), 1579-1586.

Wu, Q. (2010a). A hybrid-forecasting model based on Gaussian support vector machine and chaotic particle swarm optimization. *Expert Systems with Applications*, *37*(3), 2388–2394. doi:10.1016/j.eswa.2009.07.057

Wu, Q. (2010b). Power load forecasts based on hybrid PSO with Gaussian and adaptive mutation and Wv-SVM. *Expert Systems with Applications*, *37*(1), 194–201. doi:10.1016/j.eswa.2009.05.011

Wu, Q. (2011). Hybrid forecasting model based on support vector machine and particle swarm optimization with adaptive and Cauchy mutation. *Expert Systems with Applications*, *38*(8), 9070–9075. doi:10.1016/j.eswa.2010.11.093

Wu, Y., Tang, Y., Han, B., & Ni, M. (2015). A topology analysis and genetic algorithm combined approach for power network intentional islanding. *International Journal of Electrical Power & Energy Systems*, *71*, 174–183. doi:10.1016/j.ijepes.2015.02.030

Xia, L., & Du Taihang, T. S. (2011). A fuzzy logic variable hysteresis band current control technique for three phase shunt active filter. *Proc. IEEE Conf. CASE* (pp. 1-4).

Xu, L., & Chow, M.-Y. (2006). A classification approach for power distribution systems fault cause identification. *IEEE Transactions on Power Systems*, *21*(1), 53–60.

Xu, L., Chow, M.-Y., & Taylor, L. S. (2007). Power Distribution Fault Cause Identification with Imbalanced Data Using the Data Mining Based Fuzzy Classification E-Algorithm. *IEEE Transactions on Power Systems*, *22*(1), 164–171. doi:10.1109/TPWRS.2006.888990

Xu, Z., Jia, Q.-S., Guan, X., & Shen, J. (2013). Smart Management of Multiple Energy Systems in Automotive Painting Shop. *IEEE Transactions on Automation Science and Engineering*, *10*(3), 603–614. doi:10.1109/TASE.2012.2236554

Y. Y. Hsu, C.C. Liu & C. T. Huang, (1988). Application of Power System Stabilizers and Static VAR Compensators on a Longitudinal Power System. *IEEE Trans. on Power Systems*, 3(4).

Yamini, B., & Selvi, D. V. (2010, November). Cloud virtualization: A potential way to reduce global warming. Proceedings of the 2010 conference on Recent Advances in Space Technology Services and Climate Change RSTSCC (pp. 55-57). IEEE. doi:10.1109/RSTSCC.2010.5712798

Yang, G.-H., & Li, V. O.-K. (2010). Energy Management System and Pervasive Service-Oriented Networks. *Proceedings of the2010 First IEEE International Conference on Smart Grid Communications (SmartGridComm)* (pp. 1–6). doi:10.1109/SMARTGRID.2010.5622001

Yang, B., Vittal, V., & Heydt, G. T. (2006). Slow-Coherency-Based Controlled Islanding: A Demonstration of the Approach on the August 14, 2003 Blackout Scenario. *Power Systems. IEEE Transactions on*, *21*(4), 1840–1847. doi:10.1109/TPWRS.2006.881126

Yanzhi Wang, Xue Lin, Pedram, M. (2014). Coordination of the Smart Grid and Distributed Data Centers: A Nested Game-Based Optimization Framework. *Proceedings of the 5th IEEE Annual Power & Energy Society Innovative Smart Grid Technologies Conference to Showcase Smart Grid Investment Grant Projects.*

You, H., Vittal, V., & Xiaoming, W. (2004). Slow coherency-based islanding. *Power Systems. IEEE Transactions on*, *19*(1), 483–491. doi:10.1109/TPWRS.2003.818729

Younis, O., & Fahmy, S. (2004). Distributed Clustering for Scalable, Long-Lived Sensor Networks. *Mobile Computing, 3*(4), 366–379. doi:10.1109/TMC.2004.41

Youssef, K. H., & Wahba, M. A., Yousef, Hasan A., Sebakhy, O.A. (2008). A new method for voltage and frequency control of stand-alone self-excited induction generator using PWM converter with variable DC link voltage. *Proceedings of theAmerican Control Conference 2008* (pp. 2486-2491). doi:10.1109/ACC.2008.4586864

Yu, Y., Estrin, D., & Govindan, R. (2001). *Geographical and Energy-Aware Routing: a recursive data dissemination protocol for Wireless Sensor Networks* (Technical Report UCLA-CSD TR-01-0023). UCLA Computer Science Department.

Yuwono, R. Y. (2009). X10 Protocol Man Machine Interface Implementation using Labview. *Journal of Applied Sciences*, *9*(19), 3562–3568. doi:10.3923/jas.2009.3562.3568

Zabin, F., Misra, S., Woungang, I., Rashvand, H. F., Ma, N. W., & Ali, M. A. (2008). REEP: Data-centric, energy-efficient and reliable routing protocol for wireless sensor networks. *IET Communications*, *2*(8), 995–1008. doi:10.1049/iet-com:20070424

Zarei, B., Zeynali, M., & Nezhad, V. M. (2010). Novel cluster Based Routing Protocol in Wireless Sensor Network. *International Journal of Computer Science Issues*, *7*(4), 32–36.

Zhang, H., Wang, Z., Gao, F., & Zhou, D. (2012). A New Method for the Daily Electricity Consumption Forecast of Energy Intensive Enterprise. *Proceedings of the2012 Asia-PacificPower and Energy Engineering Conference APPEEC* (pp. 1–4). doi:10.1109/APPEEC.2012.6307482

Zhao Tan, Peng Yang, Nehorai, A. (2014). An Optimal and Distributed Demand Response Strategy With Electric Vehicles in the Smart Grid. *IEEE Transactions on Smart Grid*, 5(2), 861 – 869.

Zhao, B., Guo, C.X., & Cao, Y.J. (2005). A Multiagent-Based Particle Swarm Optimization Approach for Optimal Reactive Power Dispatch. *IEEE Transactions on Power Systems*, *20*(2), 1070–1078. doi:10.1109/TPWRS.2005.846064

Zhao, H. X., & Magoulès, F. (2010). Parallel support vector machines applied to the prediction of multiple buildings energy consumption. *Journal of Algorithms & Computational Technology*, *4*(2), 231–249. doi:10.1260/1748-3018.4.2.231

Zhao, P., Suryanarayanan, S., & Simoes, M. G. (2013). An Energy Management System for Building Structures Using a Multi-Agent Decision-Making Control Methodology. *IEEE Transactions on Industry Applications*, *49*(1), 322–330. doi:10.1109/TIA.2012.2229682

Zhou, D., Gao, F., Guan, X., Chen, Z., Li, S., & Lu, Q. (2004). Daily electricity consumption forecast for a steel corporation based on NNLS with feature selection. *Proceedings of the2004 International Conference on Power System Technology PowerCon '04* (*Vol. 2*, pp. 1292–1297). doi:10.1109/ICPST.2004.1460201

Zhou, A., Qu, B.-Y., Li, H., Zhao, S.-Z., Suganthan, P. N., & Zhang, Q. (2011). Multiobjective evolutionary algorithms: A survey of the state of the art. *Swarm and Evolutionary Computation*, *1*(1), 32–49. doi:10.1016/j.swevo.2011.03.001

Zhou, Q., Shirmohammadi, D., & Liu, W.-H. E. (1997). Distribution Feeder Reconfiguration for Service Restoration and Load Balancing. *IEEE Transactions on Power Systems*, *12*(2), 724–729. doi:10.1109/59.589664

Zhuang, Y., & Menzies, R. W. (1995). Effect of DC Capacitance of a STATCOM on the dynamic performance at a weak HVDC terminal. Proc of IEEE WESCANEX '95 (pp. 289-293).

Zhu, Y. L., Yang, Y. H., Hogg, B. W., Zhang, W. Q., & Gao, S. (1994). An expert system for power systems fault analysis. *IEEE Transactions on Power Systems*, *9*(l), 503–509.

About the Contributors

Smita Shandilya is an eminent scholar and energetic researcher with excellent teaching and research skills. She achieved excellent result in all the subjects she has taught till date. She has over 20 quality research papers in international & national journals & conferences to her credits. She has delivered several invited talks in national seminars of high repute. Her research interests are Power System Planning and Smart Micro Grids. She is one of the core members of the research and development section of her Institute. She is also involved in various projects like the establishment of Center of Excellence (CoE) on Smart Grid in the Institute, Establishment of Training cum Incubator centre on Solar Systems under Institute-Industry Collaboration.

* * *

Abdelkader Abdelmoumene obtained his PhD from the Institute of Electrical Engineering and Electronics (Ex-INELEC), University of Boumerdes in 2015. He received also both Electromechanical Engineer and Magister degrees in Electrical Engineering from the Faculty of Oil and Chemistry at the University of Boumerdes, Algeria, in 2006 and 2010 respectively. He joined the Engineering Division of the National Company of Great Petroleum Works (a subsidiary of Sonatrach Group) in 2008. His areas of interest include Power System Protection, Monitoring and Control, Safety and Reliability, Electromechanical systems and High Voltage Engineering. He has authored and co-authored several journal \ conference papers.

Adnan Anwar is working with The University of New South Wales Australia. He is also a lecturer of the University of Asia Pacific. He has published more than 20 technical articles in prestigious book chapters, journals and conferences. His research interest includes Smart Grid modelling and analysis, optimization, data mining and cyber security of Smart Grid.

Hamid Bentarzi was born in Legata, Boumerdes, Algeria. He received both Electrical Engineering and Magister Degrees with honors from "Institut National d'Electricite et d'Electronique" (INELEC), Boumerdes, Algeria, in 1989 and 1992 respectively and Ph.D in Microelectronic systems from "Ecole Nationale Polytechnique" (ENP), Algiers, Algeria, in 2004. Hamid was a lecturer at INELEC, Boumerdes, Algeria, since 1991. After 1999, he has been a faculty member at the Department of Electrical and Electronic Engineering, Faculty of Engineering, University of Boumerdes, Algeria. Since 2001, he has been head of research team working in developing microelectronic systems applied to power systems in the Signal and System Laboratory, Boumerdes, Algeria. Prof. Bentarzi interests are in the fields of microelectronics, electrical protection systems and systems reliability. He has authored and co-authored over 70 technical papers. Besides, he has been a member of IEEE as well as organizing and technical committee of several conferences including WSEAS group.

Jignesh G. Bhatt received degree in Instrumentation & Control Engineering from Gujarat University, Gujarat, India, in 1997 and his Master of Technology degree in Electrical Engineering with Specialization: Measurement & Instrumentation from Indian Institute of Technology Roorkee (IIT Roorkee), Uttarakhand, India, in 2010. He served as Assistant Manager (Instrumentation) and Senior Engineer (Projects) with Public and Private Limited Companies during 1997 to 1999. Since 2000, he is with Dharmsinh Desai University, where he has been holding the position of Sr. Assistant Professor in Department of Instrumentation & Control Engineering, Faculty of Technology. He is currently pursuing his Doctoral Study (Ph.D.) as a Research Scholar in the Electrical Engineering Department, School of Technology, Pandit Deendayal Petroleum University, Gandhinagar, India. He also serves as Principal Investigator and Coordinator for the Sponsored Research and Community Development Projects funded by State and Central Governments. He has authored research publications in reputed national-international referred journals-conferences, contributed book chapters and technical reports, delivered Invited Expert Lectures and guided M.Tech. Dissertations. He serves as Reviewer and Editorial Board Member to reputed peer-reviewed international journals-conferences such as IEEE, Elsevier, IET, IFAC, etc. His research interests include Automation Systems, Building Automation, e-Governance, e-Learning, Instrumentation, Process Control, Solar Energy, Solar City, Smart City, Smart Grid and Wireless Sensor Networks with special focus on industrial and real-life applications.

Jose Luis Calvo-Rolle was born in A Coruna, Spain, in 1974. He received the M.S. and Ph.D. degrees in Industrial Engineering from the University of Leon, Leon, Spain, in 2004, and 2007, respectively. He is an Associate Professor of Automatic Control and the head of Industrial Engineering Department, Faculty of Engineering, University of A Coruna, Spain. His main research interests have been centered in applying expert system technology to the diagnosis and control systems and in intelligent training systems for control engineering, optimization and education.

José-Luis Casteleleiro-Roca received the B.S. from University of Coruna in 2003, the M.S. in Industrial Engineering from the University of Leon in 2012, and now he is a Ph.D student in the University of Coruna. He has been Technical Engineer in the Spanish Navy since 2004 (in the Arsenal from Ferrol); he worked in Missiles workshop for 11 years, performing tests and certifying the SM-1 and SM-2 missiles from the Spanish Navy. Now, he works in the fire weapon control workshop, performing the maintenance and repairing different weapons in the F-100 type vessels. Since 2014, he is also part of the teaching and research staff of the UDC as a part-time associate professor in the Industrial Engineering Department. He is co-author of 5 publications in journals indexed in JCR, 7 contributions in international scientific conferences and more than 10 contributions in national conferences. He is reviewer in several international journals and member of different committees in scientific conferences (program committee and organization committee) as IDEAL, HAIS and SOCO. His main research interests have been centered in applying expert system technologies to the diagnosis and control systems and in intelligent systems for control engineering and optimization.

Monika Dixit is presently pursuing a M. Tech. from SIRT, RGTU, Bhopal. She has completed her B. E. from NIIST, RGTU, Bhopal in 2010.

Manuel Domínguez González received his PhD degree in Engineering from the University of Oviedo (Spain) in 2003. He is a professor of automatic control and the head of the SUPPRESS research group and the Institute of Automatic Control (AAC-IAF) at the University of León (Spain). His research interests include control engineering, data mining, industrial process monitoring, and virtual/remote laboratories for higher education in engineering.

Neeti Dugaya is an Assistant Professor of the EX Dept. of SIRTS.

Shaimaa Ahmed El-said is now an Assistant Professor in the Electronics and Communications Department, Faculty of Engineering, Zagazig University, Egypt. She received her MSc degree in Network Security in 2006 and PhD in Multimedia Security in 2011 from the Faculty of Engineering, Zagazig University, Egypt. Her current research interests include wireless networks, mobile computing, routing security, data mining, image processing, medical imaging, multimedia security and soft computing techniques. She has authored books and has published many research papers in international journals and conference proceedings.

Luis Alfonso Fernández Serantes was born in Viveiro, Spain, in 1992. He received the BSc. degree in Electrical Engineering from the University of A Coruña, A Coruña, Spain, in 2014. His main research interests have been centred in applying expert system technology to the diagnosis and control systems and in power electronic systems for the development of new devices.

José A. Gámez received his M.S. degree in Computer Science in 1991, and his Ph.D. degree in Computer Science in 1998, both from the University of Granada, Spain. He joined the Department of Computer Science at the University of Castilla-La Mancha (UCLM) in 1991, where he is currently a Full Professor. His main research interests include probabilistic reasoning, Bayesian networks, evolutionary algorithms, machine learning and data mining. In these topics Dr. Gamez has edited five books and five special issues of international journals. He is the (co)author of more than one hundred papers in journals, books, and refereed international conferences.

Aboul Ella Hassanien (Abo) is a Professor at Cairo University, Faculty of Computers and Information, IT Department and the chair of the technology center of blind and visual impaired people. Abo received his B.Sc. with honours in 1986 and M.Sc degree in 1993, both from Ain Shams University, Faculty of Science, Pure Mathematics and Computer Science Department, Cairo, Egypt. On September 1998, he received his doctoral degree from the Department of Computer Science, Graduate School of Science & Engineering, Tokyo Institute of Technology, Japan. He has authored/coauthored over 160 research publications in peer-reviewed reputed journals, book chapters and conference proceedings. He has served as the program committee member of various international conferences and reviewer for various international journals. He has received the excellence younger researcher award from Kuwait University for the academic year 2003/2004. He has guest edited many special issues for international scientific journals. He has directed many funded research projects. He is the editor and co-editor for more than 20 books in the area of rough computing, computational intelligence, social networks, bioinformatics, and E-commerce. His research interests include, rough set theory, wavelet theory, computational intelligences, medical image analysis, information security, intelligent environment and multimedia processing.

Ganesh Neelakanta Iyer is currently working with Progress Software Development as QA Architect. He is also a Visiting Faculty at IIIT, Hyderabad and an Adjunct Professor at BVRIT, Narsapur. He has received his Bachelors degree in Computer Science and Engineering with University first rank from Mahatma Gandhi University, Kerala, India in 2004 and Masters and PhD degrees from National University of Singapore in 2008 and 2012 respectively. He has several years of Industry experience at Progress Software Hyderabad, Sasken Communication Technologies and NXP semiconductors, Bangalore. His research interests include Cloud computing, Game theoretic approaches, Distributed computing, aPaaS architectures, Software Quality Analysis and Cloud Testing methodologies. He has several international publications including book chapters, journals and conferences. He serves as the chair for several international conferences. He is also a member of the IEEE-CS Industry outreach committee for 2014. He has also delivered several workshops/seminars/invited talks in various research and industry forums as well as in academics. He is also an expert in performing Kathakali, a traditional Indian dance.

Girish Neelakanta Iyer is currently working with Aryanet Institute of Technology as Assistant Professor in the Department of Computer Science and Engineering. He has received his Bachelor's degree in Computer Science and Engineering from Anna University, Trichy, Tamil Nadu, India in 2011 and a Masters in Network and Internet Engineering from Karunya University, Coimbatore, India with University Third Rank, in 2014. His research interests include Cloud computing, Hadoop, Big Data Analytics, Distributed computing, Network Security. He has several international publications including journals and conferences. He has also presented many papers at national level technical symposiums.

Farkhondeh Jabari was born in 1989 in Iran. She received her B.Sc. and M.Sc degrees in Electrical Power Systems Engineering from University of Tabriz, Tabriz, Iran in 2012 and 2014, respectively. She is currently a PhD student with the Faculty of Electrical and Computer Engineering, University of Tabriz, Tabriz, Iran. Her research interests include power system stability and control, power system defensive splitting, power system uncertainties, design and scheduling of combined cooling, heating and power systems.

Swati Jaiswal graduated in 2012 from RGPV, Bhopal, in computer science engineering. She was awarded a gold medal in M.Tech. She has a total of 6 years teaching experience as an assistant professor. She has published 7 international papers to date. She is currently associated with SKNSITS, Lonavala.

Suman Jonnalagadda has graduated in B.Tech Computer Science Engineering from Jawaharlal Nehru Technological University, Hyderabad, Telangana, India and Post Graduation in M.Tech Computer Science Engineering from Acharya Nagarjuna University, Guntur, Andhra Pradesh, India. He is currently working as an Assistant Professor in the Department of Computer Science Engineering at B.V. Raju Institute of Technology, Narsapur, Telangana, India. He has more than 7 years of teaching and industry experience from reputed engineering colleges. He has published 5 research papers in various journals and conference proceedings. He is also the faculty representative member of IEEE & CSI in the BVRIT College. His area of research interest is in Cloud Computing.

Pattabhi Mary Jyosthna has received B.Tech and M.Tech degrees in computer Science and Engineering from Jawaharlal Nehru Technological University, Hyderabad. She is pursuing a Ph.D in Computer Science and Engineering at VelTech Dr. RR & Dr. SR Technical University, Chennai. Her research interests include Cloud Computing and Data Security. She has 10 years of Teaching experience from reputed engineering colleges. She is currently working as Assistant Professor in the Department of Computer Science and Engineering at B.V. Raju Institute of Technology, Narsapur, Telangana, India.

Amir Manzoor holds a bachelor's degree in engineering from NED University, Karachi, an MBA from Lahore University of Management Sciences (LUMS), and an MBA from Bangor University, United Kingdom. He has many years of diverse professional and teaching experience working at many renowned national and internal organizations and higher education institutions. His research interests include electronic commerce and technology applications in business. He is a member of Chartered Banker Institute of UK and Project Management Institute, USA.

Jesus Martínez-Gómez received the MSc. and PhD degrees in Computer Science in 2008 and 2011, respectively, from the University of Castilla-La Mancha, Spain. He is currently member of the Intelligent Systems and Data Mining research group in the Albacete Research Institute of Informatics I3A and post doc researcher in the University of Alicante. He has been part time lecture in telecommunications in the Cuenca polytechnics school. His research includes robotics, place classification, artificial intelligence, multimodal human-robot interaction, computer vision, and android development.

Behnam Mohammadi-Ivatloo received the BSc degree in electrical engineering from University of Tabriz, Tabriz, Iran, in 2006, and the MSc and PhD degrees from the Sharif University of Technology, Tehran, Iran, in 2008 and 2012, all with honours. He is currently an assistant professor with Faculty of Electrical and Computer Engineering, University of Tabriz, Tabriz, Iran. His main area of research is economics, operation and planning of intelligent energy systems in a competitive market environment.

Juan Aurelio Montero Sousa was born in A Coruna, Spain, in 1965. He received a degree in History from the University of Santiago de Compostela, A Coruña, Spain, and in 1999 a degree in Law from UNED, Spain. He is a teacher in UNED (National University of Distance Education) Spain. He is completing a doctoral thesis about the electricity market and his main research interests have been centered the electricity market, renewable energy, efficiency and education.

Asmaa Osama is an electrical engineer. She graduated from the Faculty of Engineering, Zagazig University, Egypt. Her current research interests include wireless networks, mobile computing, routing and soft computing techniques. She has published many research papers in international journals and conference proceedings.

Kashika Dubey Pathak is a budding academician and researcher in the field of Computer Science and Engineering. She has qualified GATE in year 2010 and pursued her M.Tech in CSE from SATI College, Vidisha. Her area of interests include Image Processing, Software Engineering and Networking. She has achieved appreciable results in all the subjects taught till date. She has also been awarded as "Best Teacher" by SRIJAN Committee in year 2015, for achieving 100% result in RGPV University Examinations. She is EMC certified professional and guided software projects. She is also an author of the book "Soft Computing & Applications" published in year 2016.

Satya Prakash has completed post graduation in 2012 from the National Institute of Technology, Bhopal (MANIT), India. Graduated from SIRT, Bhopal in the year 2007. Has a total of 7 years of experience in Academics as an Assistant Professor. Currently is attached to SKNSITS, Lonavala, in the Electrical Department.

Sajad Najafi Ravadanegh was born in Iran, in 1976. He received the B.Sc. degree from the University of Tabriz, Tabriz, Iran, and the M.Sc. and Ph.D. degrees from the Amirkabir University of Technology, Tehran, Iran, in 2001, 2003, and 2009, respectively, all in electrical engineering. He is currently an Assistant Professor with the Electrical Engineering Department, Azarbaijan Shahid Madani University, Tabriz, where he is responsible for the Smart Distribution Grid Research Laboratory. His current research interests include smart distribution networks and microgrids optimal operation and planning, power system stability and control, power system controlled islanding, optimization algorithms applications in power systems, nonlinear dynamic, and chaos. He has authored/co-authored of over 50 technical papers.

Joyanta Kumar Roy is a prominent entrepreneur and educationist. He was the founder Chairman of M/S System Advance Technologies Pvt. Ltd., Kolkata. He did his PhD (Technology) from the Department of Applied Physics, Calcutta University and executed a number of projects related to control, automation and instrumentation in several engineering sectors. He has worked for many educational institutes as Principal and Dean. Presently, he is associated with MCKV Institute of Engineering as Dean (Research and Consultancy), looking after institutional management of research. Beside this portfolio, he is also working now as Departmental Head of Electronic and Communication Engineering. He is writer, Technical speaker and researcher in inter-disciplinary field of Science and Technology. He received several awards and published many articles in journals and conferences. He is a senior member IEEE, EC member IET, Fellow of the IWWWA and Fellow of IETE. His present research interests include, multifunction sensors, sensor data mining, sensor data fusion and biomedical signal analysis.

Dipu Sarkar was born in West Bengal, India on 20th January 1979. He graduated in Electrical Engineering from University of Kalyani, India in 2003 and obtained the Master Degree in Electrical Engineering with specialization in Electrical Power Systems from the University of Calcutta in 2007. He did his PhD at Bengal Engineering and Science University, Howrah, India presently known as IIEST in the year of 2013. He is an Associate Dean (Research) and an Asst. Professor in the Department of Electrical Engineering, National Institute of Technology, Nagaland. He has nearly 11 years of teaching experience. He has more than 12 national and international journal and conference papers.

Vivek Saxena was graduated in B.E Electrical & Electronics Engg in 2007. He did his Master of Engineering in Power System in from University Institute of Technology affiliated to Rajiv Gandhi Proudyogiki Vishwavidyalaya, Bhopal in 2012. He has 4 years & 9 months of working experience that includes 1 year 6 months in Automation Industry & 3 years 3 month in teaching undergraduate program. He is currently working as an Asst. Professor at SKNSITS, Lonavala. His research interest includes Power Systems & Power Electronics.

Heresh Seyedi was born in Iran, 1979. He received his B.Sc., M.Sc. and Ph.D. degrees of electrical engineering in 2001, 2003 and 2008, respectively, all from the University of Tehran, Iran. Dr. Seyedi is currently an associate professor with the faculty of Electrical and Computer Engineering, University of Tabriz, Iran. His areas of interest include digital protection of power systems, wide area protection and power system dynamics.

Toshi Sharma is a research scholar at MANIT, Maulana Azad National Institute of Technology.

Jayadeep Srikakolapu born in Andhra Pradesh. He did his B.tech in Electrcal and Electronics from B V C engineering college affiliated to JNTU Kakinada and completed Masters in Power Electronics from SGSITS Indore. Currently, he is working as an assistant professor in BVC Engineering College, Odalarevu.

Pallavi Thakkur is an Assistant Professor at SIRTS, Bhopal.

Vivek Venkobarao is an innovative engineering problem solver with over 30 publications, 6 patent applications under examination and experience in advanced product engineering leadership and individual contributor roles. Co Author in the Handbook of Research on Emerging Technologies for Electrical Power Planning, Analysis, and Optimization" "Renewable Energy" from an reputed international publisher Invited Reviewer - IEEE R10 conferences - Inderscience Journals - Transseller Journals - Elsevier Publishers

Gonzalo Vergara received a BSc in Theoretical Physics in 1995 and a MSc in Applied Physics in 1996 from the University of Valencia, Spain. In 2015, he obtained a PhD from the University of Castilla-La Mancha, Spain. He is a researcher in the Intelligent Data Analysis Laboratory (IDAL) group at the University of Valencia and collaborates with the Intelligent Systems and Data Mining research group at the Albacete Research Institute of Informatics (I3A). His work is about the application of machine learning methods. His current research interests are in the areas of data mining and machine learning, in particular, neural networks and ensemble methods.

Index

I

L

M

N

O

P

Q

R

S

T

V

W

www.ingramcontent.com/pod-product-compliance
Lightning Source LLC
LaVergne TN
LVHW081250011026
849162LV00006BA/1324